English–Spanish and Spanish–English Glossary of Geoscience Terms

Diccionario Inglés–Español y Español–Inglés de Términos de Geociencias

Gary L. Prost

CRC Press
Taylor & Francis Group
Boca Raton London New York

CRC Press is an imprint of the
Taylor & Francis Group, an **informa** business

CRC Press
Taylor & Francis Group
6000 Broken Sound Parkway NW, Suite 300
Boca Raton, FL 33487-2742

ISBN-13: 978-9-056-99562-1 (pbk)

Publisher's Note

The publisher has gone to great lengths to ensure the quality of this reprint but points out that some imperfections in the original copies may be apparent.

Cover artwork: *"Towers at Short Creek, Vermillion Cliffs," by William Henry Holmes, plate XIX in Clarence Edward Dutton's "The Physical Geology of the Grand Cañon District," U.S. Geological Survey, Second Annual Report (Washington, DC, 1882).*

British Library Cataloguing in Publication Data

Prost, G. L.
English–Spanish and Spanish–English glossary of geoscience terms = Diccionario inglés–español y español–inglés de términos de geociencias
1.English language – Dictionaries – Spanish 2.Spanish language – Dictionaries – English 3.English language – Technical English – Glossaries, vocabularies, etc.
4.Spanish language – Technical Spanish – Glossaries, vocabularies, etc. 5.Geology – Terminology
I.Title
423.6'1

English–Spanish and Spanish–English Glossary of Geoscience Terms

---

Diccionario Inglés–Español y Español–Inglés de Términos de Geociencias

To Adam, Benjamin and Elizabeth

A Adan, Benjamin e Isabela

CONTENTS

PREFACE

The purpose of this glossary is to provide a ready reference to those in the geosciences with the need to translate from English to Spanish or vice versa. English-speaking countries cover most of two continents (North America and Australia) and England, whereas Spanish-speaking countries cover another entire continent (Central and South America and much of the Caribbean) as well as Spain. With the continuing growth of commerce between Spanish- and English-speaking countries, it is increasingly important to have a clear and precise understanding of what is being said in the technical literature as well as in technical presentations and day-to-day transactions.

Prior to this work there has been no comprehensive English–Spanish / Spanish–English glossary of geotechnical terms. The terms and phrases collected here apply to all of the geosciences, including geology (and its many branches, such as stratigraphy, structure, geochemistry, paleontology, and remote sensing), geophysics, petroleum exploration and production, mining, hydrology, engineering geology (including civil engineering and surveying), and environmental geology. This reference is meant to assist both the professional and student in the earth sciences. In addition, business and banking terms and phrases used in annual reports, proposals, finance, and commodities markets are included for those in the business community. Finally, common business and science terms covering legal, climatic, geographic, and computer terminology and units of measure are included for completeness.

An attempt has been made to include all of the most commonly used terms, as well as many that are not so common. However, since the technical as well as popular languages are constantly changing, no work can be all-inclusive. This should be considered a work in progress. It is hoped that this glossary will contribute to clear communication, a better understanding, and closer working relationships among geoscientists, engineers, and businessmen in both cultures.

PREFACIO

El propósito de este Diccionario es el de proporcionar una referencia práctica a aquellos científicos de las geociencias que tienen la necesidad de traducir del inglés al español y viceversa. Además de Inglaterra, los países de habla inglesa cubren la mayor parte de dos continentes (Norteamérica y Australia) mientras que, exceptuando Brasil, los países de habla hispana cubren Centroamérica, Sudamérica, España y la mayor parte del Caribe. Dado el continuo aumento del comercio entre los países de ambas lenguas, cada vez es más urgente la necesidad de entender con exactitud y precisión lo que se está comunicando en la literatura y las exposiciones técnicas tanto como en toda transacción cotidiana.

No se conocen antecedentes de diccionarios de términos de geociencias inglés–español y español–inglés de la envergadura del presente trabajo. Este diccionario está orientado para servir tanto a profesionales como a estudiantes en las ciencias de la Tierra. Los términos y las frases recogidas aquí provienen de todas las geociencias, incluyendo la geología (y sus numerosas ramas, tales como estratigrafía, geología estructural, geoquímica, paleontología, y teledetección), la geofísica, la exploración y producción del petróleo, la minería, la hidrología, la geología ingenieril (incluidas ingeniería civil y agrimensura), y la geología ambiental. Además se incluyen términos y expresiones comerciales y bancarias usadas en informes anuales, propuestas, finanzas y mercados, de tal modo que también la comunidad financiera pueda aprovecharlo. Finalmente, se han incluido como complemento términos legales, climáticos, geográficos, y de computación, comunes en el mundo de los negocios y las ciencias, asi como unidades de medida.

Se ha intentado incluir todos los términos más comunmente usados, asi como también otros no tan comunes. Sin embargo, dado que tanto el lenguaje técnico como el popular está cambiando constantemente, ningun trabajo puede incluirlo todo. Por esta razón, este debe considerarse un trabajo en evolución. Se espera que este diccionario contribuya a una comunicación más clara, a un entendimiento mejor, y relaciones laborales más estrechas entre los geocientíficos, ingenieros, y hombres de negocios pertenecientes a ambas culturas.

EXPLANATORY NOTES

The following abbreviations are used to assist the user with grammar:

f.	feminine noun	a.	adjective
m.	masculine noun	vb.	verb
adv.	adverb	p.	preposition

When a word can act both as noun and adjective, such as the names of geologic epochs, both abbreviations will be shown. Adjectives will be shown only in the masculine form. Some words, like "microcline," can have both masculine and feminine forms: "microclino" and "microclina." Both are given.

When terms are unique or specific to a country or region, the following abbreviations are used:

And	Andes	Ec	Ecuador
Arg	Argentina	Méx	Mexico
Bol	Bolivia	Par	Paraguay
CA	Central America	Pe	Peru
Car	Caribbean	PR	Puerto Rico
Ch	Chile	Sp	Spain
Col	Colombia	UK	United Kingdom
Cu	Cuba	US	United States
DR	Dominican Republic	Vz	Venezuela

When the same spelling has more than one meaning, synonyms will be separated by commas, while different meanings will be separated by semicolons. Some examples:

grade *f.* (slope) pendiente, cuesta, rampa; *m.* gradiente, declive, sesgo; (quality) *m.* grado, clase; *f.* calidad

grade *vb.* (to classify) calificar, clasificar, graduar, tasar; (to level) nivelar, emparejar, allanar, aplanar, enrasar

plot *m.* solar; gráfico; *f.* parcela; impresión

sound *f.* (geog.) sonda, boca; *m.* boquerón, estrecho o brazo de mar; (noise) sonido, ruido

ingreso *m.* entrance; earnings, revenue

manto *m.* bed, layer, stratum; nappe; stratabound mineral deposit; mantle

mesa *f.* mesa, table mountain; principal facet on a precious stone

If the meaning or synonym is ambiguous, the field of endeavor will be shown parenthetically, i.e., **aguja** *f.* needle; small vein (min.). The following list contains those fields used in this glossary:

aeron.	aeronautics
astron.	astronomy
biol.	biology
botan.	botany
chem., *quím.*	chemistry
com.	commerce
comp.	computation and computers
elect.	electricity and instrumentation
geog.	geography and topography
geol.	geology
hydr.	hydrology
legal	legal and law
meas.	measurements
mech.	mechanics
min.	mining
naut.	nautical terms
paleo.	paleontology
perf.	drilling
petrol.	petrology and petrography
phys., *física*	physics
ref.	refining and smelting
seis.	seismic and geophysics
survey, *levan.*	surveying

Terms are sometimes spelled differently depending on the country; e.g., Albian can be spelled Albense, Albiense, or Albiano. If the various spellings are sufficiently close, as in this example, they will be given together in one citation. Otherwise, they will appear as separate entries. In some cases a term can only be translated as a phrase; e.g., "along strike" translates as "a lo largo del rumbo." In such cases the entire phrase will be given. Often a word is taken directly from one language into the other. "Arroyo" is a proper English word taken originally from the Spanish, but with a slight change of meaning. On the other hand, "avoirdupois" is taken directly from the French, and "fenster" directly from the German.

Alphabetization follows English and Spanish norms. For example, Spanish contains three extra letters, ch (which follows c), ll (which follows l), and ñ (which follows n). Numbers in the English–Spanish section use Spanish norms, e.g., 5.000 and 0,214. In the Spanish–English section they follow English norms, e.g., 5,000 and 0.214.

EXPLICACION

Las siguientes abreviaturas se usan para asistir al usuario con la gramática:

f.	sustantivo femenino	a.	adjetivo
m.	sustantivo masculino	vb.	verbo
adv.	adverbio	p.	preposición

Cuando una palabra es tanto sustantivo como adjetivo, tal como los nombres de las épocas geológicas, figurarán ambas abreviaturas. Los adjetivos se indicarán solo en el género masculino. Algunas palabras, tales como "microcline" en inglés, pueden admitir géneros tanto masculino como femenino en espanol: en este caso se dan ambos, "microclino" y "microclina."

Cuando un término es unico o de uso específico a un país o región, se usarán las siguientes abreviaturas:

And	Andes	Ec	Ecuador
Arg	Argentina	Méx	México
Bol	Bolivia	Par	Paraguay
CA	América Central	Pe	Perú
Car	Caribe	PR	Puerto Rico
Ch	Chile	Sp	España
Col	Colombia	UK	Inglaterra
Cu	Cuba	US	Estados Unidos
DR	República Dominicana	Vz	Venezuela

Cuando el termino tiene más de un significado, los sinonimos se separan con comas, mientras que los significados diferentes se separan con punto y coma. Ejemplos:

grade *f.* (slope) pendiente, cuesta, rampa; *m.* gradiente, declive, sesgo; (quality) *m.* grado, clase; *f.* calidad

grade *vb.* (to classify) calificar, clasificar, graduar, tasar; (to level) nivelar, emparejar, allanar, aplanar, enrasar

plot *m.* solar; gráfico; *f.* parcela; impresión

sound *f.* (geog.) sonda, boca; *m.* boquerón, estrecho o brazo de mar; (noise) sonido, ruido

ingreso *m.* entrance; earnings, revenue

manto *m.* bed, layer, stratum; nappe; stratabound mineral deposit; mantle

mesa *f.* mesa, table mountain, principal facet on a precious stone

Si el significado o sinónimo es ambiguo, el campo de mayor uso se presenta entre paréntesis, por ejemplo, **aguja** *f.* needle; (min.) small vein. La siguiente lista contiene los campos de las ciencias que cubre este diccionario:

aeron.	aeronáutica
astron.	astonomía
biol.	biología
botan.	botánica
chem., quím.	química
com.	comercio
comp.	computación y computadoras
elect.	electricidad e instrumentación
geog.	geografía y topografía
geol.	geología
hydr.	hidrología
legal	legal
meas.	mediciones
mech.	mecánica
min.	minería
naut.	náutica
paleo.	paleontología
perf.	perforación
petrol.	petrología y petrografía
phys., física	física
ref.	refinación y fundición
seis.	sísmica y geofísica
survey, levan.	levantamiento de mapas y agrimensura

Algunos términos pueden sufrir ligeras variaciones de acuerdo al país que se considere; por ejemplo, Albian puede verse escrito como Albense, Albiense, o Albiano. Si las distintas variantes son bastante similares entre sí, como en este caso, se darán juntas en una sóla cita. De lo contrario, aparecerán en referencias separadas. A veces, un término puede traducirse sólo como una expresión; por ejemplo, "along strike" es traducido como "a lo largo del rumbo." En estos casos se citará la frase entera. Muchas veces un idioma toma directamente una palabra del otro. "Arroyo" es una palabra originaria del español, que ha sido asimilada por el inglés con un pequeño cambio en el significado. Análogamente, "avoirdupois" está tomada del francés, y "fenster" directamente del alemán.

La alfabetización sigue las normas del inglés y del español. Por ejemplo, el español tiene tres letras extras, ch (que sigue a la c), ll (después de l), y ñ (que sigue a la n). En inglés la "ch" se encuentra dentro de la "c," y la "ll" dentro de la "l." Para las cifras numéricas en la sección inglés–español se siguen las normas del español, e.g., 5.000 y 0,214. En la sección español–inglés se siguen las normas del inglés, e.g., 5,000 y 0.214.

ACKNOWLEDGMENTS

Many colleagues contributed references and personal lexicons to this glossary.

Among them I would like to thank: Antenor Aleman, Claudio Bartolini, Patricia Beach, Art Berman, John Blickwede, Burke Burkart, David Crowe, Peter Drobeck, Samuel Eguiluz, Eric Fullmer, José Galicia, Francisco Giovanelli, Cristy Harrison, Claudio Manzolillo, Randall Marrett, Anna Morales, Salvador Ortuño Arzate, Joshua Rosenfeld, Matt Trout, Victor Vega, Peter Wang, and Greg Wessel.

Special thanks are directed to my teacher and good friend Maria Cecilia Chaboudy.

AGRADECIMIENTOS

Muchos de mis colegas han contribuido con referencias y léxicos propios para este diccionario.

Entre ellos quisiera agradecer a: Antenor Aleman, Claudio Bartolini, Patricia Beach, Art Berman, John Blickwede, Burke Burkart, David Crowe, Peter Drobeck, Samuel Eguiluz, Eric Fullmer, José Galicia, Francisco Giovanelli, Cristy Harrison, Claudio Manzolillo, Randall Marrett, Anna Morales, Salvador Ortuño Arzate, Joshua Rosenfeld, Matt Trout, Victor Vega, Peter Wang, y Greg Wessel.

Mis agradecimientos mas efusivos están dirigidas a mi maestra y buena amiga Maria Cecilia Chaboudy.

GLOSSARY OF GEOSCIENCE TERMS

ENGLISH - SPANISH

A

Aalenian *m. & a.* Aalenense, Aaleniense, Aaleniano

abandon *vb.* abandonar, desamparar

abandoned *a.* abandonado, suspendido || ***abandoned meanders*** *m.* meandros abandonados || ***abandoned well*** *m.* pozo abandonado

abandonment *m.* abandonamiento, desamparo, desistimiento, abandono

abandonments *m.* abandonos

aberrant *a.* aberrante

aberration *m.* error; *f.* aberración

abiogenesis *f.* abiogénesis

abioglyph *m.* abioglifo

abiotic *a.* abiótico

ablation *f.* ablación, desgaste

ablution *f.* ablución

Abney level *m.* nivel Abney, clinómetro

abnormal *a.* anormal

abnormality *f.* anormalidad

above *p.* por encima de, sobre de; *a.* sobre

abrade *vb.* bruñir, desgastar, raer, raspar, pulverizar; (chem., quím.) roer

abraded *a.* desgastado

abrasion *f.* abrasión || ***abrasion resisting*** *a.* resistente a la abrasión

abrasive *a.* abrasivo, rayante, corroyente

abrupt *a.* abrupto; (geog.) barrancoso, escarpado, quebrado

abscissa *f.* abscisa

absence *f.* ausencia

absolute *a.* absoluto || ***absolute age*** *f.* edad absoluta || ***absolute permeability*** *f.* permeabilidad absoluta || ***absolute value*** *m.* módulo, valor absoluto || ***absolute zero*** *m.* cero absoluto

absorb *vb.* absorber, amortiguar

absorbency *f.* absorbencia

absorbent *a.* absorbente

absorber *m.* absorbedor, amortiguador

absorption *f.* absorción, imbibición || ***absorption coefficient*** *m.* coeficiente de absorción || ***absorption factor*** *m.* factor de absorción || ***atmospheric absorption*** *f.* absorción atmosférica || ***ground absorption*** *f.* absorción del suelo || ***preferential absorption*** *f.* absorción preferencial

abstract *m.* resumen, sumario

abundance *m.* abasto; *f.* abundancia, afluencia

abundant *a.* abundante, afluente

abut *vb.* empalmar, ensamblar, empotrar || ***abut against*** *vb.* empotrar en, apoyarse en

abutment *m.* contrafuerte, refuerzo, empotramiento, estribo, lindero

abyss *m.* abismo; *f.* sima

abyssal *a.* abisal, abísico, abismal, insondable

acadialite *f.* acadialita

Acadian *m. & a.* Acadiense, Acadiano, Acadico, Jabaliense, Jabaliano

acanthite *f.* acantita

acanthopore *m.* acantóporo

accelerate *vb.* acelerar

acceleration *f.* aceleración || ***gravitational acceleration*** *f.* aceleración de gravedad

accelerometer *m.* acelerómetro

access *m.* acceso

accessibility *f.* accesibilidad

accessible *a.* accessible, visitable

accessory *m.* accesorio || ***accessory element*** *f.* pieza accesoria || ***accessory minerals*** *m.* minerales accesorios

accident *m.* acontecimiento fortuito, accidente

accidental *a.* accidental

acclimatization *f.* aclimatación

acclivity *m.* contrapendiente

accommodate *vb.* acomodar

accommodated *a.* acomodado

accommodation *f.* acomodación || ***accommodation space*** *m.* espacio disponible

account *f.* cuenta, *m.* informe financiero (o bancario) || ***account balance*** *m.* corte de caja || ***accounts payable*** *f.* cuentas por pagar || ***accounts due*** *f.* cuentas por cobrar

account *vb.* dar cuenta, explicar

accountant *m.* contador

accounting *f.* contabilidad

accreted *a.* acrecionado || ***accreted terrane*** *m.* terreno acrecionado

accretion *f.* acrecencia, acreción, yuxtaposición, acumulación; *m.* aumento, acrecentamiento, aluvionamiento, acrecimiento

accretionary *a.* acrecional || ***accretionary prism*** *f.* prisma acrecionaria || ***accretionary ridge*** *f.* cresta acrecionaria

accrue *vb.* acrecentar, crecer, aumentar

accumulate *vb.* acumular

accumulated *a.* acumulado

accumulation *f.* acumulación; cumulo

accuracy *f.* exactitud, precisión, fidelidad

accurate *a.* correcto, exacto

acenaphthylene *m.* acenaftileno

acetate *m.* acetato

acetic *a.* acético

acetone *f.* acetona

acetylene *m.* acetileno || ***acetylene torch*** *m.* mechero de acetileno, quemador de acetileno; *f.* antorcha de acetileno

achavalite *f.* achavalita

achondrite *m.* acóndrito

achroite *f.* acroíta

achromatic *a.* acromático

acicular *a.* acicular || ***acicular gold*** *m.* [Col.] alfiler de oro

acid *a.* agrio, ácido

acid *m.* ácido || ***acid bottle*** *f.* botella de ácido || ***acid inhibitor*** *m.* supresor de ácido, inhibidor de ácido || ***acid meter*** *m.* acidómetro || ***acid-proof*** *a.* antiácido, a prueba de ácido || ***acetic acid*** *m.* ácido acético || ***boric acid*** *m.* ácido bórico || ***carbolic acid*** *m.* ácido carbólico, ácido fénico, phenol || ***carbonic acid*** *m.* ácido carbónico || ***cyanic acid*** *m.* ácido ciánico || ***formic acid*** *m.* ácido fórmico || ***hydrochloric acid*** *m.* ácido clorhídrico, ácido hidroclórico, ácido muriático || ***hydrofluoric acid*** *m.* ácido fluorhídrico, fluorhidrato || ***isobutyric acid*** *m.* ácido isobutírico, ácido isobutílico || ***muriatic acid*** *m.* ácido muriático || ***nitric acid*** *m.* ácido nítrico || ***oxalic acid*** *m.* ácido oxálico || ***phosphoric acid*** *m.* ácido fosfórico || ***sulfuric acid*** *m.* ácido sulfúrico || ***terephthalic acid*** *m.* ácido tereftálico

acidic *a.* persilíceo

acidic lava *f.* lava ácida

acidification *f.* acidificación

acidify *vb.* acidificar

acidity *f. & a.* acidez

acidize *vb.* acidificar, acidular

acidizing *f.* acidulación

acknowledgment *m.* reconocimiento

aclinic *a.* aclínico

acmite *f.* acmita, aegerina, aegirina

acoustic *a.* acústico || ***acoustic impedance*** *f.* impedancia acústica || ***acoustic log*** *m.* perfil acústico

acquisition *f.* adquisición

acre *m.* acre (4,047 × 10^{-3} km. cuad.) || ***acre foot*** *m.* acrepie (1,233 m^2)

acreage *f.* superficie en acres, extensión en acres

across *adv.* a través; *p.* a través de

acrylic *m.* acrílico

act of God *m.* caso fortuito

actifier (gas) *m.* activador (de gas)

actinolite *f.* actinolita, actinota, waldheimita; *m.* actinoto

action *f.* acción || ***galvanic action*** *f.* acción galvánica

activate a well *vb.* activar el pozo

activated *a.* activado

activation *f.* activación || ***activation energy*** *f.* energía de activación, energía potencia

active *a.* activo || ***active fault*** *f.* falla activa || ***active volcanism*** *m.* vulcanismo activo

activity *f.* (chem., quím.) actividad

acute *a.* agudo

acute-angled *a.* acutángulo, ángulo agudo

acyclic *a.* acíclico

adamantine *a.* adamantino, adiamantado, diamantino

adamellite *f.* adamelita

adamite *f.* adamina

adapt *vb.* adaptar, conformar

adaptability *f.* adaptabilidad

adaptation *f.* adaptación

adaptor *m.* adaptador || ***pump adaptor*** *m.* adaptador de bomba, conector de bomba, ensamblador de bomba || ***casing adaptor*** *m.* adaptador de tubería, acoplador de tubería

add *vb.* agregar, añadir || ***add to*** *vb.* aumentar

additive *m.* aditivo

adductor *a.* oclusor, aductor || ***adductor muscle*** *m.* músculo aductor, músculo oclusor

adhere *vb.* adherirse, pegarse, aficionarse

adhesion *f.* adhesión, adherencia

adhesive *m.* adhesivo

adiabatic *a.* adiabático

adit *m.* acceso, brocal, socavón, tiro de recueste; *f.* boca, bocamina, entrada, bartolina, galería, contramina

adjacent *a.* adyacente, adjunto, contiguo

adjust *vb.* ajustar, compensar, regular

adjusted *a.* ajustable

adjusted ajustado

adjusted stream *f.* corriente ajustada

adjuster *m.* ajustador

adjustment *m.* ajuste; *f.* compensación, regulación

admissible *a.* admisible

admissibility *f.* admisibilidad, tolerancia

admix *vb.* mezclar

admixture *m.* adicional, adicionante, agregado coloidal; *f.* admixtura, comezcla

adobe *m.* adobe; *f.* [Col.] cangagua

adolescent *a.* adolescente

adsorb *vb.* adsorber

adsorbent *a.* adsorbente

adsorption *f.* adsorción

adsorptive *a.* adsorbente

adularia *f.* adularia; *m.* feldespato nacarado

ad valorem *adv.* conforme a su valor, ad valórem

advance *m.* avance || ***advance boring*** *m.* barreno de reconocimiento || ***advance heading*** *f.* galería de avance || ***advance of the sea*** *m.* avance de mar; *f.* transgresión || ***advance on wages*** *m.* adelanto; *f.* [Pe.] alanoca || ***glacial advance*** *m.* avance de glaciar

advanced *a.* adelantado, avanzado

advantage *f.* ventaja || ***competitive advantage*** *f.* ventaja competitiva

advertise for bids *vb.* anunciar la licitación

advertisement *m.* aviso, anuncio; *f.* propaganda, advertencia

advice *m.* consejo; *f.* recomendación

aegirine *f.* egirina

aegirite *f.* egirita

aenigmite *f.* enigmatita

aeolian *a.* eólico

aeon *m.* eón

aerate *vb.* aerear

aeration *f.* aeración, ventilación

aerator *m.* aerador, aereador

aereated *a.* aireado

aerial *a.* aéreo || ***aerial geology*** *f.* aerogeología || ***aerial mapping*** *f.* aerofotogrametría, levantamiento aéreo || ***aerial mosaic*** *m.* mosaico de fotografías aéreas || ***aerial photography*** *f.* aerofotografía, fotografía aérea || ***aerial photomosaic*** *m.* fotomosaico aéreo || ***aerial tram*** *f.* aerovía de cable, [Arg.] alambre; *m.* andarivel, telesferico

aerobic *a.* aeróbico

aerodynamic *a.* aerodinámico

aeromagnetic *m.* aeromagnético

aeromagnetics *f.* aeromagnetría, aeromagnetometría

aerosol *m.* aerosol

affect *vb.* impactar, afectar

affinity *f.* afinidad

affluent *m.* afluente, tributario

after *a.* poste, posterior; *adv.* después; *p.* después de, según

aftershock *m.* sismo final, temblor secundario; *f.* réplica de un terremoto

Africa *f.* África

Aftonian *m. & a.* aftonense

agate *f.* ágata || ***dark green agate*** *f.* prasma

age *f.* edad || ***absolute age*** *f.* edad absoluta

agency *f.* agencia

agent *m.* agente || ***catalytic agent*** *m.* agente catalítico || ***flocculating agent*** *m.* agente floculador || ***mineralizing agent*** *m.* agente mineralizador || ***reducing agent*** *m.* agente reductor

agglomerate *m.* aglomerado

agglomerate *vb.* aglomerar, arremolinarse

agglomeration *f.* aglomeración, arremolinamiento

aggradation *f.* agradación, adición por sedimentación

aggrade *vb.* agradar, acumular, elevarse

aggregate *m.* agregado || ***graded aggregate*** *m.* agregado escalonado, agregado graduado, agregado gradado || ***run-of-mine aggregate*** *m.* agregado tal come sale, agregado caótico

aggregate *vb.* agregar, aglutinar, acumular

aging *m.* envejecimiento

agitate *vb.* agitar

agitation *f.* agitación

agitator *m.* agitador

Agnostozoic *m. & a.* Agnostozoico

agonic line *f.* línea de no declinación

agreement *m.* acuerdo, convenio; ajuste

agricultural *a.* de cultivo

A-horizon *m.* horizonte A

aikinite *f.* aciculita

air *m.* aire || ***airborne*** *a.* transportado por aire, aéreo || ***air compressor*** *m.* compresor de aire || ***air-cooled*** *a.* enfriado por aire || ***air drive*** *m.* inyección de gas, empuje por aire, desplazamiento por aire || ***air filter*** *m.* depurador de aire || ***air gun*** *m.* cañón de aire; *f.* fuente de aire, fuente de poder || ***air injection*** *m.* inyección de aire || ***air lift*** *m.* extracción por aire || ***air shaft*** *m.* pozo de ventilación || ***airtight*** *a.* hermético, sellado || ***air vent*** *m.* respiradora || ***bad air*** *m.* aire viciado o contaminado || ***compressed air*** *m.* aire comprimido || ***exhaust air*** *m.* aire extraido, aire viciado || ***free air*** *m.* aire libre || ***fresh air*** *m.* aire puro

airphoto *f.* fotografía aérea, aerofotografía || ***high oblique airphoto*** *f.* fotografía aérea obliqua con horizonte sensible || ***low oblique airphoto*** *f.* fotografía aérea obliqua sin horizonte sensible || ***vertical airphoto*** *f.* fotografía aérea vertical

airplane *m.* avión

akmolith *m.* acmolito

alabandite *f.* alabandina

alabaster *m.* alabastro, alabastrino, tecali

alaskite *f.* alasquita, alaskita

albedo *m.* albedo

albertite *f.* albertita

Albian *m. & a.* Albense, Albiense, Albiano

albite *f.* albita; *m.* feldespato sódico, sodaclaso

albitization *f.* albitización

albumin *f.* albúmina

alcohol *m.* alcohol || ***ethyl alcohol*** *m.* alcohol etílico || ***methyl alcohol*** *m.* alcohol metílico

alexandrite *f.* alejandrita, alejandrina

algae *f.* algas, alga

algal *a.* algino, de algas, algáceo || ***algal reef*** *m.* arrecife algino

algicide *m.* algicida

alginite *f.* alginita

Algomian *m. & a.* Algomiano, Algomiense

Algonkian, Algonquin *m. & a.* Algónquico, Algonkiano, Agnostozóico || ***lower Algonkian, Algonquin*** *m. & a.* Kalaviense

algorithm *m.* algoritmo

alias *m.* (comp.) alias

aliasing (seis.) *f.* señal de artefacto

alidade *f.* alidada

align *vb.* alinear, centrar, enderezar

alignment *m.* alineamiento; *f.* alineación

alike *a.* parecido, conforme a, similar, semejante

aliphatic *a.* alifático

aliquat *f.* alícuota

alkali *a.* álcali || ***alkali basalt*** *m.* basalto alcalino || ***alkali flat*** *m.* llano alcalino || ***alkali metal*** *m.* metal alcalino || ***alkali rhyolite*** *f.* riolita alcálica (o alcalina) || ***alkali trachyte*** *f.* traquita alcálica (o alcalina)

alkaline *a.* alcalino || ***alkaline earth*** *m.* alcalino térreo

alkalinity *f.* alcalinidad

alkaloid *m.* alcaloide

alkane *m.* alcano

alkene *m.* alqueno

alkyl *a.* alquilo, alquino

alkylation *f.* alcoilación, alquilación, alquinación

allanite *f.* allanita, ortita, yttro-ortita || ***impure allanite*** *f.* wasita || ***lithium allanite*** *f.* weissigita

allochthon *m.* alóctono

allochthonous *a.* alóctono

allogenic *a.* alógeno

allomorphism *m.* alomorfismo

allophane *f.* alófana, zirlita

allotment *m.* repartimiento, concesión

allotriomorphic *a.* alotriomorfo, anihédrico, alotriomórfico

allotropic *a.* alotropía, alotrópico

allotropy *f.* alotropía

allotype *m.* alotipo

allowable *m.* permitido, contingente, admisible, permisible

allowance *f.* tolerancia

alloy *f.* aleación, liga, aligación, aquilatación; *m.* aleaje

alloy *vb.* alear, ligar, aquilatar

alluvial *a.* aluvial, aluvional || ***alluvial apron*** *f.* cubierta aluvial; *m.* frente de deyección || ***alluvial fan*** *m.* abanico aluvial || ***alluvial plain*** *f.* llanura aluvial, planicie aluvial || ***alluvial terrace*** *m.* terraplén (o terraza) aluvial

alluviation *f.* acumulación aluvial; *m.* aluvionamiento

alluvium *f.* aluvión, cobijadura, llena, sobresabana; *m.* acarreo, chorroborro, aterramiento, deslava, [Arg.] aluvionario, [And.] huaico, [Méx.] deslave

almanac *m.* almanaque

almandine, almandite *f.* almandina, almandita

almost *a.* casi

alnoite *f.* karbonatita alnoíta

along *a.* a lo largo del || ***along bedding*** a lo largo de la estratificación || ***along strike*** a lo largo del rumbo, a rumbo

alpha particle *f.* partícula alfa

alpine *a.* alpestre, alpino

Alpine *m.* & *a.* Alpino, Alpinense || ***Alpine glaciation*** *f.* glaciación de montaña

altaite *f.* altaita

alteration *f.* alteración

altered *a.* alterado || ***altered rock*** *f.* roca alterada; *m.* revenidero

altimeter *m.* altímetro

altimetric *a.* altimétrico, altímetro

altitude *f.* altitud, altura, cota, elevación

altocumulus *m.* altocúmulo

alum *m.* alumbre || ***alum-bearing*** *a.* alúmbrico || ***alum quarry*** *f.* alumbrera

alumina *f.* alúmina

aluminite *f.* werhemannita

aluminous *a.* alumbroso, aluminoso

aluminum *m.* aluminio || ***aluminum-bearing*** *a.* aluminífero

alunite *f.* alunita, alumbre de piedra

alunogen *f.* queramohalita

alveolar *a.* alveolar

alveole, alveolus *m.* alvéolo

amalgam *f.* amalgama, pella, pasta || ***gold amalgam*** *f.* auromalgama || ***gold or silver amalgam*** *f.* (min.) pasta de oro o plata

amalgamate *vb.* amalgamar

amalgamated *a.* amalgamado

amalgamation *f.* amalgamación, azogería || ***hot amalgamation process*** *m.* [Méx.] beneficio de cazo, azogue || ***cold amalgamation process*** *m.* beneficio de patio

amalgamator *m.* cazeador

amarantite *f.* amarantita

amazonite *f.* amazonita

amber *m.* ámbar, cárabe, succino

ambient *a.* ambiental, ambiente

ambligonite, amblygonite *f.* ambligonita

amausite *m.* petrosilex

amethyst *f.* amatista; *m.* [Méx.] chichicle, chichitle

amide *f.* amida

amine *f.* amina

amino acid *m.* aminoácido

amitosis *f.* amitosis

ammonia *m.* amoníaco

ammonite *f.* amonita

ammonium *a.* amonio

amoeba *f.* ameba

amorphous *a.* amorfo

amortization *f.* amortización

amount *f.* cantidad ‖ ***total amount*** *m.* monto

ampelite *f.* ampelita

amphibian *m.* anfibio

amphibious *a.* anfibio

amphibole *m.* anfíbol; *f.* anfíbola

amphibolite *f.* anfibolita, afanita ‖ ***amphibolite schist*** *m.* esquisto de anfibolita

amphibolization *f.* anfibolización

amphitheater *m.* anfiteatro

amplify *vb.* amplificar

amplitude *f.* amplitud

amygdaloid *m.* amígdaloide

amygdaloidal *a.* amígdaloide

amygdule *f.* amígdala

anabergite *f.* anabergita

anaerobic *a.* anaeróbio, anaeróbico

anal *a.* anal

analcime, analcite *f.* analcima, analcita, zeolita dura

analog *m.* analógico

analogy *m.* & *a.* análogo

analysis *m.* & *f.* análisis ‖ ***cost analysis*** *m.* análisis de costo ‖ ***elemental analysis*** *m.* análisis elemental ‖ ***isotopic analysis*** *m.* análisis isotópico ‖ ***qualitative analysis*** *m.* análisis cualitativo ‖ ***quantitative analysis*** *m.* análisis cuantitativo

analyst *m.* & *f.* analista

analytical *a.* analítico

analyze *vb.* analizar

analyzer *m.* analizador

anamorphism *m.* anamorfismo

anastomose *vb.* anastomosarse, entrenzarse

anastomosing *a.* anastomosado

anatase *f.* anatasa, wiserina

anatexis *m.* anatexis; *f.* anatexia

anatexite *f.* anatexita

anauxite *f.* anauxita

anaxial *a.* anaxil

ancient *a.* antiguo

andalusite *f.* andalusita, andalucita ‖ ***altered andalusite*** *f.* westanita

Andean *a.* Andino

andesine *f.* andesina

andesite *f.* andesita

andorite *f.* webnerita, sundita

andradite *f.* andradita

anelastic *a.* anelástico, inelástico, sin elasticidad

anelasticity *f.* anelasticidad, inelasticidad

anemometer *m.* anemómetro

anemone *f.* anémona

aneroid barometer *m.* barómetro aneroide

angiosperm *f.* angiosperma

angle *m.* ángulo ‖ ***acute angle*** *m.* ángulo agudo ‖ ***angle of dip*** *m.* inclinación, buzamiento, echado ‖ ***angle of friction*** *m.* ángulo de fricción ‖ ***angle of incidence*** *m.* ángulo de incidencia ‖ ***angle of intersection*** ángulo de intersección ‖ ***angle of polarization*** ángulo de polarización ‖ ***angle of refraction*** ángulo de refracción ‖ ***angle of repose*** *m.* ángulo de reposo, ángulo de talud natural, talud de reposo ‖ ***angle of slope*** *m.* ángulo de pendiente ‖ ***critical angle*** *m.* ángulo crítico ‖ ***extinction angle*** *m.* ángulo de extinción ‖ ***final cutoff angle*** *m.* ángulo final de truncamiento ‖ ***fundamental step-up angle*** *f.* inclinación de las rampas en el primer escalón || ***incidence angle*** *m.* ángulo de incidencia || ***initial cutoff angle*** *m.* ángulo inicial de truncamiento || ***initial step-up angle*** *m.* ángulo de escalón inicial || ***optic angle*** *m.* ángulo óptico || ***reflection angle*** *m.* ángulo de reflexión || ***right angle*** *m.* ángulo recto || ***rotation angle*** *m.* ángulo de rotación || ***shear angle*** *m.* ángulo de cizalladura || ***obtuse angle*** *m.* ángulo obtuso

anglesite *f.* anglesita, [Pe.] asta de venado

Ångström, angstrom *m.* ángstrom (10^{-8} cm)

angular *a.* angular, anguloso || ***angular shear*** *m.* cizalla angular || ***angular unconformity*** *f.* discordancia angular

angularity *f.* angulosidad

anhedral *a.* anhedral, alotriomórfico

anhydride *m.* anhídrido, anhidro

anhydrite *f.* anhidrita, carstenita, karstenita

anhydrous *a.* anhidro, anhídrico

aniline *f.* & *a.* anilina

animal *m.* animal

anion *m.* anión

Anisian *m.* & *a.* Anisiense, Anisiano, Virgloriano, Virgloriense

anisometric *a.* anisométrico

anisotropic *a.* anisótropo, anisotropico

anisotropy *f.* anisotropía

ankerite *f.* ankerita, anquerita

annabergite *f.* annabergita; *m.* niquelocre

anneal *vb.* recocer, destemplar

annealed *a.* recocido

annealing *f.* recocción || ***annealing furnace*** *m.* horno de recocción

annelid *f.* anélido

annotation *f.* anotación

announcement *m.* aviso, anuncio

annual *a.* anual

annular *a.* anular || ***annular drainage*** *m* drenaje (o avenamiento) anular

annulus *m.* ánulo, espacio anular

anode *m.* ánodo, polo positivo

anomalous *a.* anómalo

anomaly *f.* anomalía || ***free air gravity anomaly*** *f.* anomalía de aire libre || ***gravity anomaly*** *f.* anomalía gravimétrica || ***magnetic anomaly*** *f.* anomalía magnética

anorogenic *a.* anorogénico, anorogenético

anorthite *f.* anortita; *m.* feldespato calizo

anorthoclase *f.* anortoclasa

anorthosite *f.* anortosita, plagioclasita || ***corundum anorthosite*** *m.* korund-anortosita

anoxic *a.* anóxico, sin oxígeno

antarctic *a.* antártico, antartido

Antarctica *f.* Antártica, Antartida

antecedent *a.* antecedente || ***antecedent stream*** *m.* río antecedente

antenna *f.* antena

anterior *a.* anterior

anthophyllite *f.* antofilita

anthracene *m.* antraceno

anthracite *f.* antracita, hullita, hulla || ***anthracite-bearing*** *a.* antracitoso || ***anthracite coal*** *m.* carbón antracita

anthraconite *f.* antraconita

anti-alias filter *m.* filtro antialias

anticlinal *a.* anticlinal || ***anticlinal axis*** *m.* eje anticlinal, [Méx.] charnela || ***anticlinal closure*** *m.* cierre anticlinal || ***anticlinal crest*** *f.* cresta del anticlinal || ***anticlinal nose*** *f.* nariz anticlinal

anticline *m.* anticlinal || ***asymmetric anticline*** *m.* anticlinal asimétrico || ***breached anticline*** *m.* anticlinal fracturado, anticlinal abierto, anticlinal desmantelado || ***doubly overturned anticline*** *m.* anticlinal cornudo, anticlinal abanico || ***hanging wall anticline*** *m.* anticlinal colgado (o colgante) || ***inverted anticline*** *m.* anticlinal volcado || ***overturned anticline*** *m.* anticlinal recostado (o recumbente, invertido, o volcado) || ***recumbent anticline*** *m.* anticlinal acostado (o recostado, o recumbente) || ***symmetric anticline*** *m.* anticlinal simétrico

anticlinorium *m.* anticlinorio

antidune *f.* antiduna

antiform *f.* antiforma

antifreeze *m.* & *a.* anticongelante

antigorite *f.* antigorita, williamsita, zermattita

antimonite *f.* antimonita

antimony *m.* antimonio, [Bol.] alcohol || ***antimony-bearing*** *a.* antimoniado

antioxidant *m.* antioxidante

antiperthite *f.* antipertita

antiripple *f.* antióndula, antirizadura

antirust *m.* & *a.* anticorrosivo, antioxidante

antitaxial fibrous veins *f.* venas fibrosas antitaxiales

antithesis *f.* antítesis

antithetic *a.* antitético || ***antithetic fault*** *m.* falla antitética

antitoxin *f.* antitoxina

antlerite *f.* antlerita, vernadskita

anus *m.* ano

apatite *f.* apatita; *m.* apatito || ***carbonate apatite*** *f.* podolita, tavistockita

aperture *f.* abertura, apertura; *m.* orificio

apex *m.* ápice, vértice; *f.* cúspide

aphanite *f.* afanita

aphanitic *a.* afanítico

aphotic *a.* afótico

apical *a.* apical

API (American Petroleum Institute) gravity *f.* gravedad API (Instituto Americano del Petróleo)

aplite *f.* aplita

apophyllite *f.* apofilita, albina || ***green apophyllite*** *m.* xylochloro

apophysis *f.* apófisis

Appalachian *a.* Apalache, Apalacheano

apparent *a.* aparente || ***apparent dip*** *m.* buzamiento aparente, echado aparente

apparatus *m.* aparato, aparejo

appearance *f.* apariencia, aparición; *m.* aspecto

appendage *m.* apéndice

appendix *m.* apéndice

application *f.* aplicación; (com.) propuesta, solicitud

apply *vb.* aplicar, solicitar

apportion *vb.* proratear

appraisal *f.* tasación, apreciación, justipreciación, tasa, evaluación; *m.* avalúo, aforamiento, aforo, aprecio

appraise *vb.* aforar, apreciar, avaluar, justipreciar

appraised value *m.* justiprecio

appraiser *m.* justipreciador, perito

appreciate *vb.* apreciar

appreciation in value *f.* valorización

approach *m.* método

approximation *f.* aproximación

apron *m.* delantal, mandil, cona aluvial

Aptian *m.* & *a.* Aptense, Aptiense, Aptiano

aqua fortis *m.* ácido nítrico

aqua regia *f.* agua regia

aquamarine *f.* aguamarina; *m.* berilio verdemar

aquapulse *m.* acuapulso

aquatic *a.* acuático

aqueduct *m.* acueducto

aqueous *a.* acuoso, ácueo

aquiclude *f.* acuiclusa; *m.* acuicludo, estrato impermeable

aquifer *m.* acuífero; *f.* capa acuífera, capa freática, napa || ***artesian aquifer*** *m.* acuífero artesiano || ***linear aquifer*** *m.* acuífero lineal || ***radial aquifer*** *m.* acuífero radial

Aquitanian *m.* & *a.* Aquitaniense, Aquitaniano

aquitard *m.* acuitardo, estrato impermeable

aragonite *m.* aragonito; *f.* aragonita, chimborazita, zeyringita

arbitrate *vb.* arbitrar

arbitration *f.* arbitración

arc *m.* arco || ***arc-shaped vein*** *f.* veta arqueada || ***island arc*** *m.* arco de islas, arco isla, arco insular || ***volcanic arc*** *m.* arco volcánico

Arcadian *m.* & *a.* Arcadiense, Arcadiano

arch *m.* arco; *f.* bóveda || ***natural arch*** *m.* arco natural, portal

archaeocyathid *m.* arqueociátido

archaeologist *m.* arqueólogo

archaeology *f.* arqueología

arched *a.* abovado

Archean *m.* & *a.* Arqueano, Arcáico

arched *a.* arqueado, curvado

Archeozoic *m.* & *a.* Arqueozoico

archipelago *m.* archipiélago

architecture *f.* (geol.) arquitectura || ***landscape architecture*** *f.* arquitectura paisajista || ***sequence architecture*** *f.* arquitectura de la secuencia

arctic *m. & a.* ártico

arcuate *a.* arqueado

area *f.* área, ámbito, región, zona, comarca, sector || ***area balance*** *m.* balanceo por área, balance de áreas || ***drainage area*** *f.* área de drenaje || ***exploration area*** *f.* área de exploración || ***recreation area*** *f.* área recreativa || ***unproven area*** *f.* área no probada

areal *a.* areal || ***areal geology*** *f.* geología superficial, geología regional

areation *f.* aireación, aeriación

arenaceous *a.* arenoso, arenáceo, aglutinado, arenisco

Arenigian *m. & a.* Arenigense, Arenigiano

arenite *f.* arenita, arenisca || ***quartz arenite*** *f.* arenita de cuarzo

areole *f.* aréola

arête *f.* arista

argentiferous *a.* argentífero

argentite *f.* argentita, acanelado, argirita, argirosa, plata agria, plata dúctil, plata gris, plata vítrea, plata negra, polvorilla; *m.* bodoque, metal polvorillo, [Méx.] azulaque

argillaceous, argillic *a.* arcilloso, argiloso, pelolítico

argillite *f.* argilita

argon *m.* argón

Argovian *m. & a.* Argoviense, Argoviano

arid *a.* árido, desértico

aridity *f.* aridez

Arkansan *m. & a.* Arkansiense, Arkansiano

arkose *f.* arcosa

arkosic *a.* arcósico

arm *m.* brazo || ***arm of the sea*** *m.* brazo de mar || ***cantilever arm*** *m.* brazo volado || ***control arm*** *m.* brazo de gobierno || ***lever arm*** *m.* brazo de palanca

armored clay ball *m.* rodado de arcilla acorazado

aromatic *a.* aromático

arrange *vb.* arreglar

arrangement *f.* disposición, distribución; *m.* arreglo, dispositivo

array *f.* configuracion, formación, matriz; *m.* orden, arreglo

arrester (flame) *m.* chispero, parallamas

arrival time *f.* llegada; *m.* tiempo de llegada, arribo

arroyo *f.* quebrada, barranca; *m.* barranco

asdic *m.* sonar, asdic

arsenate *m.* arseniato

arsenic *m.* arsénico

arsenide *m.* arseniuro

arsenopyrite *f.* arsenopirita, pirita arsenical; *m.* hierro arsenical, hierro blanco, mispíquel

artesian *a.* artesiano, surgente || ***artesian spring*** *f.* fuente artesiana || ***artesian water*** *f.* agua artesiana || ***artesian well*** *m.* pozo artesiano

arthropod *m.* artrópodo

articulate, articulated *a.* articulado

articulation *f.* articulación

artifact *m.* artifacto, artificio

artificial lift *f.* elevación artificial

asbestos *m.* asbesto, amianto || ***tremolite-actinolite asbestos*** *f.* zillerita

ascent *m.* ascenso; *f.* trepada, subida

aseismic *a.* asísmico

asexual *a.* asexual

ash *f.* ceniza, escoria; *a.* cinerítico || ***ash flow*** *f.* alud de cenizas, cinerita, nube ardiente || ***ash flow tuff*** *m.* sillar || ***fly ash*** *m.* hollín || ***volcanic ash*** *f.* toba volcánica (o de cenizas)

Ashgillian *m. & a.* Ashgilliense, Ashgilliano

Asia *f.* Asia

aspect *m.* aspecto, vertiente; *f.* fase

asphalt *m.* asfalto, betún, [Méx.] chapopote, [Arg., Cu.] chapuro; *f.* brea, brea mineral, grahamita, malta || ***asphalt base crude*** *m.* crudo de base asfáltica || ***asphalt-bearing*** asfaltífero

asphaltene *m. & a.* asfalteno

asphaltic *a.* asfáltico

asphaltite *f.* asfaltita, grahamita

asphyxiate *vb.* asfixiar

assay *m.* ensayo, ensaye, aquilatamiento, contraste; *f.* aquilatación, evaluación, titulación || ***assay button*** *m.* pallón || ***assay house*** *m.* casa del apartado

assay *vb.* ensayar; acrisolar, aquilatar, quilatar

assayer *m.* ensayador, quilatador

assemble *vb.* armar, montar, ensamblar

assemblage *f.* asociación, conjunto || ***fossil assemblage*** *f.* asociación de fósiles

assembly *f.* instalación, montura

assess *vb.* evaluar

assessment *m.* aforo; *f.* evaluación

assets *m.* activos, bienes || ***capital assets*** *m.* bienes de capital || ***liquid assets*** *m.* activos circulantes

assignment *m.* encargo

assimilate *vb.* asimilar

assimilation *f.* asimilación, digestión

associate *m. & f.* socio

associated *a.* asociado

association *f.* asociación

assorted *a.* variado

assumption *f.* suposicion

astatic *a.* astático

asterated sapphire *f.* asteria

asterism *m.* asterismo

asteroid *m.* (paleo.) asteroideo; asteroide, (astron.) planetoide

asthenosphere *f.* astenosfera

Astian *m. & a.* Astiense, Astiano

astrakanite *f.* astracanita

astrolabe *m.* astrolabio

astronomic(al) *a.* astronómico

astronomy *f.* astronomía

asymmetric(al) asimétrico, anórtico || ***asymmetric(al) anticline*** *m.* anticlinal asimétrico || ***asymmetric(al) fold*** *m.* pliegue asimétrico

asymmetry *f.* asimetría

asymptote *f.* asíntota

asymptotic *a.* asintótico

asynchronous *a.* asíncrono, asincrónico

atacamite *f.* atacamita, remolenita

atectonic *a.* atectónico

atlas *m.* atlas

atmosphere *f.* atmósfera

atmospheric *a.* atmosférico

atoll *m.* atolón, atol

atom *m.* átomo

atomic *a.* atómico || ***atomic number*** *m.* número atómico || ***atomic weight*** *m.* peso atómico

at random *m. & a.* al azar

attach *vb.* unir, juntar, fijar, atar

attached *a.* adherido, fijo, agregado

attachments *m.* accesorios, anexos

attack *m.* ataque || ***bacterial attack*** *m.* ataque bacterial

attenuate *vb.* atenuar

attenuation *f.* atenuación

attitude *f.* disposición, posición; *m.* comportamiento

attraction *f.* atracción

attrition *m.* desgaste

attritus *m.* atrito

Aturian *m. & a.* Aturiense, Aturiano

atypical *a.* atípico

audio frequency *f.* audiofrecuencia

audit *f.* revision

audit *vb.* revisar

auganite *f.* auganita

augen gneiss *m.* gneis ojiforme, augengneis, neis de ojos, neis glandular || ***augen gneiss structure*** *f.* estructura ojiforme

auger *f.* barrena, broca, catadora, tienta; *m.* taladro

augite *f.* augita

aulacogen *m.* aulacógeno

aureole *f.* aureola

aurichalcite *f.* auricalcita

auriferous *a.* aurífero

austenite *f.* austenita

Australia *f.* Australia

authiclastic *a.* auticlástico

authigenesis *f.* autigénesis

authigenic *a.* autígeno, autigénico

autochthon *m.* autóctono

autochthonous *a.* autóctono

autoclast *m.* autoclasto

autoclastic *f.* autoclástico

autoclave *f.* autoclave

autocorrelation *f.* autocorrelación

autogenetic *a.* autógeno

autolith *m.* autolito

automation *f.* automatización

autometamorphism *m.* autometamorfismo

automorphic *a.* automórfico

autopore *m.* autóporo

autumn *m.* otoño

Autunian *m. & a.* Autuniense, Autuniano

autunite *f.* autunita

auxiliary *a.* auxiliar || ***auxiliary minerals*** *m.* minerales auxiliares

avalanche *f.* avalancha; *m.* alud, [And.] volcán || ***ice avalanche*** *m.* alud de hielo

aventurine *f.* aventurina, venturina

average *a.* promedio

avicularium *m.* aviculario

avoirdupois *m.* avoirdupois (1 libra = 0,4536 kg) || ***avoirdupois weight*** *m.* peso avoirdupois

awl *f.* lezna

axe *f.* hacha || ***mine axe*** *f.* ademadora

axial *a.* axial, axil, del eje || ***axial angle*** *m.* ángulo de charnela || ***axial plane*** *m.* plano axial || ***axial plane cleavage*** *m.* clivaje del plano axial

axillary *a.* axilar

axinite *f.* axinita, yanolita

axis *m.* eje || ***axis of symmetry*** *m.* eje de simetría || ***crystallographic axis*** *m.* eje cristalográfico || ***fold axis*** *m.* eje de pliegue || ***maximum and minimum principal axes*** *m.* ejes principales máximo y mínimo || ***optic axis*** *m.* eje óptico || ***principal strain axis*** *m.* eje principal de deformación

azimuth *m.* acimut, azimut, rumbo

azimuthal *a.* acimutal, azimutal, zenital

azoic *m. & a.* azóico

azurite *f.* azurita, malaquita azul, chessylita, quesilita; *m.* cobre azul

B

B horizon *m.* horizonte B del perfil del suelo

babbitt *m.* metal blanco, metal de antifricción, babbitt

back *a.* trasero, dorso; *adv.* posterior, atrás

backdeep *f.* fosa posterior, posfosa, trasfosa

backfire *f.* contraexplosión

backfill *m.* relleno, terraplén, rehincho

backfill *vb.* rellenar, rehinchar

backflow *f.* contracorriente; *m.* contraflujo

background *m.* fondo; antecedente

backhoe *m.* retroexcavador; *f.* pala mecánica, zanjadora

backlimb *m.* flanco trasero, flanco posterior

backoff joint *f.* unión desprendible

backpack *f.* mochila

backplug a well *m.* retrotaponamiento

backpressure *f.* contrapresión

backreef *m.* postarrecife, tras arrecife

backshore *f.* playa alta, retroplaya, ribera posterior, zona de supramarea; *a.* post-litoral

backsight *f.* nivelada de atrás; retrolectura retrovisual, visual inversa, visual atrás

backslope *f.* menor pendiente, pendiente posterior; *m.* talud exterior

backstop of a critical wedge *m.* contratope

backstroke *f.* carrera de retroceso del émbolo

backthrust *f.* retrofalla, retrofalla inversa, retrocabalgadura

back up *vb.* (comp.) respaldar

backwash *f.* resaca; *m.* retrolavado, retroceso

backwater *m.* remanso

bacteria *f.* bacteria

bad air *m.* aire viciado

badlands *f.* tierras malas, malpaís; *m.* [And.] huaiqueria

baffle *m.* deflector, desviador, desvío, tabiquc; *f.* chicana

bag *m.* saco

baguio *m.* baguío

bahada *f.* bajada

bail *f.* asa, manija; (com.) fianza

bail *vb.* achicar

bailer *m.* achicador; *f.* cuchara || ***sand bailer*** *m.* achicador de arena

bailing *m.* achique, cuchareo, baldeo

Bajocian *m. & a.* Bajocense, Bajociense, Bayociense, Bajociano, Bayociano

bake *vb.* hornear

bakelite *f.* bakelita

baking *f.* cocción

balance *f.* balanza, balancín, báscula; *m.* equilibrio, balance || ***analytical balance*** *f.* balanza analítica || ***area balance*** *m.* balanceo por área || ***balance beam*** *m.* balancín || ***balance due*** *m.* alcance, monto por cobrar || ***balance sheet*** *m.* corte de caja || ***balance weight*** *m.* contrapeso || ***credit balance*** *m.* saldo acreedor || ***debit balance*** *m.* saldo deudor || ***energy balance*** *m.* balance de energía || ***line-length balance*** *m.* balance por larga, balanceo por la longitud de líneas estratales || ***material balance*** *m.* balance de materiales || ***net balance*** *m.* saldo líquido || ***torsion balance*** *f.* balanza de torción

balance *vb.* balancear, contrapesar, (com.) saldar

balanced *a.* balanceado || ***balanced sections*** *m.* transectos balanceados; *f.* secciónes balanceadas

balas ruby *m.* balaje, rubí balas, rubí espinela

ball and seat valve *f.* válvula de bola y asiento

ball and socket joint *f.* articulación esférica

ballast *m.* lastre, balasto, balastro, balastre; *f.* zahorra

ball bearing *m.* cojinete de bolas

balled bit *m.* trépano empastado

ball grinder *f.* amoladora de bolas

ball gudgeon *m.* cojinete, muñón esférico

ball joint *f.* unión esférica

ball valve *f.* válvula de bola

balsam *m.* bálsamo

band (e.g., of fields, of reefs) *f.* faja, liga, banda, lista, franja, tendencia || ***band brake*** *m.* freno de banda || ***bandpass filter*** *m.* filtro pasabandas || ***band saw*** *f.* sierra sin fin || ***band width*** *f.* anchura de banda

banded *a.* rayado, bandeado, cinteado, fajeado

banding *a.* bandamiento, bandeamiento

bank *m.* banco, bajío; *f.* orilla, ribera || ***bank lending funds to mines*** *m.* banco de avíos || ***bank note*** *m.* billete || ***sand bank*** *f.* restinga || ***sloping bank*** *m.* ribazo

bankrupt *a.* insolvente, quebrado

bankrupt *vb.* hacer bancarrota

bankruptcy *f.* quiebra

bar *f.* barra; palanca, alzaprima; (phys., física) bara; *m.* alfaque; (ingot) riel || ***bay mouth bar*** *f.* barrera litoral || ***beach bar*** *f.* barra litoral, barrera litoral, restinga || ***longshore bar*** *f.* barra prelitoral, barrera prelitoral || ***river bar*** *f.* barra de río || ***sand bar*** *f.* barra de arena; *m.* tómbolo

barbed drainage *m.* diseño de avenamiento en lengüeta

barchan, barkhan *f.* barján, barcana, barchan

barge *f.* barcaza, bacha, gabarra; *m.* lanchón, chalán, planchón || ***drilling barge*** *f.* gabarra de perforación || ***pipelaying barge*** *f.* barcaza tiendetubos

barite *f.* barita, baritina, wolnina; *m.* chekao, espato pesado, wolnym, [Bol.] cachí

barium *m.* bario

barnacle *m.* cirrípedo, bárlano || ***stalked barnacle*** *m.* percebe, escaramujo

barograph *m.* barógrafo

barometer *m.* barómetro

barometric *a.* barométrico

barred basin *f.* cuenca obstruida

barrel *m.* barril, tonel, tambor (equivalente a 158,98 litros petróleo) || ***barrel of oil equivalent (BOE)*** *m.* barriles de petróleo equivalente || ***barrel of oil per day (BOPD)*** *m.* barriles de petróleo por día

Barremian *m. & a.* Barremense, Barremiense, Barremiano

barren *a.* no productivo, pobre, estéril, seco, bobo, calvo, infecundo || ***barren rock*** *f.* borra, borrasca; *m.* tepetate || ***barren vein material*** *f.* [Bol., Pe.] broza

barrier *f.* barrera; valla || ***barrier bar*** *f.* barrera litoral; *m.* banco barrera || ***barrier beach*** *m.* banco, banco barrera, bajío, cordón litoral; *f.* barra, restinga || ***barrier complex*** *m.* complejo de barrera || ***barrier reef*** *f.* arrecife barrera || ***barrier island*** *m.* islote litoral

barter *m.* treque; *f.* trueque, permutación

barter *vb.* trocar, permutar

Bartonian *m. & a.* Bartoniense, Bartoniano

barysphere *f.* barisfera

basal *a.* basal, básico || ***basal conglomerate*** *m.* conglomerado basal

basalt *m.* basalto || ***basalt flow*** *f.* colada basáltica, efusión de basalto || ***basalt plateau*** *f.* meseta basáltica; *m.* trapp || ***columnar basalt*** *m.* basalto prismático

basaltic *a.* basáltico

basanite *f.* basanita

base *f.* base; *m.* zócalo || ***base level*** *m.* nivel de referencia, base de erosión || ***base line*** *f.* línea de base, base de triangulación || ***base map*** *m.* plan de base, mapa de referencia || ***base metal*** *m.* metal de base || ***base mineral index*** *m.* índice de los minerales básicos || ***base of weathering*** *f.* base del suelo intemperizado || ***base station*** *f.* estación de base

basement rock *m.* basamento, sótano, zócalo, sócalo; *a.* basal || ***basement rock complex*** *m.* complejo de base, complejo basal

basic *a.* (quím., chem.) básico; (petrol.) básico, máfico || ***basic rocks*** *f.* rocas basicas, [Arg.] basitas

basin *f.* cuenca, hondonada; pileta, pila, alcubilla, hoya, torca, cubeta, [Méx.] alberca; *m.* cuenco, estanque, vaso, [Arg.] tajamar || ***backarc basin*** *f.* cuenca postarco || ***barred basin*** *f.* cuenca obstruida || ***basin floor*** *m.* piso de cuenca || ***basin floor fan*** *m.* abanico de piso de cuenca || ***closed basin*** *f.* cuenca cerrada || ***catchment basin*** *f.* cuenca colectora || ***drainage basin*** *f.* cuenca de drenaje o hidrográfica || ***internal basin*** *m.* bolsón || ***piggyback basin*** *f.* cuenca acuesta, cuenca transportada || ***sedimentary basin*** *f.* cuenca (o cubeta) sedimentaria || ***starved basin*** *f.* cuenca empobrecida

basis *m.* base

basket *f.* cesta, taza, canasta; *m.* cesto, casquillo || ***basket bit*** *f.* barrena de cesto

bassanite *f.* vibertita

bastnesite *f.* weibyeíta

batch *f.* tanda; (ref.) cochada, [Pe.] templa; *m.* lote || ***batch distillation*** *f.* destilación por cochadas || ***batch processing*** *m.* (ref., comp.) proceso por lotes || ***batch vaporization*** *f.* vaporización por lotes

batholith *m.* batolito

Bathonian *m. & a.* Batonense, Batoniense, Batoniano

bathyal *a.* batial

bathymetric *a.* batimétrico

bathymetry *f.* batimetría

bathypelagic *a.* batipelágico

bathysphere *f.* batisfera

battery *m.* tanque, grupo; acumulador; *f.* batería; pila || ***tank battery*** *f.* batería de tanques

Baumé gravity *f.* densidad Baumé

bauxite *f.* bauxita, cliachita, wocheinita

bay *f.* bahía, anconada, angra, cala, ensenada; *m.* ancón, caney || ***bay bar*** *f.* espiga || ***bay head delta*** *f.* delta de cabeza de bahia || ***bay mouth bar*** *f.* barra de desembocadura de bahía

bayonet gauge *m.* indicador tipo bayoneta

bayou *m.* brazo pantanoso, brazo del río, ríachuelo, ríacho

BCF *m.* mil millones de pies cúbicos

beach *f.* playa || ***barrier beach*** *f.* playa de barrera || ***beach ridge*** *f.* restinga || ***raised beach*** *f.* playa levantada, playa costera

beacon *f.* baliza; *m.* radiofaro

bead-like *a.* perliforme

beaker *m.* cubilete, bidón; *f.* copa, bombona, redoma

beam *f.* viga; *m.* travesaño; balancín; (phys., física) rayo || ***balance beam*** *m.* balancín || ***beam hanger*** *f.* barra de suspensión || ***beam pump*** *f.* bomba de balancín || ***walking beam*** *m.* balancín || ***working beam*** *m.* balancín

Beaman stadia arc *m.* arco Beaman

bear *vb.* sostener, resistir

bearing *f.* chumacera, marcación, orientación; *m.* cojinete; rumbo, arrumbamiento || ***ball bearing*** *m.* cojinete de bolas, bolitas de cojinete || ***bearing metal*** *m.* babbitt || ***roller bearing*** *m.* cojinete de rodillos

Becke line *f.* linea de Becke

bed (geol.) *f.* capa, estratificación, cama, camada, hebra, vena; *m.* lecho, estrato, horizonte, jazido, yacimiento, banco || ***bed-length balancing*** *m.* balanceo por la longitud de líneas estratales || ***bed load*** *m.* acarreos || ***bed plane slip*** *m.* deslizamiento capa-a-capa || ***bed set*** *m.* conjunto de estratos || ***bottomset bed*** *f.* capa basal || ***carrier bed*** *f.* via de migración, capa de migración, formación conductor || ***coal bed*** *f.* capa de carbón || ***cross bed*** *m.* estrato cruzante || ***foreset bed*** *f.* capa frontal; *m.* estratos frontales || ***key bed*** *f.* capa clave, capa guía; *m.* estrato índice || ***marker bed*** *f.* capa guía, capa clave || ***overturned bed*** *f.* capa invertida o tumbada || ***red bed*** *f.* capa roja || ***river bed*** *m.* cauce, fondo, alveo || ***thick bed*** *m.* banco || ***topset bed*** *f.* capa dorsal, capa superior

bedded *a.* estratificado || ***thick-bedded*** *f.* estratificación espesa (o potente) || ***thin-bedded*** *f.* estratificación delgada

bedding *f.* estratificación; *m.* manteo || ***backset bedding*** *f.* estratificación inclinada inversa || ***bedding plane*** *m.* plano de estratificación; *f.* superficie de estratificación || ***convolute bedding*** *f.* estratificación intraplegada (o convoluta) || ***cross bedding*** *f.* estratificación cruzada, diagonal, o entrecruzada; *m.* estratos entrecruzados || ***graded bedding*** *f.* estratificación gradada, graduada, o gradual || ***laminar bedding*** *f.* estratificación laminar || ***lenticular bedding*** *f.* estratificación lenticular || ***massive bedding*** *f.* estratificación masiva || ***wavy bedding*** *f.* estratificación ondulítica o ondulada

bedrock *m.* lecho de roca, estrato de roca, subestrato, basamento; *f.* roca viva

bedsets *m.* conjuntos de estratos || ***shingled bedsets*** *m.* conjuntos de estratos escalonados (o imbricados)

beerbachite *f.* beerbachita, beerbaquita

behave *vb.* conducirse, comportarse

behavior *m.* comportamiento, funcionamiento

behead *vb.* captar, capturar, decapitar

behind *adv.* atrás, poste; *a.* posterior; *p.* per atrás

belemnite *f.* belemnita

belemnoid *m.* belemnoideo

below *p.* por abajo de, debajo de; *adv.* yuso, bajo, abajo, debajo

belt (mountain) *m.* cinturón, faja, banda, cinta, zona, cordillera || ***belt drive*** *m.* mando por correa || ***conveyor belt*** *f.* banda transportadora || ***fold belt*** *m.* cinturón de pliegues, cinturón plegado || ***mineral belt*** *f.* zona metalífera, faja metalífera || ***mountain belt*** *m.* cinturón de montañas; *f.* sierra, cordillera || ***volcanic belt*** *m.* cinturón volcánico; *f.* cadena volcánica

bench *m.* banco, (min.) antepecho, bancal; *f.* banqueta, terraza de roca, bancada, banca || ***bench level*** *m.* nivel de banco

benchmark *f.* cota de referencia, acotación; *m.* banco de cota fija, punto de referencia, machote, [Vz.] botalón

bend *m.* doblez; *f.* curva || ***expansion bend*** *f.* curva de dilatación, curva de expansión || ***river bend*** *m.* recodo; *f.* retuerta

bend *vb.* combarse, doblarse

bending *f.* dobladura, encorvadura, flexión || ***bending stress*** *m.* esfuerzo de flexíon

beneath *adv.* & *p.* subyacente

benefit *m.* beneficio, bienes; *f.* merced, utilidad

benitoite *f.* benitoíta

bent *a.* acombado, doblado

benthic, benthonic *a.* bentónico

bentonite *f.* bentoníta

bentonitic *a.* bentonítico

benzene *m.* benceno; *f.* bencina

benzol *m.* benzol

benzylacetylene *m.* bencilacetileno

bergschrund *f.* rimaya, grieta circence

berm *f.* berma; *m.* bancal || ***summer berm*** *f.* plataforma de verano || ***winter berm*** *f.* plataforma de invierno

Berriasian *m.* & *a.* Berriasiense, Berriasiano

berthierite *f.* bertierita
beryl *m.* berilo, berillo, berilio || ***cesium and lithium beryl*** *f.* worobieffita
beryllium *m.* berilio
be subject to *vb.* sufrir
beta particle *f.* partícula beta
beta ray *m.* rayo beta
better *a.* mejor, superior
bevel *m.* bisel, chaflán
bevel *vb.* biselar, chaflanar
beveled *a.* biselado, truncado
BHT (bottom hole temperature) *f.* temperatura del fondo
bias *m.* al sesgo; *f.* oblicuidad
biased *a.* sesgo, sesgado
biaxial *a.* biaxial, diaxial, biáxico
bibb *m.* grifo
bicarbonate *m.* bicarbonato
bichambered *a.* bilocular
bichloride *m.* bicloruro
bichromate *m.* bicromato
biconcave *a.* bicóncavo
biconical *a.* bicónico
biconvex *a.* biconvexo
bicyclodecane *m.* biciclodecano
bicyclohexane *m.* biciclohexano
bicyclononane *m.* biciclononano
bid *f.* licitación || ***bid round*** *f.* licitación
bid *vb.* licitar
bidder *m.* postor
bifid *a.* bífido, bifurcado
bifurcate *a.* bifurcado
bight *f.* ensenada, curva, cala; *m.* seno
bihexagonal *a.* bihexagonal
bilateral *a.* bilateral
bilge *f.* sentina
bill *f.* factura, cuenta || ***bill of exchange*** *f.* letra de cambio || ***bill of lading*** *m.* conocimiento de embarque
bill *vb.* pasar la cuenta, cobrar
bimodal *a.* bimodal
bin *f.* celda, tolva
binary *a.* binario || ***binary code*** *m.* código binario
bind *vb.* ligar
binder *m.* aglomerante, ligador, aglutinante
bindstone *f.* costras formadas por organismos
binocular *a.* binocular
binoculars *m.* binóculo, gemelos, binoculares
binomial *a.* binominal
biochemical *a.* bioquímico
bioclast *m.* bioclasto
bioclastic *a.* bioclástico
biofacies *m.* biofacies
biogenesis *f.* biogénesis
biogenic *a.* biogénico, biogenético
bioherm *f.* bioherma
biolith *f.* biolita, zoolita
biologist *m.* biólogo
biology *f.* biología
biomarker *m.* biomarcador
biomass *f.* biomasa
bioremediation *f.* bioremediación
biosparite *f.* biosparita
biosphere *f.* biósfera
biostratigraphic *a.* bioestratigráfico || ***biostratigraphic unit*** *f.* unidad bioestratigráfica
biostratigraphy *f.* bioestratigrafía
biostromal *a.* biostromal
biostrome *f.* biostroma
biota *f.* biota
biotic *a.* biótico
biotite *f.* biotita, mica negra
bioturbated *a.* bioperturbado, bioturbado
bioturbation *f.* bioturbación, bioperturbación
biozone *f.* biozona
bipyramid *m.* dipirámide, bipirámide
bipyramidal *a.* dipirámide, bipirámidal
bipolarity *f.* bipolaridad
bird foot delta *f.* delta pie de pájaro
birefringence *f.* birrefringencia
birefringent *a.* birrefringente
bisect *vb.* bisecar, bisectar
bisectrix, bisector *f.* bisectriz

bismuth *m.* bismuto || ***bismuth carbonate or oxide*** *f.* bismuto ocre

bismuthinite, bismuthite *f.* bismutinita, bismutina, bismutita

bisulfate *m.* bisulfato

bit *f.* (perf.) barrena, boca de barrena, [Col.] broca, fresa; *m.* (perf.) trépano; (comp.) digito binario || ***auger bit*** *f.* barrena espiral || ***bit basket*** *m.* cesto de barrena || ***bit breaker*** *m.* desconectador de barrena || ***bit cone*** *m.* cono de barrena || ***cable drilling bit*** *f.* barrena percutente || ***disk bit*** *f.* barrena de discos || ***jet bit*** *f.* barrena de chorro || ***roller bit*** *f.* barrena de rodillos || ***rotary bit*** *f.* barrena rotatoria || ***rotary disk bit*** *f.* barrena giratoria de disco

bitetragonal *a.* bitetragonal

bitumen *m.* betumen, betún, bitumen, [Ec.] copey; azulaque

bituminization *f.* bitumenización

bituminous *a.* bituminoso, bituminífero || ***bituminous coal*** *m.* carbón bituminoso || ***bituminous shale*** *f.* lutita bituminosa

bivalent *a.* bivalente

bivalve *m. & a.* bivalvo

bixbyite *f.* bixbyita, partridgeita, sitaparita

blade *f.* aleta

black hole *m.* agujero negro

blanket *f.* manta, sábana; *m.* manto || ***blanket deposit*** *f.* manta || ***blanket-like*** *a.* manteado || ***blanket sand*** *m.* manto de arena

blast *f.* voladura, explosión, dinamitación, ráfaga, detonación; *m.* disparo, soplo, chorro, cohete || ***blast furnace*** *m.* alto horno, horno de fundación || ***blast hole*** *f.* perforación para voladura; *m.* chocolón

blast *vb.* pegar, volar, estallar

blaster *m.* detonador, fulminante

blasting *f.* voladura, dinamitación || ***blasting cap*** *m.* fulminante; *f.* cápsula explosiva, cápsula

blastoid *m.* blastoideo

blastogranitic *a.* blastogranítico

blastopore *m.* blastóporo

blastoporphyritic *a.* blastoporfirítico

bleach *vb.* descolorar, descolorir, blanquear

bleached *a.* descolorido, blanqueado

bleaching *m.* blanqueo, descoloramiento

bleed off *vb.* purgar

bleed point *m.* punto de purga

bleeder *m.* grifo de purga

bleeder valve *m.* purgador

bleeding *m.* flujo

blend *f.* mezcla, mixtura

blend *vb.* mezclar

blended *a.* mezclado

blender *f.* mezcladora

blending *a.* mezclante

blind *a.* ciego, sin salida, persiana || ***blind load*** *f.* veta ciega || ***blind thrust*** *f.* falla ciega, cabalgadura con traza oculta || ***blind zone*** *f.* zona ciega

blister *f.* ampolla

blizzard *f.* nevisca, nevasca

block *m.* bloque, motón; *f.* garrucha, polea || ***block caving*** *f.* explotación por socavación y derrumbe || ***block diagram*** *m.* diagrama en bloque || ***fault block*** *m.* bloque de falla || ***slide block*** *f.* placa de deslizamiento

block out (min.) *vb.* bloquear, cubicar, delinear

blocked out reserves *f.* reservas comprobadas

blocky *a.* blocoso, en forma de bloques

blödite *f.* warthita

blood *f.* sangre

bloodstone *f.* calcedonia roja y verde, dionisia, sanguinaria; *m.* heliotropo

bloom *f.* (botan.) flor, floración || ***bloom of oil*** *f.* fluorescencia del petróleo

blow, blow away, blow off *vb.* soplar

blowback *m.* soplo inverso

blower *m.* soplador, ventilador, insuflador

blowoff valve *f.* válvula de descarga o de escape; *m.* purgador

blowout *m.* reventón; *f.* erupción, surgencia; (geog.) depresión eólica

blowout preventer (BOP) *f.* válvula de seguridad o de emergencia; *m.* impiderreventones, cierre de seguridad, preventor

blowpipe, blowtorch *m.* soplete

blueprint *f.* copia azul, heliografía; *m.* fotocalco azul

blueprint *vb.* fotocalcar

bluff *m.* risco, morro, acantilado, barranco; *f.* barranca

board of directors *f.* junta directiva; *m.* consejo directivo

boat *m.* bote, barco; *f.* lancha || ***life boat*** *m.* bote salvavidas

bodenbenderite *f.* bodenbenderita

body *m.* cuerpo, armazón || ***extrusive body*** *m.* cuerpo efusivo || ***igneous body*** *m.* cuerpo ígneo || ***intrusive body*** *m.* cuerpo intrusivo || ***magmatic body*** *m.* cuerpo magmático || ***ore body*** *m.* cuerpo de mena, cuerpo mineralizado, cuerpo mineral

bog *m.* pantano, aguazal, cenegal, fangal, lodazal, marjal, [PR] cienagal, [Arg., Ch.] agualotal; *f.* ciénaga, [Méx.] zoquetal || ***bog iron*** *f.* goethita, limonita || ***bog manganese*** *f.* acerdesa

boghead coal *m.* [Sp.] bogue

boil *vb.* hervir

boiler *f.* caldera, boiler

boiling *f.* ebullición, cocción || ***boiling point*** *m.* punto de ebullición

Bollandian *m. & a.* Bollandiense, Bollandiano

bolson *m.* bolsón

bolt *m.* perno, tornillo, bulón

bolt *vb.* atornillar, empernar, bulonar

bomb *f.* (geol.) bomba; (phys., física) bomba

bonanza *f.* bonanza

bond *m.* trabazón, ligazón, grado de afinidad, enlace; (com.) bono; *f.* adhesión, (chem., quím.) aligación, ligación, (com.) fianza || ***atomic bond*** *m.* enlace atómico

bond *vb.* adherir, ligar

bondsman *m.* fiador

bone *m.* hueso || ***bone coal*** *m.* carbón argiláceo || ***bone-colored*** *a.* ahuesado || ***bone hard*** *a.* ahuesado

Bononian *m. & a.* Bononiense, Bononiano

bonus *m.* bono; *f.* prima || ***cash bonus*** *m.* bono de caja; *f.* bonificación

bony *a.* óseo, huesoso, ososo, huesudo

book *m.* libro

bookkeeper *m.* tenedor de libros

bookkeeping *f.* teneduría de libros

boom *m.* aguilón, botalón

booster *m.* reforzador, aumentador de presión || ***booster pump*** *f.* bomba reforzadora || ***booster station*** *f.* estación auxiliar

boot *m.* manguito

BOP (blowout preventer) *f.* válvula de seguridad o de emergencia; *m.* impiderreventones, cierre de seguridad, preventor

BOPD (barrels of oil per day) *m.* barriles de petróleo por día

boracite *f.* boracita

borate *m.* borato

borax *f.* bórax, borato de sosa, zala || ***borax deposit*** *f.* [Arg., Bol., Ch.] boratera || ***refined borax*** *m.* atíncar

border *f.* frontera, comarca, orla; *m.* borde, cuadro, marco, canto, margen, reborde

bordering *a.* limítrofe, rayano

borderland *f.* país marginal; *m.* borde continental

bore *f.* perforación; *m.* taladro, agujero; calibre, diámetro interior, pozo

bore *vb.* perforar, taladrar, barrenar, agujerear, sondear, sondar

boreal *a.* boreal, ártico, relativo al Norte

bored *a.* perforado

borehole *m.* taladro, barreno, sondeo, pozo, hueco

borer *f.* barrena de cateo

borescope *m.* calibrescopio

boric *a.* bórico

boride *m.* boruro

boring *f.* perforación de reconocimiento, trepa, perforación de cateo; *m.* sondeo, sondaje, trepado; *a.* perforante, taladrador

bornite *f.* bornita, erubescita; *m.* cobre abigarrado, mineral "pavo real"

boron *m.* boro

bort *m.* diamante negro

boss *m.* jefe, capataz, sobrestante, colero, mandón; (geol.) macizo, intrusivo; *f.*

(min.) protuberancia || ***shift boss*** *m.* minero de cuarto

bostonite *f.* bostonita

botryogen *f.* quetenita

botryoidal *a.* botrioide, arriñonado

bottleneck *m.* gollete de botella, paso angosto, atascamiento, cuello de botella

bottlenecking *f.* reducción del diámetro

bottom *m.* fondo; pie; base, hondo, pique || ***bottom dwelling*** *a.* viviente del fondo || ***bottom hole pressure*** *f.* presión del fondo del pozo || ***bottom hole temperature (BHT)*** temperatura del fondo del pozo || ***bottom land*** *f.* tierra de aluvión, vega || ***bottom set·bed*** *f.* capa basal || ***sea bottom*** *m.* fondo del mar

bottomless *a.* insondeable

bottoms *m.* residuos, asientos

bottomset(s) *m.* cimiento sedimentario || ***bottomset beds*** *f.* capas basales

boudin *f.* estructura de salchicha; *m.* boudín

boudinage *f.* estructura moniliforme, boudinage

Bouguer anomaly *f.* anomalía de Bouguer

Bouguer correction *f.* corrección de Bouguer

boulangerite *f.* boulangerita

boulder *m.* bloque, guijarro grande, peñasco, canto, canto rodado, bolón, cabezón, cancho, cuartón, morillo; pedrusco; *f.* peña, galga || ***boulder clay*** *f.* till, arcilla pedregosa

bounce cast *m.* calco de roce

bouncing pin *m.* indicador de golpeteo; *f.* aguja indicadora

boundary *m.* lindero, límite, linde, contacto; *f.* comarca, frontera || ***boundary marker*** *m.* balate || ***claim boundary*** *m.* lindero || ***constant pressure boundary*** *m.* límite de presión constante || ***no-flow boundary*** *f.* barrera sin flujo o impermeable || ***plate boundary*** *m.* límite de placa

boundstone *f.* biolitita

bournonite *f.* bournonita, wolchita

box canyon *m.* cajón

box core *m.* testigo cajón

Boyle's Law *f.* ley de Boyle

brace *f.* puntal, riostra; *m.* jabalcón || ***vertical brace*** *f.* riostra

brace *vb.* apuntalar, arriostrar, entibar, acodar, acodalar, apear, jabalconar

braced *a.* ligado, reforzado

bracing *m.* acodalamiento, acodamiento, apeo, refuerzo

brachial *a.* braquial

brachidium *m.* braquidio

brachiole *m.* braquiolo

brachiophore *m.* braquióforo

brachiopod *m.* brachiópodo, braquiópodo

bracing *f.* traba, puntal; *m.* refuerzo, ademado

brackish *a.* salobre, fluviomarino, salado

brackishness *f.* salobridad

bradenhead *m.* cabezal con prensaestopa

Bradfordian *m.* & *a.* Bradfordiense, Bradfordiano

braided stream *m.* curso ramificado; *f.* corriente ramificada

braidplain *f.* llanura trenzada, llanura de curso ramificado

brake freno, retranca

branch *f.* bifurcación, ramificación; rama; *m.* brazo || ***branch line*** *m.* ramal || ***branch point*** *m.* punto de ramificación || ***branch vein*** *f.* rama; *m.* ramal || ***secondary branch*** *m.* ramal

branch *vb.* bifurcarse, ramificarse, acometer, ramalearse

branched *a.* bifurcado, ramificado

branchial *a.* branquial

branching *f.* ramificación; *a.* ramificado, bifurcado

brand *f.* marca de fábrica

brass *m.* latón, bronce

braunite *f.* braunita

bravoite *f.* bravoita

breach *vb.* aportillar, fracturar, abrir

breached (anticline) *a.* (anticlinal) fracturado, abierto, descopetado

breadth *f.* anchura; *m.* ancho

break *f.* rotura, fractura, hendidura; interrupción, cuarteadura, quebraja, quebradura, quiebra, ruptura; *m.* cuarteo, rompimiento

break *vb.* romper, fracturar, interrumpir, desconectar, cuartear, quebrantar, quebrantarse, quebrar, quebrarse || ***break ground*** *vb.* arrancar, derribar, [Méx.] tumbar || ***break ore*** *vb.* [Méx.] tumbar

breaking ground *m.* arranque, derribo

breaker *m.* rompiente

breaking ground *m.* abataje

break in slope *f.* ruptura de pendiente

breakout *vb.* desenroscar tubería

breakthrough *m.* rompimiento || ***water breakthrough*** *f.* penetración del agua, llega, llegada

breakwater *f.* rompeolas, tajamar

breccia *f.* brecha || ***fault breccia*** *f.* brecha de falla || ***tectonic breccia*** *f.* brecha tectónica || ***volcanic breccia*** *f.* brecha volcánica

brecciated *a.* brechiforme, brechoide, brechoso, brechado

brecciation *f.* brechación

breeze *f.* brisa

breunerite *f.* walmstedtita

breviceracone, brevicone *m.* brevicono

brick *m.* ladrillo || ***clay brick*** *m.* adobe || ***fire brick, insulating brick*** *m.* ladrillo refractario

brickyard *f.* adobera, adorbería, ladrillera

bridge *m.* puente

Brigg's standard *f.* (perf.) especificaciones Brigg

bright *a.* pulido, claro, bruñido, pulimentado, brillante || ***bright stock*** *m.* aceite pesado refinado

brightness *m.* brillo

bright spot *m.* punto brillante

brilliance *f.* brillantez

brine *f.* salmuera, agua salada

briny *a.* salobre

British Thermal Unit (BTU) *f.* unidad térmica británica

brittle *a.* frágil, quebradizo, quebradizado, agrio, rígido; deleznable, quebrajoso, quebrantable

brittleness *f.* fragilidad

broad *a.* ancho

brochantite *f.* brochantita, kamarezita, warringtonita

brochure *m.* desplegable, folleto

broggerite *f.* broggerita

broken *a.* quebrado, roto

(to be) broken *vb.* quebrarse, romperse

broker *m.* corredor de bolsa

brokerage *f.* casa de comisión

bromargyrite *f.* bromargira, bromargirita

bromide *m.* bromuro

bromine *m.* bromo

bromlite *f.* bromlita

bronze *m.* bronce || ***bronze age*** *f.* edad de bronce

bronzite *f.* broncita

brook *m.* arroyo, arroyuelo; *f.* acequia, cañada, quebrada

brookite *f.* brookita, jurinita, piromelana

brown coal *m.* lignito, carbón pardo

brucite *f.* brucita

Brunton compass *f.* brújula Brunton

bryozoan *m.* briozoario, briozoo

BSI (British Standard Institute) *m.* Instituto Británico de Normas

BTU *f.* unidad térmica británica

bubble *f.* burbuja || ***bubble point*** *m.* punto de burbujeo || ***bubble pressure*** *f.* presión de burbujeo || ***bubble pulse*** *m.* (seis.) pulso de burbuja

buck *vb.* engrasar roscas, triturar minerales con martillo

bucket *m.* balde, cangilón, capacho, [Ch.] canguilón; *f.* cubeta || ***bucket fill*** *f.* cubetada || ***dragline bucket*** *m.* balde de arrastre; *f.* cubetadraga || ***mine bucket*** *m.* balde de extracción

bucking *f.* trituración

buckle *m.* (fold) pliegue bucle; (clasp) abrazadera, hebilla

buckle *vb.* doblarse

buckling *m.* pandeamiento, encorvamiento

budding *f.* gemación

budget *m.* presupuesto

buff *m.* color de ante

buffer *m.* tope, paragolpes; (polish) pulidor; (chem., quím.) tampón; *f.* (comp.) memoria intermedia (o compensadora)

bug (comp.) *m.* error; *f.* falla en software o hardware

build *vb.* construir || ***build a mill*** *vb.* plantar

builder *m.* constructor

build-up *f.* masa, estructura, barrera

bulge *f.* comba, protuberancia

bulged *a.* pando, hinchado

bulk *m.* volumen, macizo; *f.* masa || ***in bulk*** *a.* al granel, al por mayor || ***bulk modulus*** *m.* módulo de volumen || ***bulk sale*** *f.* venta al granel || ***bulk storage plant*** *f.* almacén

bulldozer *f.* empujadora niveladora, escrepa de empuje, oruga, "caterpílar"

bullion *m.* metal en barra; *f.* [Ch., Pe., Bol.] piña

bull wheel *m.* torno de herramientas

bulwark *m.* baluarte

bulwarked *a.* abaluartado

bundled *a.* atado, unido

bunker fuel *m.* combustóleo para barcos

Bunsen burner *m.* quemador Bunsen

bunsenite *f.* buncenita

buoy *f.* boya, baliza

buoyancy *f.* flotación, flotabilidad

burden *m.* cargo; gravamen

Burdigalian *m. & a.* Burdigaliense, Burdigaliano

bureaucracy *f.* burocracia

Bureau of Standards *m.* negociado de normas de EUA

burial *m.* entierro, sepelio

buried *a.* enterrado, sepultado; soterraño, soterrado, subterráneo || ***buried hill*** *m.* morro del subsuelo

burn *vb.* quemar, abrasar, arder, inflamar, incendiar

burner *m.* quemador, mechero || ***Bunsen burner*** *m.* quemador Bunsen

burrow *f.* excavación, madriguera, conejera, bioturbación; *m.* agujero

burrow *vb.* agujerear, excavar

burrowed *a.* agujereado, excavado || ***burrowed texture*** *f.* textura agujereada

burrowing *a.* excavador

burst *vb.* reventar, estallar

bury *vb.* enterrar, sepultar, soterrar

bushing *m.* buje, casquillo, manguito

business *m.* comercio, negocio; *f.* empresa

bustamite *f.* bustamita

butadiene *m.* butadieno

butane *m.* butano

butene *m.* buteno

butlerite *f.* butlerita

butte *f.* mesa, hamada; *m.* monte aislado, cerro testigo

button *m.* botón

buttress *m.* contrafuerte, macho, machón

butts *m.* residuos

butyl acetate *m.* acetato butílico

butylacetylene *m.* butilacetileno

butylbenzene *m.* butilbenceno

butylene *m.* butileno, buteno

buy *vb.* comprar

buyer *m.* comprador

bypass *m.* desvío; *f.* desviación, derivación || ***bypass zone*** *f.* zona desviada

bypass *vb.* desviar

byproduct *m.* subproducto, producto secundario

bysmalith *m.* bismalito

byte *m.* byte

bytownite *f.* bitounita

C

C horizon *m.* horizonte C del perfil del suelo

cable *m.* cable; *f.* [Vz.] guaya || ***cable tool*** *f.* herramienta para cable, herramienta para perforación a percusión || ***cable way*** *f.* cablevía

cadastral *a.* cadastral

cadastre *m.* catastro

cadmium *m.* cadmio

cage *m.* (min.) ascensor

cairn *m.* mojón

cake *f.* incrustación

cake *vb.* conglutinarse, endurecer, cuajarse

caked *a.* incrustado

caking *m.* (perf.) encostramiento

Calabrian *m. & a.* Calabriense, Calabriano, Calabrés

calamine *f.* calamina, zinconisa; *m.* óxido de cinc, zinc hidrocarbonatado, zinc oxidado silicífero

calaverite *f.* calaverita

calc-alkaline *a.* calcalcalino

calcarenite *f.* calcarenita

calcareous *a.* calcáreo, calizo, calar, calero, calichoso

calcic *a.* cálcico

calcification *a.* calcificación

calcify *vb.* calcificar

calcigranite *m.* kalkgranito

calcilutite *f.* calcilutita

calcination *f.* calcinación, tostación, tostadura

calcine *vb.* calcinar, quemar, reverberar

calciner *m.* calcinador

calcirudite *f.* calcirudita

calcisiltite *f.* calcilimolita

calcisponge *f.* calciesponja

calcite *f.* calcita, zeyringita; *m.* espato calcáreo, espato cálcico, espato calizo, espato de cal, [Ch.] cachí

calcitic *a.* calcítico

calcium *m.* calcio || ***calcium chloride*** *m.* cloruro de calcio || ***calcium hypochlorite*** *m.* hipoclorito cálcico

calculate *vb.* calcular, computar, derivar

calculation *f.* calculación, tasación; *m.* cálculo, cómputo || ***rough calculation*** *m.* tanteo

caldera *f.* caldera

Caledonian *m. & a.* Caledoniano, Calidoniense

caledonite *f.* caledonita

calf wheel *m.* torno de las tuberías

caliber *m.* calibre

calibrate *vb.* calibrar

calibrated *a.* calibrado

calibration *f.* calibración, ajuste

caliche *m.* caliche, talpetate, [Méx.] tepetate; *f.* nitratina, nitratita

californite *f.* californita

caliper *m.* calibrador, calibre

call for bids *m.* llamado a licitación

Callovian *m. & a.* Caloviense, Calloviense, Caloviano, Calloviano

CALM (Catenary Anchor Leg Mooring) *m.* monopodio de carga con ancla catenaria

calorie *f.* caloría

calorimeter *m.* calorímetro

calorimetry *f.* calorimetría

calving *f.* calva

calycal *a.* del cáliz

calyx *m.* cáliz

cam *f.* leva

camera *f.* cámara || ***aerial camera*** *f.* cámara aérea || ***camera lucida*** *f.* cámara clara || ***mapping camera*** *f.* cámara cartográfica

camarinid *m.* camerínido

Cambrian *m. & a.* Cámbrico, Cambriano

camp *m.* campamento || ***mining camp*** *m.* campamento o campo minero

camp *vb.* acampar

campaign *f.* campaña

Campanian *m. & a.* Campaniense, Campaniano

campanulate *a.* acampanado, campanudo

camphane *m.* canfeno

camptonite *f.* camptonita

campylite *f.* campilita

can *f.* vasija, lata; *m.* bote

Canada balsam *m.* bálsamo del Canadá

Canadian *m. & a.* Canadiano, Canadiense

canal *f.* acequia, caz; *m.* canal, [Méx.] colagón || ***diversion canal*** *m.* canal de derivación

cancrinite *f.* cancrinita

candle power *m.* poder iluminante

cannel coal *f.* ampelita; *m.* [Sp.] canel

canopy *m.* baldaquin, dosel, toldo || ***salt canopy*** *m.* baldaquin de sal

cantalite *f.* cantalita

cantilever *m.* cantilever, voladizo, volado || ***cantilever arm*** *m.* brazo volado || ***cantilever beam*** *f.* viga voladiza, viga acartelada

canyon *m.* cañón, cajón, cañadon, barrancón, congosto, hocino; *f.* cañada

cap *f.* capa, tapa; *m.* cubierto; (perf.) detonador; casquete || ***blasting cap*** *m.* fulminante, casquillo; *f.* cápsula explosiva || ***cap rock*** *f.* roca que cubre, roca de cubierta, capa protectora; *m.* capuchón; estrato de cobertura, techo || ***gas cap*** *m.* casquete de gas || ***ice cap*** *m.* manto glacial

capability *f.* capacidad

capable *a.* capaz

capacitance *f.* capacidad eléctrica

capacitor *m.* capacitor; condensador

capacity *f.* capacidad || ***design capacity*** *f.* capacidad calculada || ***flow capacity*** *f.* capacidad de flujo || ***heat capacity*** *f.* capacidad térmica || ***production capacity*** *f.* capacidad de producción || ***rated capacity*** *f.* capacidad estimada o nominal; *m.* potencial nominal

cape *m.* cabo; *f.* punta, angla

capillary *a.* capilar || ***capillary action*** *f.* capilaridad || ***capillary fringe*** *f.* franja capilar; orla freática || ***capillary water*** *f.* agua capilar

capillarity *f.* capilaridad

capital (com.) *m.* capital || ***capital assets*** *m.* bienes de capital, activos, capital activo, capital físico || ***capital stock*** *f.* acciones de capital || ***investment capital*** *m.* capital de inversión || ***venture capital*** *m.* capital arriesgado, capital de riesgo, capital invertido || ***working capital*** *m.* capital activo, capital de trabajo

capitalize on *vb.* aprovechar, capitalizar

capped *a.* (perf.) tapado; (geol.) cubierto

capping a well *vb.* controlar (o tapar) la erupción de un pozo

capstan *m.* cabrestante

capture *f.* captación, captura || ***stream capture*** *f.* captación de río

capture *vb.* capturar, atrapar, coger, pescar

car *m.* carro || ***car load*** *f.* vagonada || ***flat car*** *m.* carro de plataforma || ***hand car*** *m.* carro de mano || ***mine car*** *m.* carro minero || ***tank car*** *m.* carro cisterno, carro tanque

Caradocian *m. & a.* Caradociense, Caradociano

carapace *m.* carapacho, caparazón

carat *m.* quilate (equivalente a 205 mg)

carbene *m.* carbeno

carbide *m.* carburo

carbohydrate *m.* carbohidrato

carbolic acid *m.* ácido carbólico, phenol

carbon *m.* carbono; carbón || ***carbon chain*** *f.* cadena de hidrocarburos || ***carbon dioxide*** *m.* ácido carbónico, bióxido de carbono, bióxido carbónico || ***carbon-14*** *m.* carbono 14 || ***carbon–hydrogen ratio*** *f.* relación carbono–hidrógeno || ***carbon monoxide*** *m.* óxido carbónico, monóxido de carbono || ***carbon ratio*** *f.* proporción de carbono || ***carbon steel*** *m.* acero al carbono || ***carbon tetrachloride*** *m.* tetracloruro de carbono || ***fixed carbon*** *m.* carbón fijo || ***organic carbon*** *m.* carbono orgánico || ***total organic carbon (TOC)*** *m.* carbono orgánico total (COT)

carbonaceous *a.* carbonáceo, carbonoso

carbonate *m.* carbonato; *f.* caliza; *a.* carbonatado

carbonation *f.* carbonación

carbonatite *f.* carbonatita, karbonatita

carbonatization *f.* carbonatación

carbonic *a.* carbónico || ***carbonic acid*** *m.* ácido carbónico || ***carbonic-rich water*** *f.* acídula

Carboniferous *m. & a.* Carbonífero, Carbónico

carbonization *f.* carbonización

carborundum *m.* carborundo

carboxyl group *m.* grupo carboxilo

carburate *vb.* carburar

cardinal points *m.* puntos cardinales

care *f.* solicitud; *m.* cuidado

cargo *m.* cargamento, flete; *f.* carga

carina *f.* carena, quilla

carinate *a.* carenado, aquillado

Carlsbad twinning *f.* macla de Carlsbad

carlstainite *f.* karlsteinita

carnallite *f.* carnalita

carnelian *f.* carneola

Carnian *m. & a.* Karniense, Karniano, Carniense, Carniano

carnotite *f.* carnotita

carpoid *m.* carpoideo

carrier *m.* conductor, portador; transportador, acarreador || ***carrier bed*** *f.* via de migración, capa de migración, formación conductor

carry *vb.* llevar, cargar, portar

carryover *m.* exceso

cartel *m.* gremio

cartilage *m.* cartílago

cartographer *m.* cartógrafo

cartographic *a.* cartográfico

cartography *f.* cartografía

cartridge *m.* cartucho || ***blasting cartridge*** *m.* cartucho de voladura

carve *vb.* grabar, tallar, labrar, esculpir

carved *a.* labrado

carving *f.* labra

cascade *f.* cascada

cascadite *f.* kalicamptonita

cased-off well *m.* pozo entubado

case history *m.* antecedente de un caso, historial clínico, caso histórico, caso práctico

cash *m.* efectivo, pago anticipado, contado || ***cash on delivery (COD)*** *m.* cobro contra entrega, cobro o devolución || ***cash receipts*** *f.* entrada

casing *f.* tubería de revestimiento o de ademe, cañería aisladora o de entubación; *m.* casco, forro, revestidor, tubo de ademe || ***casing adapter*** *m.* adaptador para cañería de entubación || ***casing a well*** *f.* entubación || ***casing cutter*** *m.* cortatubos || ***casing head gas*** *m.* gas húmedo || ***casing head pressure*** *f.* presión en la cabeza de cañería || ***casing perforator*** *m.* perforador de tubos || ***casing plug*** *m.* tapón de tubería || ***casing point*** *f.* profundidad de entubación || ***casing pressure*** *f.* presión interior de la tubería || ***casing pump*** *f.* bomba para cañería de entubación || ***casing scraper*** *m.* raspatubos || ***casing shoe*** *f.* zapata de cementación de la tubería || ***casing splitter*** *m.* tajatubos, rajatubos || ***casing string*** *f.* sarta de revestimiento || ***casing conductor*** *m.* tubo conductor || ***intermediate casing*** *f.* tubería intermedia || ***surface casing*** *f.* primera tubería || ***well casing*** *m.* [Méx.] ademado

cask *m.* casco

cassiterite *f.* casiterita, cal de estaño; *m.* óxido de estaño

cast *m.* calco; *f.* impronta || ***cast iron*** *m.* hierro fundido, hierro colado, fundición || ***flute cast*** *m.* turboglifo intersepto || ***load cast*** *m.* calco de carga || ***scour cast*** *m.* turboglifo

cast *vb.* fundir, moldear, colar, vaciar

casting *f.* fundición

castings *f.* bolita fecal, coprolito

cataclasis *f.* cataclasis

cataclastic *a.* cataclástico

cataclysm *m.* cataclismo

catalog *m.* catálogo

catalysis *f.* catálisis

catalyst *m.* catalizador

catalytic *a.* catalítico || ***catalytic alkylation*** *f.* alquilación catalítica || ***catalytic cracking*** *m.* cräcking catalítico || ***catalytic dehydrogenation*** *f.* deshidrogenación catalítica || ***catalytic desulfurization*** *f.* desulfuración catalítica || ***catalytic medium*** *m.* agente catalítico

catalyze *vb.* catalizar

cataphoresis *f.* cataforesis

cataract *f.* catarata; *m.* rabiones, raudal

catastrophic *a.* catastrófico

catastrophism *m.* catastrofismo

catchment area *f.* zona de captación, cuenca imbrífera

catenary *f.* catenaria

cathead *m.* torno auxiliar

cathode *m.* cátodo, polo negativo || ***cathode ray*** *m.* rayo catódico

cathodic *a.* catódico || ***cathodic protection*** *f.* protección catódica

cation *f.* catión || ***cation exchange*** *m.* intercambio de bases

cat's eye *m.* ojo de gato; *f.* asteria

caucasite *f.* kaukasita

cauldron *m.* caldero || ***cauldron subsidence*** *m.* hundimiento en caldero

caulk, calk *vb.* recalcar, calafatear

causeway *m.* dique terraplén

caustic *a.* cáustico, optálico

cave *f.* abra, caverna, gruta, cueva, ajaquefa, bóveda, covacha, chingana, sima; *m.* antro, socavón || ***vertical cave*** *m.* [Méx.] sótano

cave *vb.* derrumbarse, hundirse

caved *a.* derrocado, hundido, derrumbado, desplomado || ***caved ground*** *m.* caídos

cave-in *m.* derrumbe, hundimiento, revenimiento, desplome

cavern *f.* caverna, bóveda, cueva, abra, gruta; *m.* antro, salón

cavernous *a.* cavernoso

caving *m.* derrumbamiento, derrumbe, hundimiento; *f.* [Bol.] aiza

cavings *m.* derrumbes

cavity *f.* cavidad, vacuola; abra, oquedad; *m.* hueco || ***sulfur-lined cavity*** *m.* [Pe., Ec.] máchay

cay *m.* cayo, islote

ceiling *m.* cielo raso, techo

celadonite *f.* kmaita

celestial sphere *f.* esfera celeste

celestite *f.* celestita, celestina

cell *f.* célula, celdilla; celda; pila, elemento

cellular *a.* celular

cellulose *f.* celulosa

Celsius scale *f.* escala Celsius

cement *m.* cemento; aglutinante

cement *vb.* cementar, aglutinar

cementation *f.* cementación

cemented *a.* cementado

cementer *m.* cementador

cementing *f.* cementación

Cenomanian *m. & a.* Cenomaniano, Cenomaniense, Cenomanense

Cenozoic *m. & a.* Cenozoico, Caenozoico, Cainozoico, Kainolítico

cenozone *f.* cenozona

center *m.* centro || ***center counter*** *m.* contador central || ***center line*** *f.* línea de centro

center *vb.* centrar

centigrade *m.* centígrado

centigram *m.* centígramo

centiliter *m.* centilitro

centimeter *m.* centímetro

centimeter-gram-second (cgs) *m.* centímetro-gramo-segundo (cgs)

centipoise *m.* centipoise

centistoke *m.* centistoke

central *a.* central

Central America *f.* América Central

centrifugal *a.* centrífugo

centrifuge *f.* centrífuga

centripetal *a.* centrípeto

cephalon *m.* escudo cefálico, céfalo, cefalón

cephalopod *m.* cefalópodo

cephalothorax *m.* cefalotórax

ceramic *a.* cerámico

cerargyrite *f.* cerargirita, querargira, querargirita, plata córnea, plata córnea blanca, plata parda; *m.* metal blanco

ceratite *f.* ceratita

cerite *f.* ocroíta

cerium *m.* cerio

certificate *m.* certificado, título || ***certificate of weights and measures*** *m.* certificado de pesas y medidas

certification *f.* certificación

cerussite *f.* cerusa, cerusita, albayalde; *m.* carbonato plomizo, plomo blanco

cesium *m.* cesio

ceylonite *f.* zeylanita

chabazite *f.* chabasita, chabasia, adipita, zeolita cúbica

Chadronian *m. & a.* Chadroniano, Chadroniense

chain *f.* (geog.) cadena; (meas.) cadenada; *m.* rosario || ***chain structure*** *f.* estructura en cadena || ***chain survey*** *m.* levantamiento en cadena

chain *vb.* cadenar

chalcanthite *f.* calcantita, alcaparrosa azul; *m.* vitriolo de cobre

chalcedony *f.* calcedonia, carneola, jenzschita || ***blue chalcedony*** *f.* zafirita

chalcocite *f.* calcocita, calcosina, redruthita; *m.* sulfuro de cobre

chalcopyrite *f.* calcopirita, chalcopirita, pirita de cobre, pirita cobriza; *m.* abronzado, bronce, coribronce, sotlanque

chalcostibite *f.* wolfsbergita

chalcotrichite *f.* calcotriquita

chalk *f.* creta, tiza, greda, toba; *m.* buro, gis

chalky *a.* gredoso, cretoso, calcáreo

chalybite *f.* siderita, calibita, hierro espático

chamber *f.* cámara, caja; columna; *m.* (min.) anchurón, chiflón, salón, [Pe.] bovedón || ***combustion chamber*** *f.* cámara de combustión || ***chamber and pillar*** *a.* explotación por cámaras y pilares

chamosite *f.* chamoisita

Champlainian *m. & a.* Champlainiano, Champlainiense

chance *m. & a.* azar; *f.* probabilidad; oportunidad

change *m.* cambio, vuelto

change *vb.* cambiar, transformar || ***change direction*** *vb.* torcer || ***change strike*** *vb.* torcer el rumbo

channel *f.* canal, acequia; ranura, canaleta, agojía, vaguada, zanja, zubía, [Sp.] alfagra; *m.* cauce, álveo, furo || ***channel bar*** *f.* barra de cauce || ***channel fill*** *m.* depósito de relleno de cauces || ***channel sampling*** *m.* muestreo por canaleta

channeling *f.* canalización

chaotic *a.* caótico

chaparral *m.* matorral, chaparro

character *m.* carácter, rasgo

characteristic *a.* característico

characteristic *f.* característica; *m.* rasgo

characterize *vb.* caracterizar

charcoal *m.* carbón de madera, carbón

charge *f.* carga || ***hydrocarbon charge*** *f.* carga de hidrocarburo || ***smelter charge*** *f.* parva

charge *vb.* cargar; pasar la cuenta

charged *a.* cargado

Charles' law *m.* ley de Charles

Charmouthian *m. & a.* Charmoutiense, Charmoutiano, Charmoutense

Charnian *m. & a.* Charniano, Charniense

charnockite *m.* charnockíto; *f.* charnoquita

chart *f.* carta, gráfica; *m.* esquema, diagrama, mapa marina, ábaco, gráfico, registro || ***flow chart*** *m.* organigrama, ordinograma || ***nautical chart*** *f.* carta náutica || ***organization chart*** *m.* ordinograma

charting *f.* cartografía

chasm *f.* sima, fossa, grieta; *m.* abismo

chassignite *m.* chassignito

chatter mark *f.* marca de rechinamiento

Chattian *m. & a.* Chattiense, Chattiano

Chautauquan *m. & a.* Chautauquiense, Chautauquiano

Chazyan *m. & a.* Chaziense, Chaziano

check *f.* comprobación, verificación

check *vb.* comprobar, verificar, revisar

checkshot survey *m.* tiro de velocidad

chelae *f.* quelas

chelicerate *m.* quelicerado

chelidon *f.* quelidonia

chemical *a.* químico || ***chemical affinity*** *f.* afinidad química || ***chemical analysis*** *m.* análisis químico || ***chemical composition*** *f.* composición química || ***chemical compound*** *m.* compuesto químico || ***chemical engineer*** *m.* ingeniero químico || ***chemical reaction*** *f.* reacción química

chemist *m.* químico

chemistry *f.* química

Chemungian *m. & a.* Chemungiense, Chemungiano

chernozem *m.* chernozem

chert *m.* pedernal, chert, horsteno, sílex; *f.* lidita, piedra, córnea

chertification *f.* chertificación

chessylite *f.* chessylita, azurita

Chesterian *m. & a.* Chesteriense, Chesteriano

chevron fold *m.* pliegue cabrío, pliegue chevrón

chiastolite *f.* quiastolita

Chideruan *m. & a.* Chideruano, Chideruense

chief *m.* jefe || ***crew chief*** *m.* jefe de cuadrilla, jefe de tripulación || ***party chief*** *m.* jefe de grupo

chief *a.* principal || ***chief executive officer*** *m.* jefe ejecutivo oficial

chilidium *m.* quilidio

chill *vb.* acerar, enfriar

chilled contact *f.* margen enfriada (o congelada)

chimney *f.* chimenea, tronera; *m.* canal de erupción

China clay *m.* caolín; *f.* arcilla china

chip *f.* astilla, astilladura, lasca || ***chip sampling*** *m.* muestreo por picado

chip *vb.* desmenuzar, descantillar

chisel *m.* formón, cincel, cortafierro, cortafrío, cortahierro; *f.* acocadera, tajadera || ***stonemason's chisel*** *m.* puntero; *f.* uñeta

chisel *vb.* cincelar

chitin *f.* quitina

chitinous *a.* quitinoso

chiton *m.* quitón

chladnite *f.* chladnita

chloracetone *f.* cloroacetona

chloanthite *f.* cloantita

chlorate *m.* clorato

chloric *a.* clórico

chloride *m.* cloruro || ***polyvinyl chloride (PVC)*** *m.* cloruro de polivinilo || ***vinyl chloride*** *m.* cloruro de vinilo

chlorinated hydrocarbons *m.* hidrocarburos clorinados

chlorination *f.* clorinación

chlorine *m.* cloro || ***chlorine process ore treatment*** *m.* beneficio por cloruración

chlorite *f.* clorita || ***swelling chlorite*** *f.* pseudoclorita

chloritic *a.* clorítico

chloritization *f.* cloritización

chloritoid *m.* cloritoide

chloroanaline *f.* cloroanilina

chloroform *m.* cloroformo

chlorophane *f.* clorofana

chlorophyll *f.* clorofila

choke *m.* estrangulador, obturador || ***adjustable choke*** *m.* estrangulador graduable

choke *vb.* estrangular

choked *a.* estrangulado

choker *m.* estrangulador, reductor

chondrite *f.* condrita; *m.* chondrito

chondritic *a.* chondrítico, chóndrulo

chondrodite *f.* condrodita

chondrule *m.* cóndrulo

choose *vb.* escoger, seleccionar

chordate *m.* cordado

Christmas tree *m.* (perf.) arból de conexiones, arból de navidad, arból de válvulas, cabezal

chromate *m.* cromato

chromatic *a.* cromático || ***chromatic aberration*** *f.* aberración cromática || ***chromatic dispersion*** *f.* dispersión cromática

chromatography *f.* cromotografía

chrome *m.* cromo

chromite *f.* (petrol.) cromita, cromoferrita, siderocroma; *m.* (chem., quím.) cromito, hierro cromado

chromitite *f.* cromitita

chromium *m.* & *a.* cromo || ***chromium steel*** *m.* acero al cromo

chromosome *f.* cromosoma

chronograph *m.* cronógrafo

chronolite *m.* cronolito

chronological *a.* cronológico

chronology *f.* cronología

chronometer *m.* cronómetro

chronostratigraphic *a.* cronoestratigráfico

chrysene *m.* criseno

chrysoberyl *m.* crisoberilo; *f.* cimofana

chrysocolla *f.* crisócola

chrysolite *f.* crisolita; *m.* crisolito

chrysoprase *f.* crisoprasa

chrysotile *m.* crisotilo; *f.* zermattita

churchite *f.* weinschenkita

churn *vb.* batir, agitar

churn drilling *f.* perforación a percusión

churned *a.* agitado

chute *f.* caída, cárcava, chorrera; *m.* conducto, criadero, chorreadero, saetín || ***loading chute*** *f.* alcancía || ***ore chute*** *m.* resbaladero, rumbadero; *f.* tolva

cianuration *f.* cianuración

cilium *m.* cilio

cinder *f.* & *a.* ceniza volcánica, ceniza; cinerítico || ***cinder cone*** *m.* cono de cenizas, cono cinerítico

cinnabar *m.* cinabrio, bermellón; *f.* cinabarita, paria, [Pe.] llimpi

cinnamon stone *f.* hessonita, piedra canela, essonita; *m.* jacinto

circle *m.* círculo || ***antarctic circle*** *m.* círculo polar antártico || ***arctic circle*** *m.* círculo polar ártico || ***great circle*** *m.* gran círculo

circuit *m.* circuito

circular *a.* circular

circulate *vb.* circular

circulated *a.* circulado

circulating *a.* circulante || ***circulating pump*** *f.* bomba de circulación

circulation *f.* circulación

circumference *f.* circunferencia

cirque *m.* circo; *f.* zanoga

cirriped *m.* cirrípedo

cirrocumulus *m.* cirrocúmulo

cirrostratus *m.* cirroestrato

cirrus *m.* cirro

cistern *f.* cisterna

citrine *f.* citrina

clad *a.* enchapado

claim *f.* reclamación; *m.* reclamo, pedimento || ***claim corner*** *m.* mojón; *f.* mojonera || ***claim title*** *m.* título de propiedad || ***mining claim*** *f.* pertenencia, propiedad minera

clam *f.* almeja

clamp *f.* abrazadera; grapa, grampa, agarradera

clarain *f.* clarita

clarifier *m.* clarificador

clarify *vb.* clarificar, aclarar

clasp *f.* abrazadera

class *f.* clase; *m.* tipo

class *vb.* clasificar

classifiable *a.* clasificable

classification *f.* clasificación, tipificación; *m.* esquema

classified *a.* clasificado; proprietario

classify *vb.* clasificar, graduar

clast *m.* clasto, grano, fragmento

clastic *a.* clástico, fragmentoso || ***clastic dike*** *m.* dique clástico

clathrate *m.* metano congelado bajo del piso del mar, hidrato de gas

clause *f.* cláusula

clausenite *f.* klausenita

clausthalite *f.* claustalita

clay *f.* arcilla, argila, argilla; *m.* barro, jaboncillo || ***boulder clay*** *f.* arcilla pedregosa, tilíta || ***calcareous clay*** *m.* leche de tierra || ***clay ball*** *m.* rodado de arcilla || ***clay brick or soil*** *m.* adobe || ***clay drape*** *f.* capa de arcilla; *m.* manto de arcilla || ***clay gall*** *f.* inclusión de arcilla || ***clay pit*** *m.* andurrial, barrial, barrero; *f.* barrera, gredal || ***clay deposit*** *m.* barrero || ***fire clay*** *m.* [Cu.] barro refractorio || ***potter's clay*** *f.* arcilla figulina; *m.* barro de olleros, gres || ***prodelta clay*** *f.* arcilla frente deltaico || ***refractory clay*** *f.* arcilla refractoria || ***sapropellic clay*** *f.* arcilla sapropélica || ***volcanic clay*** *f.* moya

clayey *a.* arcilloso, argiloso, barroso, gredoso

claystone *f.* arcillita, argilolita, [Vz.] arcilita

clean *a.* puro, neto, limpio

clean *vb.* limpiar

cleaner *m.* limpiador

cleaning *f.* limpieza

cleanout bailer *f.* cuchara limpiapozos

clear land *vb.* desmontar, [Ec.] zocallar, [Col.] descapotar

clearance *m.* espacio libre, despejo, luz; *f.* tolerancia

clearing *f.* abra; *m.* desmonte

cleat *f.* (clasp) abrazadera; (coal) fractura

cleavable *a.* hendible

cleavage *m.* clivaje, crucero; *f.* esquistocidad || ***axial plane cleavage*** *m.* clivaje de plano axial || ***crenulation cleavage*** *m.* clivaje de crenulación, crucero ondulado || ***flow cleavage*** *m.* clivaje de flujo || ***fracture cleavage*** *m.* clivaje de fractura || ***pencil cleavage*** *m.* clivaje de lápiz, crucero astilloso || ***slaty cleavage*** *m.* clivaje pizarreño, clivaje pizarroso, crucero verdadero || ***slip cleavage*** *m.* clivaje de transposición

cleave *vb.* hender

cleaving *m.* hendimiento

cleft *f.* abertura

clevis *m.* grillete, clavijero

cliachite *f.* bauxita, cliachita

client *m.* cliente

cliff *m.* acantilado, cantil, risco, barranco, frontón, precipicio, derrumbadero, despeñadero, paredón, peñasco; *f.* bufa, escollera, llambria, pared

climate *m.* clima

climatic *a.* climático

climatology *f.* climatología

climax *m.* clímax, cima

climb *f.* trepada

climb *vb.* subir, trepar, escalar

climbing ripples *f.* entrecruzamiento de óndulas sobrepuestas, ondulitas superpuestas

clinker *f.* escoria

clinochlore *m.* clinocloro

clinoclase *f.* clinoclasa, afanesa

clinoenstatite *f.* clinoenstatita

clinoform *f.* clinoforma

clinograph *m.* clinógrafo

clinohumite *f.* clinohumita

clinometer *m.* clinómetro, inclinómetro, eclinómetro

clinopyroxinite *f.* clinopiroxenita

clinozoisite *f.* clinozoisita

clintonite *f.* xantofilita

clip *f.* abrazadera, grapa; *m.* sujetador

clockwise *a. & adv.* en el sentido de las agujas (o manacillas) del reloj, en sentido cronométrico

cloanthite *f.* cloantita

clod *m.* terrón

clog *f.* obstrucción; *m.* atascamiento

clog *vb.* taponar, obturar, tapar

clogged *a.* obstruído, taponado, obturado

close *a.* próximo, cerca

close *vb.* cerrar

closed *a.* cerrado

close-up *f. & a.* toma (o vista) de primer plano; vista de cerca

closure *m.* cierre estructural, cierre yacimiento

cloud *f.* nube || ***ash cloud*** *f.* nube de cenizas || ***cloud point*** *m.* punto de enturbamiento, punto de opacidad, punto de neblina; *f.* temperatura de precipitación de sólidos en petróleo || ***squall cloud*** *f.* nube de chubascos || ***thunder cloud*** *f.* nube de tormenta

cloudiness *f.* nebulosidad, nubosidad, opacidad

cloudy *a.* nublado

cluster *m.* grupo

clutch *m.* embrague

coagulant *m.* coagulante

coagulate *vb.* coagular, conglutinarse, cuajarse

coagulation *f.* coagulación

coagulator *m.* coagulador

coal *m.* carbón; *f.* hulla || ***anthracite coal*** *f.* antracita, hullita; *a.* antracitoso || ***ash coal*** *f.* albertita || ***asphaltic coal*** *f.* albertita || ***bituminous coal*** *m.* carbón bituminoso; *f.* hulla || ***brown coal*** *m.* carbón pardo, lignito || ***cannel coal*** *f.* ampelita || ***coal-bearing*** *a.* hullífero || ***coal bed*** *f.* hullera || ***coal bin*** *f.* carbonera || ***coal dust*** *m.* cisco || ***coal gas*** *m.* gas de carbon || ***coal measure*** *m.* estrato carbonífero || ***coal mine*** *f.* carbonera, hullera || ***coal mining machine*** *f.* rozadora || ***coal miner*** *m.* hullero || ***coal oil*** *f.* kerosina || ***coal tar*** *f.* brea de alquitrán, brea de carbón, brea de hulla; *m.* alquitrán de hulla || ***coking coal*** *m.* carbón aglomerante, carbón aglutinante, carbón de cok, carbón graso, carbón de piedra pasada || ***dull coal*** *m.* carbón oscuro, carbón mate || ***hard coal*** *f.* hornaguera || ***humic coal*** *m.* chameolito || ***screened coal*** *f.* grancilla || ***semi-bituminous coal*** *m.* carbón semibituminoso || ***unprocessed coal*** *m.* todouno

coal-bearing *a.* carbonífero

coalbed methane *m.* gas grisu; metano de carbón

coalesce *vb.* unirse, juntarse, juntar, soldarse, conglutinarse, fundirse

coalescence *f.* coalescencia, soldadura

coalification *f.* incarbonización, carbonización

coaly *a.* carbonoso

coarse *a.* grueso, gordo, basto, brusco, bruto, recio, tosco, áspero || ***coarse-grained*** de grano grueso

coarsen *vb.* engrosar(se)

coarseness *f.* tosquedad

coarsening upward *a.* engrosamiento hacia arriba, granocreciente hacia arriba

coast *f.* costa, orilla; *m.* litoral ‖ ***emergent coast*** *f.* costa emergente

coastal *a.* costero, costanero, litoral ‖ ***coastal plain*** *f.* planicie costera, llanura costera o costanera; *m.* llano costero

coastline *f.* línea de costa

coated *a.* revestido, encapado

coating *m.* baño, recubrimiento, revestimiento

cob *vb.* (min.) chancar, quebrar

cobalt *m.* cobalto ‖ ***cobalt bloom*** *f.* eritrita ‖ ***cobalt oxide*** *m.* zafre

cobaltic *a.* cobáltico

cobaltite *f.* cobaltita, cobaltina

cobber *m.* pepenador

cobble *m.* guijarro, guija, canto, canto grande, chino, canto rodado mediano, bolón, chinarro, chinazo, morillo, guijón; *f.* peña, china

Coblencian, Coblentzian, Coblenzian *m. & a.* Coblenciense, Coblenciano, Coblentziano, Coblentziense, Koblenzense

coccolith *m.* cocolito; *f.* cocolita

coccolithic limestone, cocconite *f.* kokkonita

code *m.* código; *f.* (legal) clave

codify *vb.* codificar

codimer *m.* codímero

coefficient *m. & a.* coeficiente ‖ ***coefficient of elasticity*** *f.* coeficiente de elasticidad ‖ ***coefficient of expansion*** *f.* coeficiente de expansión ‖ ***correlation coefficient*** *f.* coeficiente de correlación; *m.* razón de correlación

coelenterate *m.* coelentéreo, celenterado

coeloma *f.* celoma

coenenchyme *f.* cenénquima

coenosteum *m.* cenósteo

coffer dam *f.* represa (o presa) encofrada

coffinite *f.* cofinita

cognate inclusion *f.* inclusión relacionada

cogwheel ore *f.* bournonita

coherence *f.* coherencia

coherent *a.* coherente

cohesion *f.* cohesión

cohesive strength *f.* resistencia cohesiva

coil *f.* vuelta, espira, espiral, circunvolución; (elect.) bobina; *m.* arrollamiento, rollo

coiled *a.* enrollado, enroscado, bobinado

coincidence *f.* coincidencia

coke *m.* cok, coque

coke *vb.* coquificar, coquizar

coking *f. & a.* coquificación, coquización ‖ ***coking plant*** *f.* coquería

col *m.* collado, col, cuello, tollón

cold *a.* frío

colemanite *f.* colemanita

collapse *m.* colapso, derrumbamiento, desplome, hundimiento, derrumbe; *f.* caída completa

collapse *vb.* aplastarse, derrumbarse

collapsed pipe *m.* tubo caído, tubo apachado

collapsible *a.* plegable, plegadizo

collapsing *a.* aplastamiento

collar *m.* collar, cuello, collarín, anillo, aro, brocal; *f.* argolla ‖ ***drill collar*** *m.* cuello de tubería o de perforación ‖ ***shaft collar*** *f.* boca de mina ‖ ***well collar*** *f.* boca de pozo

collateral *m.* resguardo

collect *vb.* recoger, recolectar, muestrear, juntar

collection *f.* colección

collections *m.* cobros

collector *m.* colector; receptor, receptor de ondas

collide *vb.* chocar

colliery *f.* hullera

collision *m.* choque

collimate *vb.* colimar

collimation *f.* colimación

collimator *m.* colimador

colloid *m. & a.* coloide, gelatinoso

colloidal *a.* coloidal, coloide, coloideo, gelatinoso

colloform *m.* coloformo

colloidal *a.* coloidal

collophane *m.* colófano, colofano; *f.* colofanita, naaurodita

colluvial *a.* coluvial

colluvium *f.* coluvión

colony *f.* colonia

colophonite *f.* colofonita

color *m.* color

colorimeter *m.* colorímetro

colorimetric *a.* colorimétrico

colorless *a.* incoloro

columbite *f.* colombita

columella *f.* columela

columellar *a.* columelar

column *f.* columna; *m.* pilar || ***geologic column*** *f.* columna geológica || ***oil column*** *f.* columna de petróleo || ***stratigraphic column*** *f.* columna estratigráfica

columnar *a.* columnar || ***columnar basalt*** *m.* basalto prismático, basalto columnar || ***columnar jointing*** *f.* disyunción prismática, junta prismática || ***columnar section*** *f.* columna estratigráfica

Comanchean *m. & a.* Comancheano, Comanchense

comatulid *m.* comatúlido

combination *f.* combinación

combine *vb.* conjuntar, combinar

combined *a.* conjuntado, combinado

combustible *m. & a.* combustible

combustion *f.* combustión

comet *f.* cometa

command *m.* (comp.) comando, mandato

commerce *m.* comercio

commercial *a.* comercial

commingle *vb.* mezclarse

commingled *a.* mezclado

comminute *vb.* moler

commitment *m.* compromiso; *f.* obligación

commodities *m.* productos, géneros, frutos; *f.* mercaderías, producciones

common *a.* común || ***common depth point (CDP)*** *m.* punto común profundo, punto de reflejo común || ***common stock*** *m.* acciones ordinarias

community *f.* comunidad

compact *vb.* consolidar, compactar

compact *a.* compacto, apretado

compacting *f.* compactación

compaction *f.* compacción, compactación, consolidación, [PR, Vz.] compacidad || ***differential compaction*** *f.* compacción diferencial

compactness *f.* compacidad

company (Co.) *f.* compañía (Cía.), empresa || ***company name*** *m.* razón social || ***holding company*** *f.* compañía matriz

comparison *f.* comparación

compass *f.* brújula, compás || ***compass rose*** *f.* rosa de los vientos

compensation *f.* compensación || ***isostatic compensation*** *f.* compensación isostática

competence *f.* competencia

competent *a.* competente, capaz

competition *f.* competencia

competitor *m.* competidor

compilation *f.* compilación, recopilación

compile *vb.* compilar, recopilar

compiler *m.* recopilador

complaint *f.* reclamación

complementary *a.* complementario

complete *a.* todo, entero, total

completion *f.* terminación, finalización; (perf.) *f.* terminación; *m.* acabamiento || ***dual completion*** *f.* terminación a dos zonas || ***multiple, multizone completion*** *f.* perforación a múltiples zonas, terminación de zonas múltiples

complex *m.* complejo, paquete; *a.* complejo || ***basement complex*** *m.* complejo de basamento || ***metamorphic core complex*** *m.* complejo nuclear metamórfico || ***sedimentary complex*** *m.* complejo de estratos || ***thrust complex*** *m.* complejo de cabalgamiento

complexity *f.* complejidad

component *m.* componente

composite *a.* compuesto || ***composite volcano*** *m.* cono compuesto, estratovolcán

composition *f.* composición

compound *m. & a.* compuesto, combinado, mezclado

composite *a.* compuesto || ***color composite*** *f.* composición de color

comprehensive *a.* comprensivo

compress *vb.* comprimir

compressed *a.* comprimido, apretado

compressible *a.* compresible

compressibility *f.* compresibilidad

compression *f.* compresión

compressional waves *f.* ondas compresionales

compressive *a.* compresivo || ***compressive force*** *f.* fuerza compresiva || ***compressive strength*** *f.* resistencia al aplastamiento, resistencia compresiva

compressor *m.* compresor

computation *m.* cómputo, cálculo; *f.* computación

compute *vb.* calcular, computar

computer *m.* computador, ordenador; *f.* computadora, computista || ***personal computer (pc)*** *m.* computador personal, ordenador doméstico, pc

concave *a.* cóncavo

concealed *a.* tapado, oculto, disimulado, cubierto

concentrate *m.* concentrado, asiento

concentrate *vb.* concentrar, apurar, nuclear, [Méx.] despajar

concentrated *a.* concentrado

concentrates *f.* gandinga; *m.* reconcentrados || ***crushed concentrates*** *f.* ñakha || ***rich concentrates*** *m.* polvillos, concentrados

concentration *f.* concentración; contenido || ***concentration tailings*** *m.* despaje, colas

concentrator *m.* concentrador

concentric *a.* concéntrico

concept *m.* concepto || ***exploration concept*** *m.* play, concepto geológico

concession *f.* concesión, patente; *m.* derecho || ***concession holder*** *m.* concesionario || ***to grant a concession*** *vb.* concesionar

conchoidal *a.* concoide, concoideo

concordance *m.* acordamiento; *f.* concordancia

concordant *a.* concordante

concordantly *adv.* concordantamente

concrete *m.* hormigón, concreto || ***reinforced concrete*** *m.* cemento armado, concreto reforzado

concretion *f.* concreción

concurrent *a.* concurrente, coexistente

condensate *m.* condensado, gas licuado

condensation *f.* condensación || ***retrograde condensation*** *f.* condensación retrógrada

condense *vb.* condensar

condensed *a.* condensado || ***condensed section*** *f.* sección condensada

condenser *m.* (petrol., ref.) condensador

conditions *f.* condiciones || ***market conditions*** *f.* condiciones del mercado

conductance *f.* conductancia

conduction *f.* conducción, transmisión

conductive *a.* conductivo, conductible

conductivity *f.* conductividad, conductibilidad || ***thermal conductivity*** *f.* conductibilidad térmica

conductor *m.* conductor

conduit *m.* acueducto, conducto, aguaducho, caño, tubo conducto, [Méx.] colagón

cone *m.* cono || ***alluvial cone*** *m.* cono aluvial, cono de deyección || ***cinder cone*** *m.* cono de cenizas || ***composite cone*** *m.* cono compuesto || ***cone in-cone structure*** *f.* estructura cono entre cono, estructura cono en cono, estructura conos coaxiales || ***lava cone*** *m.* cono de lava || ***pyroclastic cone*** *m.* cono piroclástico || ***water cone*** *m.* cono de agua

confidence level *m.* coeficiente (o nivel) de confianza

configuration *f.* configuración

confined pressure *f.* presión confinada

confining pressure *f.* presión de confinamiento

confirm *vb.* confirmar

confirmation *f.* confirmación

confluence *f.* confluencia

confluent *a.* confluente

conform *vb.* conformar

conformable *a.* concordante

conformal *a.* ortomórfico

conformity *f.* concordancia, acordancia, (geol.) acordanza; *m.* ortomorphismo

conglomerate *m.* conglomerado; *f.* pudinga
conglomeratic *a.* conglomerático, conglomerado
congruent *a.* congruente, congruo
congruous *a.* congruente
Coniacian *m. & a.* Coniaciense, Coniaciano
conical *a.* cónico, acuminado
conic *a.* cónico || ***conic projection*** *f.* proyección cónica
conifer *f.* conífera
coning (water) *f.* conificación del agua
conjugate *m. & a.* conjugado
connate *a.* innata || ***connate water*** *f.* agua innata, agua congénica, agua fósil
connect *vb.* enlazar, conectar, juntar, unir, ensamblar
connection *f.* conexión, unión, junta; *m.* acoplamiento, enchufe, enlazamiento
connecting rod *f.* biela
conodont *m.* conodonto
consanguineous *a.* consanguíneo
consequent *a.* consecuente || ***consequent stream*** *m.* curso consecuente
conservation *f.* conservación, reparación
consignee *m.* consignatorio, depositario
consignment *f.* consignación, partida
consignor *m.* consignador
consistency *f.* consistencia
consistent *a.* consistente
console *f.* consola
consolidate *vb.* consolidar
consolidated *a.* consolidado, compactado
consolidation *f.* consolidación
consortium *m.* consorcio
constant *f.* constante || ***conversion constant*** *f.* constante de conversión || ***dielectric constant*** *f.* constante dieléctrica || ***elastic constant*** *f.* constante elástica
constellation *f.* constelación
constituent *m.* componente, constituyente, elemento, ingrediente
constraints *f.* restricciones
constriction *f.* constricción
construct *vb.* construir
construction *f.* construcción
constructional *a.* constructivo, por acumulación
constructive *a.* constructivo
consultant *m.* asesor
consume *vb.* consumir
consumer *m.* consumidor
consumption *m.* consumo
contact *m.* contacto || ***contact metamorphism*** *metamorfismo* de contacto || ***contact zone*** *f.* zona de contacto || ***gas–oil contact*** *m.* contacto gas-petróleo (o gas–aceite) || ***gas–water contact*** *m.* contacto gas–agua || ***gradational contact*** *m.* contacto de transición, contacto gradual || ***oil–water contact*** *m.* contacto petróleo-agua (o agua-aceite); *f.* zona aceite–agua || ***sedimentary contact*** *m.* contacto sedimentario || ***sharp contact*** *m.* contacto neto (claro, o abrupto) || ***tectonic contact*** *m.* contacto tectónico
contain *vb.* contener
container *m.* recipiente
contaminate *vb.* contaminar, impurificar
contaminated *a.* contaminado, impurificado
contamination *f.* contaminación, impurificación
contemporaneous *a.* contemporáneo, simultáneo
content *m.* contenido
context *m.* contexto
contiguous *a.* vecino
continent *m.* continente
continental *a.* continental || ***continental borderland*** *m.* borde continental || ***continental drift*** *f.* deriva continental || ***continental margin*** *f.* margen continental || ***continental rise*** *m.* borde de plataforma continental; *f.* ondulación continental || ***continental shelf*** *f.* plataforma continental || ***continental shield*** *m.* escudo continental, núcleo continental || ***continental slope*** *f.* pendiente continental, talud continental
contingency *m.* contingencia, imprevisto
continuity *f.* continuidad
continuous *a.* continuo || ***continuous phase*** *f.* fase continua

contorted *a.* retorcido, torcido, contorsionado

contortion *f.* contorsión

contour *m.* contorno; *f.* curva de nivel, isoípsa || ***contour line*** *f.* curva de nivel, isohipsa, isoípsa || ***contour map*** *m.* plano topográfico o nivelado, mapa acotado || ***depth contour*** *f.* isobata

contouring *f.* acotación; *m.* delineamiento || ***parallel contouring*** *f.* acotación paralela

contract *m.* contrato, destaje, destajo || ***contract work*** *m.* [Bol.] destaje, destajo || ***production-sharing contract*** *m.* contrato de producción compartida

contract *vb.* contraerse

contraction *f.* contracción, disminución

contractor *m. & f.* contratista, asentista

contrast *m.* contraste

contribution *f.* contribución

control *m.* control || ***automatic control*** *m.* control automático || ***control point*** *m.* punto fijo, punto de comprobación, punto de control || ***geodetic control*** *m.* control geodético || ***ground control*** *m.* mando de tierra

control *vb.* controlar, gobernar, regular

controlled *a.* controlado, regulado

controller *m.* regulador

convection *f.* convección, contravección || ***convection current*** *f.* corriente de convección

convention *m.* convenio; *f.* convención

conventional *a.* convencional, corriente, común

converge *vb.* confluir, juntarse, converger

convergence *f.* convergencia || ***evolutional convergence*** *f.* convergencia evolutiva

convergent, converging *a.* convergente

conversion *f.* conversión || ***conversion constant*** *f.* constante de conversión || ***conversion table*** *f.* tabla de conversión || ***conversion unit*** *f.* unidad de conversión

converted *a.* convertido

converter *m.* convertidor

conveyor *m.* transportador mecánico

convex *a.* convexo; abombado

convolute, convoluted *a.* enrollado, convoluto || ***convoluted bedding*** *f.* estratificación convoluta || ***convoluted folding*** *m.* intraplegamiento

convolution *f.* convolución

cool *a.* fresco

cool *vb.* enfriar

coolant *m.* líquido enfriador, líquido refrigerante

cooler *m.* enfriador, refrigerador

cooling *m.* enfriamiento || ***cooling coil*** *m.* serpentín enfriador || ***cooling tower*** *f.* torre de enfriamiento || ***evaporative cooling*** *m.* enfriamiento por evaporación

coombe, coomb, combe *m.* valle estrecho

coordinate *f.* coordenada || ***coordinate axes*** *m.* ejes de coordenadas || ***coordinate paper*** *m.* papel cuadriculado || ***coordinate plane*** *f.* coordenada plana

coordination *f.* coordinación

coordinator *m.* coordinador

copalite *f.* copalita; *m.* copal

copiapite *m.* acije; *f.* janosita, knoxvillita

copolymer *m.* copolímero

copolymerization *f.* copolimerización

copper *m.* cobre, alambre; [Pe.] anta || ***copper-bearing*** *a.* cuprífero, aerífero, cobrizo || ***copper carbonate*** *m.* cardenillo || ***copper carbonate-stained*** *a.* acardenillado || ***copper colored*** *a.* acaparrosado, acobrado || ***copper mine*** *f.* mina de cobre, chacra || ***copper ores*** *f.* mena de cobre; *m.* [Méx.] abronzado || ***copper oxide*** *m.* cobre roso, óxidos de cobre || ***copper pyrite*** *f.* calcopirita; *m.* cobre amarillo || ***copper sulfate*** *m.* ferrete, lipis, vitriolo azul || ***copper sulfide*** *f.* copaquira, chalcosina || ***disseminated native copper*** *f.* [Bol.] barilla || ***native copper*** *m.* cobre virgen, cobre nativo

copperas *f.* colpa; *m.* vitriolo de marto

coprolite *m.* coprolito

copy *f.* copia

copyright *m.* título de propiedad, derecho de autor

coquina *f.* coquina

coral *m. & a.* coral; *a.* coralino, coralígeno, [Vz.] corálico || ***coral reef*** *m.* arrecife coralino

coraliferous *a.* coralífero

coraliform *a.* coraliforme

Corallian *m. & a.* Coraliense, Coraliano

coralline *a.* coralino

cord *f.* cuerda

cordierite *f.* cordierita, dicroita

cordillera *f.* cordillera

core *m.* núcleo, cuesco, [Arg.] testigo, [Col.] corazón; *f.* muestra || ***core barrel*** *m.* sacatestigos, sacanúcleos, tubo portatestigos; *f.* sacamuestras || ***core bit*** *f.* barrena sacamuestras; *m.* cortanúcleos || ***core catcher*** *m.* atrapanúcleos, recogemuestras || ***core drill*** *f.* barrena sacanúcleos; *m.* cortanúcleos || ***core puller*** *m.* extractor de testigo || ***core pusher*** *m.* expulsanúcleos, expulsatestigos || ***core sample*** *f.* muestra de núcleo, muestra de pared || ***earth's core*** *m.* núcleo terrestre || ***earth's inner core*** *m.* nife, núcleo interior; *f.* nifesima || ***earth's outer core*** *m.* núcleo exterior || ***oriented core*** *m.* núcleo orientado || ***sidewall core*** *m.* núcleo lateral, núcleo de pared

core *vb.* sacar nucleos, corazonar, extraer núcleos, nuclear

cored *a.* nucleado

corer *m.* muestreador de núcleos

coring *m.* desmuestre

cornice *f.* cornisa

corollary *m. & a.* corolario

corona *f.* corona

corporate name *m.* razón social

corporation (Inc.) *f.* sociedad anónima (SA), empresa, companía; *m.* gremio

corrasion *f.* corrasión, remoción

correct *a.* correcto

corrected *a.* corregido

correction *f.* corrección, rectificación || ***datum correction*** *f.* corrección al nivel de referencia || ***drift correction*** *f.* corrección de deriva || ***free-air correction*** corrección de aire libre || ***phase correction*** *f.* corrección de fase || ***terrain correction*** *f.* corrección topográfica

correlate *vb.* correlacionar, poner en correlación, amarrar

correlation *f.* correlación || ***correlation coefficient*** *f.* coieficiente de correlación; *m.* razón de correlación || ***stratigraphic correlation*** *f.* correlación estratigráfica

correlative conformity *f.* conformidad correlativa

correlogram *m.* correlograma

correspond *vb.* corresponder

corrode *vb.* corroer, desgastar, oxidar, roer

corrosion *f.* corrosión; desgaste, roedura

corrosive *a.* corrosivo, corroyente

corrugated *a.* acanalado, corrugado

corsite *f.* corsita, napoleonita

corundum *m.* corindón, carborundo, corundo, espato adamantino, rubí oriental

coseismic *a.* cosísmico

cosmic *a.* cósmico || ***cosmic mass*** *f.* masa cósmica || ***cosmic rays*** *m.* rayos cósmicos

cosmogenetic *a.* cosmogenético

cosmology *f.* cosmología

cost *m.* costo, coste, gasto || ***cost efficient*** *a.* costo eficiente || ***development cost*** *m.* costos de desarrollo || ***extraction cost*** *m.* costo de extracción || ***intangible cost*** *m.* costo intangible || ***lifting cost*** *m.* costo de extracción || ***startup cost*** *m.* gasto de iniciación de operación

costean, costeen *f.* trinchera

cotter pin *m.* pasador hendido; *f.* chaveta hendida

coulée *m.* coulée; *f.* colada, zanja de desagüe

coulomb *m.* culombio

coulsonite *f.* vanado-magnetita

count *m.* recuento || ***point count*** *m.* conteo de puntos

counter *m.* contador || ***Geiger counter*** *m.* contador geiger

counteract *vb.* contrarrestrar

counterbalance *m.* contrapeso, equilibrio

counterbore *m.* abocardo de fondo plano, contrapozo

counterclockwise *a.* contrario a las agujas del reloj, contrario al orden cronométrico, en sentido contrario a las manacillas del reloj

counterflow *f.* contracorriente

counterseptum *m.* contra-septo, septo antípoda

counterweight *m.* contrapeso

country *m.* país || ***country rock*** *m.* respaldo; *f.* piedra bruta, roca viva, roca de caja, [Méx.] armadura

countryside *m.* paisaje

couple *vb.* unir, ensamblar, engranar, acoplar

coupler *m.* acoplador; *f.* unión

coupling *m.* acoplamiento, acoplador, manguito; *f.* unión || ***hydraulic coupling*** *m.* acoplador hidráulico

course *m.* curso; (survey, levan.) tiro

covalent *a.* covalente

covariance *f.* covarianza

cove *f.* ensenada, caleta, angra, abra, cala, moldura de rompiente; *m.* ancón, nicho de rompiente

covellite *f.* covellina, covellita

cover *f.* tapa, cubierta, cubertura, máscara, sábana; *m.* cubierto || ***cover glass*** *m.* cubreobjeto || ***cover rock*** *f.* roca de cubierta || ***cover slip*** *m.* cubreobjeto

cover *vb.* tapar, cubrir || ***to be covered by sand cover*** *vb.* arrambarse

coverage *f.* cobertura, superficie cubierta

covered *a.* cubierto, tapado

covering *f.* cobertura; *m.* recubrimiento

crab *m.* (biol.) cangrejo; (aeron.) desviación

crack *f.* grieta, hendidura; fractura, agrietadura, cuarteadura, fisura, quebraja, quebradura, quiebra, raja, rajadura; *m.* cuarteo || ***hairline crack*** *f.* grieta capilar || ***mud crack*** *f.* grieta de desecación || ***thermal crack*** *f.* termoclasa

crack *vb.* (ref.) craquear, desintegrar; grietarse, (mech.) fracturar, fisurarse, quebradizar, cuartear, desquebrajar, fraccionar, hender, quebrajar, quebrantar, rajar, romper

cracked *a.* cuarteado, rajado, agrietado, estrellado, hendido, grietado, grietoso, resquebrajo || ***cracked by drought*** *a.* abertal

cracking *f.* desintegración, fisuración, fraccionación, (ref.) decomposición térmica; *m.* fraccionamiento, grieteo, hendimiento, (ref.) crácking, [Arg.] craqueo || ***catalytic cracking*** *m.* cráсking catalítico || ***thermal cracking*** *m.* cráсking térmico; *f.* disociación térmica

Cracoian *m. & a.* Cracoiense, Cracoiano

crag *m.* risco, cueto, roquedo, peñasco; *f.* peña

craggy *a.* arriscado

crandallite *f.* seudowavelita, pseudowavelita

crane *f.* grúa

crankcase oil *m.* aceite para el cárter

crankshaft *m.* cigüeñal

crash *m.* choque; (comp.) *vb.* caer

crater *m.* cráter

craton *m.* cratógeno, kratógeno, núcleo

credit *m.* crédito, auxilio

creditor *m.* acreedor

creek *m.* arroyo, caño, arroyuelo, ríachuelo; *f.* boca, cala, caleta, ría, quebrada, ensenada

creep *f.* reptación; *m.* derrubio, flujo, resbalamiento, movimiento paulatino del terreno, deslizamiento lento, [Bol., Ch.] estero || ***power law creep*** *f.* ley de potencia de reptación || ***soil creep*** *f.* reptación del suelo

creeping *a.* reptante

crenulated *a.* crenulado

crenulation *m.* arrugamiento, crenulación

creodont *m.* creodonto

crescent *a.* en media luna

crest *f.* cresta, cima, cuchilla, cumbre; *m.* crestón, copete, vértice

Cretaceous *m. & a.* Cretácico, Cretáceo

crevasse *f.* grieta, grieta de glaciar, crevasse, rimaya || ***crevasse splay*** *m.* abanico de desembalse

crevice *f.* grieta; abertura, covacha, fisura, hendidura

crew *f.* brigada, cuadrilla de trabajadores, personal, dotación, tripulación, [Méx.] paro; *m.* turno, personal, equipo || ***day crew*** *f.* parada de hacienda, paro de obra

crib *m.* (perf.) enchufe de lodo

cribate *a.* cribado

cribbing *m.* entibado

cribwork *m.* (min.) ademado de cajón

crinkle *f.* crenulación, corrugación, ondulación

crinkled *a.* óndulado, corrugado

crinkle mark *f.* seudoóndula

crinoid *m.* crinoideo, crinoide

cristianite *f.* kristianita

cristobalite *f.* cristobalita

criteria *m.* criterio

critical *a.* crítico || ***critical wedge*** *f.* cuña crítica

crocidolite *f.* crocidolita

crocoite *f.* crocoisita, crocoicita, crocoísa, crocoita; *m.* plomo rojo

crooked *a.* encorvado || ***crooked hole*** *m.* agujero curvo o torcido; *f.* pata de perro || ***crooked pipe*** *m.* tubo doblado o curvado

crop out *vb.* aflorar

crossbedding *f.* estratificación cruzada, estratificación diagonal, capas o estratos entrecruzados

crosscorrelation function *f.* función de croscorrelación

crosscut *m.* crucero, socavón crucero, transverso; *f.* cruzada, labor através, travecaja, traviesa

crosscut *a.* atrevesado, cruzado, transversal, transverso

crosscut *vb.* atrevesar, cruzar

crosscutting *m.* & *a.* cruzamiento; cruzante || ***crosscutting relationship*** *f.* relación de corte

cross folds *m.* pliegues cruzados

cross hairs *m.* hilos, retículos

cross lamination *f.* laminación cruzada, láminas cruzadas

crossline *f.* (seis.) línea ortogonal, línea perpendicular

crossopterygian *m.* crosopterigio

crossover *m.* cruce

crossplot *m.* gráfico ____ contra ____

cross ripples *f.* rizaduras, ondulitas u ondulaciones de interferencia

cross section *f.* sección, sección cruzada; *m.* corte transversal, corte, transecto, perfil || ***stratigraphic cross section*** *m.* corte estratigráfico || ***structural cross section*** *m.* corte estructural

cross spread *m.* & *a.* despliegue de cruz

cross stratification *f.* estratificación cruzada

crowding *m.* agolpamiento

crowbar *f.* barrena, palanca; *m.* [Ch.] chuzo

crown *f.* copa del árbol

crown block *m.* caballete portapoleas

crow's nest *f.* (perf.) plataforma de corona, plataforma de seguridad

crucible *m.* crisol, tasco; *f.* pileta

cruciform *a.* cruciforme

current *f.* corriente

crude *m.* & *a.* crudo; bruto || ***crude oil*** *m.* petróleo crudo; *f.* nafta || ***heavy crude*** *m.* crudo pesado || ***light crude*** *m.* crudo liviano || ***light-sweet crude*** *m.* crudo liviano dulce || ***raw crude*** *m.* crudo virgen || ***sour crude*** *m.* crudo agrio, crudo sulfuroso || ***sweet crude*** *m.* crudo dulce || ***waxy crude*** *m.* crudo parafínico || ***weathered crude*** *m.* petróleo oreado

crumble *vb.* desmoronarse, desmenuzar(se), desleznar

crumbly *a.* desleznable, desmoronado

crumple *vb.* arrugar

crumpling *m.* arrugamiento

crush *vb.* aplastar, bocartear, chancar, machacar, machucar, majar, quebrantar, romper, triturar

crushed *a.* chancado, aplastado, ñuto

crusher *m.* bocarte, coche, machador, quebrador, quebrantador, machucador, triturador; *f.* cochina, chancadora, quebradora, trituradora || ***rock crusher*** *m.* rompepiedras

crushing *m.* aplastamiento, machacamiento, machaqueo, majamiento, quebrantador, triturador; *f.* molienda, trituración || ***crushing machine*** *m.* machador || ***crushing of ore*** *f.* chanca; *m.* requiebro || ***crushing strength*** *f.* resistencia a la compresión

crust *f.* corteza, corteza terrestre, costra, capa || ***continental crust*** *f.* corteza continental || ***oceanic crust*** *f.* corteza oceánica

crustal decoupling *m.* desacople cortical

cryoconite *m.* crioconito; *f.* kryokonita

cryogenic plant *f.* planta criogénica

cryolite *f.* criolita

cryptic *a.* criptógeno

cryptocrystalline *a.* criptocristalino

cryptomagmatic *a.* criptomagmático

cryptomere *f.* criptómera

cryptovolcanic *a.* criptovolcánico

crystabolite *f.* cristabalita

crystal *m.* cristal || ***crystal lattice*** *f.* red cristalina, enrejado cristalino || ***crystal system*** *m.* sistema cristalográfica

crystalline *m.* cristalino

crystallinity *f.* cristalinidad

crystallization *f.* cristalización

crystallize *vb.* cristalizar

crystallizing *a.* cristalizador, cristalizante

crystalloblastic *a.* cristaloblástico

crystallographic *a.* cristalográfico

crystallography *f.* cristalografía

crystobalite *f.* cristobalita

CSO (complete shutoff) *m.* cierre completo

ctenidium *m.* ctenidio, branquia

cubanite *f.* chalmersita

cube *m.* cubo

cubic *a.* cúbico || ***cubic meters*** *m.* metros cúbicos || ***cubic feet*** *m.* pies cúbicos

cuesta *f.* cuesta

Cuisian *m.* & *a.* Cuisense, Cuisiano

culm *f.* hulla menuda

culmination *f.* culminación

culture *f.* cultura; *m.* cultivo

culvert *m.* carcamo; *f.* alcantarilla

cumberlandite *f.* comberlandita

cummingtonite *f.* cummingtonita

cumulative *a.* cumulativo, acumulado || ***cumulative production*** *f.* producción acumulada

cumulite *m.* cumulito

cumulonimbus *m.* cumulonimbo

cumulophyric *a.* cumulofírico

cumulus *m.* cúmulo

cup *f.* copa

cupel *m.* tascotermocupla

cupola *f.* cúpula, bóveda; *m.* cubilote

cupric *a.* cúprico

cupriferous *a.* cobrizo, cuprífero

cuprite *f.* cuprita; *m.* óxido de cobre || ***red cuprite*** *f.* ziguelina

cuprous *a.* cuproso, cuprifero

currency *f.* moneda corriente, moneda acuñada, divisas, moneda legal; *m.* efectivo || ***hard currency*** *f.* divisas fuertes

current *f.* corriente; correntada; *m.* corrental || ***alternating current*** *f.* corriente alterna || ***convection current*** *f.* corriente de convección || ***current account*** *f.* cuenta corriente || ***density current*** *f.* corriente de densidad || ***direct current*** *f.* corriente continua || ***equatorial current*** *f.* corriente ecuatorial || ***eddy current*** *f.* corriente reversa o parásita; *m.* remolino, cabezón || ***longshore current*** *f.* corriente litoral || ***rip current*** *f.* corriente de rompiente || ***seasonal current*** *f.* corriente estacional || ***turbidity current*** *f.* corriente de turbidad o turbidez, corriente turbia o turbidítica

cursor *m.* cursor

curvature *f.* curvatura, encorvado

curve *f.* curva || ***decline curve*** *f.* curva de declinación || ***production curve*** *f.* curva de producción

curved *a.* encorvado

curvilinear *a.* curvilíneo

cuselite *f.* kuselita

cusp *f.* cúspide

cuspate *a.* en forma de cúspide, arqueado

customer *m.* cliente

customs *f.* aduana || ***customs duty*** *m.* derechos de aduana

cut *m.* porcentaje de agua y sólidos en suspensión en el análisis de petróleos; corte, rebaje, tajo; *f.* cortadura, raja, rajadura, incisión; *a.* labrado, rajado || ***cut and fill*** *a.* desmonte y relleno, relleno de cauce || ***cut-off*** *a.* minimo || ***final cut*** *m.* corte final || ***open cut*** *m.* rajo, tajo abierto

cut *vb.* cortar, (stone) cantear, picar; (gems) abrillantar, labrar, rajar

cutinite *f.* cutinita

cutter *m.* (perf.) cortatubos || ***stone cutter*** *m.* pedrero

cutaway view *m.* corte transversal

cuticle *f.* cutícula

cutoff *m.* corte; minimo || ***cutoff line*** *f.* línea de truncamiento || ***cutoff point*** *m.* punto de truncamiento || ***cutoff valve*** *f.* válvula de cierre automático

cutting stone or gems *f.* labra, talla; *m.* tallado

cuttings *f.* (perf.) muestras de canal, virutas, cortaduras; *m.* cortes de pozo, ripios

cuttlefish *m.* calamar

cyanide *m.* cianuro || ***cyanide process ore treatment*** *m.* beneficio por cianuración

cyanotrichite *f.* namacualita

cycle *m.* ciclo || ***erosional cycle*** *m.* ciclo de erosión

cyclic, cyclical *a.* cíclico

cycling *f.* recirculación

cycloalkane *m.* cicloalcano

cyclobutane *m.* ciclobutano

cyclobutene *m.* ciclobuteno

cyclodont *m. & a.* ciclodonto

cycloheptane *m.* ciclohepteno

cyclohexane *m.* ciclohexano

cyclohexene *m.* ciclohexeno

cyclone *m.* ciclón || ***cyclone separator*** *m.* separador ciclonal

cyclonic *a.* ciclónico

cycloparaffins *m.* hidrocarburos cicloparafínicos

cyclopentane *m.* ciclopentano

cyclothem *f.* ciclotema

cyclotron *m.* ciclotrón

cylinder *m.* cilindro

cylindrical structure *f.* estructura cilíndrica

cymophane *f.* cimofana; *m.* crisoberilo

cyprite *f.* ciprita

cyrenoid *a.* cirenoide

cyrtoceracone, cyrtocone *m.* cirtocono

cystic *a.* cístico

cystoid *m.* cistoideo

cytology *f.* citología

cytoplasm *f.* citoplasma

cytosome *f.* citosoma

D

dacite *f.* dacita

dacitic *a.* dacítico

dactylethra *m.* dactíletros

dactylioceracone *m.* dactilioceracono

daily *a.* diario, por día

dale *m.* vallecito

dam *f.* (hydr.) cortina, presa, represa, (geol.) barrera; *m.* rebosadero, cerramiento, embalse, dique || ***arch dam*** *m.* dique de arco || ***buttress dam*** *m.* dique de machones || ***earthfill dam*** *m.* dique de terraplén; *f.* presa de tierra y escollera

dam *vb.* represar, tranear, embalsar, trinchar

damage *m.* perjuicio, daño; *f.* avería || ***formation damage*** *m.* daños a formación

damburite *f.* damburita

dame coiffée *m.* pilar de tierra, pináculo

damourite *f.* damourita, talcita

damp *a.* húmedo, mojado

damp, dampen *vb.* amortiguar, hacer menos vibraciones; humedecer, humectar, mojar

damped *a.* amortiguado

damper *m.* registro, regulador de tiro; amortiguador

damping *m.* amortiguamiento || ***electromagnetic damping*** *m.* amortiguamiento electromagnético

dampness *f.* humedad

danaite *f.* danaita

danburite *f.* danburita

Danian *m. & a.* Danense, Daniense, Daniano, Danés, Jorgense, Jorgiano

darcy *m.* darcy, medida de permeabilidad

darkroom *m.* cuarto oscuro

Darwin glass *f.* queenstownita

data *m.* datos; *f.* información || ***data base*** *m.* banco de datos, base de datos || ***data entry person*** *m. & f.* capturista || ***data***

gathering *f.* recopilación de datos || ***data management*** *f.* gestión de datos; *m.* manejo de datos || ***data processing*** *m.* proceso de datos || ***data processor*** *m.* ordenador, procesador de datos || ***data set*** *m.* conjunto de datos || ***relational data*** *f.* información relacional

date *f.* fecha; datación

date *vb.* fechar, datar

dating *a.* fechamiento

datolite *f.* datolita

datum *m.* datum, marco de referencia, nivel de comparación, dato, horizonte guía, nivel de referencia, plano de nivel, plano de referencia; *f.* cota de referencia, cota de comparación

davidite *f.* ufertita

day *m.* día || ***day work*** *m.* jornal || ***work day*** *f.* jornada

dead *a.* muerto, apagado || ***dead oil*** *m.* aceite muerto || ***dead weight*** *m.* peso muerto

deadman *m.* macizo de anclaje; punto de atadura

deaerator *m.* desaereador, extractor de aire

dealer *m.* agente comercial, distribuidor

deasphalted oil *m.* petróleo desasfaltado

deasphalting *f.* desasfaltación, separación de asfalto

debanding *f.* delineación

debit *m.* pasivo || ***debit balance*** *m.* saldo deudor, saldo acargo

debottleneck *vb.* desembotellar

debris *f.* escombrera, deyección; *m.* escombros, despojos, detritos || ***debris flow*** *m.* flujo de escombros (o de detritos)

debt *f.* deuda; *m.* pasivo

debug *vb.* depurar, eliminar fallos

debugging *f.* depuración

debutanize *vb.* debutanizar

decalcification *f.* descalcificación

decalcify *vb.* descalcificar

decane *m.* decano

decant *vb.* decantar

decantation *f.* decantación

decanter *f.* ampolla de decantación

decapitating *m.* descabezamiento

decapod *m.* decápodo

decarbonate *vb.* descarbonatar, decarburar

decarbonation *f.* descarburación

decarbonizing *f* decarbonización

decarburation *f.* descarburación

decay *f.* descomposición, putrefacción, decaimiento

decay *vb.* descomponerse, pudrirse, podrirse

decelerate *vb.* desacelerar

deceleration *f.* desaceleración, deceleración

decibel *m.* decibel

deck *f.* placa, bandeja; *m.* piso

declination *f.* declinación

decline *m.* descenso || ***decline curve*** *f.* curva de declinación

declivity *m.* declive, pendiente, talud; derrame; *f.* inclinación, ladera

decollement *f.* estructura de despegue; *m.* nivel de despegue, despegue

decompose *vb.* descomponer

decomposed *a.* descompuesto, podrido

decomposition *f.* descomposición

decompress *vb.* descomprimir

decompression *f.* descompresión

deconvolution *f.* deconvolución, desconvolución

decoupling *m.* desacople

decrease *vb.* disminuir, reducir, rebajar

decrease *f.* reducción, disminución, rebaja

deduce *vb.* deducir

deduction *f.* deducción, rebaja

deed *m.* título de propiedad

deep *a.* profundo, hondo

deep *m.* fondo, abismo; *f.* sima || ***back deep*** *f.* fosa posterior || ***deep-seated*** *a.* emplazado profundo || ***foredeep*** *f.* fosa anterior

deepen *vb.* ahondar, profundizar, zahondar

deeply *adv.* hondamente, profundamente

defect *m.* defecto, vicio; *f.* imperfección

deferred *a.* diferido, suspendido

defective *a.* defectuoso

define *vb.* definir

deflation *f.* deflación, denudación eólica

deflect *vb.* desviar

deflection *f.* desviación, deflexión

deflector *m.* desviador

deform *vb.* deformar

deformation *f.* deformación || ***penetrative deformation*** *f.* deformación penetrante

deformed *a.* deformado

degas *vb.* desgasificar

degasser *m.* desgasificador

degassification, degassing *f.* desgasificación

deglaciation *m.* desenglazamiento

degradation *f.* degradación

degrade *vb.* degradar; disminuir; rebajar

degree *m.* grado, graduado || ***bachelor's degree*** *f.* licenciatura || ***degree of freedom*** *m.* grado de libertad || ***master's degree*** *f.* maestría || ***doctor's degree*** *m.* doctorado

dehumidify *vb.* deshumedecer

dehydrate *vb.* deshidratar

dehydration *f.* deshidratación

dehydrator *m.* deshidratador; deshidratante

dehydrogenation *f.* deshidrogenación

deisobutanizer *m.* desisobutanizador

delay *m.* retardo, retraso || ***delay time*** *m.* tiempo de retardo (o de atraso)

delayed *a.* atrasado, retardado || ***delayed ignition*** *m.* encendido atrasado (o retardado)

deleterious *a.* deletéreo

delicate *a.* delicado

delimit *vb.* delimitar, acotar, demarcar, deslindar

delimitation *f.* deslindamiento

delineate *vb.* delinear, quedar determinado

deliver (sediment) *vb.* aportar (sedimentos); entregar

deliverability *f.* entrega

delivery *m.* reparto, de reparto, conducto; *f.* entrega, de entrega, descarga, salida

dell *f.* cañada

delta *m.* delta || ***constructive delta*** *m.* delta constructivo || ***delta front*** *m. & a.* frente deltáico || ***delta plain*** *f.* intradelta || ***delta switching*** *m.* cambio del delta || ***destructive delta*** *m.* delta destructivo

deltaic *a.* deltáico

delthirium *m.* deltirio

deltidium *m.* deltidio

deltohedron *m.* deltoedro

deluge, deluvium *m.* diluvio

demagnetize *vb.* desmagnetizar, desimantar, desimanar

demagnetization *f.* desimanización, desimantación, desmagnetización, demagnetización

demagnetized *a.* desimantado, desmagnetizado, demagnetizado

demand *f.* demanda

demand *vb.* demandar, exigir

demarcate *vb.* demarcar

demarcation *f.* demarcación, deslinde; *m.* deslindamiento

demethanizer tower *f.* torre desmetanizadora

demodulation *f.* demodulación

demoiselle *m.* pilar de tierra, pináculo

demonstrate *vb.* manifestar, demostrar

demosponge *f.* demoesponja

demulsification *f.* desmulsificación

demulsifier *m.* desmulsificador

demulsify *vb.* desmulsificar

denature *vb.* desnaturalizar

dendrite *f.* dendrita

dendritic *a.* dendrítico, dendriforme, dendroide, dendroideo

dendroid *a.* dendroide, dendriforme

dense *a.* compacto, denso, cargado, espeso, pesado

densitometer *m.* densitómetro

density *f.* densidad || ***density current*** *f.* corriente de densidad

dental lamella *f.* lámina dental

dentate *m.* dentado, dentellado

dent *f.* hendidura

dentation *f.* dentación

denticle *m.* dentículo

dentition *f.* dentición

denudation *f.* denudación, remoción; *m.* lavaje

denude *vb.* denudar, desnudar

denuded oil *m.* petróleo despojado de fracciones livianas

deodorize *vb.* desodorizar

deoil *vb.* desaceitar

deoiling *f.* descaptación

department *m.* departamento || ***purchasing department*** *m.* departamento de compras || ***sales department*** *m.* departamento de ventas

dependent *a.* dependiente

deplete *vb.* agotar

depleted *a.* agotado, empobrecido

depletion *m.* agotamiento, empobrecimiento; *f.* depleción

deployment *m.* despliegue

depocenter *m.* depocentro

depolarization *f.* despolarización; *a.* despolarizante

deposit *m.* depósito, yacimiento, criadero || ***mineral deposit*** *m.* venero || ***ore deposit*** *m.* depósito mineral

deposit *vb.* depositar, yacer, [Ch.] aconcharse

deposited *a.* depositado; *adv.* yacente

deposition *f.* deposición, depositación, acumulación; *m.* depósito

depositional *a.* depositado, deposicional

depot *m.* almacén

depreciation *f.* depreciación

depression *f.* depresión, hondonada, hoya, hoyada, torca; *m.* abajor

depropanizer *m.* despropanizador

depth *f.* profundidad; *m.* hondo || ***depth gauge (gage)*** *m.* calibre de profundidad || ***depth migration*** *f.* migración en profundidad || ***focal depth*** *f.* profundidad focal || ***water depth*** *m.* brazaje

deranged drainage *m.* avenamiento anárquico

derivative *m.* derivado || ***first derivative*** *f.* primera pendiente

dermal *a.* dérmico

derrick *f.* torre de perforación (o de extracción, o de sondeo), cabria, grúa; *m.* faro || ***derrick boom*** *m.* botalón || ***derrick man*** *m.* farero, torrero, [Vz.] encuellador, [Méx.] chango

desaggregation *f.* desagregación

desalinate *vb.* desalar, desalinizar

desalination *f.* desalación, desalinización

desalter *m.* desalador

desander *m.* desarenador

descend *vb.* abajar, bajar

descendant *m.* descendiente

descendence *f.* descendencia

descending *a.* descendente

descent *f.* descendencia, caída, bajada; pendiente, inclinación; *m.* descenso

describe *vb.* describir

description *f.* descripción

desert *m.* desierto, yermo; *f.* paramera || ***desert pavement*** *m.* adoquín || ***desert varnish*** *m.* barniz del desierto

desiccant *m. & f.* desecante

desiccate *vb.* desaguar, desecar

desiccated *a.* desecado, resquebrajo

desiccating *m.* desecamiento; *a.* desecante

desiccation *f.* desecación

desiccator *m.* desecador

design *m.* diseño, dibujo, plano || ***design capacity*** *f.* capacidad calculada

design *vb.* trazar, diseñar, proyectar, idear, dibujar, bosquejar

desilicification *f.* desilicificación

desilicify *vb.* desilicificar

desintegrate *vb.* desintegrar

desintegration *f.* desintegración

desma *f.* desma

desorption *f.* deadsorción, desabsorción

destriping *f.* delineación

destructional basin *f.* cuenca de erosión

destructional forms *f.* formas de erosión, formas de destrucción

destructive *a.* destructivo

desulfurization *f.* desulfurización, desulfuración

detach *vb.* despegar, desprender

detachment surface *m.* despegue, superficie de despegue

detail *m.* detalle

detection *f.* detección

detector *m.* detector || ***detector spread*** *m.* despliegue de detectores || ***gas detector*** *m.* detector de gas || ***hot-wire detector*** *m.* detector térmico

detergent *m. & a.* detergente
deterioration *f.* deterioración
determinant *m. & a.* determinante
determination *f.* determinación
determine *vb.* determinar
detonate *vb.* explotar
detonation *f.* detonación, explosión
detonator *m.* fulminante, detonador
detour *m.* desvío; *f.* desviación
detrital, detritic *a.* detrítico, petroclástico
detritus *m.* detrito, escombros
deuteric *a.* deutérico
deuterium *m.* deuterio, hidrógeno pesado
deuteropore *m.* deuteróporo
develop *vb.* aprovechar, desarollar, explotar
developed *a.* desarrollado || ***developed field*** *m.* campo desarrollado, yacimiento en explotación
development *m.* desarrollo; fomento, aprovechamiento; *f.* producción, explotación
deviate *vb.* desviarse
deviated *a.* desviado || ***deviated hole*** *m.* pozo desviado
deviation *f.* desviación, deriva, pata de perro; *m.* desvío
device *m.* aparato, mecanismo, ingenio, (comp.) dispositivo
devitrification *f.* desvitrificación
devitrify *vb.* desvitrificar
devoid *a.* desprovisto, carente
Devonian *m. & a.* Devónico, Devoniano
dewater *vb.* desecar, desaguar, achicar
dewatering *m.* achique, desagüe; *f.* (min.) ziquitumba
dewax *vb.* desparafinar
dewaxing *f.* desparafinación
dew point *m.* punto de rocío, punto de niebla, punto de condensación
dextral *a.* diestro, dextroso, dextral
diabase *f.* diabasa, dolerita || ***olivine diabase*** *m.* quine
diachronous *a.* diacrónico
diagenesis *f.* diagénesis
diagenetic *a.* diagenético
diagnose *vb.* diagnosticar
diagnosis *f.* diagnosis
diagnostic *m. & a.* diagnóstico
diagnositics *f.* diagnóstica
diagolite *f.* diagolita
diagonal *a.* diagonal
diagram *m.* diagrama, croquis, esquema, bosquejo; *f.* gráfica || ***block diagram*** *m.* diagrama de bloques || ***flow diagram*** *m.* diagrama de circulación, diagrama de elaboración, diagrama de flujo || ***phase diagram*** *m.* diagrama de fases || ***pie diagram*** *m.* diagrama de sectores || ***rose diagram*** *m.* diagrama roseta
dial *m.* cuadrante; *f.* esfera
dialitic *a.* dialítico
dialogite *f.* dialogita
dialysis *f.* diálisis
diamagnetic *a.* diamagnético
diamantine *a.* diamantino
diameter *m.* diámetro
diamictite *f.* simictita
diamond *m.* diamante; xanthokón || ***black diamond (bort)*** *m.* diamante negro || ***cut diamond*** *m.* diamante tallado || ***diamond-bearing*** *a.* diamantífero || ***diamond chip*** *f.* chispa || ***diamond cutter*** *m.* diamantista || ***diamond drill*** *f.* barrena con punta de diamante || ***diamond-like*** *a.* diamantado, diamantino || ***diamond merchant*** *m.* diamantista || ***diamond-shaped*** *a.* romboidal || ***uncut diamond*** *m.* naife
diaphanous *a.* diáfano
diaphragm *m.* diafragma; *f.* membrana
diaphtoresis *m.* diaftóresis
diapir *m.* diapiro, domo
diapiric fold *m.* acis
diapirism *m.* diapirismo
diapsid *m. & a.* diápsido
diaspore *m.* diásporo; *f.* diáspora
diastem *f.* diastema
diastrophic *a.* diastrófico
diastrophism *m.* diastrofismo, tectonismo; *f.* diastroma
diatom *f.* diatomea
diatomaceous *a.* diatomáceo || ***diatomaceous earth*** *f.* tierra diatomácea, harina fósil, infusoriolita

diatomic *a.* diatómico

diatomite *f.* diatomita, tierra de diatomeas, trípoli, tripolita

diatreme *f.* diatrema; *m.* tubo de explosión

dibranchiate *m.* & *a.* dibranquial, dibranquiado

dichotomous *a.* dicotómico

dichroic *a.* dicroico

dichroism *m.* dicroísmo

dichroite *f.* dicroita

dickite *f.* dickita

dicrasite *f.* plata antimonial

dictyonine *m.* dictionino

dicyclic *a.* dicíclico

diductor *a.* separador, diductor

die *vb.* expirar, morir

dielectric constant *f.* constante dieléctrica

diesel *m.* & *a.* diesel

diethyl *a.* dietilo

difference *f.* diferencia

differential *a.* diferencial || ***differential compaction*** *f.* compactación diferencial, consolidación || ***differential equation*** *f.* ecuación diferencial || ***differential erosion*** *f.* erosión diferencial || ***differential pressure*** *f.* presión diferencial || ***differential thermal analysis*** *m.* análisis térmico diferencial

differentiate *vb.* diferenciar, distinguir

differentiated *a.* distinguido, diferenciado, diasquístico

differentiation *f.* diferenciación || ***magmatic differentiation*** *f.* diferenciación magmática

diffract *vb.* difractar

diffraction *f.* difracción || ***phantom diffraction*** *f.* difracción fantasma || ***x-ray diffraction*** *f.* difracción radiográfica

diffractive *a.* difractivo, difrangente

difractometer *m.* difractómetro

diffuse *a.* difundido, extendido

diffuse *vb.* difundir

diffuser *m.* difusor

diffusion *f.* difusión, propagación

diffusivity *f.* difusividad

dig *vb.* cavar, excavar, socavar la tierra, hoyar || ***dig ditches*** *vb.* zanjar, zanjear || ***dig under*** *vb.* socavar

digenite *f.* digenita

digest *vb.* digerir

digger *m.* cavador

digging *f.* ahondadura, cava, cavadura; *m.* ahondamiento, ahonde

digit *m.* dígito || ***binary digit*** *m.* dígito binario

digital *a.* digital || ***digital processing*** *m.* tratamiento digital

digitate, digitated *a.* digitado

digitization *f.* digitización, digitalización

digitize *vb.* digitizar, digitalizar

digitized *a.* digitizado

dihedral *a.* diedro

dihedron *m.* diedro

dihexagonal *a.* dihexagonal

dike *m.* dique, filón; malecón, gang, macho, reborde, xenofiro; *f.* [Bol.] pantalla || ***dike swarm*** *m.* enjambre de diques || ***ring dike*** *m.* dique anular

dilate *vb.* dilatar

dilated *a.* dilatado

dilation *f.* dilatación

diligence *f.* solicitud

dilute *vb.* desleir, diluir, deshacer, disolver; atenuar, aguar

dilute, diluted *a.* atenuado; diluído, desleído, disuelto

dilution *f.* dilución, desleimiento

diluvial *a.* diluvial

diluvium *m.* diluvión

Dimantian *m.* & *a.* Dimantiense, Dimantiano

dimension *f.* dimensión

dimethyl *a.* dimetilo

dimorphic *a.* dimorfo

dimorphism *m.* dimorfismo

dimyrian *a.* dimiario

dinoflagellate *m.* dinoflagelado

dinosaur *m.* dinosaurio

diode *m.* diodo

diolefin *f.* diolefina

diopside *m.* diópsido; *f.* diópsida

dioptase *f.* dioptasa

diopter *f.* dioptra

diorite *f.* diorita, sabanera

dioxide *m.* dióxido || ***silicon dioxide*** *m.* dióxido de sílice || ***sulfur dioxide*** *m.* dióxido de sulfuro

dip *m.* buzamiento, hundimiento, recuesto, tendido, [Méx.] echado; *f.* pendiente, [Arg., Ch.] inclinación, caída || ***apparent dip*** *m.* buzamiento aparente || ***dip domain*** *m.* dominio de inclinación || ***dip slip fault*** *f.* falla transversal de buzamiento || ***dip slope*** *f.* pendiente estructural, pendiente de buzamiento || ***forward dip*** *f.* inclinación frontal || ***rear dip*** *f.* inclinación dorsal || ***regional dip*** *m.* buzamiento regional || ***steep dip*** *m.* buzamiento empinado || ***up dip*** *m.* buzamiento arriba

dip *vb.* buzar, inclinarse, recostar

diphenyl *a.* difenilo

dipleurula *f.* dipléurula

diplobrathid *m.* diplobrátido

diplograptid *m.* diplográptido

diploid *m.* diploedro

diploma *m.* título

Diplopoda *m.* diplópodos

diplopore *m.* diplóporo

dipmeter *m.* inclinómetro

dipole *m.* dipolo, bipolo

dipping *a.* buzante, inclinado, hundido

dipyramid *f.* bipirámide, dipirámide

dipyritization *f.* werneritización

direction *f.* orientación, dirección; *m.* sentido, rumbo, azimut

directional *f.* direccional || ***directional drilling*** *f.* perforación de desviación controlada, perforación dirigida, perforación direccional

dirt *m.* suelo; *f.* tierra; basura, viruta, pegadura; suciedad; *m.* barro, mugre || ***dirt pile*** *m.* terrero

dirty sandstone *f.* arenisca sucia

dirty work *m.* trabajo bruto, trabajo sucio

disaggregation *f.* desagregación

discard *vb.* quitar, descartar, desechar

discharge *m.* caudal, derrame; desagüe, gasto, disparo; *f.* descarga, expulsión, educción

discharge *vb.* descargar, descartar

discern *vb.* discernir

discoasterid *m.* discoastérido

discocyclinid *m.* discociclínido

discolored *a.* descolorido

disconformity *f.* disconformidad, discordancia paralela

disconnect *vb.* desconectar

discontinuity *f.* discontinuidad || ***Mohorovicic (Moho) discontinuity*** *f.* discontinuidad de Mohorovicic

discontinuous *a.* discontinuo

disc-shaped *a.* discoidal

discordance *f.* discordancia

discordant *a.* discordante

discount *m.* descuento; *f.* rebaja || ***discount rate*** *f.* tasa de descuento

discount *vb.* rebajar

discounted future net cash flow *m.* flujo neto futuro de caja descontado

discover *vb.* descubrir, alumbrar

discovered *a.* descubierto, descubridor

discovery *m.* hallazgo, descubrimiento || ***discovery well*** *m.* pozo descubridor

discriminate *vb.* discernir, discriminar

discrimination *m.* discernimiento; *f.* discriminación

disengage *vb.* desacoplar

disharmonic *a.* disarmónico, inarmónico

disintegrate *vb.* desagregar, desintegrar, disgregar

disintegrated *a.* desagregado, desintegrado

disintegration *f.* desintegración, disgregación

disk *m.* disco || ***disk drive*** *f.* (comp.) unidad de disco || ***floppy disk*** *m.* (comp.) disco flexible, diskette || ***magnetic disk*** *m.* disco magnético

diskette *m.* diskette

disk-shaped *a.* discoidal

dislocation *f.* dislocación, perturbación; *m.* botamiento || ***dislocation glide and climb*** *m.* dislizamiento y subida de dislocaciones

dislodge *vb.* desalojar, desplazar
dismantle *vb.* desmantelar
disoriented *a.* desorientado
disperse *vb.* dispersar
dispersed *a.* dispersado
dispersing agent *m.* agente dispersador
dispersion *f.* dispersión
disphenoid *m.* disfenoide
displace *vb.* desplazar, dislocar, reemplazar
displaced *a.* desplazado, dislocado
displacement *m.* desplazamiento, movimiento, rechazo || ***differential displacement*** *m.* desplazamiento diferencial
display *f.* representación, representación visual || ***liquid crystal display*** *f.* pantalla de cristal líquido
display *vb.* desplegar, presentar
disposal *f.* disposición, eliminación
dispose *vb.* desechar, eliminar
disposition *f.* disposición
disrupt *vb.* reventar, romper
disrupted *a.* dislocado
dissected *a.* cortado, encañado || ***dissected peneplain*** *f.* penillanura encañada (o disectada o cortada)
disseminate *vb.* diseminar
disseminated *a.* diseminado
dissemination *m.* diseminación
dissertation *f.* tesis doctoral, disertación
dissipate *vb.* disipar
dissociation *f.* disociación
dissolution *f.* disolución || ***dissolution surface*** *f.* aspillera
dissolve *vb.* disolver, descoagular, descuajar, desleir
dissolved *a.* desleído, disuelto
dissolving *f.* disolución
distal *a.* distante, lejos, lejano
distance *f.* distancia || ***focal distance*** *f.* distancia focal
distension *f.* distensión
disthene *f.* distena
distill *vb.* destilar
distillate *m.* destilado, condensado || ***straight run distillate*** *m.* destilado íntegro
distillation *f.* destilación || ***batch distillation*** *f.* destilación por cochadas, destilación por lotes, refinación por lotes || ***flash distillation*** *f.* destilación relámpago, destilación por expansión instantánea || ***fractional distillation*** *f.* destilación fraccionada || ***steam distillation*** *f.* destilación al vapor || ***vacuum distillation*** *f.* destilación al vacío
distilled water *f.* agua destilada
distillery *f.* distilería
distinct *a.* distinto
distinctive *a.* distintivo, característico, diferencial, distinguido, distinto
distinguish *vb.* diferenciar, distinguir, discernir
distort *vb.* deformarse, retorcerse, distorcionarse
distortion *f.* distorsión, deformación
distributable *a.* repartible
distributary *m.* distributario, distributivo, brazo; *a.* distributivo
distribute *vb.* distribuir, repartir
distribution *f.* distribución
distributor *m.* distribuidor
district *m.* distrito; *f.* comarca, intendencia || ***mining district*** *f.* comarca minera, marca; *m.* distrito minero
disturbance *f.* perturbación, disturbio
disturbed *a.* perturbado
disulfide *m.* bisulfuro
disymmetry *f.* disimetría
ditch *f.* zanja, cuneta de desagüe, acequia, cárcava, hoya, trinchera; *m.* foso, acueducto || ***ditch digger*** *m.* zanjeador; *f.* zanjadora || ***drainage ditch*** *m.* charquero || ***irrigation ditch*** *m.* cauce || ***small ditch*** *f.* tijera
ditetragonal *a.* ditetragonal
diurnal *a.* diurno, diario
divalent *a.* divalente, bivalente
dive *f.* inmersion
dive *vb.* sumergirse, bucear; (geol.) buzar
diver *m.* buzo
divergence *f.* divergencia
divergent *a.* divergente, radiado
diversification *f.* diversificación

diversify *vb.* diversificar

diversion *f.* diversión, desviación; *m.* desvío

divert *vb.* desviar, rodear

divest *vb.* desinvertir, vender

divestment *f.* desinversión, venta

divide *m.* (geog.) divorcio de aguas; *f.* crestería, divisoria

dividend *m.* dividendo

divining rod *f.* varita de rabdomante

division *f.* división

DO (dissolved oxygen) *m.* oxígeno disuelto

dock *m.* muelle, dique

dockyard *m.* astillero

document *m.* documento; *f.* carta

dodecahedron *m.* dodecahedro

dodecane *m.* dodecano

dog *m.* detenedor, trinquete

Dogger *f.* Jura pardo

doghouse *f.* (perf.) caseta, caseta del cabrestante

dogleg *m.* (perf.) doblez, pata de perro

dogtooth spar *m.* dientes de perro

dolerite *f.* dolerita || ***olivine dolerite*** *f.* estelión || ***quartz dolerite*** *f.* kongadiabasa

doldrums *f.* calmas ecuatoriales

doline, dolina *f.* dolina

dolomite *f.* dolomita, (petrol.) dolomía; *m.* espato bruno, espato romboidal || ***dolomite with cinnabar*** *f.* ziguelina

dolomitic *a.* dolomítico

dolomitization *f.* dolomitización

dolomitize *vb.* dolomitizar

domain *m.* dominio || ***forward dip domain*** *m.* dominio de inclinación frontal || ***rear dip domain*** *m.* dominio de inclinación dorsal

domal *a.* dómico

dome *m.* domo; bóveda, techo, abombamiento, cimborrio, tumor; *f.* cúpula || ***salt dome*** *m.* domo salino; *f.* cúpula salina

Domerian *m. & a.* Domeriense, Domeriano

domestic *a.* doméstico || ***domestic market*** *m.* mercado interno

dominant *a.* dominante

domino *m.* dominó

doodlebug *m.* (seis.) geófono

dope *f.* grasa para roscas

Doppler shift *m.* efecto Doppler

dormant *a.* latente

dorsal *a.* dorsal

DOT (directional orientation tool) *f.* herramienta de orientación direccional

double *m. & a.* doble, gemelo || ***double section*** *f.* sección doble

doubt *f.* incertidumbre, duda

down *adv.* abajo || ***down building*** *m.* diapirismo; *f.* construcción hacia abajo || ***down dip*** *m.* echado abajo; buzamiento abajo || ***down draft*** *m.* tiro hacia abajo || ***down grade*** *m.* abajadero, declive || ***down hole*** *a.* hoyo abajo || ***downlap*** *m.* traslapo hacia abajo || ***bidirectional downlap*** *m.* traslapo hacia abajo bidireccional || ***down pour*** *m.* chaparrón, chubasco || ***down slope*** *a.* cuesta abajo, descendente || ***down slope ripple*** *f.* óndula descendente || ***down spout*** *m.* bajante; *f.* bajada pluvial || ***downstream*** *m.* río abajo, agua(s) abajo, corriente abajo || ***downthrown*** *a.* movido hacia abajo || ***downthrown fault block*** *m.* labio bajo, bloque bajo; *f.* caída || ***down time*** *m.* tiempo pasivo || ***down warp*** *m.* alabeo hacia abajo, alabeo cóncavo

Downtonian *m. & a.* Downtoniense, Downtoniano

dowsing *f.* rabdomancia

draft *m.* (texto) tiro; (aire) succión; corriente || ***first, rough draft*** *m.* tiro, borrador

draft *vb.* dibujar, ilustrar

drafting *m.* diseño, dibujo

draftsman *m.* dibujante, delineante, delineador

drag *vb.* arrastrar

drag *m.* arrastre; *f.* traba, oposición, resistencia, gancha || ***drag fold*** *m.* pliegue de arrastre, pliegue de falla || ***dragline*** *f.* dragalina, dragalínea, excavadora de arrastre, draga || ***drag mill*** *m.* (min.) arrastre

drain *m.* desagüe, desaguadero, dren, vertedero, (min.) cárcamo, desaguador || ***drain cock*** *m.* purgador

drain *vb.* desaguar, purgar, drenar, escurrirse, achicar, agotar, avenar, desecar, vaciar, sanear

drainage *m.* drenaje, desagüe, saneamiento, avenamiento, desagüe, escurrimiento || ***annular drainage*** *m.* drenaje anular || ***dendritic drainage*** *m.* drenaje dendrítico || ***depletion-drive drainage*** *m.* drenaje por expansión || ***deranged drainage*** *m.* drenaje desordenado || ***drainage area*** *m.* vertiente || ***drainage divide*** *m.* divorcio de aguas; *f.* crestería, parteaguas || ***gas-drive drainage*** *m.* drenaje por expansión de gas || ***gravity drainage*** *m.* drenaje por gravedad || ***internal drainage*** *m.* drenaje centrípetro || ***net drainage*** *m.* drenaje efectivo || ***radial drainage*** *m.* drenaje radial || ***trellis drainage*** *m.* drenaje acostillado

drained *a.* saneado

draining (of a mine, well, etc.) *m.* achicamiento, avenamiento, desagüe, desecamiento; *f.* purga

drape *m.* manto

drape *vb.* cubrir, depositar sobre, yacer sobre

draw *m.* arroyo; *f.* quebrada

draw *vb.* dibujar; arrastrar; aspirar, ilustrar, trazar

drawdown *m.* (hydr., perf.) descenso de nivel

draw down *vb.* bajar, bajarse

drawing *m.* dibujo, compás; *f.* ilustración

draw-works *m.* motores del cabrestante || ***draw-works shed*** *f.* caseta del cabrestante

dredge *f.* draga, puntal

dredge *vb.* dragar

dredger *m.* dragador

dredging *m.* dragado, dragaje

dregs *f.* heces, zurrapa; *m.* (petróleo) aceitón

dress *vb.* (min.) preparar minerales || ***dress a bit*** *vb.* reparar un trépano || ***dress stone*** *vb.* [Méx.] castigar, labrar

dressing (stone) *f.* labra, talla; *m. & a.* tallado

dried *a.* desecado

drift *m.* tendencia; (glacial) material de acarreo, (min.) paso, socavón a hilo de veta, tunel, [Bol.] camino; *f.* (geol.) deriva || ***continental drift*** *f.* deriva continental, traslación continental || ***drift correction*** *f.* correción de deriva || ***drift indicator*** *m.* indicador de desviación || ***drift meter*** *m.* desviómetro || ***glacial drift*** *m.* material de acarreo glacial, arcillas glaciales

drift *vb.* derivar, derivarse

drifting *m.* derivando

drill *m.* perforador; taladro, abocardo, barreno; *f.* barrena, perforadora, sonda, taladradora || ***cable drill*** *f.* barrena de cable || ***churn drill*** *f.* [Vz.] chompa || ***core drill*** *f.* barrena sacamuestras, barrena sacanúcleos, barrena sacatestigos, barrena tubular || ***drill bit*** *m.* taladro, trépano || ***drill hole*** *m.* barreno, pocillo, pozo, agujero; *f.* perforación || ***drill hole terminated prematurely*** *m.* pingo || ***drill pipe*** *f.* tubería de perforación || ***drill rig*** *f.* perforadora; *m.* taladro, [Vz.] sancocho || ***drill ship*** *m.* barco perforador, buque de perforación || ***drill stem test (DST)*** *m.* examen de pruebas del contenido de la formación por media de la tubería de perforación || ***drill string*** *f.* columna de sondeo, columna de perforación, tubería de perforación || ***hand drill*** *f.* barra de mano, barrena, barrena de mina || ***rotary drill*** *f.* barrena rotatoria

drill *vb.* perforar, barrenar, sondear, sondar, taladrar

drillable *a.* perforable

drillability *f.* perforabilidad

drilled *a.* perforado; horadado

driller *m.* perforador, taladrador || ***inexperienced driller*** *m.* novato, inexperto

driller's log *m.* informe de perforación

drilling *f.* perforación, trepa; *m.* sondaje, sondeo, trepado; *a.* perforador, perforante, taladrador || ***cable drilling*** *f.* perforación con cable || ***core drilling*** *f.* perforación a núcleos, perforación con extracción de testigos || ***diamond drilling*** *f.* perforación con diamante || ***directional drilling*** *f.* perforación dirigida, perforación direccional, perforación guiada || ***drilling barge*** *f.* gabarra de perforación || ***drilling mud*** *m.* lodo de perforación || ***drilling rate*** *m.* grado de penetración || ***exploratory drilling*** *f.* perforación de cateo, perforación de reconocimiento, perforación exploratoria || ***percussion drilling*** *f.* perforación a percusión || ***rotary drilling*** *f.* perforación por rotación, perforación rotativa || ***slim-hole drilling*** *f.* perforación de diámetro reducido || ***submarine drilling*** *f.* perforación bajo agua || ***wildcat drilling*** *f.* perforación de ensayo

drip *vb.* gotear

dripstone *f.* travertina

drive *m.* empuje; *f.* impulsión, presión, propulsión || ***disk drive*** *f.* (comp.) unidad de disco || ***gas drive*** *m.* empuje por gas || ***gravity drive*** *f.* impulsión por gravedad || ***miscible drive*** *m.* desplazamiento por fase miscible || ***primary drive*** *m.* flujo natural || ***solution gas drive*** *f.* impulsion por gas disuelto || ***water drive*** *m.* empuje hidrostático; *f.* impulsión por agua

drive *vb.* (min.) dar cuele, impulsar, avanzar

drizzle *f.* llovizna

drizzle *vb.* paramear, lloviznar

drop *f.* gota; caída, baja, bajada; *m.* descenso

drop *vb.* bajar, caer

dross *f.* escoria, horrura, impurezas, desechos

drought *f.* sequía

drown *vb.* ahogar, sumergir

drowned *a.* ahogado, sumergido || ***drowned platform*** *f.* plataforma ahogada

drum *m.* tambor

drumlin *m.* drumlin

druse, drusy *m.* laque; *f.* drusa; *a.* en drusa, de quartz

dry *a.* seco, árido || ***dry dock*** *m.* dique seco || ***dry gas*** *m.* gas seco || ***dry hole*** *m.* pozo seco || ***dry ice*** *m.* hielo seco

dry *vb.* secar, desecar, sanear

drying *m.* desecamiento; *f.* desecación; *a.* desecante

DSDP (Deep Sea Drilling Project) *m.* proyecto de perforación en aguas profundas

DST (Drill Stem Test) *m.* examen de pruebas del contenido de la formación por medio de la tubería de perforación

duct *m.* conducto, tubo

ductile *a.* dúctil, maleable

ductility *a.* ductilidad, maleabilidad, flexibilidad

dufrenite *f.* kraurita

dull *a.* mate, opaco

dumortierite *f.* dumortierita

dump *m.* lugar donde se echan y amontonan las basuras, acopio, botadero, vaciadero, vertedero, terrero; *f.* escombrera || ***memory dump*** *m.* vuelco de la memoria

dump *vb.* descargar, volcar, botar, descartar, vaciar, voltear

dumping *f.* vaciada; *m.* vuelco

dune *f.* duna, aas, [Méx.] arenal; *m.* médano || ***barchan dune*** *f.* duna falciforme || ***climbing dune*** *f.* duna trepadora || ***dune field*** *m.* medanal || ***longitudinal dune*** *f.* duna longitudinal || ***parabolic dune*** *f.* duna parabólica || ***sand dune*** *f.* duna de arena || ***seif dune*** *f.* duna seif || ***transverse dune*** *f.* duna transversal

dunite *f.* dunita

duplex *m. & a.* dúplex, doble, gemelo, duplo, dúplice

durability *f.* durabilidad

durain *f.* durita

duration *f.* duración

duricrust *f.* duricostra

dusk *m.* obscurecido, fusco, anochecer

dust *m.* polvo || ***dust cap*** *f.* tapa guardapolvo || ***dust proof*** *a.* a prueba de polvo || ***dust seal*** *m.* sello gaurdapolvo || ***fine dust*** *m.* polvillo

dusty *a.* turbio

duster *m.* pozo seco, pozo improductivo

dutchman *m.* raigón

duty *m.* servicio, trabajo, rendimiento

dwarf *m.* enano

dwt (deadweight tons) *f.* toneladas de peso muerto

dye *f.* tintura, tinta; *m.* tinte, colorante

dying out *f.* desaparición

dynamic *a.* dinámico || ***dynamic range*** *f.* dinámica

dynamics *f.* dinámica

dynamite *f.* dinamita

dinamiting *f.* dinamitación

dynamo *m.* dinamo

dynamometamorphism *m.* dinamometamorfismo, metamorfísmo dinámico

dynamometer *m.* dinamómetro

dyne *f.* dina

dysodont *m. & a.* disodonto

dystrophic *a.* distrófico

E

early *a.* temprano

earnings *m.* ingresos, rendimiento; *f.* ganancias

earth *f.* tierra; *m.* suelo, terrero; globo terrestre; *a.* terrero || ***alkaline earth*** *f.* tierra alcalina || ***diatomaceous earth*** *f.* tierra de diatomeas, tierra diatomácea (o diatomeas) || ***earth sciences*** *f.* geociencias, ciencias de la tierra || ***Fuller's earth*** *f.* tierra de batán (o de Fuller) || ***loose earth*** *f.* tierra floja || ***rare earth*** *f.* tierra rara || ***red earth*** *m.* terrojo

earthen *a.* terrestre

earthfill reservoir or dam *f.* represa de tierra; *m.* terraplén

earthly *a.* terrestre

earthquake *m.* terremoto, temblor, sismo, [Vz.] cimbrozano; *f.* sacudida sísmica || ***distant earthquake*** *m.* telesismo

earthwork *m.* terraplén

earthy *a.* terroso, térreo

easement *f.* descarga, servidumbre

east *m.* este, naciente, oriente

eastern *a.* oriental

ebb *m.* flujo, reflujo; *f.* yusente || ***ebb tide*** *f.* marea menguante, bajada, marea baja, vaciamar, vaciante; *m.* bajamar, jusente; *a.* jusente, yusente

ebonite *f.* ebonita

eccentric *a.* excéntrico

eccentricity *f.* excentricidad

echo *m.* eco || ***echo sounder*** *f.* ecosonda || ***echo sounding*** *m.* sondeo acústico

echosounder *f.* ecosonda; *m.* ecosondador

echinoderm *m.* equinodermo

echinoid *m.* equinoideo

eclipse *m.* eclipse

ecliptic *f.* & *a.* eclíptica, eclíptico

eclogite *f.* eclogita

ecology *f.* ecología

economic *a.* económico || ***economic geology*** *f.* geología económica

economics *f* écónómica

economy *f.* economía

ecosystem *m.* ecosistema

ectoderm *m.* ectodermo

ectoplasm *f.* ectoplasma

eddy *m.* remolino, contraflujo; *f.* contracorriente

edelforsite *f.* zeolita roja

edenite *f.* edenita

edge *m.* borde, reborde, chaflán, margen, canto, labio; *f.* orilla || ***edge waters*** *f.* aguas marginales || ***edge zone*** exosarco || ***leading edge*** *m.* borde de ataque || ***trailing edge*** *m.* borde posterior

edit *vb.* editar, revisar

edrioasteroid *m.* edrioasteroideo

eduction *f.* educción, descarga

effect *m.* efecto

effective *a.* eficaz, efectivo

effectiveness *f.* eficacia

efficiency *f.* eficiencia; eficacia; rendimiento || ***overall efficiency*** *m.* rendimiento total || ***sweep efficiency*** *f.* eficacia de barrido

efficient *a.* eficaz, eficiente

efficiently *adv.* eficazmente

efflorescence *f.* eflorescencia

efflorescent *a.* eflorescente

effort *m.* esfuerzo

effluent *m.* efluente, escurrente; *f.* aguas residuales

effusion *f.* efusión

effusive *a.* efusivo

Eifelian *m.* & *a.* Eifeliense, Eifeliano

einsteinium *m.* einstenio

eject *vb.* expulsar, arrojar, expeler, lanzar, aventar

ejecta *f.* deyección, eyección

ejector *m.* eyector || ***steam ejector*** *m.* eyector a vapor

ejectum *m.* eyecto

elaborate *vb.* elaborar

elastic *a.* elástico || ***elastic coefficient*** *m.* coeficiente elástico || ***elastic deformation*** *f.* deformación elástica || ***elastic discontinuity*** *f.* descontinuidad elástica || ***elastic flow*** *m.* flujo elástico; *f.* fluencia elástica || ***elastic limit*** *m.* limite elástico || ***elastic modulus*** *m.* coeficiente de elasticidad || ***elastic rebound*** *m.* rebote elástico

elasticity *f.* elasticidad

elastomer *m.* elastómero

elaterite *f.* elaterita

electric *a.* eléctrico || ***electric conductivity*** *f.* conductividad eléctrica || ***electric drive*** *m.* accionamiento eléctrico || ***electric field*** *m.* campo eléctrico || ***electric induction*** *f.* inducción eléctrica || ***electric log*** *m.* registro, registro eléctrico, perfil eléctrico || ***electric prospecting*** *f.* prospección eléctrica || ***electric resistivity*** *f.* resistividad eléctrica

electrical *a.* eléctrico

electrician *m.* electricista

electricity *f.* electricidad

electrochemical *a.* electroquímico

electrochemistry *f.* electroquímica

electrode *m.* electrodo

electrofiltration *f.* electrofiltración

electrogalvanic *a.* electrogalvánico

electrolysis *f.* electrólisis, electrolización

electrolyte *m.* electrolito

electrolytic *a.* electrolítico

electromagnet *m.* electroimán

electromagnetic *a.* electromagnético

electromagnetism *m.* electromagnetismo

electrometallurgy *f.* electrometalurgía

electromotive force *f.* fuerza electromotriz

electron *m.* electrón

electronegative *a.* electronegativo

electronic *a.* electrónico

electronics *f.* electrónica

electropositive *a.* electropositivo

electrorefining *f.* refinación eléctrica

electrostatic *a.* electroestático

electrum *m.* electro; *f.* plata mixta

element *m.* elemento || ***matrix element*** *m.* elemento matricial || ***trace element*** *elemento* trazo, oligoelemento

elevate *vb.* levantar, altearse, alzarse

elevated *a.* elevado

elevation *f.* elevación, cota, altura, cotación; *m.* alzamiento, altitud; alto; nivel || ***elevation mark*** *f.* acotación || ***regional elevation*** *f.* nivel regional

ellipse *f.* elipse || ***strain ellipse*** *m.* elipsoide de deformación || ***infinitessimal strain ellipse*** *m.* elipsoide de deformación infinitesimal

ellipsoid *m.* elipsoide

ellipsoidal *a.* elipsoidal

elliptic, elliptical *a.* elíptico

elongate *vb.* alargar

elongated *a.* alargado

elongation *f.* elongación, prolongación; *m.* alargamiento

elutriation *f.* elutriación

eluvial *a.* eluvial

eluviation *f.* eluviación

eluvium *m.* eluvión

emanation *f.* emanación

emanometer *m.* emanómetro

embankment *m.* terraplén, zopetero; *f.* valla, terraza

embayment *f.* ensenada, bahía; *m.* embalse

embed *vb.* empotrar, incrustar

embedded *a.* empotrado, incrustado, [Méx.] ahogado

embolite *f.* embolita, ortobromita, plata córnea verde, plata verde

embrittlement *a.* friabilidad, fragilidad

embryo *m.* embrión

embryology *f.* embriología

embryonic *a.* embrionario, en embrión

emerald *f.* esmeralda; *m.* [Col.] cañutillo

emerge *vb.* brotar, alumbrar, emerger, nacer

emergence *f.* emersión, salida; *m.* levantamiento

emergency *f.* emergencia

emergent *a.* emergente

emery *m.* esmeril

emission *f.* emisión

empirical *a.* empírica

emplace *vb.* emplazar

emplacement *m.* emplazamiento

employ *vb.* emplear || ***employ by the day*** *vb.* jornalar, jornalear

employee *m.* empleado

employer *m.* empleador, patrón, dueño

employment *m.* empleo; *f.* ocupación, aplicación

emptiness *m.* vacío

empty *a.* vacío

empty *vb.* vaciar

Emscherian *m. & a.* Emscheriense, Emscheriano

Emsian *m. & a.* Emsiense, Emsiano

emulsification *f.* emulsificación

emulsified *a.* emulsionado

emulsifier *f.* emulsionador, emulsivo

emulsify *vb.* emulsificar

emulsion *f.* emulsión

enamel *m.* esmalte

enantiomorphic *a.* enantiomórfico

enargite *f.* enargita

encase *vb.* encastrar, embutir, envolver

encased *a.* encapado, envuelto

enclose *vb.* abrazar, encerrar

enclosed *a.* cubierto, encerrado, blindado; adjunto, encerrado

enclosure *m.* cercamiento, recinto; cercado

encode *vb.* codificar

encrinite *f.* encrinita

encroachment *f.* intrusión

encrust *vb.* incrustar

encrustation *f.* incrustación

encrusted *a.* incrustado

encumbrance *m.* gravamen

end *m.* extremo, fin || ***heavy end*** *m.* corte pesado final || ***light end*** *m.* corte liviano final

endemic *a.* endémico

endexine *f.* endexina

endlichite *f.* endliquita, endlichita

endobiontic *a.* endobióntico

endogenesis *f.* endogénesis

endogenetic *a.* endógeno, endogenético

endogenous *a.* endógeno

endoglyph *m.* endoglifo

endomorphic *a.* endomórfico

endomorphism *m.* endomorfísmo

endoplasm *f.* endoplasma

endorse *vb.* endosar

endorsement *m.* endoso

endosketleton *m.* endoesqueleto

endosperm *f.* endosperma

endothermal, endothermic *a.* endotérmico

ends *m.* productos livianos de petróleo

endurance *f.* resistencia, duración

endure *vb.* soportar

en echelon *a.* en echelon, en escalón, escalonado, en rosario || ***en echelon faults*** *f.* fallas escalonadas || ***en echelon folds*** *m.* pliegues escalonados

energy *f.* energía, fuerza, potencia || ***activation energy*** *f.* energía potencial, energía de activación || ***energy-deficient*** *f.* energía deficiente || ***free energy*** *f.* energía libre || ***kinetic energy*** *f.* energía cinética || ***mechanical energy*** *f.* energía mecánica || ***radiant energy*** *f.* energía radiante || ***thermal energy*** *energía* térmica

engine *m.* motor; *f.* máquina

engineer *m.* ingeniero (Ing.), maquinista || ***chemical engineer*** *m.* ingeniero químico || ***mechanical engineer*** *m.* ingeniero mecánico || ***reservoir engineer*** *m.* ingeniero de yacimientos

engineering *f.* ingeniería

engrave *vb.* grabar

enhance *vb.* realzar

enhanced *a.* mejorado

enhancement *m.* mejoramiento, realce

enlarge *vb.* ampliar, amplificar, aumentar

enlargement *m.* agrandamiento, aumento, ensanchamiento; *f.* expansión, ampliación

enlighten *vb.* iluminar

enrich *vb.* enriquecer; aumentar

enrichment *m.* enriquecimiento || ***secondary enrichment*** *m.* enriquecimiento secundario

enstatite *f.* enstatita

enterprise *m.* negocio

enthalpy *f.* entalpía

entrained *a.* arrastrado || ***entrained air*** *m.* aire arrastrado

entrainment *m.* arrastre de líquidos en vapores

entrance *f.* entrada, boca; *m.* acceso, ingreso

entrenched *a.* encajonado

entropy *f.* entropía; peso térmico

environment *m.* ambiente, medioambiente, cercanías || ***depositional environment*** *m.* ambiente de deposición

environs *m.* alrededores; *f.* cercanías

enzyme *f.* enzima

Eocambrian *m.* Eocámbrico, Eocambriano

Eocene *m. & a.* Eoceno, Eocénico

eolian *a.* eólico, eolio

eolianite *f.* eolianita

eon *m.* eón

Eotvos unit *f.* unidad Eotvos

Eozoic *m. & a.* Eozoico

epeiric *a.* epeírico, epírico, epicontinental

epeirogenesis *m.* epeirogénesis, epeirogenia

epeirogenetic, epeirogenic *a.* epeirogenético, epeirogénico

epeirogeny *f.* epirogenia, epeirogenia

ephebic *a.* efébico

ephemeral *a.* efimero

epicenter *m.* epicentro

epiclastic *a.* epiclástico

epicontinental *a.* epicontinental

epidermis *m.* epidermo; *f.* epidermis

epidosite *f.* epidosita

epidote *m.* epidoto; *f.* epidota

epigenesis *m.* epigénesis

epigenetic *a.* epigenético, xenogénito, xenogenuoso

epiglyph *m.* epiglifo

epipelagic *a.* epipelágico

epiphysis *f.* epífisis

epiphyte *f.* epifita

epiphytic *a.* epifito

epiplanktonic *a.* epiplanctónico

epipodite *m.* epipodito

epirostrum *m.* epirostro

episode *m.* evento, episodio

epistilbite *f.* orizita

epistome *f.* epístoma

epitheca *f.* epiteca

epithermal *a.* epitermal

epivalve *f.* epivalva

epizonal *a.* de la epizona

epizone *f.* epizona

epoch *f.* época

epoxy *f.* resina, epoxia

Epsom salt *f.* sal de Epsom, sal amarga

epsomite *f.* epsomita, sal amarga

equation *f.* ecuación || ***wave equation*** *f.* ecuación de onda

equator *m.* ecuador

equatorial *a.* ecuatorial

equigranular *a.* equigranular

equilateral *a.* equilátero

equilibrium *m.* equilibrio || ***dynamic equilibrium*** *m.* equilibrio dinámico || ***equilibrium constant*** *m.* factor constante de equilibrio || ***unstable equilibrium*** *m.* equilibrio inestable

equinox *m.* equinoccio || ***autumnal equinox*** *m.* equinoccio de otoño || ***spring equinox*** *m.* equinoccio de primavera

equip *vb.* habilitar

equipment *m.* equipo, útiles; *f.* habilitación

equipoise *m.* equilibrio; contrapeso

equipotential *a.* equipotencial, piezométrico || ***equipotential surface*** *f.* superficie equipotencial

equity *m.* (com.) propio, capital || ***equity investment*** *m.* capital de inversión || ***shareholder equity*** *m.* capital propio de los accionistas

equivalence, equivalency *f.* equivalencia

equivalent *a.* equivalente

equivalve *m.* equivalvo

era *f.* era

erect *vb.* montar, eregir, construir, levantar

erg *m.* (phys., física) ergio; (geog.) areg, arej

erode *vb.* erosionar, desgastar, carcomar, denudar, derrubiar, deslavar, gastar, roer

eroded *a.* desgastado

erodible *a.* deslavable, erosionable, gastable

erosion *f.* erosión, desgaste, ablación, abrasión, degradación, denudación, deslava, roedura; *m.* (del río) desmoronamiento || ***chemical erosion*** *f.* erosión química || ***erosion surface*** *f.* superficie erosional || ***mechanical erosion*** *f.* erosión mecánica

erosional remnant *m.* isleo cobijante, remanente erosional

erosive *a.* erosivo

erratic *a.* errático ǁ ***glacial erratic*** *m.* bloque errático

erratic *m.* canto errático

error *m.* error ǁ ***mean square error*** *m.* error cuadrático medio ǁ ***rounding error*** *m.* error de redondeo ǁ ***truncation error*** *m.* error de truncamiento

eruption *f.* erupción, efusión

eruptive *m.* eruptivo, volcánico

erythrite *f.* eritrita, eritrina

escape *f.* fuga; *m.* derrame, escape

escape *vb.* escapar

escarpment *m.* acantilado, risco, escarpe, peñasco, resbalo; *f.* escarpa

esker *m.* esker

esmeraldite *f.* nordfieldita, northfieldita

esophagus *m.* esófago

essence *f.* esencia

essential *a.* esencial

essexite *f.* esexita

essonite *f.* essonita, piedra canela, jacinto

establish *vb.* establecer

ester *m.* ester

esterellite *f.* esterelita

estimate *f.* estimación, tasa, apreciación; *m.* aprecio, cálculo ǁ ***rough estimate*** *m.* tanteo

estimate *vb.* estimar, avaluar, tantear

estimation *f.* estimación

estimator *m.* estimador

estuarine *a.* estuarino

estuary *m.* estuario, estero; *f.* boca del río, ría

etch *vb.* atacar con ácido

etching *m.* ataque químico

ethane *m.* etano; dimetilo

ethanol *m.* etanol; alcohol etílico

ether *m.* éter

ethyl *m.* etilo

ethylene *m.* etileno, eteno

ettringite *f.* woodfordita

euclase *f.* euclasa

eucryptite *f.* eucriptita

eucrystalline *f.* eucristalina

eugeosyncline *f.* eugeosinclinal

euhedral *a.* euhedral, euhédrico

eulamellibranch *f.* eulamelibranquia

eulytite *f.* bismutoblenda

eupelagic *a.* eupelágico

euphotic *a.* eufótico

euphotide *f.* eufótida

Eurasia *f.* Eurasia

Europe *f.* Europa

europium *m.* europio

euryapsid *m.* & *a.* euriápsido

euryhaline *a.* eurihalino

eurypterid *m.* euriptérido

eustasy *f.* eustasia

eustatic *a.* eustático

eutaxitic *a.* eutaxítico

eutectic *f.* eutéctica; *a.* eutéctico

eutrophic *a.* eutrófico

euvitrian *m.* xylovitrian, xylovitriano

euxenic *a.* euxénico

evacuate *vb.* evacuar, vaciar

evacuated *a.* evacuado

evacuation *f.* evacuación

evaluate *vb.* evaluar, valorar

evaluation *f.* evaluación

evaporate *vb.* evaporar, evaporarse, vaporar, (salmuera) cuajar

evaporation *f.* evaporación, vaporización ǁ ***evaporation tank*** *f.* pila

evaporative cooling *m.* enfriamiento por evaporación

evaporator *m.* evaporador

evaporite *f.* evaporita, roca salina

evaporitic *a.* evaporítico

evapotranspiration *f.* evapotranspiración

even *a.* liso, par, llano, regular, parejo

event *m.* evento

evolution *f.* evolución

evolutionary *a.* evolutivo

exact *a.* exacto, preciso

exaggerated scale *f.* escala exagerada

exaggeration *f.* exageración

exam *m.* examen

examination *f.* examinación, indagación; *m.* examen

examine *vb.* examinar, reconocer, recorrer, tantear

example *m.* ejemplo

excavate *vb.* excavar; abarrancar, acavernar, cavar, zapar, [Col., Vz.] banquear

excavated *a.* excavado; abarranco

excavating *a.* excavador

excavation *f.* excavación, ahondadura, cava, hoya hondonada; *m.* ahonde, ahondamiento, hornacho, socavón, tajo, apique

excavator *m.* cavador; *f.* excavadora, zapadora || ***bucket-wheel excavator*** *f.* excavadora de rueda de cangilones

excess *m. & a.* exceso || ***excess profit*** *f.* excedente de ganancias, exeso de utilidades

exchange *m.* cambio, intercambio; treque || ***foreign exchange*** *f.* (com.) divisas

exchanger *m.* intercambiador || ***heat exchanger*** *m.* (inter)cambiador de calor; termopermutador

excrescence *f.* excrecencia

excursion *f.* excursión

exempt *vb.* franquear

exemption *f.* exención, franquicia

exfoliate *vb.* exfoliar, exfoliarse, desconcharse

exfoliation *f.* exfoliación

exhalant *m. & a.* escurrente

exhalation *f.* exhalación, emanación

exhaust *m.* tubo de escape; escape || ***exhaust gases*** *m.* gases del escape

exhaust *vb.* descargar, vaciar, agotar

exhausted *a.* agotado

exhibit *f.* exposición

exhibit *vb.* mostrar, manifestar || ***exhibit indications of ore*** *vb.* pintar

exhibition *f.* exhibición

exhumation *f.* exhumación

exhumed *a.* exhumado

exine *f.* exina

exinite *f.* exinita

existence *f.* existencia

exogenesis *f.* exogénesis

exogenetic *a.* exogenético

exogenous *a.* exógeno

exoglyph *m.* exoglifo

exomorphic *a.* exomórfico

exomorphism *m.* exomorfismo

exopodite *m.* exopodito

exoseptum *m.* exosepto

exoskeleton *m.* exoesqueleto

exosphere *f.* exósfera

exothermal, exothermic *m.* exotérmico

exotic *a.* alóctono, exótico, arrastrado

expand *vb.* extender, expandir, ensanchar, abrir, dilatar

expandable *a.* expandible

expanded *a.* extendido, dilatado || ***expanded section*** *f.* sección extendida

expander *m.* expansor, ensanchador || ***tube, pipe expander*** *m.* abocinador de tubos, ensanchatubos, expandidor de tubos

expanding *a.* expandible, ensanchable

expansion *f.* expansión, dilatación || ***adiabatic expansion*** *f.* expansión adiabática || ***isothermal expansion*** *f.* expansión isotérmica || ***expansion joint*** *f.* junta de dilatación o expansión || ***expansion line*** *f.* tubería de dilatación || ***expansion loop*** *f.* curva de dilatación || ***expansion ratio*** *f.* relación de expansión || ***expansion ring*** *m.* anillo de expansión

expectation *f.* expectativa

expedition *f.* campaña, expedición

expel *vb.* expulsar

expenditures *m.* gastos || ***capital expenditures*** *m.* gastos de capital

expense *m.* gasto, costo, coste || ***fixed expense*** *m.* gasto fijo

experiment *m.* experimento

experiment *vb.* investigar, experimentar

experimental *a.* experimental

expert *m. & a.* experto, perito || ***mining expert*** *m.* güiris

expertise *f.* pericia; *m.* expertaje

expiration *m.* vencimiento; *f.* (min.) caducación

expire *vb.* caducar, expirar, vencerse

expired *a.* caducado, vencido

explain *vb.* explicar, ilustrar

explanation *f.* explicación

explode *vb.* estallar, explotar

exploit *vb.* explotar, aprovechar

exploitation *f.* explotación; *m.* desarrollo, laboreo, aprovechamiento, disfrute

exploration *f.* exploración, búsqueda, busca; *m.* reconocimiento, reconociendo; (min.) cateo

explorationist *m.* explorador

exploratory *a.* exploratorio

explore *vb.* explorar, investigar, buscar, reconocer, (min.) catear

explored *a.* explorado, buscado, (min.) cateado

explorer *m.* explorador, investigador, buscador

explosion *f.* explosión, detonación; *m.* disparo, reventón

explosive *m.* explosivo

exponent *m.* exponente

export *f.* exportación

export *vb.* exportar

exporter *m.* exportador

exposed *a.* expuesto || ***to be exposed*** *vb.* aflorar(se), estar expuesto

exposure *m.* afloramiento; *f.* exposición

express *vb.* representar, expresar

expression *f.* expresión || ***topographic expression*** *f.* expresión topográfica

expropriate *vb.* expropiar

expropriation *f.* expropiación

expulsion *f.* expulsión

exsolution *f.* exsolución

extend *vb.* amplificar, extender, alargar, ampliar, prorrogar

extension *f.* extensión, distensión, prolongación || ***extension fracture*** *f.* fractura de extensión || ***time extension*** *f.* prórroga

extensional *a.* distensivo, extensional

extent *f.* extensión

exterior *m. & a.* exterior

external *a.* externo, exterior

extinct *a.* extinto, extinguido, apagado

(to become) extinct *vb.* extinguirse

extinction angle *m.* ángulo de obscurecimiento (o extinción)

extinguish *vb.* apagar, extinguir

extinguished *a.* apagado

extra *a.* extra

extract *m.* extracto

extract *vb.* extraer, (min.) apartar, desvenar, quitar, sacar

extractor *m.* extractor || ***drill extractor*** *m.* sacabarrena || ***tool extractor*** *m.* extractor de herramientas

extraction *f.* extracción, obtensión, saca; *m.* apartamiento mineral || ***solvent extraction*** *f.* extracción por solvente

extractor *m.* extractor

extraneous *a.* extraño

extrapolate *vb.* extrapolar

extrapolation *f.* extrapolación

extrude *vb.* estirar

extruded *a.* estirado

extrusion *m.* estiramiento por presión; *f.* efusión, erupción, extrusión

extrusive *m. & a.* extrusivo, efusivo

exude *vb.* exudar, resumir

eye *m.* ojo; *f.* ojal, argolla || ***cat's eye (agate)*** *m.* ojo de gato || ***eye base*** *f.* base ocular, distancia interpupilar || ***eye of the storm*** *m.* ojo de huracán

eyepiece *m.* lente ocular, ocular

F

fabric *m. & a.* petrofábrico; *f.* textura; tela; tejido

face *f.* cara, superficie, faz; *m.* frente, rostro || ***cleavage face*** *f.* cara del crucero || ***crystal face*** *f.* faceta || ***mine face*** *m.* frente de corte || ***working face*** *m.* frontón, tajo; *f.* faz de laboreo

face *vb.* (piedra) carear

facet *f.* faceta || ***crystal facet*** *f.* cara del cristal, faceta || ***principal facet*** *f.* mesa

faceted spur *m.* contrafuerte facetado

facies *f.* facies

facility *f.* planta, facilidad

facing *f.* capa de refuerzo; *m.* revestimiento, forro

fact *m.* hecho, dato

factor *m.* factor || ***factor of safety*** *m.* factor de seguridad

factory *f.* fábrica, planta

Fahrenheit *m.* grados Fahrenheit

failure *f.* (mech.) cesión; quiebra; falla mecánica, rotura; *m.* fracaso, defecto

fair *a.* regular, normal

fall *f.* caída, inclinación, baja; *m.* descenso || ***fall of sealevel*** *f.* bajada del nivel del mar; *m.* descenso del nivel del mar, decremento del nivel del mar || ***waterfall*** *m.* salto

fall *vb.* bajar, abajar, caerse, ceder, descender

false *a.* falso, aparente || ***false bottom*** *m.* piso falso, fondo falso

Famennian *m. & a.* Famenniense, Famenniano

family *f.* familia

fan *m.* abanico, ventilador || ***alluvial fan*** *m.* abanico aluvial || ***basin floor fan*** *m.* abanico de piso de cuenca || ***fan-shaped*** *a.* flabeliforme || ***submarine fan*** *m.* abanico submarino || ***talus fan*** *m.* abanico de deyección, abanico de talud

fang *m.* diente, colmillo

fanglomerate *m.* fanglomerado; *f.* raña

fasciole *m. & f.* fascíolo, fascíola

fathom *f.* braza, brazada; equivale a 1,67 m

fatigue *f.* fatiga; *m.* agotamiento || ***fatigue limit*** *m.* límite de resistencia

fault *f.* falla; paraclasa, dislocación, quiebra; *m.* botamiento, resbalón, salto || ***antithetic fault*** *f.* falla antitética || ***boundary fault*** *f.* falla marginal, falla limitante || ***branching fault*** *f.* falla ramificada || ***composite fault*** *f.* falla compuesta || ***detachment fault*** *f.* falla de despegue || ***dip-slip fault*** *f.* falla de buzamiento, falla según buzamiento, falla transversal de buzamiento || ***drag fault*** *f.* falla de arrastre || ***en echelon fault*** *f.* fallas al tres bolillo, fallas en escalón || ***fault block*** *f.* gleba tectónica || ***fault gouge*** *f.* harina de falla; *m.* detrito de falla || ***fault heave*** *m.* rechazo horizontal de la falla || ***fault line scarp*** *f.* barranca de falla con erosión || ***fault scarp*** *f.* barranca de falla, escarpa de falla || ***fault sliver*** *f.* escame || ***fault throw*** *m.* salto, rechazo vertical de la falla || ***fault zone*** *f.* zona de falla || ***gravity fault*** *f.* falla de gravedad || ***growth fault*** *f.* falla de crecimiento || ***high-angle fault*** *f.* falla de alto ángulo || ***inferred fault*** *f.* falla inferida, falla probable || ***keystone fault*** *f.* falla en clave de arco || ***listric fault*** *f.* falla lístrica || ***low-angle fault*** *f.* falla de bajo ángulo || ***normal fault*** *f.* falla normal, falla de ascenso, falla de gravidad, falla gravitacional || ***reverse fault*** *f.* falla inversa, falla de descenso, falla invertida || ***rotational fault*** *f.* falla giratoria, falla por rotación, falla pivotal || ***scissors fault*** *f.* falla pivotal, falla de tijera || ***sealing fault*** *f.* falla sellante || ***strike-slip fault*** *f.* falla lateral, falla transcurrente, falla de transcurencia, falla corriente, falla de rumbo, falla longitudinal de rumbo, falla rumbodeslizante, falla transversal de rumbo || ***synthetic fault*** *f.* falla sintética || ***tear fault*** *m.* desgarramiento; *f.* falla de desgarre || ***thrust fault*** *f.* cabalgadura, falla colgante, falla de altoempuje, falla acostada, falla de corrimiento, falla de escurrimiento, falla inversa, falla por empuje; *m.* acabalgamiento, cabalgamiento, sobrecorrimiento, corrimiento || ***transform fault*** *f.* falla transformante || ***underthrust fault*** *f.* falla de bajoescurrimiento || ***wrench fault*** *f.* falla lateral, falla transcurrente, falla de transcurrencia

fault *vb.* dislocar, fallar

faulted *a.* dislocado, fallado

faulting *m.* fallamiento

faulty *a.* defectuoso

fauna *f.* fauna

faunal *a.* fáunico, faunístico || ***faunal assemblage*** *f.* asociación de fauna || ***faunal zone*** *f.* zona faunística

favorable *a.* favorable, enderezado

fayalite *f.* fayalita

feasible *a.* practicable, viable, factible

feature *m.* rasgo, característica

fecal pellet *f.* pella fecal

fee *m.* honorario, costo

feedback *f.* retroalimentación ‖ ***feedback loop*** *m.* bucle de realimentación

feeder *m.* alimentador, tubo de alimentación ‖ ***feeder dike*** *m.* dique alimentador, dique fuente ‖ ***feeder stock*** *m.* tronco alimentador ‖ ***feeder waveguide*** *f.* guía de ondas de alimentación

feedstock *f.* carga, materia prima

fees *m.* honorarios

feldspar *m.* feldespato

feldspathic *a.* feldspático

feldspathoid *m.* feldspatoide

fell *f.* colina; *m.* monte

felsenmeer *f.* cancha; *m.* campo de bloques

felsic, felsitic *a.* félsico, felsítico

felsite *f.* felsita

female *f.* mujer, (botan., biol.) hembra; *a.* femenino

femic *a.* fémico

fen *f.* tierra pantanosa; *m.* pantano

fenestrae *f.* cavidades milimétricas rellenas de calcita

fenestrate *a.* fenestrado

fenster *f.* ventana, ventana tectónica; *m.* miradero

ferberite *f.* ferberita

fergusite *f.* ferrogusita

fergusonite *f.* fergusonita

fermentation *f.* fermentación

fern *m.* helecho

ferrate *m.* ferrato

ferric *a.* férrico

ferricopiapite *f.* challantita

ferromagnesian *m. & a.* ferromagnesiano, ferromagnésico

ferromagnetic *a.* ferromagnético

ferronickel *m.* ferroníquel; *f.* josefinita

ferrosilite *f.* ferrosilita

ferrous *a.* ferroso, férreo

ferruginous *a.* ferruginoso, ferrugíneo

fersmannite *f.* fersmannita

festoon cross-bedding *f.* estratificación cruzada festoneada

fetch *f.* búsqueda y carga; area fuente

fetid *a.* hediondo, fétido, mal olor

fiber *f.* fibra, vetita ‖ ***carbon fiber*** *f.* fibra de carbono

fibrolite *f.* fibrolita

fibrous *a.* fibroso, ahebrado, bacilar

fiducial mark *f.* marca fiduciaria

field *m.* campo ‖ ***field book*** *f.* libreta; *m.* cuaderno de campo ‖ ***field geology*** *f.* geología de campo ‖ ***field of view*** *m.* campo de observación, campo de un microscopio ‖ ***field party*** *f.* brigada de campo ‖ ***field work*** *m.* trabajo de campo ‖ ***electromagnetic field*** *m.* campo electromagnético ‖ ***gold field*** *m.* campo aurífero ‖ ***gravitational field*** *m.* campo de gravedad ‖ ***kame field*** *m.* campo de kames ‖ ***magnetic field*** *m.* campo magnético ‖ ***oil field*** *m.* campo del petróleo, campo petrolero, campo petrolífero

figure *f.* figura, ilustración

filament *m.* filamento; *f.* fibra

filamentous *a.* filamentoso

file *m.* archivo, fichero; *f.* lima ‖ ***file card*** *f.* ficha

file *vb.* registrar

filibranch *a.* filibranquiado

filiform *a.* filiforme

fill *m.* relleno, terraplén, llenado

fill *vb.* llenar; rellenar, terraplenar, atibar, colmarse

filled-to-spill *a.* a punto de derrame

filler *m.* relleno

filling *m.* llenado, relleno

film *f.* película, membrana, mica ‖ ***film overlay*** *f.* sobrepuesta en película ‖ ***interfacial film*** *f.* película interfacial ‖ ***oil film*** *f.* película de aceite

filter *m.* filtro, filtrador ‖ ***directional filter*** *m.* filtro direcional

filter *vb.* filtrar, depurar

filtering *f.* coladura; *a.* filtrante, filtrado

filtrate *m. & a.* filtrado; *m.* revoque

filtration *f.* filtración, percolación

fin *f.* (geog.) aleta; *m.* nervio

final *a.* terminal

finance *f.* finanzas; *m.* asuntos

finance *vb.* aviar, financiar

financial *a.* rentístico, monetario, financiero, financista, económico || ***financial statement*** *m.* balance; informe de cuentas

financier *m.* aviador, financista

financing *f.* financiación

find *m.* hallazgo, descubrimiento

find *vb.* descubrir, alumbrar, encontrar

fine *f.* multa (com., legal)

fine *a.* fino

fine *vb.* afinar(se)

fine-grained *a.* de grano fino

fineness *f.* fineza, finura; *m.* tenor

fines *m.* finos; ripios, polvo menudo que pasa por los filtros

fining upward *adv.* afinamiento hacia arriba, granodecreciente hacia arriba

finish *vb.* acabar, terminar

finite differences *f.* diferencias finitas

finite elements *m.* elementos finitos

fiord, fjord *m.* fiordo

fire *m.* fuego || ***fire brick*** *m.* ladrillo refractario || ***fire clay*** *f.* arcilla infusible; *m.* tasco || ***fireproof*** *a.* ignífugo, a prueba de fuego

fire *vb.* (perf.) disparar, prender fuego, encender

firing *m.* fogueo; encendimiento

firm *a.* firme

firn *m.* firn; *f.* neviza || ***firn*** *m. & a.* nevado || ***firn ice*** *m.* geloide

first aid *m.* primer auxilio || ***first aid kit*** *m.* botiquín

first arrival *f.* primeras ondas que llegan

firth *m.* valle sumergido, brazo de mar, fiordo

fiscal *a.* fiscal || ***fiscal stamp*** *m.* timbre fiscal || ***fiscal year*** *m.* año fiscal

fish *m.* pescado, labor de salvamento

fish *vb.* pescar, rescatar

fish-tail bit *f.* barrena cola de pescado

fishing *f.* pesca, rescate || ***fishing basket*** *f.* cesta de pesca || ***fishing tool*** *m.* pescador; *f.* pinza

fissile *a.* físil, hendible, rajadizo, firme

fissility *f.* fisilidad

fission *f.* fisión || ***fission tracks*** *f.* huellas de fisión, trazas de fisión

fissionable *a.* fisionable

fissure *f.* fisura, grieta, hendidura, abertura, abra, agrietadura, cuarteadura, quebrada, sima, quebradura, quiebra, quebrantadura, rajadura; *m.* barranco, cuarteo || ***fissure eruption*** *f.* erupción lineal || ***fissure flow*** *f.* efusión por grietas || ***fissure vein*** *f.* veta de relleno de fisuras || ***fissure vent*** *f.* grieta de erupción

fissuring *m.* grieteo

fit *vb.* ajustar, montar, empalmar, adaptar

fitted *a.* ajustado

fitter *m.* ajustador

fittings *f.* piezas, conexiones; *m.* accesorios, aditamientos || ***pipe fittings*** *m.* accesorios para tubería

five spot flooding system *m.* sistema de inyección de agua a cuatro pozos para producir petroleo por presión hidráulica por conducto de un quinto pozo; arreglo de cinco pozos

fix *vb.* fijar; asegurar, asentar, empotrar; reparar, arreglar

fixed *a.* empotrado, fijo; arreglado, reparado || ***fixed carbon*** *m.* carbón estable; carbón fijo || ***fixed bed catylist*** *m.* catalizador de lecho fijo

fixing agent *m.* fijador

fjord *m.* fiordo

flabelliform, flabellate *a.* en forma de abanico

flag *f.* (survey, levan.) bandera; *m.* señalizador

flagellate *m.* flagelado

flagellum *m.* flagelo

flaggy *a.* lajoso

flagstone *m.* [Méx.] cuartón, lastrón; *f.* chanca, laja lastra, loncha, losa

flake *f.* laminilla, escama, hojuela; *m.* casquito

flaky *a.* escamoso; floculento

flame *f.* llama, flama || ***flame arrester*** *f.* parallamas; *m.* chispero || ***flame structure*** *f.* estructura de llama

flange *f.* brida, pletina; oreja, pestaña, cabeza, cuerda, pata, rama; *m.* platillo; base, patín; reborde, resalte; cordón

flange *vb.* rebordear, embridar, bordear, bridar

flanged *a.* alado

flange-like *a.* alado, con forma de ala

flank *m.* flanco, costado, lado; *f.* ladera, ala

flank *vb.* flanquear

flanked *a.* flanqueado

flare *f.* quema, tea; *m.* destellador, mechero

flaring *a.* ensanchado

flaser *a.* flaser || ***flaser structure*** *f.* estructura flaser, estructura de flujo escamosa, láminas de escamas rodeadas de un cuerpo lenticular

flash *m.* relámpago; destello; resplandor; fusilazo || ***flash distillation*** *f.* destilación por expansión instantánea || ***flash point*** *m.* punto de deflagración

flask *m.* frasco, matraz; *f.* botella, redoma

flat *a.* plano, llano, parejo || ***flat land*** *m.* campo raso; *f.* sabana || ***flat-lying*** *a.* horizontal || ***flat pebble*** *m.* guijarro aplanado

flatiron *f.* plancha

flatten *vb.* aplanar, aplastar, asombrarse

flattened *a.* (geog.) plano, ñato, [Arg.] chato

flattening *m.* achatamiento, aplanamiento, aplastamiento

flaw *m.* defecto, vicio; *f.* imperfección, grieta, mancha, quebraja || ***flaw in an emerald*** *f.* yerba

flecked *a.* abigarrado

flexibility *f.* flexibilidad

flexible *a.* flexible

flexure, flexural fold *f.* flexura, flexión

flexural *a.* flexural || ***flexural slip*** *m.* deslizamiento

flexural *f.* cizalla flexural

flight line *f.* línea de vuelo

flint *m.* pedernal, chert, lumbre, petrosilex, sílex, [Méx.] tecpatl; *f.* piedra, córnea

float *m.* (talud) flotador; acarreo, acarreos, rodado

float *vb.* flotar

floating *a.* flotante || ***floating-point arithmetic*** *f.* aritmética de coma flotante

flocculation *f.* floculación

flocculating agent *m.* agente floculador

flocculent *m.* floculento

floe *m.* banco de hielo, témpano, bandejón

flood *a.* creciente

flood *f.* inundación, crecida, avenida, aguada, arriada, riada, venida, [Ch.] creces, [Arg.] avalancha; *m.* aguaducho, creciente, diluvio || ***flash flood*** *f.* inundación repentina || ***flood plain*** *m.* lecho de crecimiento, área inundable, lecho mayor, talud de ribera, plano aluvial; *f.* vega, planicie de inundación || ***flood-prone*** *a.* alagadizo, anegable || ***flood tide*** *f.* pleamar || ***flood surface*** *f.* superficie de inundación

flood *vb.* inundar, anegar, aguarse, alagar, apantanar, arriarse, llenar, sofocarse, [Ch.] alagunar

flooded *a.* inundado, ahogado, alagado, crecido, lleno, (min.) sofocado || ***easily flooded*** *a.* inundable || ***flooded mine or excavation*** *m.* abrevadero

flooding *m.* crecimiento; *f.* inundación || ***flooding surface*** *f.* superficie de inundación

floor *m.* piso, suelo, bajo; *f.* planta, baja

flora *f.* flora

florencite *f.* koivinita

floscelle *m.* floscelo

flotation *f.* flotación, flotabilidad

flour *a.* en polvo

flow *m.* flujo, corriente, caudal, derrame, gasto; *f.* descarga, colada, corredura || ***cash flow*** *m.* flujo de caja, flujo de efectivo || ***future cash flow*** *m.* flujo futuro de caja || ***debris flow*** *m.* flujo de escombros, flujo de escombra || ***elastic flow*** *f.* fluencia elástica || ***flow cast*** *m.* calco de flujo || ***flow capacity*** *m.* aforo || ***flowchart*** *m.* diagrama de flujo, organigrama, ordinograma || ***flow cleavage*** *f.* foliación de flujo || ***flow diagram*** *m.* diagrama de flujo || ***flow folding*** *m.* plegamiento de flujo || ***flow into*** *vb.* afluir || ***flow layering*** *f.* camadas de flujo || ***flow line*** *f.* línea de flujo || ***flowmeter*** *m.* medidor de caudal || ***flow rate*** *f.* velocidad del flujo; *m.* real de fontanero || ***flow structure*** *f.* textura fluidal, estructura de flujo || ***heat flow*** *m.* flujo térmico || ***laminar flow*** *m.*

flujo laminar || *lava flow* m. flujo de lava; f. corriente de lava || *mud flow* m. flujo de fango, flujo de lodo || *plastic flow* m. flujo plástico || *turbulant flow* m. flujo turbulento || *viscous flow* m. flujo viscoso

flow, flow into vb. fluir, correr, desembocar, dimanar

flowage m. metamorfismo plástico

flowing a. corriente, fluente || *flowing pressure* f. presión de flujo

flowing well m. pozo surgente

flowstone f. incrustación calcárea, travertina

fluctuate vb. fluctuar, variar

fluctuating a. fluctuante

fluctuation f. fluctuación; m. balance

flue m. tubo, flus; tragante, humero, escape; f. tronera || *flue gas* m. gas de chimenea

fluid m. fluído; f. agüilla, aguosidad; a. fluente, flúido || *frac fluid* m. flujo de fracturación || *mineralizing fluid* m. flúido de mineralización; f. flúido madre || *wetting fluid* m. flúido humedo, flúido humectante

fluidity a. fluidez

fluidizing f. fluidificación

flume m. tubo separador; caño, cárcamo, conducto saetín; f. caz, acequia, canal

fluoborite f. nocerita

fluocerite f. tysonita

fluorescence f. fluorescencia

fluorescent a. fluorescente

fluoride m. fluoruro

fluorine f. fluorina; m. flúor

fluorite f. fluorita, fluorina; m. espato de flúor || *cerian fluorite* f. yttrocerita

fluoroscope m. fluoroscopio

fluorspar flux m. espalto

flushed a. lavado || *flushed zone* f. zona lavada

flute f. estría || *flute cast* m. turboglifo

fluted a. acanalado, estriado, canalado

fluvial a. fluvial

fluvioglacial a. fluvioglacial

fluviomarine a. fluvial marino, fluviomarino

flux m. fundente, flujo, fluidificante, metal de ayuda || *smelting flux* m. magistral

fluxion m. flujo; f. fusión; fluxión, derivada

flysch m. flysch

flywheel m. volante

foam f. espuma || *foam rubber* f. espuma de caucho, espuma de hule

focal a. focal || *focal depth* f. profundidad focal || *focal length* f. longitud focal, distancia focal || *focal plane* m. plano focal

focus m. foco, enfoque; epicentro || *buried focus* m. foco cubierto || *coarse focus* m. tornillo macrométrico || *fine focus* m. tornillo micrométrico || *velocity focus* m. foco de velocidad

focus vb. enfocar, focalizar

fog f. niebla, neblina

fold m. pliegue, plegamiento, doblez; (seis.) cubrimiento, duplicidad || *arcuate fold* m. pliegue arqueado || *asymmetric fold* m. pliegue asimétrico || *broad fold* m. pliegue suave, pliegue amplio || *buckle fold* m. pliegue bucle, pliegue circular || *chevron fold* m. pliegue chevron, pliegue cabrío || *closed fold* m. pliegue cerrado || *competent fold* m. pliegue armónico || *concentric fold* m. pliegue concéntrico || *detachment fold* m. pliegue de despegue || *diapiric fold* m. pliegue diapírico, diapiro || *disharmonic fold* m. pliegue disarmónico || *dome fold* m. pliegue abovedado || *drag fold* m. pliegue de arrastre || *elongate fold* m. pliegue elongado (o largo o alargado) || *en echelon fold* m. pliegue escalonado || *fan fold* m. pliegue abanico || *fault-bend fold* m. pliegue flexurado por falla || *fault-propagation fold* m. pliegue por propagación de la falla || *flexural fold* f. flexura, flexión || *fold belt* m. cinturón plegado || *forced fold* m. pliegue forzado || *gentle fold* m. pliegue suave || *imbricate folds* m. pliegues imbricados || *inverted fold* m. pliegue volcado (o invertido, o tumbado) || *isoclinal fold* m. pliegue isoclinal, pliegue cerrado || *kink fold* m. pliegue kink || *normal fold* m. pliegue parado || *overturned fold* m. pliegue invertido (o volcado, o tumbado) || *parallel fold* m. pliegue paralelo || *plunging fold* m.

pliegue buzante || ***recumbent fold*** *m.* pliegue acostado (o recostado, yacente, o recumbente) || ***sheath fold*** *m.* pliegue de cubierta || ***similar fold*** *m.* pliegue semejante o similar || ***small fold*** *f.* arruga || ***symmetric fold*** *m.* pliegue simétrico o recto || ***tight fold*** *m.* pliegue apretado, pliegue angosto || ***upright fold*** *m.* pliegue parado

fold *vb.* plegar, arrugar

folded *a.* plegado, arrugado, doblado

folding *m.* plegamiento, plegado; *f.* arrugación, plegadura || ***chevron folding*** *m.* plegamiento en chevron || ***cross folding*** *m.* plegamiento entrecruzado || ***en echelon folding*** *m.* plegamiento en escalón || ***fault-bend folding*** *m.* plegamiento por flexión de fallas || ***fault-propagation folding*** *m.* plegamiento por propagación de fallas || ***flexural folding*** *m.* plegamiento verdadero || ***flexural flow folding*** *m.* plegamiento por flujo || ***flexural slip folding*** *m.* plegamiento por flexión y deslizamiento || ***micro folding*** *m.* plísamiento

foliated *a.* foliado, hojoso

foliation *f.* foliación, esquistocidad, exfoliación, foliatura

follow-up *vb.* perfeccionar, dar seguimiento

fool's gold *f.* pirita

foot *m.* pie, 30,48 cm || ***foot pound*** *m.* librapié

foot-activated *adv.* accionado por pedal

foothill *f.* estribación; ladera, piedemonte, falda de montaña; *m.* pie de montaña

footpound *f.* libra-pié

footwall *m.* bloque hundido, bloque yacente, bloque bajo, arrastre, bajo, piso, reliz del bajo; *a.* yusano, yusero || ***footwall flat*** *m.* plano basal || ***footwall ramp*** *f.* rampa basal

foraminifera *m. & a.* foraminífero

force *f.* fuerza, potencia, energía; *m.* esfuerzo || ***buoyant force*** *f.* fuerza de flotación || ***centrifugal force*** *f.* fuerza centrífuga || ***compressive force*** *f.* fuerza compresora

forced *a.* forzado

forceful injection *f.* inyección forzada

ford *m.* vado, paso

forecast *m.* pronóstico; *f.* previsión

forecast *vb.* pronosticar

foredeep *f.* antecubeta, fosa anterior, prefosa

foreground *m.* primer plano

foreign exchange *f.* divisas

foreland *m.* antepaís, cabo, promontorio; *f.* región frontera, zona más externa, tierra adentro || ***foreland basin*** *f.* antefosa

forelimb *m.* flanco anterior

foreman *m.* capataz, sobrestante, mayordomo, jefe, campero, capitán, colero, mandón, obrajero, [Vz.] caporal || ***blasting foreman*** *m.* pegador

forereef *m.* antearrecife

foreset beds *m.* estratos (o capas) frontales

fore-shock *m.* sismo preliminar, temblores previos, aviso sísmico, sismo precursor

foreshore *m.* litoral entre mareas, sublitoral, zona de marea

foresight *f.* visual adelante

forest *m.* bosque; *f.* selva

forward modeling *m.* modelaje iterativo, modelaje hacia adelante

forfeit *vb.* (min.) caducar

forfeiture *f.* (legal) pérdida legal por incumplimiento

forge *f.* fragua, forja

forge *vb.* fraguar, forjar

forged *a.* fraguado

forging *f.* forjadura; pieza forjada; *m.* fraguado

fork *f.* bifurcación, confluencia

fork *vb.* bifurcarse

forked *a.* ahorquillado

forking *f.* bifurcación

form *f.* forma, figura

form *vb.* conformar, formar

(to be) formed *vb.* formarse

formaldehyde *m.* formaldehido

format *m.* formato

format *vb.* formatear

formation *f.* formación || ***formation damage*** *f.* daños a formación || ***formation tester*** *m.* verificador de la formación

formic *a.* fórmico

formula *f.* fórmula

forsterite *f.* forsterita

fortune *m.* caudal; *f.* fortuna, suerte

forwarding agent *m.* agente de fletes, agente expedidor

fosse, foss *f.* fosa

fossil *m.* fósil || ***fossil assemblage*** *f.* asociación de fósiles || ***guide fossil*** *m.* fósil guía, fósil índice, fósil indicador

fossiliferous *a.* fosilífero

fossilize *vb.* fosilizar

(to become) fossilized *vb.* fosilizarse

fossilization *f.* fosilización

fossula *f.* fósula

foul *vb.* ensuciar, manchar

fouling *m.* ensuciamiento

(to be) found *vb.* disponerse, encontrarse, hallarse

foundation *m.* base, cimiento, fundamento, fundación

foundry *f.* fundición

fountain *f.* fuente

four-wheel drive *m.* cuatro por cuatro, impulsión por cuatro ruedas, de doble, tracción

fowlerite *f.* fowlerita

frac *vb.* fragmentar || ***frac gradient*** *m.* gradiente de fractura || ***frac job*** *f.* fragmentación, fracturación || ***frac fluids*** *m.* flúidos de fractura

fractile *f.* fractil

fraction *f.* fracción, proporción

fractional *a.* fraccionario || ***fractional distillation*** *f.* destilación fraccionada

fractionate *vb.* fraccionar

fractionating column *f.* torre de destilación fraccionaria

fractionation *m.* fraccionamiento; *f.* fraccionación

fractionator *f.* torre fraccionadora; *m.* fraccionador

fracture *f.* fractura, rotura, agrietadura, disyunción, quebradura, quebrantadura, quiebra, grieta, hendidura; *m.* quebrantamiento, rompimiento || ***en echelon fracture*** *f.* fractura escalonada || ***extension fracture*** *f.* fractura de extensión || ***fracture cleavage*** *m.* clivaje de fractura || ***fracture plane*** *m.* plano de fractura || ***fracture zone*** *f.* área de fracturación || ***sigmoidal fracture*** *f.* fractura sigmoidea

fracture *vb.* fracturar, agrietar, agrietarse, desquebrajar, quebrajar, quebrantar, quebrantarse, quebrar, quebrarse, romper

fractured *a.* fracturado, quebradizado, quebrado, quebradizo

fracturing *m.* agrietamiento, fracturamiento; *f.* fracturación || ***artificial fracturing*** *m.* torpedeamiento

fracturing *vb.* agrietarse

fragile *a.* frágil, quebradizo, quebrajoso

fragment *m.* fragmento, pedazo; *f.* lasca

fragment *vb.* fragmentar

fragmentation *f.* fragmentación

frame *m.* armazón, marco, bastidor; *f.* estructura, tirantería, armazón || ***skid frame*** *m.* armazón de corredera, bastidor de corredera

framework *m.* marco, patrón, arreglo; *f.* armadura, armazón, marca, trama

franchise *f.* franquicia

francium *m.* francio

franklinite *f.* franklinita, espinela cincífera, isofana, zincoferrita

Frasnian *m. & a.* Frasniense, Frasniano

free *a.* libre; gratis || ***free air*** *a.* de aire libre || ***free air correction*** *f.* corrección de aire libre || ***free fall*** *f.* caída libre || ***free radical*** *m.* radical libre || ***free swimming*** *a.* que nada libremente

free *vb.* liberar, franquear

freeboard *m.* francobordo, abordaje libre

freeze *vb.* congelar

freezing *a.* congelado || ***freezing point*** *m.* punto de congelación

freibergite *f.* freibergita

freight *m.* flete

frequency *f.* frecuencia || ***acoustic frequency*** *f.* frecuencia acústica || ***ground frequency*** *f.* frequencia del suelo || ***high frequency*** *f.* frequencia alta || ***low frequency*** *f.* frequencia baja || ***resonance frequency*** *f.* frecuencia de resonancia

fresh air *f.* aire puro

fresh water *m.* agua dulce

friable *a.* quebradizo, friable, desmenuzable, desleznable, labíl, quebradizado

friability *a.* friabilidad

friction *f.* fricción

frictional resistance *f.* resistencia de rozamiento, resistencia de fricción

fringe *f.* franja, margen, liga; orla; *m.* borde; fleco || ***capillary fringe*** *f.* orla freática, franja capilar

fringing reef *m.* arrecife marginal

front *a.* delantero, anterior; *m.* frente || ***cold front*** *m.* frente frío || ***stationary front*** *m.* frente estacionario || ***tectonic front*** *m.* frente tectónico || ***warm front*** *m.* frente cálido || ***wave front*** *m.* frente de onda

frontal *a.* frontero, anterior, frontal, frental

frontier *f.* frontera, zona limítrofe

frost *f.* helada, escarcha; *m.* hielo || ***frost heaving*** *f.* dilatación por congelación || ***frost polygons*** *m.* polígonos de congelación

frosted *a.* helado, deslustrado

froth *f.* espuma; burbuja

frozen *a.* congelado || ***frozen ground*** *m.* gelisol

fuchsite *f.* fucsita, fuchsita

fucoid *m.* fucoide

fuel *m.* combustible || ***fuel cell*** *f.* pila combustible || ***fuel gauge*** *m.* indicador de combustibles || ***fuel oil*** *m.* combustóleo, mazut || ***fuel*** *m.* combustible de chorro; *f.* turbosina

fugacity *f.* fugacidad

fulcrum *m.* fulcro

fulgurite *f.* fulgurita

full *a.* lleno; completo || ***full bore*** *a.* de pleno caudal

Fuller's earth *f.* galáctita, tierra de batán, arcilla de batán, arcilla decolorante, greda, greta

fulcrum *m.* fulcro

Fullonian *m. & a.* Fuloniense, Fuloniano

fumarole *f.* fumarola

fume(s) *m.* vapor, gas, vaho; *f.* (voladura) fogata

function *f.* función || ***as a function of*** en función de

function *vb.* funcionar

fund *vb.* habilitar

fund(s) *m.* fondos

fungus *m.* hongo

funiculus *m.* funículo

funnel *f.* hipónoma; *m.* embudo

furlong *m.* estadio

furnace *m.* horno, hogar, chacuaco, galemador || ***blast furnace*** *m.* horno alto, horno de fundición || ***calcining furnace*** *m.* horno de calcinación; [Sp.] calentador || ***reverberatory furnace*** *m.* comillo, galeme, hornillo || ***small furnace*** *m.* hornillo; *f.* hornaza

furnish *vb.* abastar

furrow *m.* surco

furrow *vb.* surcar, asurcar

furrowed *a.* surcado, acanalado

furrowing *m.* surcamiento

fusain *f.* fusita

fuse *m.* cohete, fusible, seguro; *f.* mecha

fuse *vb.* fundir, fusionar

fusible *a.* fusible, fundible

fusibility *f.* fusibilidad

fusinite *f.* fusinita

fusion *f.* fusión, coalescencia, soldadura

fusulinid *f.* fusulina; *m.* fusulínido

G

gab *m.* gancho

gabbro *m.* gabro || ***corundum hornblende gabbro*** *m.* korund-gabro || ***quartz gabbro*** *f.* klausenita

gabbroic *a.* gábrico, gabróico

gad *f.* cuña, piquetilla; *m.* punzón, alipús

gadolinite *f.* gadolinita, ytterbita, yttrita

gadolinium *m.* gadolino

gage, gauge *m.* manómetro, marcador, indicador, calibrador; calibre, medidor || ***strain gage, gauge*** *m.* indicador de deformación, medidor de deformación

gaging *f.* medición

gahnite *f.* gahnita, ganita, espinela cincífera

gain *f.* muesca; ganancia; *m.* beneficio

gain *vb.* ganar, vencer; propiciar

gal *m.* gal; unidad de aceleración gravimétrica

galaxite *f.* galaxita

galaxy *f.* galaxia || ***spiral galaxy*** *f.* galaxia espiral

galena *f.* galena, galenita, johnstonita; *m.* metal azul, sulfuro de plomo, (con plata) acerillo, soroche, (cúbico) alquifol, alcohol, [Bol., Pe.] carne de vaca, [Pe.] espejado

gall *f.* rozadura; *m.* raspaje || ***clay gall*** *f.* inclusión de arcilla

gallery *f.* galería, caña, rama, vía; *m.* caño, entresuelo, socavón, paso, túnel, [Bol.] camino || ***gallery roof*** *f.* [Pe.] alsa || ***main gallery*** *m.* paso real || ***parallel gallery*** *m.* paralelo || ***secondary gallery*** *m.* ramal

gallium *m.* galio

gallon *m.* galón (4,543 litros en Inglaterra; 3,785 litros en E.U.A.)

galvanic action *f.* acción galvánica

galvanized *a.* galvanizado

galvanometer *m.* galvanómetro

gamete *m.* gameto

gamma *f.* gamma; unidad de fuerza del campo magnético = 1×10^5 gauss

gamma ray *m.* rayo gamma || ***gamma ray log*** *f.* gammagrafía

gamut *f.* gama

gang *f.* brigada, cuadrilla, tanda, pandilla, [Méx.] parada

gangue *f.* ganga, borra, guija, matriz, roca madre, zafra; *m.* estéril, quijo, soroque, [Bol., Pe.] paniso, panizo

gap *m.* hiato, resquicio; desfiladero, collado, portillo, puerto, [Ch.] atravieso; *f.* garganta, angostura; brecha; barranca, hondonada, abra, caluma, estrechura, estrechez

garbage *m.* ataques; *f.* basura

garnet *m.* granate, granada || ***garnet-bearing*** *a.* granatífero

garnierite *f.* garnierita

gas *m.* gas || ***acetylene gas*** *m.* gas acetileno || ***associated gas*** *m.* gas asociado || ***aviation gas*** *f.* gasolina de aviación, gasavión || ***casinghead gas*** *m.* gas húmedo, gas de boca de pozo || ***coal gas*** *m.* gas de carbón || ***compressed natural (CNG) gas*** *m.* gas natural comprimido (GNC) || ***dry gas*** *m.* gas seco || ***exhaust gas*** *m.* gas de escape || ***flue gas*** *m.* gas de chimenea, humero || ***fuel gas*** *m.* gas combustible || ***gas analysis*** *m.* análisis de gas || ***gas burner*** *m.* quemador de gas, mechero || ***gas cap*** *f.* cúpula gasífera, capa gasífera, cresta de gas; *m.* casquete gasífero || ***gas cap drive*** *m.* empuje por gas libre || ***gas chromatography–mass spectrometry (GCMS)*** *f.* cromatografía de gas–espectrometría en masa || ***gas cloud*** *f.* nube gaseosa || ***gas condensate*** *m.* gas condensado, gas licuado || ***gas cut*** *m.* porcentaje de gas || ***gas cut mud*** *f.* inyección gasificada || ***gas detector*** *m.* detector de gas || ***gas drive*** *m.* empuje por gas en solución || ***gas leak*** *f.* pérdida de gas, fuga de gas || ***gas lift*** *f.* extracción artificial por gas || ***gas line*** *m.* gasoducto || ***gas lock*** *m.* atasco por gas || ***gas mask*** *f.* careta antigás, mascara antigás || ***gas meter*** *m.* contador de gas, medidor de gas || ***gas oil*** *m.* gasóleo || ***gas/oil contact*** *m.* contacto gas-petróleo || ***gas/oil ratio*** *m.* relacion gas-petróleo || ***gas pipeline*** *m.* gasoducto || ***gas plant*** *f.* planta de gas || ***gas pocket in a coal mine*** *m.* averno || ***gas pressure regulator*** *m.* regulador de la presión del gas || ***gas sand*** *f.* arena gasífera || ***gas scavenging*** *m.* barrido de gases de combustión || ***gas scrubber*** *m.* depurador de gas; *f.* lava de gas || ***gas separator*** *m.* separador de gas || ***gas show*** *f.* manifestación de gas || ***gas stripper*** *m.* desgasificador || ***gas trap*** *f.* trampa de gas; *m.* colector de gas || ***gas/water contact*** *m.* contacto gas-agua || ***ideal gas*** *m.* gas perfecto, gas ideal || ***inert gas*** *m.* gas inerte || ***liquified natural gas (LNG)*** *m.* gas natural licuado (GNL) || ***liquified petro-***

leum gas (LPG) m. supergás || ***natural gas** m.* gas natural, metano || ***natural gas liquids** m.* líquidos de gas natural || ***occluded gas** m.* gas ocluido || ***solution gas** m.* gas disuelto, gas en solución || ***sour gas** m.* gas ácido, gas sulfuroso, gas amargo || ***swamp gas** m.* gas de pantano || ***sweet gas** m.* gas dulce || ***unassociated gas** m.* gas libre, gas no asociado || ***waste gas** m.* gas de desecho, gas de desperdicio || ***wet gas** m.* gas húmedo

gaseous *a.* gaseoso

gash *m.* corte natural; *f.* grieta.de tracción

gasification *f.* gasificación

gasify *vb.* gasificar

gasket *f.* empaquetadura, arandela, zapatilla, guarnición

gasoline *f.* gasolina, nafta, bencina || ***casinghead, natural gasoline**, f.* gasolina natural

gas-tight *a.* hermético al gas, estanco al gas

gastrolith *m.* gastrolito

gastropod *m.* gasterópodo

gather *vb.* recoger, acumular; (seis.) *f.* familia || ***CDP gathers** f.* familias CDP

gathering line *f.* cañería colectora

gathering station *f.* estación recolectora

gauge, gage *f.* vara, sonda; *m.* escandallo, manómetro, marcador, indicador, medidor || ***gauge pressure** f.* presión manométrica || ***gauge table** m.* cuadro de medición || ***pressure gauge** m.* manómetro || ***standard gauge** f.* plantilla normal || ***strain gauge** m.* medidor de deformación || ***stream gauge** m.* hidrómetro, manómetro

gauge *vb.* aforar, medir, graduar

gauging pole *f.* varilla graduada

gauss *m.* gaussio, gauss; unidad de intensidad magnética reemplazada desde 1938 por Oersted = 100,00 gammas

gaylussite *f.* gaylussita; *m.* urao

gear *m.* engranaje, engrane

Gedinnian *m. & a.* Gedinniense, Gedinniano

gedrite *f.* gedrita

gehlenite *f.* velardenita

geiger-counter *m.* contador geiger

geiser *m.* géiser

gel *m.* gel; *f.* gelatina

gel *vb.* gelificarse

gelatinous *a.* gelatinoso

gelling agent *m.* aditivo gelatinizador

gem *f.* gema || ***genuine gem** a.* gema oriental

gemmology *f.* gemología

gene *m.* gen

general *a.* general

generalized section *m.* pérfil esquemático

generate *vb.* generar

generation *f.* generación

generator *m.* generador

generic *a.* genérico

genesis *f.* génesis

genetic *a.* genético

genetics *f.* genética

genotype *m.* especie tipo, genotipo

gentle *a.* amplio, suave

genus *m.* género

geoanticline *f.* geoanticlinal, anticlinal regional

geocentric *a.* geocéntrico

geochemical *a.* geoquímico

geochemist *m.* geoquímico

geochemistry *f.* geoquímica

geochronologic, geochronological *a.* geocronológico

geochronology *f.* geocronología

geocronite *f.* geocronita

geode *f.* géoda, bocarrena

geodepression *f.* geodepresión

geodesy *f.* geodesía

geodetic *a.* geodético

geodetics *f.* geodésica

geodynamic *a.* geodinámico

geognosy *f.* geonosia

geographer *m.* geógrafo

geographic *a.* geográfico

geography *f.* geografía

geohydrology *f.* geohidrología

geoid *m.* geoide

geologic *a.* geológico || ***geologic column*** *f.* columna geológica, sección geológica || ***geologic map*** *m.* mapa geológico, carta geológica || ***geologic section*** *f.* columna geológica, sección geológica || ***geologic setting*** *m.* ámbito geológico || ***geologic time*** *m.* tiempo geológico

geological *a.* geológico

geologist *m.* geólogo || ***well site geologist*** *m.* geólogo de pozo

geology *f.* geología, geognosia || ***well site geology*** *f.* geología de pozo

geomagnetic *a.* campo geomagnético, geomagnético

geometric *a.* geométrico

geometrically *adv.* geométricamente

geometry *f.* geometría

geomorphic *a.* geomórfico

geomorphology *f.* geomorfología

geophone *m.* geófono; sismómetro, micrófono, receptor || ***geophone spread*** *f.* haz de geófonos

geophysical *a.* geofísico

geophysicist *m.* geofísico

geophysics *f.* geofísica

geo-reference *f.* geo-referencia

geo-referenced *a.* geo-referenciado

Georgian *m. & a.* Georgiense, Georgiano

geoscience *f.* geociencia

geoscientist *m. & f.* geocientífico

geosuture *f.* geosutura

geosyncline, geosynclinal *f. & a.* geosinclinal

geotectonics *f.* geotectónica

geothermics *f.* geotérmica

geothermal, geothermic *a.* geotérmico || ***geothermal, geothermic gradient*** *f.* pendiente geotérmica

gerontic *a.* senil, geróntico

gersdorffite *f.* gersdorfita

geyser *m.* géiser

geyserite *f.* geiserita

giant *a.* gigante

gibbsite *f.* gibbsita, yesita

gigantism *m.* gigantismo

gill *f.* agalla, branquia

gilsonite *f.* gilsonita

gimbal *m.* balancín de la brújula

Givetian *m. & a.* Givetiense, Givetiano

glabella *f.* glabela

glacial *a.* glacial, glaciario || ***glacial cirque*** *m.* circo glaciario; *f.* zanoga || ***glacial drift*** *m.* acarreos de glaciar, derrubios glaciarios || ***glacial erratic*** *m.* bloque errático || ***glacial outlet*** *f.* boca de glaciar

glaciated *a.* aborregado, glaciado

glaciation *f.* glaciación

glacier *m.* glaciar, helero, ventisquero || ***alpine glacier*** *m.* glaciar alpino || ***continental glacier*** *m.* glaciar continental || ***hanging glacier*** *m.* glaciar colgante, glaciar colgado || ***mountain glacier*** *m.* glaciar de montaña || ***rock glacier*** *m.* glaciar de roca, glaciar rocoso || ***valley glacier*** *m.* glaciar de valle

glaciofluvial *a.* glacifluvial

glaciolacustrine *a.* glacilacustre

glaciology *f.* glaciología; *m.* glaciarismo

glance *m.* lustre

glass *m.* vidrio || ***Pyrex glass*** *m.* vidrio pyrex || ***volcanic glass*** *m.* vidrio volcánico

glassy *a.* vidrioso, vítreo

glaubertite *f.* glaubertita

glauconite *f.* glauconita, glauconia

glauconitic *a.* glauconítico

glaucophane *f.* glaucofana, glaucodot

glaze *m.* lustre

glen *m.* hocino

glide *m.* deslizamiento, escurrimiento || ***glide plane*** *m.* plano de escurrimiento, plano de deslizamiento

global *a.* global

globe *m.* globo

globigerina *f.* globigerina

globular *a.* globular, globuloso, globoso

globule *m.* glóbulo

glomeroporphyritic *a.* glomeroporfirítico

glory hole *m.* chupón, rajo abierto

gloss *m.* brillo, lustre

Glossopteris *m.* glosópteris

glossy *a.* lustroso

gloves *m.* guantes

glow *f.* incandescencia

glue *f.* cola, goma; *m.* pegamento

glycerine *f.* glicerina

glycol *m.* glicol

glyptolith *m.* gliptolito

gmelinite *f.* gmelinita, gmelita

gneiss *m.* gneis, neis ‖ ***augen gneiss*** *m.* gneis ojiforme, neis glandular, gneis de ojos

gneissic *a.* gnéisico, néisico, sarabiado

goethite *f.* goetita, xantosiderita; *m.* hierro acicular

goggles *f.* gafas protectoras, antiparras

gold *m.* oro, [Bol., Pe.] cori ‖ ***fine gold*** *m.* obrizo, oro puro ‖ ***of fine gold (monedas)*** *a.* zahén ‖ ***free gold*** *m.* oro libre ‖ ***gold-bearing*** *a.* aurífero ‖ ***gold bullion*** *m.* oro en barras ‖ ***gold cake*** *m.* tejón ‖ ***gold dust*** *f.* arena de oro; *m.* oro en polvo ‖ ***gold ingot*** *f.* palacra, palacrana ‖ ***gold mine*** *f.* arrugia ‖ ***gold nugget*** *m.* aborto de oro, papa; *f.* palacra, palacrana, pepita ‖ ***gold pan*** *f.* gamella, lavadora; *m.* lavador, platón, plato para lavar mineral ‖ ***gold plate*** *m.* tejón ‖ ***gold rush*** *f.* fiebre del oro ‖ ***gold speck*** *f.* chispa ‖ ***massive gold*** *m.* choclo de oro ‖ ***native gold*** with bismuth *f.* wismutaurita ‖ ***placer gold*** *m.* oro de aluvión (o de placer) ‖ ***vein gold*** *m.* oro de veta, oro filoniano

golden *a.* dorado, de oro, áureo, áurico

gonangium *m.* gonangio

goniatite *f.* goniatita

goniometer *m.* goniómetro

gorge *f.* garganta, angostura, barranca, cañada, quebrada, caluma, hoz, [Sp.] alfagra; *m.* barranco, cañadón, cañón, desfiladero, barrancón, carcavón, cajón, congosto, hocino

goslarite *m.* acije blanco

gossan *m.* gossan; *m.* rebosadero, sombrero de alteración, [Méx.] amarillo, [Bol., Pe.] paco; *f.* espuma

gooseneck *m.* tubo en S, cuello de ganso; *f.* conexión en S

Gotlandian *m.* & *a.* Gotlándico

gouge *m.* jaboncillo, [Bol., Pe.] panizo, paniso; *f.* arcilla gredosa, gubia

gouge mark *f.* marca lunada

government *m.* gobierno

governor *m.* regulador, controlador

graben *f.* zanja geológica, grabenfosa tectónica, fosa, graben, cuenca por fallas, fosa tectónica, hoya de falla, región de hundimiento

gradation *f.* graduación; *m.* paso gradual

gradational *a.* gradual

grade *f.* (geog.) pendiente, cuesta, rampa; *m.* gradiente, declive, sesgo, [Ch., Col.] cuelga; (meas.) *m.* grado, clase, contenido; *f.* calidad

grade *vb.* (to classify) calificar, clasificar, graduar, tasar

grade *vb.* (to level) nivelar, explanar, emparejar, allanar, aplanar, enrasar

graded *a.* aplanado, gradado ‖ ***graded bed*** *f.* capa gradada ‖ ***graded stream*** *m.* río en equilibrio

grader *f.* niveladora, perfiladora

gradient *f.* pendiente, inclinación; *m.* gradiente, grado ‖ ***geothermal gradient*** *m.* gradiente geotérmico ‖ ***hydraulic gradient*** *m.* gradiente hidráulico, gradiente piezométrico ‖ ***temperature gradient*** *m.* gradiente geotérmico

grading *f.* gradación

gradual *a.* gradual

graduated *a.* graduado ‖ ***graduated cylinder*** *f.* probeta graduada

graduation *f.* graduación; licenciatura

grahamite *f.* grahamita, asfalto

grain *m.* grano; *f.* textura, finura ‖ ***gold grains*** *f.* granalla

grainstone *f.* caliza granular

grainy *a.* granoso

gram *m.* gramo

granite *m.* granito, kaligranito, kalkgranito, kamgranito; *f.* berroqueña, ortoclasita, [Pe.] ala de mosca ‖ ***alkali granite*** *f.* karlsteinita ‖ ***biotite granite*** *f.* kristianita ‖ ***granite wash*** *m.* lavado de granito; *f.* arcosa ‖ ***graphic granite*** *f.* runita ‖ ***rapakivi granite*** *f.* wiborgita ‖ ***sanidine granite*** *f.* kaukasita

granitic *a.* granítico

granitization *f.* granitización

granitoid *m.* granitoide; *a.* granitoideo

granoblastic *a.* granoblástico

granodiorite *f.* granodiorita

granophyre *m.* granipórfido, granofiro, pegmatofíro

granophyric *a.* granofírico, pegmatofírico

grant *f.* concesión, merced, patente; *m.* derecho

grant *vb.* concesionar

grantee *m.* concesionario

grantor *m.* otorgante

granular *a.* granular, granudo, granoso

granularity *f.* granosidad

granulated *a.* granulado

granulation *f.* granulación; *m.* granulado

granule *m.* gránulo; *f.* guija

granulite *f.* granulita, kodurita

granulose *a.* granuloso

graph *f.* gráfica; *m.* esquema gráfico, ábaco

graphed *a.* graficado

graphic *m.* & *a.* gráfico

graphics *f.* gráfica

graphite *m.* grafito, plombagina, alquifol, carbón, lápiz, plomo negro || ***graphite mine*** *m.* lapizar

graphitic *a.* grafítico

graptolite *m.* graptolito

grass *f.* yerba, pasta

grasshopper counterbalance *m.* contrapeso oscilante articulado

grasslands *f.* pradería

grate, grating *f.* reja

graticule *f.* cuadrícula

grating *f.* rejilla, parilla, reja, verja; *m.* emparrillado, rejado

gravel *f.* grava, almendrilla, guija, zaborra, [Cu.] granza; *m.* cascajo, casquijo, granzón, guijo, pedregullo, peñascal, rodado, tucurubay || ***fine gravel*** *f.* gravilla; *m.* sábulo || ***gravel bank*** *m.* banco de grava, casquijo || ***gravel bed*** *m.* cascajero || ***gravel deposit*** *m.* cascajal, cascajar, cascajo, lleira || ***gravel packing*** *m.* relleno de cascajo || ***gravel pit*** *m.* cascajal, cascajar; *f.* gravera || ***lag gravel*** *f.* gravilla de desierto

gravelly *a.* cascajoso, guijoso

gravimeter *m.* gravímetro

gravimetric *a.* gravimétrico

gravimitry *f.* gravimetría

gravitation *f.* gravitación

gravitational *a.* gravitativo, gravitacional || ***gravitational attraction*** *f.* acción de la gravedad; atracción de la gravedad; acción gravitativa || ***gravitational constant*** *m.* constante de gravitación || ***gravitational field*** *m.* campo de gravedad

gravity *f.* gravedad || ***API gravity*** *f.* gravedad API || ***apparent gravity*** *f.* gravedad aparente || ***gravity anomaly*** *f.* anomalía gravimétrica || ***gravity feed*** *f.* alimentación por gravedad || ***specific gravity*** *f.* gravedad específica; *m.* peso específico

grayscale *f.* gama de tonos de gris

graywacke, greywacke *f.* grauvaca, grauwaca, usamerita; vaciagris

grazing land *f.* tierra de pastoreo

grease *f.* grasa

greasy *a.* grasoso

greenhorn *m.* bronco

greenockite *f.* greenockita, greenoquita, blenda cadmífera, xantocroíta; *m.* sulfuro de cadmio

greensand *f.* arena glauconítica

greenschist *m.* esquisto clorítico

greenstone *f.* piedra verde, wallerita; tierra verde

Greenwich mean time *f.* hora de Greenwich

greisen *m.* greisen, gres

grid *m.* enmallamiento; *f.* malla, red, cuadrilla, rejilla, cuadrícula, reja || ***seismic grid*** *m.* enmallamiento sísmico; *f.* malla sísmica || ***survey grid*** *m.* enmallamiento agrimensura

grid *vb.* cuadricular, mallar

gridded *a.* cuadriculado

grind *vb.* moler, amolar, afilar, esmerilar, majar, pulir, triturar

grinder *f.* moledora, amoladora; esmeriladera, rectificadora, muela, afiladora; *m.* moledor, triturador, molino

grinding *f.* amoladura, molienda, trituracion; *m.* majamiento

grindstone *f.* piedra de amolar, muela; *m.* afilador

grip *f.* agarradera

grit *f.* arena, arenisca gruesa, gravilla, arenilla, zaborra; *m.* cascajo

gritty *a.* arenoso, sabuloso

grizzly (screen) *f.* criba

groin *m.* muro de defensa

grommet *m.* anillo protector, orificio blindado

groove *m.* surco; *f.* acanaladura, estría; ranura, muesca || ***groove cast*** *m.* calco de surco

groove *vb.* acanalar

grooved *a.* ranurado; acanalado, canalado

grooving *m.* acanalamiento

gross *a.* gruesa; (com.) bruto || ***gross income*** *f.* (com.) rentas brutas || ***gross production*** *f.* producción en bruto || ***gross thickness*** *m.* espesor bruto || ***gross weight*** *m.* peso bruto

grossularite *f.* grosularia, grosularita

grotto *f.* gruta, sima

ground *f.* tierra; *m.* suelo, pique; terreno || ***ground level*** *m.* nivel del terreno || ***ground mass*** *f.* pasta mineral, matriz; *m.* base vidriosa || ***ground water*** *m.* agua subterránea || ***ground work*** *f.* infraestructura || ***rocky ground*** *m.* pedriscal, pedragal, tierra pedregosa

ground *a.* ñuto

grounded circuit *m.* circuito a tierra

grounding *m.* (elect.) contacto a tierra

group *m.* grupo; *f.* haz || ***group interval*** *m.* (seis.) intervalo de grupo

grout *m.* revestimiento

grove *f.* arboleda; *m.* socavón

growth *m.* crecimiento || ***growth fault*** *f.* falla de crecimiento || ***growth rate*** *f.* proporción de crecimiento

grubstake *m.* avío, suministro, abastecimiento; *f.* provisión

grubstake *vb.* aviar, abastecer, aprovisionar

grunerite *f.* grunerita

grus *m.* grus

guano *m.* guano, guanero; *f.* covadera

guarantee *f.* garantía; (com.) fianza

guarantee *vb.* garantizar, garantir

guard *m.* centinela, guardián

guard *vb.* guardar

gudgeon *m.* muñon, gorrón, macho

guide *f.* guía || ***guide book*** *f.* libreta guía; *m.* manual

guidelines *f.* líneas-guía, líneas de anclaje

guild *m.* gremio

gulch *m.* barranco, arroyo; *f.* barranca, hondonada, quebrada

gulf *m.* golfo || ***Gulf of Mexico*** *m.* Golfo de México

gullied *a.* arroyado

gully *f.* quebrada, cárcava, ranura, hondonada, barranca, torrentera, [Méx.] barranquilla; *m.* surco, arroyo, badén, barranco, cajón

gully *vb.* arroyar

gum *f.* goma

gumbo *m.* lodo gelatinoso, gumbo; *f.* arcilla plástica

gummy *a.* pegajoso

gun *f.* escopeta; *m.* fusil, cañón || ***gunpowder*** *f.* pólvora || ***perforating gun*** *m.* perforador de balas

gunite, gunnite *f.* gunita

gusher *m.* pozo brotante, eruptivo, o surgente

gusset *m.* codo de hierro, hierro angular de refuerzo

gust *f.* racha, ráfaga

gutter *f.* acequia, alcantarilla, cuneta; *m.* caño, desagüe

guyot *m.* guyot

gymnosperm *f.* gimnosperma

gypsiferous *a.* yesífero, yesoso

gypsum *m.* yeso, aljez, aljor; *f.* tiza || ***gypsum deposit*** *m.* aljezar || ***gypsum kiln*** *f.* yesería || ***gypsum mine*** *m.* yesal, yesar, yesera || ***gypsum quarry*** *m.* aljezar

gyroceracone *m.* girocono

gyroscope *m.* giroscopio

gyroscopic *a.* giroscópico

H

habit *m.* hábito

habitat *m.* hábitat || ***wildlife habitat*** *m.* hábitat de la fauna

hachured *a.* sombreado

hachures *m.* sombreados; *f.* hachuras

hackly *a.* astilloso, ganchudo

hadal *a.* hadal

hade *m.* buzamiento; *f.* inclinación

hail *m.* granizo

haiweeite *f.* ranquilita

half *a.* medio || ***half-life*** *f.* media vida

halide *m.* haluro

halite *f.* halita, sal, sal común, sal gema

halleflinta *f.* haleflinta

halmyrogenic *a.* halmirógeno

halloysite *f.* halloisita, haloisita

halo *m.* halo; *f.* corona || ***hydrocarbon halo*** *m.* halo de hidrocarburos

halogen *m.* & *a.* halógeno

haloid *m.* haloideo

halokinesis *f.* lalocínesis

halophyte *f.* halófila

halotrichite *f.* halotriquita

hammada *f.* hamada; *m.* desierto de piedra

hammer *m.* martillo; macho; mazo || ***rock hammer*** *f.* almadaneta, almádena, almaína || ***sledge hammer*** *f.* porra || ***stonecutter's hammer*** *f.* alcotana

hammer *vb.* golpetear, bronquear, machar, martillar

hand-actuated *adv.* accionado manualmente

handbook *f.* libreta guía; *m.* manual

hand lens *f.* lupa

hand sample *f.* piedra de mano

hanging *a.* colgante || ***hanging valley*** *m.* valle colgante || ***hanging wall*** *m.* bloque levantado, bloque alto, alto, reliz del alto, techo, techo de galería; *f.* pendiente, tapa || ***hanging wall anticline*** *m.* anticlinal colgado (o colgante) || ***hanging wall flat*** *m.* plano colgado (o colgante) || ***hanging wall ramp*** *f.* rampa colgada

haplopore *m.* haplóporo

haplozoan *m.* haplozoo

harbor *m.* puerto

hard *a.* duro || ***hard coal*** *m.* carbón mineral || ***hard currency*** *f.* divisas fuertes || ***hard drive*** *m.* disco duro || ***hard ground*** *f.* costras endurecidas || ***hard hat*** *m.* casco, sombrero protector || ***hard rock*** *f.* rocas cristalinas, [Pe.] ala de mosca

harden *vb.* acerar, endurecerse

hardening *m.* endurecimiento, temple

hardness *f.* dureza, (agua) crudeza

hardpan *f.* tosca, tierra endurecida, tortilla, [Ec.] canahua; *m.* caliche, [Méx.] tepetate

hardware *f.* ferretería; dotación física; *m.* equipo físico, componentes electrónicos, (comp.) hardware

harmonic *a.* armónico

harsh *a.* áspero

hartite *f.* josena

harzburgite *f.* harzburgita

hastingsite *f.* hastingsita

hat *m.* sombrero || ***hard hat*** *m.* sombrero protector, casco

hauerite *f.* pirita de manganeso

haul *vb.* arrastrar

haulage *m.* arrastre

hauling *m.* transporte, acarreo, arrastre

hausmannite *f.* hausmannita

Hauterivian *m.* & *a.* Hauteriviense, Hauteriviano

hauynite, hauyne *f.* haüyna

hawser *m.* cable de remolque

haystack *m.* cerro almiar, cerro pepino

hazard *m.* peligro, riesgo; azar

haze *f.* niebla, bruma

head *f.* cabeza, boca del pozo, cabezal; *m.* jefe de un grupo || ***hydraulic head*** *f.* altura de flúido; carga hidráulica, altura de carga, altura piezométrica, caída, tonga

header *f.* cabezal, cabecera; *m.* colector-cabezal || ***header record*** *m.* registro de cabecera

headframe *m.* castillete, puente, saetín, torre, torre de extracción (o de perforación, o de sondeo)

heading *m.* frente, rayo de barrido

headland *m.* promontorio

headquarters *f.* central, cuartel general, oficina central

headward erosion *f.* erosión retrocedente

headwaters *f.* aguas de cabecera, cabecera; *m.* naciente, nacimiento, ojo de agua

health *f.* salud

heap *m.* montón, rimero, [Car.] volcán; *f.* pila, tonga

heap *vb.* amontonar

hearth *m.* hogar

heat *m.* calor; *f.* calefacción || ***heat capacity*** *f.* capacidad térmica || ***heat content*** *m.* contenido de calor || ***heat exchanger*** *m.* cambiador de calor, permutador de calor, termopermutador || ***heat flow*** *m.* flujo térmico (o calórico) || ***heat of combustion*** *m.* calor de combustión || ***heat of fusion*** *m.* calor de fusión || ***heat of vaporization*** *m.* calor de vaporización || ***heat-producing*** *a.* calorífero, calorífico || ***heat transfer*** *f.* transferencia de calor || ***heat-treated*** *a.* tratado al calor || ***heat treatment*** *m.* tratamiento térmico || ***latent heat of fusion*** *m.* calor latente de fusión || ***specific heat*** *m.* calor específico || ***waste heat*** *m.* calor perdido

heat *vb.* caldear, calefaccionar, calentar

heater *m.* calentador

heath *m.* brezal

heating *f.* calda, calefacción; *m.* calentamiento

heave *f.* elevación; *m.* alzamiento, rechazo horizontal

heavy *a.* pesado, denso, cargado || ***heavy duty*** *a.* para trabajo pesado || ***heavy minerals*** *m.* minerales pesados || ***heavy oil*** *m.* petróleo crudo, aceite espeso, aceite pesado; *f.* brea mineral

hectare *f.* hectárea, hectara

hectorite *f.* hectorita

hedenbergite *f.* hedenbergita

height *f.* altura, cima, elevación; *m.* alto, colmo

helical, helicoidal *a.* helicoidal, hélico

heliodor *m.* heliodoro

heliotrope *m.* heliotropo

heliport *m.* helipuerto

helium *m.* helio

helix *f.* hélice

helmet *m.* casco, casco de seguridad

Helvetian *m. & a.* Helveciense, Helvetiense, Helveciano, Helvetiano

hematite *f.* hematita; *m.* oligisto, hierro oligisto, campanil, rubio || ***specular hematite*** *f.* hematita especular; *m.* oligisto

hematitic *a.* hematítico

Hembergian *m. & a.* Hembergiano, Hembergiense

hemera *f.* hemera

hemichordate *m.* hemicordado

hemimorphic *a.* hemimórfico

hemimorphite *f.* hemimorfita

hemipelagic *a.* hemipelágico

hemiseptum *m.* hemisepto

hemisphere *m.* hemisferio

hemispheric *a.* hemisférico

heptane *m.* hepteno

Hercynian *m. & a.* Herciniano

hercynite *f.* hercinita

heredity *f.* herencia

hermetic *a.* hermético, a prueba de aire

herringbone *a.* espina de pez || ***herringbone structure*** *f.* estructura espigada

Hertz *m.* hercio

herzenbergite *f.* kolbeckina

hessite *f.* hessita

hessonite *f.* hessonita, essonita

hetairite *f.* wolftonita, zinkhausmannita

heteroblastic *a.* heteroblástico

heterochronous *a.* heterocrónico

heterogeneity *f.* heterogeneidad

heterogeneous *a.* heterogénio

heterodont *m. & a.* heterodonto

heteromorphism *m.* heteromorfismo

heteropod *m.* heterópodo

heterostrophy *f.* heterostrofia

heterotopic *a.* heterotópico

heterotropic *a.* heterotrópico

heterozooid *m.* heterozoide

Hettangian *m. & a.* Hetangiense, Hetangiano

heulandite *f.* heulandita

hew *vb.* picar

hexacoral *m.* hexacoralario, exacoralario

hexagon *m.* hexágono

hexagonal *a.* hexágonal, exagonal, hexágono

hexahedron *m.* hexaedro

hexane *m.* hexano

hexene *m.* hexeno

hexoctahedron *m.* hexoctaedro

hiatus *m.* hiato

hiddenite *f.* hiddenita

hieroglyph *m.* hieroglifo, jeroglifo

hieroglyphic *a.* jeroglífico

high *m. & a.* alto || ***high frequency*** *f.* frecuencia alta || ***high grade*** *a.* de calidad superior, de alta ley || ***high pressure*** *f.* alta presión || ***high rank*** *a.* de alto grado || ***high resolution*** *f.* resolución alta || ***high stand*** *m.* nivel de base alto; *f.* talasocracia, mar profundo || ***high tide*** *f.* pleamar, marea alta; *m.* creciente || ***high water line*** *f.* línea de pleamar || ***structural high*** *m.* alto estructural || ***topographic high*** *m.* alto topográfico

highgrade *vb.* escoger lo mejor

highlands *f.* tierras altas, región montañosa

hill *f.* loma, colina, cuesta; *m.* cerro, cerillo, alcor, altillo, alto, altozono, otero, ribazo || ***buried hill*** *m.* cerro sepultado || ***low hills*** *f.* cerrillada

hillock *f.* muela

hillside *f.* falda, ladera; *m.* flanco

hilly *a.* accidentado, cerril

hinge *f.* charnela, articulación, bisagra || ***hinge line*** *f.* línea de charnela

hinterland *m.* tras país, transpaís; zona más interior, pospaís

hire *vb.* emplear, contratar

histogram *m.* histograma

history *f.* historia || ***case history*** *f.* historial, historial clínica

hoarfrost *f.* helada blanca

hoe *f.* azada; *m.* azadón

hofmannite *f.* hofmanita

hogback *f.* crestería, cresta dentada, cresta monoclinal, cuesta, cuchilla; *m.* lomo de perro

hoist *m.* aparejo, guinche, huinche, izador; *f.* grúa, montacargas || ***inclined hoist*** *m.* mantéo

hoist *vb.* alzar, izar, mantear

hoisting *m.* acarreo, izaje

holding company *f.* compañía matriz o dominatriz

hole *m.* agujero, barreno, hueco, pozo, hoyo, foso; *f.* perforación, cavidad, fosa, hoya || ***deviated hole*** *m.* agujero desviado || ***down hole*** *a.* hoyo abajo || ***drill hole*** *m.* agujero, pocillo || ***dry hole*** *m.* pozo seco, pozo improductivo || ***kettle hole*** *f.* hoya glacial, marmita glacial || ***open hole*** *m.* pozo sin entubar, pozo franco || ***shot hole*** *m.* hoyo de disparo, hoyo de la explosión || ***slant hole*** *m.* pozo direcional, pozo dirigido || ***small drill hole*** *f.* pistola

holectypoid *m.* holectípido

hollow *m.* hueco; *f.* caverna, hoyada; *a.* ahuecado

holm *f.* isla fluvial

holmium *m.* holmio

holmquistite *f.* holmquistita

holoaxial *a.* holoaxial

Holocene *m. & a.* Holoceno, Holocénico, Reciente, Actual, posglacial

holocrystalline *a.* holocristalino, pleocristalino

holohyaline *a.* holohialino

hololeucratic *a.* hololeucrático

holophytic *a.* holofítico

holoplankton *m.* holoplancton

holotheca *f.* holoteca

holotype *m.* holotipo

homeoblastic *a.* homeoblástico

homeochilidium *m.* homeoquilidio

homeodeltidium *m.* homeodeltidio

homeomorph *m.* homeomorfo

homoclinal shifting *m.* desplazamiento homoclinal

homocline *m.* homoclinal

homogeneity *f.* homogeneidad
homogeneous *a.* homogéneo
homogenize *vb.* homogeneizar
homologous series *f.* serie homólogo
homolographic *a.* homolográfico
homotaxial *a.* homotáxico
honeycombed *a.* ojoso, muy minado
hood *m.* capuchón, capó; *f.* cubierta
hoodoo *f.* roca monigote; *m.* pilar de erosión
hook *m.* gancho, garfio; *f.* gancha
hook *vb.* enganchar
hookup *f.* conexión
hopeite *f.* zinkfilita
hopper *f.* tolva; *m.* embudo, chifladero
horizon *m.* horizonte; *f.* formación || ***A horizon*** *m.* horizonte A || ***migration horizon*** *f.* formación conductor || ***phantom horizon*** *m.* horizonte imaginario
horizontal *a.* horizontal; acamado || ***horizontal bed*** *m.* alero, estrato horizontal
horn *f.* (geog.) aguja
hornblende *f.* hornablenda, hornblenda, cornubianita, [Méx.] blenda córnea; *m.* wallerian || ***hornblende schist*** *m.* anfibolesquisto
hornblende *a.* hornabléndico || ***Mn-hornblende*** *f.* juddita
hornblendite *f.* pedrosita
hornfels *f. & a.* hornfels, hornfelsa, cornubianita, corneana, keratita
hornfelsic *a.* hornfelsítico
horn silver *f.* plata córnea, querargirita, cerargirita
horse *m.* (geol.) caballo (de piedra), caballete; *f.* roca atrapada || ***horse latitudes*** *f.* calmas subtropicales || ***horse power*** *m.* caballo de fuerza || ***horse tail*** *f.* cola de caballo
horst *f.* gleba tectónica, meseta tectónica; *m.* horst, pilar tectónico, bloque levantado
hose *f.* manguera
host *vb.* armar, servir de anfitrión
host rock *f.* roca anfitriona, formación que le circundan, roca hospedadora, [Méx.] armadura
hot *a.* caliente, calentado, calido || ***hotspot*** *f.* zona de calor pluma, efusión térmica; *m.* calefactor, punto cálido, punto candente || ***hot spring*** *f.* burga, terma, manantial termal, aguas termales; *m.* balneario
hour *f.* hora || ***man hours*** *f.* horas de trabajo
house *f.* casa || ***doghouse*** *f.* casilla, caseta || ***pump house*** *f.* casa de bombas
housing *f.* cubierta
hue *m.* color, matiz
huebnerite *f.* hubnerita
hum *m.* cerro almiar, cero pepino
humberstonite *f.* chile-loeweita
humic *a.* húmico
humid *a.* húmedo
humidify *vb.* humedecer
humidity *f.* humedad
humite *f.* humita
hummock *m.* montículo
hummocky *a.* monticulado
humus *m.* humus, mantillo
hundredweight *m.* quintal: 100 libras, 45,4 kg
hung *a.* ahorcado, colgado
huntilite *f.* chañarcillita
hureaulite *f.* palaita, wentzelita
Huronian *m. & a.* Huroniano, Huroniense
hurricane *m.* huracán, ciclón
hyaline *a.* hialino; *f.* hialina
hyalite *f.* hialita
hyalophane *f.* hialofana
hyalopilitic *a.* hialopilítico
hyalosponge *f.* hialoesponja
hybrid *m. & a.* híbrido
hydatogenic *a.* hidatógeno
hydrargillite *f.* hidrargillita
hydrargyrite *m.* óxido de mercurio
hydrate *m.* hidrato
hydrate *vb.* hidratar
hydrated *a.* hidratado
hydration *f.* hidratación
hydrator *m.* hidratador

hydraulic *a.* hidráulico || ***hydraulic gradient*** *m.* gradiente hidráulico, gradiente piezométrico || ***hydraulic head*** *f.* carga, tonga || ***hydraulic-pneumatic*** *a.* hidroneumático

hydraulics *f.* hidráulica, hidrotécnica

hydride *m.* hidruro

hydrobiology *f.* hidrobiología

hydrocarbon *m.* hidrocarburo, jugo de la tierra, petróleo || ***chlorinated hydrocarbon*** *m.* hidrocarburo clorinado || ***saturated hydrocarbon*** *m.* hidrocarburo saturado

hydrochloric *a.* clorhídrico

hydrochloride *m.* clorhidrato

hydroclast *m.* hidroclasto

hydrocracking *m.* hidrocraqueo; *f.* hidrodesintegración

hydrodialysis *f.* hidrodialisis

hydrodynamic *a.* hidrodinámico

hydrodynamics *f.* hidrodinámica

hydrofluoric *a.* fluorhídrico

hydrofrac *f.* fracturación hidráulica

hydrogen *m.* hidrógeno || ***hydrogen chloride*** *m.* hidrocloruro || ***hydrogen sulfide*** *m.* sulfuro de hidrógeno, hidrógeno sulfurado

hydrogenate *vb.* hidrogenar

hydrogenated *a.* hidrogenado

hydrogenation *f.* hidrogenación

hydrogeology *f.* hidrogeología

hydrographic *a.* hidrográfico

hydrography *f.* hidrografía

hydrologic *a.* hidrológico

hydrologist *m.* hidrólogo

hydrology *f.* hidrología

hydrolysis *f.* hidrólisis

hydrolyzer *m.* hidrolizador

hydrometer *m.* densímetro, aerómetro

hydrorhiza *f.* hidrorriza

hydrophilite *f.* clorocalcita

hydrophobic *a.* hidrófobo

hydrophilic *a.* hidrófilo

hydroplutonic *a.* ácueoigneo

hydrosilicate *m.* hidrosilicato

hydrosphere *f.* hidrósfera

hydrostatic *a.* hidrostático || ***hydrostatic head*** *f.* altura hidrostática, tonga || ***hydrostatic pressure*** *f.* presión hidrostática

hydrotheca *f.* hidroteca

hydrothermal *a.* hidrotérmico, hidrotermal

hydrous *a.* acuoso, hídrico

hydroxide *m.* hidróxido

hydroxyl *m.* hidróxilo, oxhidrilo

hydrozincite *f.* calamina, hidrocincita, zinconisa; *m.* zinc hidrocarbonatado, zinc oxidado silicífero

hydrozoan *m.* hidrozoo

hygrometer *m.* higrómetro

hygroscopic liquid *m.* líquido higroscópico

hypabyssal *a.* hipabisal, ipabismal

hyperbaric chamber *f.* cámara hiperbárica

hyperbola *f.* hipérbola

hyperbolic *a.* hiperbólico

hypergenesis *f.* retrodiagénesis

hypermelanic *a.* hipermelánico

hypersthene *f.* hiperstena; *m.* hipersteno

hypidiomorphic *a.* hipidiomorfo, subidiomórfico

hypocenter *m.* hipocentro, foco sísmico

hypochlorite *m.* hipoclorito

hypocrystalline *a.* hipocristalino

hypogene *a.* hipogénico

hyponome *m.* hiponomo

hypoparian *m.* & *a.* hipopario

hypostome *f.* hipostoma

hypothermal *a.* hipotermal

hypothesis *f.* hipótesis

hypsographic *a.* hipsográfico

hypsometer *m.* hipsómetro

hypsometric *a.* hipsométrico

hysteresis *f.* histéresis

I

I beam *f.* vigueta de sección en I

ice *m.* hielo || ***dry ice*** *m.* hielo seco || ***ice age*** *f.* edad de hielo, edad glacial || ***iceberg*** *m.* témpano de hielo, iceberg || ***icebreaker*** *m.* rompehielos || ***icecap*** *f.* capa de hielo, hielos perpetuos || ***ice field*** *m.* nevero || ***ice floe*** *m.* banco de hielo, témpano || ***ice jam*** *m.* hielo atascado || ***ice sheet*** *f.* capa de hielo; *m.* glaciar continental, helero

Iceland spar *m.* espato de Islandia

Icenian *m. & a.* Iceniano, Icenense

ichnite *f.* pisada fósil

ichnofossil *m.* ichnofósil

ichthyosaur *m.* ictiosauro

icicle *m.* carámbano

iddingsite *f.* iddingsita, oroseíta

identification *f.* identificación

idioblastic *a.* idioblástico

idiochromatic *a.* idiocromático

idiomorphic *a.* automorfo, idiomorfo, panidiomorfo

idocrase *f.* idocrasa, jacintina, wilinita

igneous *a.* ígneo || ***igneous body*** *m.* cuerpo ígneo || ***igneous rock*** *f.* roca ígnea

ignimbrite *f.* ignimbrita

ignite *vb.* inflamarse, encenderse

ignition *m.* encendido; *f.* ignición

Igualadian *m. & a.* Igualadense, Igualadiano

ijolite *f.* ijolita

ilium *m.* hueso ilíaco

illicit *a.* (legal) ilícito, ilegal

Illinoisan *m. & a.* Illinoisiano, Illinoisiense

illite *f.* illita, ilita, hidromuscovita

illuminate *vb.* iluminar

illumination *f.* iluminación

illustrate *vb.* ilustrar

illustration *f.* ilustración, figura

illuvial *a.* iluvial

ilmenite *f.* ilmenita, washingtonita

ilvaite *f.* ilvaita, lievrita, yenita

image *vb.* (seis.) registrar

image *f.* imagen || ***latent image*** *f.* imagen latente

imbed *vb.* engastar

imbedded *a.* incluído

imbibition *f.* imbibición

imbricate *vb.* imbricar

imbricated *a.* imbricado, superpuesto

imbrication *f.* imbricación

immature *a.* inmaduro

immaturity *f.* inmadurez

immersed *a.* inmerso

immersion *f.* inmersión, sumersión

immiscibility *f.* inmiscibilidad

immiscible *a.* inmiscible

impact *m.* choque, impacto; *f.* impacción

impact *vb.* impactar

impactite *f.* impactita

impedance *f.* impedancia; resistencia || ***acoustic impedance*** *f.* impedancia acústica

impel *vb.* impeler, impulsar

impeller *m.* impulsor, propulsor

imperfect *a.* imperfecto, defectuoso

impermeability *f.* impermeabilidad

impermeable *a.* impermeable

impervious *a.* impermeable

impetus *m.* ímpetu

impinge *vb.* chocar, afectar

impingement *m.* choque, tropiezo

implement *m.* implemento

implement *vb.* implementar

implementation *f.* implementación

implication *f.* inferencia

imply *vb.* denotar

import *vb.* importar

imports *f.* importación

impound *vb.* embalsar

impounding *m.* agolpamiento, endicamiento, embalse

impoverished *a.* empobrecido

impoverishment *m.* empobrecimiento

impregnate *vb.* impregnar

impregnated *a.* impregnado

impregnation *f.* impregnación
impression *f.* impresión, impronta
imprint *f.* impresión, impronta
improve *vb.* mejorar
impsonite *a.* impsonita
impulse *m.* impulso, ímpetu
impulsion *f.* impulsión
impure *a.* impuro, impurificado
impurity *f.* impureza
inaccuracy *m.* error, inexactitud
inaccurate *a.* inexacto
inactive *a.* inactivo, en reposo
inadequate *a.* inadecuado
inappropriate *a.* inapropiado
inarticulate *m. & a.* inarticulado
incandescence *f.* candencia
incandescent *a.* incandescente, candente
inch *f.* pulgada (2,54 cm)
in charge of *a.* encargado
incidence *f.* incidencia || ***incidence angle*** *m.* ángulo de incidencia || ***normal incidence*** *f.* incidencia normal || ***oblique incidence*** *f.* incidencia oblicua
incineration *f.* incineración
incipient *a.* incipiente
incised *a.* inciso, cortado || ***incised river*** *m.* río inciso || ***incised valley*** *m.* valle inciso; *f.* cañada
incision *f.* insición
inclination *f.* inclinación, pendiente, caída; *m.* declive, talud; buzamiento, recuesto
incline *m.* declive || ***shallow incline*** *f.* rampa
incline *vb.* acostar, acostarse, inclinarse, bascularse
inclined *a.* inclinado, echado || ***inclined shear*** *f.* cizalla inclinada
inclinometer *m.* inclinómetro
included *a.* incluído
inclusion *f.* inclusión; *m.* enclave || ***fluid inclusion*** *f.* inclusión de flúidos
income *m.* (com.) cobros, ingresos; *f.* rentas || ***gross income*** *f.* (com.) rentas brutas
incompetent *a.* incompetente
incongruous *a.* incongruente
inconsequent *a.* inconsecuente
incorporate *vb.* incorporar
incorporation *f.* incorporación
increase *m.* aumento
increase *vb.* aumentar
increment *m.* incremento
increment *vb.* incrementar
incrust *vb.* incrustar
incrustation *f.* incrustación, escamosión; *m.* sarro
incrusted *a.* incrustado, sarroso
incrusting *f.* incrustación; *a.* incrustante
indefinite *a.* indefinido
indemnity *f.* indemnización
indentation *f.* muesca, escotadura, melladura
indeterminable *a.* indeterminable
index *m.* índice || ***index card*** *f.* ficha || ***index fossil*** *m.* fósil índice o indicador || ***index map*** *m.* mapa índice || ***index of refraction*** *m.* índice de refracción || ***hydrogen index*** *m.* índice de hidrógeno || ***oxygen index*** *m.* índice de oxígeno || ***photo index*** *m.* índice fotográfico || ***production index*** *m.* índice de producción || ***thermal alteration index*** *m.* índice de alteración térmica
indication *f.* indicación
indications *m.* indicios
indicator *m.* indicador, medidor; *f.* guía || ***indicator mineral*** *m.* mineral indicador || ***drift indicator*** *m.* indicador de desviación
indicatrix *f.* indicatriz
indicolite *f.* indicolita, indigolita
indigenous *a.* indígena, autóctono
indirect *a.* indirecto
indisolubility *f.* indisolubilidad
indistinct *a.* indistinto
indium *m.* indio
induce *vb.* inducir
induced *a.* inducido
inducer *m.* inducidor
inducing *a.* inducidor, inductor
inductance *f.* inductancia
induction *f.* inducción

inductive *a.* inductivo
inductor *m. & a.* inductor
indurate *vb.* endurecer, indurar
indurated *a.* endurecido, consolidado, indurado
induration *f.* diagénesis, induración
industrial *a.* industrial
industry *f.* industria
inefficient *a.* ineficaz
inelastic *a.* anelástico
inert *a.* inerte || ***inert gas*** *m.* gas inerte
inertia *f.* inercia
inertinite *f.* inertinita
inexhaustable *a.* inagotable
inference *f.* inducción, inferencia
inferior *a.* inferior || ***inferior quality gem*** *a.* occidental
inferred *a.* inferido
infiltrate *vb.* infiltrar || ***infiltrate into*** *vb.* infiltrarse
infiltration *f.* infiltración
infinitesimal *a.* infinitesimal
inflated *a.* inflado
inflammable *a.* inflamable
inflate *vb.* inflar, hinchar
inflation *f.* inflación
inflection *f.* inflexión
inflow *m.* aflujo, influjo, aporte; *f.* aportación
influence *f.* influencia
influence *vb.* influir, influenciar
influx *f.* intromisión, afluencia; *m.* aflujo
information *f.* información; *m.* dato
Infracambrian *m. & a.* infracámbrico
infrared *m. & a.* infrarrojo
infrastructure *f.* infraestructura
infusible *a.* infundible
infusion *f.* infusión
ingenuity *f.* ingeniosidad, inventiva, habilidad
ingot *f.* barra de metal, plancha; *m.* lingote, bloque, riel, tejo, molde de fundición
ingredient *m.* ingrediente
ingression *f.* ingresión; *m.* ingreso
inhabit *vb.* habitar, ocupar, vivir en
inhabitable *a.* habitable
inhabitant *m.* habitante
inherent *a.* intrínseco
inheret *vb.* heredar
inherited *a.* heredado
inhibit *vb.* inhibir
inhibitor *m.* inhibidor
inhomogeneity *f.* inhomogeneidad
inject *vb.* inyectar
injected *a.* inyectado
injection *f.* inyección || ***air injection*** *f.* inyección de aire || ***center line injection*** *f.* inyección de línea media || ***foam injection*** *f.* inyección de espuma || ***gas injection*** *f.* inyección de gas || ***injection well*** *m.* pozo inyector || ***magmatic injection*** *f.* inyección magmática || ***steam injection*** *f.* inyección de vapor || ***water injection*** *f.* inyección de agua
injector *m.* inyector
injure *vb.* herir, ofender, dañar
injury *m.* perjuicio, daño
in kind *a.* en especie
inland *f. & a.* tierra adentro
inlay *f.* incrustación
inlay *vb.* incrustar
inlet *f.* toma, boca de entrada, boca de admisión, caleta, boca; *m.* estero
inlier *f.* fenster, fenestra, ventana, ventanilla; *m.* miradero, asomo, relicto interior
inner *a.* interior, interno
innocuous *a.* inofensivo
innovation *f.* innovación
in operation *a.* en funcionamiento
inorganic *a.* inorgánico, anorganógeno, ionógeno
in place *a.* en su lugar, en sitio, in situ
inosilicate *m.* inosilicato
input *m.* volumen de materia prima tratada; (comp.) datos de entrada
input *vb.* introducir datos
insect *m.* insecto
inselberg *f.* inselberg
insequent *a.* insecuente

insert *f.* inserción, guarnición
inside *a.* interior, dentro
in situ *a.* en sitio, en su lugar, ínsito, in situ
insolation *m.* insolación
insoluble *a.* insoluble
insolubility *f.* insolubilidad
inspect *vb.* inspeccionar
inspection *f.* inspección; *m.* reconocimiento
inspector *m.* inspector
instability *f.* inestabilidad
install *vb.* instalar
installation *f.* instalación, infraestructura || ***remote installation*** *f.* instalación por telemando
instantaneous *a.* instantáneo || ***instantaneous velocity*** *f.* velocidad instantánea
in stock *a.* en existencia
instruction *f.* instrucción || ***instruction book*** *m.* manual de instrucciones
instructor *m.* instructor, maestro, profesor
instrument *m.* instrumento || ***instrument calibration*** *m.* ajuste de instrumentos
instrumentation *f.* instrumentación
insular *a.* insular
insularity *f.* insularidad
insulate *vb.* aislar
insulated *a.* aislado
insulation *m.* aislamiento
insulator *m.* aislador
insurance *m.* seguro
insure *vb.* asegurar
intake *f.* toma; *m.* tubo de admisión
integrate *vb.* integrar
integrated *a.* integrado
integration *f.* integración
intensity *f.* intensidad
intensive *a.* intensivo
interbasin *f.* intercuenca
interbed, interbedding *f.* intercapa, interestratificación
interbed *vb.* interestratificar
interbedded *a.* interestratificado, entrelaminado, interyacente
intercalate *vb.* intercalar
intercalated *a.* intercalado
intercalation *f.* intercalación, interclusión
intercept *vb.* interceptar, atajar || ***intercept time*** *m.* tiempo de intercepción
interchangeable *a.* intercambiable
intercontinental *a.* intercontinental
interdigitation *f.* interdigitación
interdune *f. & a.* interduna, entreduna
interest *m.* (com.) interés, rédito; *f.* participación; rentas || ***minority interest*** *m.* (com.) interés minoritario || ***working interest*** *m.* (com.) interés en la inversión; *f.* participación del concesionario
interface *m.* acoplamiento mutuo; conexión; *f.* interfaz
interfacial *a.* interfacial || ***interfacial tension*** *f.* tensión interfacial, atracción interfacial
interference *f.* interferencia || ***interference figure*** *f.* figura de interferencia
interferometer *m.* interferómetro
interfingered *a.* interdigitado
interfingering *f.* interdigitación
interfluve *m.* interfluvio
interformational *a.* interformacional
interglacial *a.* interglacial
intergranular *a.* intergranular
intergrow *vb.* entrecrecer
intergrown *a.* entrecrecido
intergrowth *m.* entrecrecimiento, intercrecimiento
interior *a.* interno
interlaminated *a.* interlaminado, entrelaminado
interleave *vb.* interpolar, simultanear
interlobate *a.* interlobulado
interlocking *a.* que traba, que enclavija, entrelazado
intermediate *a.* intermedia || ***chemical intermediates*** *m.* productos químicos intermedios || ***intermediate mine level*** *f.* sobreguía
intermittent *a.* intermitente
intermontane *a.* intermontaño, entremontaña
internal *a.* interno

international *a.* internacional
internodal *a.* internodo
interocular distance *f.* distancia interocular
interparticle *f.* interpartícula
interpolate *vb.* interpolar
interpolation *f.* interpolación
interpret *vb.* interpretar, traducir
interpretation *f.* interpretación
interreef *m.* & *a.* interarrecife; interarrecifal
interrupt *vb.* interrumpir
interruption *f.* interrupción
intersect *vb.* cruzar, intersecarse, cortar, entrecortar
intersection, intersept *f.* intersección
interstice *m.* intersticio
interstitial *a.* intersticial
interstratification *f.* interstratificación
interstratified *a.* interestratificado
interstream *a.* intercorriente
intertidal *a.* litoral, entre mareas, intermareas
intertongued *a.* interdigitado
intertonguing *m.* engranando; *f.* borrasca de la capa, interdigitación
interval *m.* intervalo, espacio || ***group spacing interval*** *m.* intervalo de grupo
intraformational *a.* intraformacional
in transit *a.* en tránsito
intragranular *a.* intragranular
intrastratal *a.* intraestratal
intrinsic *a.* intrínseco
intrude *vb.* intruir
intruded *a.* intruido
intrusion *f.* intrusión; *m.* plutón
intrusive *a.* intrusivo
intumescence *f.* intumescencia
inundate *vb.* inundar, sumergir, anegar, anegarse, apantanar
inundated *a.* anegado, inundado
inundation of a mine *f.* aguada, inundación
invaded *a.* invadido
invasion by water, etc. *f.* invasión
inventory *m.* catastro, inventario
inverse *a.* inverso
inversion *f.* inversión
invert *vb.* invertir
invertebrate *m.* invertebrado
inverted *a.* invertido, inverso, tumbado, volcado
invest *vb.* invertir
investigate *vb.* investigar
investigation *f.* investigación
investment *f.* inversión
invoice *f.* factura, cuenta
invoice *vb.* facturar, pasar la cuenta
involute *a.* involuto
inward *a.* hacia dentro, interior, interno
iodargyrite *f.* yodirita, yodoembolita
iodate *m.* yodato
iodic *a.* yodíco
iodide *m.* yoduro, ioduro
iodine *m.* yodo, iodo
iodization *f.* yoduración
iodize *vb.* yodurar, yodar
iodized *a.* yodurado, yodado
iodobromite *f.* yodobromita
iodyrite *f.* yodargirita, yodirita, plata córnea amarilla
iolite *f.* iolita, jolita
ion *m.* ion
ionic *a.* iónico || ***ionic radius*** *m.* radio iónico
ionization *f.* ionización
ionize *vb.* ionizar
iranite *f.* khuniita
iridesce *vb.* irisar
iridescence *f.* iridiscencia, irisación; *m.* tornasol
iridescent *a.* iridiscente, iridescente, irisado, tornasolado || ***iridescent hyaline quartz*** *m.* iris
iridium *m.* iridio
iridosmium *f.* iridosmina
iron *m.* hierro, fierro || ***cast iron*** *m.* hierro colado || ***iron-bearing*** *a.* férreo, ferrífero, ferrugíneo, ferruginoso || ***iron–nickel*** *m.* hierroníquel || ***iron ore*** *m.* mineral de hierro || ***iron oxide*** *m.* óxido de hierro || ***iron stone*** *f.* roca de mineral de hierro || ***iron sulfate*** *f.* colpa || ***iron vitriol*** *m.* acije de hierro || ***iron works*** *f.* ferrería || ***scrap iron*** *m.* hierro viejo || ***wrought iron*** *m.* hierro fundido, hierro batido, hierro forjado, hierro dulce

ironstain *vb.* aherrumbrarse
irradiate *vb.* irradiar
irradiation *f.* irradiación
irreducible *a.* irreducible
irrevocable *a.* irrevocable
irrigate *vb.* regar
irrigation canal *f.* regata
ischium *m.* isquion
island *f.* isla, ínsula; *m.* isleo || ***barrier island*** *m.* islote litoral || ***island arc*** *f.* arco de islas || ***small island*** *m.* cayo
islet *m.* islote, islete, cayo; *f.* isleta, islota, islilla
isoanthracite *f.* isoantracita
isobar *f.* isobara
isobaric *a.* isobárico
isobarometric *a.* isobarométrico
isobath *f.* isobata
isobutane *m.* isobutano
isocarbon *m.* isocarbono
isochemical *a.* isoquímico
isochore *f.* isocora
isochromatic *a.* isocromático
isochrome *m.* isocromo
isochron *f.* isócrona
isochronous *a.* isócrono
isoclinal, isocline *m. & a.* isoclinal, isoclino
isoclinic *a.* isoclínico
isodont *m.* isodonto
isodynamic *a.* isodinámico
isogal *f.* isogal, isogálica
isogam *f.* isogama
isogeotherm *f.* isogeoterma
isogon *m.* isógono
isogonic *a.* isogónico, isógono
isogyre *m.* isogiro
isohaline *m. & a.* isohalino; *f.* isohalina
isohyet, isohyetal *m. & a.* isohieto, isopluvial, isoyeta
isohyet *f.* isoípsa, curva de nivel
isohypse *f.* isoípsa, curva de nivel
isolate *vb.* aislar
isolated *a.* aislado
isolation *m.* aislamiento
isolith *m.* isolito || ***isolith map*** *m.* mapa isolítico
isomagnetic line *f.* isodina, línea isomagnética
isomer *m.* isómero
isomeric *a.* isomérico
isomerism *m.* isomerismo
isomerization *f.* isomerización
isometamorphism *m.* isometamorfismo
isometric *a.* isométrico
isomorph *m.* isomorfo
isomorphic *a.* isomórfico
isomorphism *m.* isomorfismo
isomorphous *a.* isomorfo
isopach, isopachous *m.* isopaco, isopático; *f.* isopaca || ***isopach, isopachous map*** *m.* mapa isopáquico
isoparaffin *f.* isoparafina
isopentane *m.* isopentano
isopleth *f.* isopleta
isopluvial *a.* isopluvial
isopore *f.* isospora
isoprene *m.* isopreno
isoprenoid *m.* isoprenoide
isopropyl alcohol *m.* alcohol isopropílico
isostasy *f.* isostasia, isostasis
isostatic *a.* isostático || ***isostatic compensation*** *f.* compensación isostática || ***isostatic correction*** *f.* corrección isostática
isotherm *f.* isoterma
isothermal, isothermic *a.* isotermo, isotérmico
isotropism *m.* isotropismo
isotype *m.* isotipo
isotypic *a.* isotípico
isotope *m.* isótopo
isotopic *a.* isotópico
isotropic *a.* isotrópico, isótropo
isotropy *f.* isotropía
issuance *f.* emisión
isthmian *a.* ístmico
isthmus *m.* istmo
itabirite *f.* itabirita, jacutinga
italite *f.* italita

item *m.* artículo

itemize *vb.* detallar

iterate *vb.* iterar

iteration *f.* iteración

iterative *a.* iterativo

itinerary *m.* itinerario

J

jack *m.* gato, cric || ***pump jack*** *m.* gato de bombeo

jacket *f.* camisa, envoltura

Jacob's staff *f.* vara portabrújula; *m.* batón de Jacóbo, [Méx.] báculo

Jacksonian *m. & a.* Jacksoniano, Jacksoniense

jackup *f.* plataforma autoelevadiza (o fija), plataforma montada en gatos mecánicos, plataforma replegable

jacupirangite *f.* jacupirangita

jade *m.* jade, yu; *f.* nefrita, jadeita || ***green jade*** *m.* [Méx.] tecpatl

jadeite *f.* jadeita, [Méx.] chalchihuitl

jagged *a.* dentado, dentellado, mellado

jamaskite *f.* jamasquita

jamesonite *f.* jamesonita, warrenita, [Pe.] paja quemada; *m.* acerado

Janitorian *m. & a.* Janitoriano, Janitorense

jargon *m.* jargón; *f.* jerga

jarosite *f.* jarosita, utahita

jasper *m.* jaspe; *f.* dionisia, jasperita || ***agate jasper*** *f.* jaspágata || ***red jasper*** with gold-bearing pyrite *m.* zinopel

jasperization *f.* jasperización

jasperoid *a.* jasperoide

jaspilite *f.* jaspilita

jaspoid *m. & a.* jaspoide

jaw *f.* mandíbula, quijada; mordaza

jellyfish *f.* medusa

jerry can *m.* bidón

Jerseyian *m. & a.* Jerseyense, Jerseyiano

jet *m.* azabache; chorro, surtidor, jayet; jais; jet || ***jet fuel*** *m.* combustible de reacción (o de chorro); *f.* turbosina || ***jet pump*** *f.* bomba para chorro de agua || ***water jet*** *m.* chorro de agua a presión

jetty *m.* espigón, rompeolas, tajamar

jewel *f.* joya, piedra preciosa

jib crane *f.* grúa de brazo, trucha

jig *m.* gálibro, calibre, patrón

jig *vb.* cribar

jigging of ore *m.* cribadura

job *m.* trabajo, encargo; *f.* tarea, labor, [Méx.] chamba

jobber *m.* traficante, medianero, corredor

join *vb.* juntar, unir, conjuntar, empalmar, enlazar, ensamblar

joined *a.* conjuntado

joint *f.* fractura, rendija, grieta, diaclasa; junta, juntura, articulación, unión, conexión, división; *m.* acoplamiento; empalme, enchufe || ***butt joint*** *f.* junta a tope || ***compression joint*** *f.* piezoclasa || ***conjugate joints*** *f.* fracturas conjugadas || ***cooling joint*** *f.* grieta de enfriamiento || ***joint plane*** *m.* plano de diaclasa (o de fractura) || ***joint set*** *f.* juego de fracturas (o de diaclasas) || ***joint system*** *m.* sistema de diaclasas || ***joint venture*** *f.* empresa conjunta || ***lap joint*** *f.* junta de solapa || ***master joint*** *f.* diaclasa principal || ***shear joint*** *f.* diaclasa de cizalla || ***tension joint*** *f.* diaclasa de tensión

jointed *a.* articulado, diaclasado, fracturado

jointing *m.* fraccionamiento, junturamiento crucero, diaclasamiento; *f.* unión, disyunción || ***columnar jointing*** *f.* disyunción prismática || ***tabular jointing*** *f.* disyunción tabular

joule *m.* julio (2,78 × 10^{-4} watt-hora)

journal *m.* diario, periódico de noticias

journey *f.* jornada

journeyman *m.* jornalero

junction *f.* junta, unión, confluencia; *m.* empalme

junk *m.* desperdicio, ripio; *f.* chatarra

junk *vb.* abandonar

jungle *f.* selva; *m.* monte

Jupiter *m.* Júpiter
Jurassic *m. & a.* Jurásico; *f.* Jura
jut *vb.* emerger, sobresalir
jutting out *a.* emergente, sobresaliente
Juvarian *m. & a.* Juvarense, Juvariano
juvenile *a.* juvenil
juxtapose *vb.* yuxtaponer, colocar
juxtaposed *a.* yuxtapuesto
juxtaposition *f.* yuxtaposición

K

kaersutite *f.* kaersutita
kainite *f.* kainita
kakixen *f.* kakirita
kaliborite *f.* paternoita
kaligranite *m.* kaligranito, ortogranito
kalikeratophyre *m.* kalikeratófido, kaliqueratofiro
kaliphite *f.* kaliofilita
kalitrachyte *f.* kalitraquita
kame *m.* kame, kam
kammgranite *m.* kamgranito
Kansan *m. & a.* Kansaniense, Kansan
kaolin *f.* kaolinita, caolinita, tierra de porcelana; *m.* caolín, kaolín, litomargo, [Bol.] calichal
kaolinite *f.* caolinita, kaolinita, nacarita
kaolinitic *a.* kaolinítico
kaolinization *f.* caolinización, kaolinización
Karnian *m. & a.* Carniano, Carniense
karren *m.* lapiez
karst *m. & a.* carst, karst, cársico, cárstico; *f.* dolina ‖ ***karst chimney*** *m.* [Méx.] sótano ‖ ***karst tower*** *m.* pepino ‖ ***karst valley*** *f.* uvala
karstenite *f.* carstenita
karstic *a.* cárstico, kárstico
karstification *f.* carstificación, karstificación
karyokinesis *f.* cariocinesis, carioquinesis
Kaskasia *m. & a.* Kaskasia
katabatic *a.* catabático
kataforite *f.* kataforita
katamorphism *m.* catamorfismo
katatectic *a.* katatéctico
katazone *f.* catazona, katazona
katogene *m.* katogeno
katogenic *a.* katogénico, catogénico, catógeno
Kazanian *m. & a.* Kazán
keel *f.* quilla
keeled *a.* aquillado, carenado
Keewatin *m. & a.* Keewatino
keivyite *f.* keiviíta
kelly *m.* vástago de arrastre, vástago de perforación ‖ ***kelly joint*** *f.* junta kelly
kelp *f.* algas marinas
kelyphitic *a.* kelifítico, kelyfítico
kemahlite *f.* quemahlita
kenigmatite *f.* koelbingita
kentallenite *f.* quentalenita
kenyite *m.* kenyita, quenita
keramikite *f.* queramiquita
kerargyrite *f.* kerargirita
keratin *f.* queratina, ceratina
keratophyre *m.* kalikeratófido, kaliqueratofiro, keratofido, keratófiro, queratófiro, queratófido; *f.* queratoporfirita
keriotheca *f.* querioteca, cerioteca
kerite *f.* querita
kermesite *f.* pirostibina, quermesita
kernite *f.* kernita
kerogen *m.* kerógeno, kerabitumen, querógeno
kerol *m.* querol
kerosene *m.* keroseno, kerosén, aceite de carbón, queroseno, querosene, [Sp.] aceite lampante; *f.* kerosina
kerosine *f.* kerosina
kerotine *m.* queroteno
kersantite *f.* kersantita, quersantita
kesterite *f.* khinganita
ketone *f.* ketona
kettle *f.* marmita glacial; paila; *m.* caldero, hoyo glaciario

Keuper *m. & a.* Keuperiano, Kéuprico

Keweenawan *m. & a.* Keweenawiano, Keweenawense

key *f. & a.* clave; llave; (comp.) tecla; *m.* (geog.) cayo || ***key bed*** *m.* horizonte clave, estrato clave, nivel clave; *f.* capa clave

keyboard *m.* (comp.) teclado; *f.* tecla

key in, keypunch *vb.* (comp.) teclear

keystone *f.* clave del arco, clave, llave

K-feldspar *m.* k-feldespąto

khal *m.* jaki, jali

Kharkov *m. & a.* Jarkov

kick *f.* surgencia, indicación || ***pressure kick*** *f.* surgencia imprevista de presión

kick a well *vb.* arrancar un pozo, desviar un pozo

kick off *vb.* iniciar, arrancar

kieselguhr, kiselgur *m.* kieselgur, quieselguhr, quiselguhr

kieserite *f.* kieserita

kiirunvaarite *f.* kürunavaarita

kilaueite *f.* quilaveita

kill *vb.* matar || ***kill a well*** *vb.* matar (o ahogar) un pozo

kiln *m.* horno, horno de secar

kilobar *f.* kilobara

kilocalorie *f.* frigoría, kilocaloría

kilocycle *m.* kilociclo

kilogram, kilo *m.* kilogramo, kilo, quilo, quilogramo

kiloline *m.* quilolino

kiloliter *m.* kilolitro

kilometer *m.* kilómetro || ***distance in kilometers*** *m.* kilometraje

kilowatt *m.* kilovatio || ***kilowatt-hour*** *f.* kilovato hora

kimberlite *f.* kimberlita, quimberlita

Kimmerian *m.* Kimmeriano, Kimmérico

Kimmeridgian *m. & a.* Kimmeridgiense, Kimmeridgiano

kind *m.* tipo

Kinderhookian *m. & a.* Kinderhook

kinematic *a.* cinemático

kinetic *a.* cinético

kinetics *f.* cinética

kingdom *m.* (biol.) reino

kink *f.* torcedura, rectorcedura; *m.* doblez, nudo || ***kink fold*** *m.* pliegue de tipo "kink"

kinked *a.* retorcido

kips *f.* kilo-libras

kitchen *f.* (geol., chem., quím.) cocina

klaprothite *f.* wittichita

klippe *m.* isleo, klippe, peñasco

knob *m.* tornillo, botón; *f.* perilla, (geog.) protuberancia, prominencia, colina aislado

knock *m.* golpe

knock *vb.* golpear

knocking *m.* cancaneo, golpeteo, retintín; *f.* detonancia

knoll *m.* otero; *f.* colina

knot *m.* nudo

knotty *a.* nudoso

knowledge *m.* conocimiento

kodurite *f.* kodurita

kolbeckite *f.* sterrettita

komatiite *f.* komatiíta

konga diabase *f.* kongadiabasa

Konialnikian *m. & a.* Konialnikense, Konialnikiano

kornerupine *f.* kornerupina

kosmochlor *m.* kosmocloro

krennesrite *f.* krennesrita

krypton *m.* kriptón

Kungurian *m. & a.* Kongurense, Kungur, Kunguriense

kunzite *f.* kunzita

kurtosis *m.* kurtosis

kvellite *f.* quevellita

kyanite *f.* cianita, distena, zianita

kylite *f.* kilita, quilita

kyschtymite *f.* korund-anorthosita

L

lable *f.* etiqueta

labor *m.* trabajo; *f.* faena ‖ ***forced mine labor*** *m.* [Pe; Quechua] mita ‖ ***labor union*** *m.* sindicato

laboratory *m.* laboratorio

laborer *m.* obrero, operario, peón, zafrero

labradorite *f.* labradorita

laccolith *m.* lacolito

laccolithic *a.* lacolítico

lacustrine *a.* lacustre, lacunario, límnico

ladder *f.* escala, escalera

Ladinian *m. & a.* Ladiniense, Ladiniano

lag *m.* retraso, retardo, atraso

laggings *m.* revestimientos

lagoon *f.* laguna, albariza, albúfera ‖ ***coastal lagoon*** *f.* albúfera

lagoonal *a.* lagunal

lahar *m.* lahar; *f.* corriente de barro

lake *m.* lago, zafareche; *f.* laguna, [Bol., Ec.] cocha ‖ ***dry lake*** *m.* saladar ‖ ***oxbow lake*** *m.* lago en herradura, lago semilunar, brazo muerto; *f.* resaca ‖ ***salt lake*** *m.* lago salado ‖ ***small lake*** *m.* ojo, estanque, charco, chortal; *f.* charca, alberca

lamella *f.* laminilla, lámina

lamellar *a.* laminar, lamellar, laminado

lamellibranch *m.* lamelibranquio

lamelliform *a.* lameliforme

lamina *f.* lámina

laminar flow *m.* corriente laminar

laminate *vb.* laminar

laminated *a.* laminado, hojoso, lamelar, listado, bandeado

lamination *f.* laminación, lámina

lamp (miner's) *m.* candil de minero; *f.* [Méx.] cachimba

lamprophyre *m.* lamprófido, lamprófiro; *m.* klausenita, malquita

Lancastrian *m. & a.* Lancastriense, Lancastriano

lanceolate *a.* lanceolado

lancet plate *f.* lanceta

Landenian *m. & a.* Landeniense, Landeniano

land *f.* tierra; *m.* terreno, terruño ‖ ***bad lands*** *f.* tierras malas ‖ ***bottom land*** *f.* tierra de aluvión ‖ ***land forms*** *f.* formas fisiográficas ‖ ***land mass*** *f.* tierra firme, masa terrestre ‖ ***landlocked*** *a.* cercado de tierra ‖ ***land survey*** *f.* mojona; *m.* levantamiento topográfico ‖ ***low land*** *f.* tierra baja ‖ ***solid land*** *f.* tierra firme

landerite *f.* rosolita, xalostocita

landman *m.* agente de tierras

landmark *m.* mojón, rasgo sobresaliente; *f.* acotación, guarda raya, mojonera

landscape *m.* paisaje

landscaping *f.* arquitectura paisajista

landslide *m.* derrumbe, deslizamiento, desplome, deslizamiento de terreno, deslizamiento de faldeo, corrimiento, derrumbamiento, deslave, desprendimiento, lurte, resbalón, [And.] volcán, [Méx.] reliz; *f.* alud de fondo, desliz ‖ ***landslide scar*** *m.* barranco

landward-stepping *vb.* regresando hacia la tierra, de regreso a tierra

lanthanum *m.* lantano

lapilli *f.* lapilli

lapis lazuli *f.* lazurita, lapislázuli; *m.* ultramar, zafiro

Laramide *m.* Laramide

large *a.* grande, voluminoso, amplio, vasto

larsenite *f.* larsenita

larva *f.* larva

larval *a.* larval

last *a.* último

late *a.* tardio

latent heat *m.* calor latente

lateral *a.* lateral, ladero ‖ ***lateral moraine*** *f.* morena lateral, morena marginal ‖ ***lateral planation*** *m.* aplanamiento lateral ‖ ***lateral ramp*** *f.* rampa lateral ‖ ***lateral variation*** *f.* variación lateral ‖ ***left lateral*** *m.* lateral izquierdo ‖ ***right lateral*** *m.* lateral derecho

laterite *f.* laterita

lateritic *a.* laterítico ‖ ***lateritic red soil*** *f.* canga

laterization *f.* laterización

latex *m.* látex

lath *f.* latilla

latite *f.* latita

latitude *f.* latitud; *m.* alejamiento

lattice *m.* retículo, enrejado; *f.* rejilla || ***crystal lattice*** *f.* red cristalográfica

latticed *a.* enrejado

Lattorfian *m. & a.* Lattorfiense, Lattorfiano

laumontite *f.* laumontita, zeolita harinosa, zeolita de Bretaña

launch *vb.* lanzar

lawrencium *m.* laurencio

Laurentian *m. & a.* Laurentiano, Laurentiense

laurvigite *f.* laurvigita

lava *f.* lava || ***aa lava*** *f.* lava escoriácea || ***lava blister*** *f.* ampolla de lava; *m.* túmulo || ***lava flow*** *m.* flujo de lava; *f.* corriente o colada de lava, efusión de lava || ***lava tube*** *m.* tubo de lava || ***pahoehoe lava*** *f.* lava pahoehoe || ***pillow lava*** *f.* lava en almohada || ***ropy lava*** *f.* lava cordiforme

law *f.* ley; *m.* derecho || ***mining laws*** *f.* ordenanzas de minería

lawsonite *f.* lawsonita

lay *vb.* instalar, depositar || ***lay pipe*** *vb.* tender cañería

lay down *vb.* depositar

laying pipe *f.* remoción de la tubería

layer *m.* lecho, estrato, horizonte, jazido; *f.* capa, camada, cama, hebra, laja; tanda, tonga, tongada || ***key layer*** *m.* horizonte guía || ***layer parallel shear*** *f.* cizalla capa-a-capa || ***weathered layer*** *f.* capa meteorizada

layered *a.* estratificada, en capas

layering *f.* estratificación

lay off *vb.* liquidar, despedir, desemplear

layout *m.* plan; *f.* disposición, distribución

layover distortion *f.* inclinación al veedor

lay over *vb.* acostar, echarse, acostarse

lazulite *f.* lazulita, zeolita azul; *m.* espato azul

lazurite *f.* lazurita; *m.* ultramar

leach *vb.* lixiviar

leaching *f.* lixiviación

lead *m.* plomo; *f.* (phys., física, naut.) sonda || ***lead-bearing*** *a.* plomífero, plomilloso, plomoso || ***lead carbonate*** *f.* copelilla || ***lead ore*** *m.* metal plomoso, plomo || ***lead pipe*** *m.* caño de plomo || ***lead sulfide*** *m.* sulfuro de plomo || ***tetraethyl lead*** *m.* tetraetilo de plomo

lead *f.* (geol., min.) localización; *m.* prospecto, lead || ***lead line*** *m.* caño de descarga; *f.* soga de maniobras

leaden *a.* plomizo, plúmbeo

leading *a.* delantero || ***leading edge*** *m.* borde de ataque || ***leading edge technology*** *f.* tecnología de punta

leaf *f.* hoja, ala, aleta, lámina

league *f.* legua (aprox. 5572 metros)

leak *m.* derrame, escape; *f.* fuga, salida, pérdida || ***leak-proof*** *a.* a prueba de fugas

leak *vb.* derramar, gotear, perder, escurrirse

leakage *m.* derramiento, derrame, escape; *f.* fuga, filtración, pérdida

leaky *a.* agujereado

lean *a.* pobre, magro || ***lean mixture*** *f.* mezcla delgada o pobre

lease *f.* concesión, tierra arrendada; *m.* cateo, arrendamiento, permiso, arriendo, contrato; (min.) campo

lease *vb.* arrendar, alquilar

leased mine worker *m.* yanacona

leasehold *m.* inquilinato

leaseholder *m.* arrendatario

least squares best fit *f.* mínimos cuadrados linea de correlación

leave *f.* licencia; *m.* permiso

lectotype *m.* lectotipo

ledge *m.* escalón, resalto, resalte, reborde, banco; *f.* laja, cama de roca

Ledian *m. & a.* Lediense, Lediano

lees *f.* escorias, zurrapa; *m.* sedimento

leeward, lee *a.* sotavento

left *a.* izquierdo || ***left lateral*** *a.* izquierdo lateral, sinestral

leg *f.* pierna || ***dog leg*** *m.* doblez; *f.* pata de perro || ***oil leg*** *f.* zona productora de petróleo

legal *a.* legal

legend *f.* explicación; leyenda

lengenbachite *f.* jentschita

legislation *f.* legislación

length *f.* longitud, largo || ***focal length*** *f.* longitud focal || ***wavelength*** *f.* longitud de onda

lens *m.* & *f.* lente || ***concave lens*** *f.* lente cóncava || ***condenser lens*** *m.* lente de condensador || ***converging lens*** *f.* lente convergente || ***convex lens*** *f.* lente convexa || ***diverging lens*** *f.* lente divergente || ***hand lens*** *f.* lupa

lens-shaped *a.* lenticular, facoidal, en forma de lente

lenticular *a.* facoidal, lenticular || ***lenticular bedding*** *f.* estratificación lenticular || ***lenticular rock or ore body*** *f.* lenteja; *m.* lentejón

lepidoblastic *a.* lepidoblástico

lepidocrocite *f.* lepidocrocita

lepidolite *f.* lepidolita; *m.* lepidolito

lepidomelane *f.* lepidomelana

lessee *m.* arrendatario

lessen *vb.* rebajar

lessor *m.* arrendador; *f.* campista

let *vb.* alquilar, arrendar

lethal *a.* letal

letter *f.* carta || ***letter of credit*** *f.* carta de crédito

leucite *f.* leucita, anfingena, italita; *m.* leucito

leucitite *f.* vesuvianita, vesubianita, vesubiana

leucitophyre *m.* leucitrófiro

leucocratic *a.* leucocrático

levee *f.* levée; *m.* borde, dique, albardón, terraplén, malecón, dique lateral

level *a.* suave, plano; (geog.) abierto

level *m.* nivel, plano, llano; *f.* superficie || ***base level*** *m.* nivel de base, nivel de erosión local, nivel de referencia || ***confidence level*** *m.* nivel de confianza || ***detachment level*** *m.* nivel de deslizamiento, nivel de despegue || ***flood level*** *m.* nivel de creciente || ***groundwater level*** *m.* nivel hidrostático || ***level country*** *f.* campaña || ***level line*** *m.* nivel de cuerda ***level rod*** *m.* estadal || ***saturation level*** *m.* nivel de saturación || ***sea level*** *m.* nivel del mar || ***spirit level*** *m.* nivel de burbuja || ***surveyor's level*** *m.* nivel de agrimensor || ***water level*** *m.* nivel del agua

level *vb.* aplanar, nivelar, igualar, allanar, banquear, emparejar

leveled *a.* aplanado, nivelado

levelling *m.* aplanamiento, achatamiento; *f.* nivelación || ***levelling rod*** *f.* estadia || ***levelling staff*** *f.* niveleta

leveler, levelman *m.* nivelador

lever *f.* palanca

leverage *m.* palanqueo; *f.* acción de una palanca

lherzolite *f.* lherzolita

liability *f.* obligación, responsibilidad; *m.* pasivo, riesgo || ***accrued liability*** *m.* pasivo devengado || ***unlimiterd liability*** *f.* responsibilidad ilimitada

Lias, Liassic *m.* & *a.* Liás, Liásico; *f.* Jura negro

libethenite *f.* chinoita

license *m.* permiso; *f.* licencia, patente || ***exploration license*** *m.* permiso de cateo

license *vb.* licenciar

licensee *m.* concesionario

licensing *f.* expedición de licencias, autorización de licencias

lie, to be lying *vb.* yacer || ***lie, to be lying below*** *vb.* infrayacer

liebigite *f.* uranothallita

lien *m.* gravamen, embargo preventivo

life *f.* vida; duración en servicio || ***half-life*** *f.* vida media || ***lifeboat*** *m.* bote salvavidas || ***life cycle*** *m.* ciclo de la vida, ciclo vital || ***life span*** *f.* duración de la vida

lift *f.* alza, elevación; jaula de extracción || ***airlift*** *m.* extracción por aire || ***gas lift*** *m.* elevador de petróleo por presión de gas; *f.* extracción artificial por gas, surgencia por inyección de gas

lift *vb.* alzar, levantar

lifter *f.* leva, cama

lifting *f.* elevación, extracción || ***lifting cost*** *m.* costo de extracción

ligament *m.* ligamento

light *a.* (meas.) ligero, liviano; (color) claro

light *f.* luz, iluminación; *m.* faro || ***available light*** *f.* luz ambiente || ***diffused, reflected light*** *m.* trasluz || ***light table*** *f.* mesa de luz; *m.* [Méx.] tormento || ***natural light*** *f.* luz natural || ***polarized light*** *f.* luz polarizada

lightening *m.* rayo

lighthouse *m.* faro

lighting *f.* iluminación

lignite *m.* lignito

lignitic *a.* lignítico

limb *m.* lado, flanco, limbo; *f.* ala || ***back limb*** *m.* flanco posterior, flanco trasero || ***fault limb*** *m.* labio de falla || ***fold limb*** *m.* flanco del pliegue || ***forelimb*** *m.* flanco anterior

limburgite *f.* limburgita

lime *f.* cal || ***lime burner*** *m.* calero || ***lime kiln*** *f.* calera; *m.* horno de cal

limestone *f.* caliza; *m.* calcáreo, calero || ***crystalline limestone*** *f.* caliza cristalina || ***deep-weathered limestone*** *f.* chiscarra || ***limestone flux*** *f.* castina || ***limestone quarry*** *m.* calar; *f.* calera || ***oolitic limestone*** *f.* caliza oolitica || ***soft limestone*** used for bricks *m.* tabaire

limit *m.* (survey, levan.) lindero, límite, linde; *f.* comarca || ***confidence limit*** *m.* límite de confianza || ***elastic limit*** *m.* límite elástico

limit *vb.* limitar

limited *a.* limitado || ***limited partnership*** *f.* compañía en comandita, sociedad limitada

limnic *a.* límnico

limnology *f.* limnología

limonite *f.* limonita, hematita parda; *m.* hierro pardo, ocre, ocre pardo, rubio

limonitic *a.* limonitizado, limonítico

limy *a.* calizo, calcáreo, calar

line *f.* línea; fila, raya; *m.* cable; renglón || ***base line*** *f.* línea de base, línea de referencia || ***Becke line*** *f.* línea de Becke || ***coastline*** *f.* línea costera || ***contour line*** *f.* línea de nivel, curva de nivel || ***distribution line*** *m.* tubo distribuidor || ***dragline*** *m.* cable de arrastre || ***expansion line*** *f.* tubería de expansión || ***feeder line*** *m.* oleoducto tributario || ***flightline*** *f.* línea de vuelo || ***flow line*** *f.* tubería de flujo || ***fuel line*** *f.* cañería de combustible || ***gathering line*** *f.* tubería colectora, línea de captación || ***hinge line*** *f.* línea de charnela || ***international date line*** *f.* línea internacional de cambio de fecha || ***line of section*** *f.* traza de sección, traza de corte || ***line of sight*** *f.* línea de mira, visual || ***loop-gathering line*** *m.* oleoductos mallados || ***loose line*** *f.* línea de referencia, línea suelta || ***pin line*** *f.* línea alfiler, línea fijada (o fija) || ***pipeline*** *m.* oleoducto || ***plumb line*** *f.* línea de plomada || ***power line*** *f.* línea de energía eléctrica || ***reference line*** *f.* línea de referencia || ***seismic line*** *f.* línea sísmica || ***shoreline*** *f.* línea litoral, línea costera || ***shot line*** *m.* cable de explosión; *f.* línea de fuentes, línea de tiros || ***timber line*** *f.* línea de bosques || ***trunk line*** *m.* oleoducto troncal || ***wire line*** *m.* cable de alambre

lineage *m.* linaje

lineament *m.* lineamiento

linear *a.* lineal, linear

lineation *f.* lineación, alineación

lined *a.* forrado, revestido

line-length balance *m.* balanceo por larga de capa; balanceo por longitud

liner *f.* calza; *m.* forro, revestidor, tubo colador, revestimiento || ***perforated liner*** *m.* caño filtro, tubo calado, caño punzonado, tubo revestidor || ***slotted liner*** *f.* cañería perdida ranurada

lining *m.* forro, revestimiento; *f.* camisa

linguloid *a.* lingüiforme

lining *m.* revestimiento interior

link *m.* esbalón; anillo, enlazamiento

link *vb.* amarrar, corelacionar, enlazar, ligar, unir

linkage *f.* conexión, unión; *m.* empalme

linnaeite *f.* pirita de cobalto

lintonite *f.* winchellita

lip *m.* labio

liptinite *f.* liptinita

liquefaction *f.* licuefacción, licuación

liquefied petroleum gas (LPG) *m.* gas licuado de petróleo

liquid *a.* líquido || ***liquid assets*** *m.* activos circulantes

liquidate *vb.* liquidar

liquidation *f.* liquidación

liquidity *f.* liquidez

liquefiable *a.* licuable

liquification *f.* licuación, licuefacción

liquify *vb.* liquidar, descoagular, descuajar, liquifacer

liquifying *a.* descoagulante

list *f.* lista; listón, filete; tira; tabla || ***packing list*** *f.* lista de embalaje || ***price list*** *f.* lista de precios

list *vb.* enumerar, tabular

listric *a.* lístrico

listrium *m.* listrio

liter *m.* litro

litharenite *f.* litarenita

litharge *m.* litargirio, litarge; *f.* almártaga, greta

lithic *a.* lítico

lithification *f.* litificación

lithified *a.* litificado, petrificado

lithium *m.* litio

lithoclase *f.* litoclasa

lithoclast *m.* litoclasto

lithofacies *m.* litofacies

lithogenesis *m.* litogénesis

lithographic *a.* litográfico

lithologic *a.* litológico

lithology *f.* litología

lithophile *a.* litófilo

lithophysae *f.* litofisuras

lithosphere *f.* litósfera

lithostatic *a.* litostático || ***lithostatic pressure*** *f.* presión litostática

lithostratigraphic *a.* litoestratigráfico

litmus paper *m.* papel de tornasol

littoral *a.* litoral || ***littoral plain*** *m.* talud litoral

live *a.* en vivo, activo || ***live oil*** *m.* petróleo con gas || ***live steam*** *m.* vapor vivo (o directo)

lixiviation *f.* lixiviación

lherzolite *f.* lherzolita

Llandeilian *m. & a.* Llandeiliense, Llandeiliano, Llandeilo, Llandilo

Llandoverian *m. & a.* Llandoveriense, Llandoveriano, Valentiano, Valentiense

llanite *f.* llanita

Llanvirnian *m. & a.* Llanvirniense, Llanvirniano

load *f.* carga; *m.* cargo, cargamento, peso || ***bed load*** *f.* carga del lecho; *m.* gasto sólido || ***dissolved load*** *f.* carga disuelta || ***load cast*** *m.* calco de carga || ***load factor*** *m.* factor de carga || ***load on bit*** *m.* peso sobre la herramienta || ***load structure*** *f.* estructura de presión, estructura de carga || ***peak load*** *f.* carga máxima; *m.* pico de carga || ***suspended load*** *f.* carga suspendida

load *vb.* cargar

loaded *a.* cargado

loading *f.* carga; *m.* cargamento, cargando

loam *m.* barro; migajón, légamo; *f.* marga

loamy *a.* legamoso, margoso

loan *m.* préstamo, empréstito, crédito

lobate *a.* lobulado

lobe *m.* lóbulo

lobed *a.* lobulado, lobuloso

local *a.* local

locality *f.* localidad, ubicación, localización; *m.* sitio

locate *vb.* ubicar, localizar; emplazar, situar, colocar, trazar

located *a.* ubicado

(to be) located *vb.* ubicarse

location *f.* localidad, localización, ubicación, situación; *m.* sitio || ***inaccessible location*** *m.* abatón

loch *m.* lago

locus *m.* lugar geométrico

lode *m.* (min.) filón, lodo; *f.* veta, vena || ***mother lode*** *f.* veta madre

lodestone *f.* piedra imán, magnetita, calamita, mena ferruginosa

loess *m.* loes, loess; *f.* marga

log *m.* registro; *m.* perfil || ***acoustic log*** *m.* registro acústico || ***electric log*** *m.* perfil eléctrico, registro eléctrico || ***log header*** *m.* cabezal || ***log suites*** *m.* conjunto de registros || ***well log*** *m.* perfil de perforación, perfil de pozo, registro de pozo

log *vb.* registrar, anotar

logarithm *m.* logaritmo

logarithmic *a.* logarítmico

logbook *m.* diario

logging *f.* registración, diagrafía, testificación; *m.* perfilaje || ***logging unit*** *m.* mandos del cable de perforación

long *a.* largo || ***long shore drift*** *m.* desplazamiento marginal || ***long-term*** *a.* largo plazo || ***long wall mining*** *m.* tajo largo

logistics *f.* logística

longitude *f.* longitud || ***longitude line*** *m.* meridiano

longitudinal *a.* longitudinal

longshore *a.* prelitoral || ***longshore bar*** *f.* barrera prelitoral, barra prelitoral || ***longshore drift*** *f.* deriva litoral

loop *m.* polígono

loop *vb.* enlazar

looping *a.* tramo paralelo, desvío

loose *a.* suelto, flojo, inconsistente || ***loose earth*** *f.* tierra floja || ***loose line*** *f.* linea suelta

loosen *vb.* soltar

loosening *m.* desprendimiento

lophophore *m.* lofóforo

lopolith *m.* lopolito

lorenzenite *f.* ramsayita

loss *f.* (com.) pérdida || ***financial loss*** *f.* (com.) pérdidas financieras || ***mud loss*** *f.* pérdida de lodo

lost circulation *f.* circulación pérdida

lost section *f.* sección perdida

lot *m.* solar, terreno, lote; *f.* partida

Lotharingian *m.* & *a.* Lotaringiense, Lotaringiano

low *a.* bajo || ***low angle*** *a.* ángulo cerrado, ángulo bajo || ***low frequency*** *f.* baja frecuencia || ***low grade*** *a.* baja ley, pobre || ***low ground*** *m.* bajial || ***lowlands*** *f.* hondonada, tierras bajas || ***low pressure*** *f.* baja presión || ***low stand*** *m.* nivel de base bajo, bajamarea || ***low tide*** *m.* bajamar, bajante; *f.* marea baja || ***low water mark*** *m.* estiaje; *f.* bajamar

lower *a.* inferior, yusano, yusero || ***lower shoreface*** *m.* frente de playa baja

lower *vb.* abajar, bajar, rebajar

lube cut *a.* corte para lubricante

lube oil *m.* aceite lubricante

lubricant *m.* & *a.* lubricante

lubricate *vb.* engrasar, lubricar

lubricating *a.* lubricante || ***lubricating oil*** *m.* aceite lubricante

lubrication *f.* lubricación

lubricator *m.* lubricador, engrasador

lucinoid *a.* lucinoide

Ludian *m.* & *a.* Ludiense, Ludiano

Ludlovian *m.* & *a.* Ludloviense, Ludloviano

luminescence *f.* luminiscencia

luminescent *a.* luminescente

luminous *a.* luminoso

lunar *a.* lunar

lung *m.* pulmón

Lusitanian *m.* & *a.* Lusitaniense, Lusitaniano

luster *m.* lustre, brillo, lustro || ***adamantine luster*** *m.* brillo diamantino || ***dull luster*** *m.* brillo mate || ***glassy luster*** *m.* brillo vítreo || ***greasy luster*** *m.* brillo grasiento, graso, o grasoso || ***metallic luster*** *m.* brillo metálico || ***pearly luster*** *m.* brillo anacarado, brillo nacarado, brillo perlado || ***resinous luster*** *m.* brillo resinoso || ***vitreous luster*** *m.* brillo vítreo || ***waxy luster*** *m.* brillo ceroso

lustrous *a.* lustroso

lutation *f.* cementación

Lutetian *m.* & *a.* Luteciense, Luteciano

lutetium *m.* lutecio

lutite *f.* lutita

lycopod *m.* licopodio

lyssacine *a.* lisacino

lydite *f.* lidita

lying *adv.* yacente, yaciente

(to be) lying *vb.* yacer

lymnetic *a.* lacustre

M

maar *f.* diatrema; *m.* embudo de explosión, maar

maceral *m.* maceral

machine *f.* máquina

machinery *f.* maquinaria

mackinawite *f.* kansita

macroclastic *a.* macroclástico

macrocrystalline *a.* macrocristalino

macrofossil *m.* macrofósil, megafósil

macropygous *a.* macrópigo

macroscopic *a.* macroscópico

macroseism *m.* macrosísmo

macroseismic *a.* macrosísmico

macrospheric *a.* macrosférico

macrospore *f.* macrospora, megaspora

macula *f.* mácula

madrepore *f.* madrépora

Maestrichtian *m.* & *a.* Maestrichtiense, Maestrichtiano

mafic *a.* máfico, básico, ferromagnesiano, fémico, ferromagnésico

mafite *m.* mafito

Magdalenian *m.* & *a.* Magdaleniano, Magdeleniense

maghemite *f.* oximagnita

magma *f.* magma || ***magma chamber*** *f.* cámara magmática || ***parent magma*** *m.* jugo lapídeo

magmatic *a.* magmático, ortotéctico

magmatism *m.* magmatismo

magnesia *m.* leche de tierra

magnesian *a.* magnesiano

magnesiochromite *f.* picrocromita

magnesite *f.* magnesita, giobertita; *m.* carbonato magnésico; espato amargo, espato bruno

magnesium *m.* magnesio

magnet *m.* imán, magneto

magnetic *a.* magnético || ***magnetic anomaly*** *f.* anomalía magnética || ***magnetic declination*** *f.* declinación magnética || ***magnetic dip*** *f.* inclinación magnética || ***magnetic dipole*** *m.* dipolo magnético || ***magnetic field*** *m.* campo magnético || ***magnetic flux*** *m.* flujo magnético || ***magnetic gradient*** *m.* gradiente magnético || ***magnetic high*** *m.* máximum magnético || ***magnetic induction*** *f.* inducción magnética || ***magnetic intensity*** *f.* intensidad magnética || ***magnetic low*** *m.* mínimo magnético || ***magnetic moment*** *m.* momento magnético || ***magnetic north*** *m.* norte magnético || ***magnetic pole*** *m.* polo magnético || ***magnetic susceptibility*** *f.* susceptibilidad magnética || ***magnetic tape*** *f.* cinta magnética || ***vertical magnetic intensity*** *f.* intensidad magnética vertical

magnetism *m.* magnetismo || ***induced magnetism*** *m.* magnetismo inducido || ***paleo magnetism*** *m.* magnetismo fósil

magnetite *f.* magnetita, calamita, kiirunavaarita, kirunavaarita, quirunavaarita; *m.* chumpi, hierro magnético

magnetization *f.* magnetización, imanación

magnetize *vb.* magnetizar, imantar, imanar, magnetar

magnetometer *m.* magnetómetro

magnetometry *f.* magnetometría

magnification *m.* aumento; *f.* magnificación, amplificación

magnify *vb.* aumentar, ampliar, amplificar

magnifying glass *m.* lente de aumento, lente de amplificación

magnitude *f.* magnitud

main *a.* principal

mainland *f.* tierra firme; *m.* continente

maintain *vb.* mantener

maintenance *f.* conservación, manutensión; *m.* mantenimiento || ***pressure maintenance*** *m.* mantenimiento de presión; conservación de presión

make *f.* marca

make *vb.* construir, hacer, formar || ***make headway*** *vb.* adelantar || ***make hole*** *vb.* perforar, progresar con la perforación

maker *m.* fabricante, constructor

malachite *f.* malaquita, malquita; *m.* cobre verde

malacostracan *m.* malacostráceo

male *m.* macho, masculino

malic *a.* málico

malleability *f.* maleabilidad

malleable *a.* maleable

mallet *f.* maza, maceta; *m.* martillo, mazo

Malm *f.* Jura blanco, Malm

maltha *f.* brea mineral

mammal *m.* mamífero

mammillary *a.* mamilar

manager, management *f.* gerente; *m.* jefe, administrador, gerencia

mandible *f.* mandíbula

manganandalusite *f.* viridina

manganate *m.* manganato

manganese *m.* manganeso || ***manganese oxide*** *f.* manganesa; *m.* óxido de manganeso, [Col.] canturrón

manganic *a.* mangánico

manganiferous *a.* que contiene manganeso

manganite *f.* manganita

manganous *a.* manganoso

mangrove *m.* mangle || ***mangrove swamp*** *m.* manglar

manifest *vb.* manifestar, aparentar

manifold *m.* múltiple, tubo distribuidor

manipulation *f.* manipulación

manner *m.* modo; *f.* manera

manometer *m.* manómetro

manpower *f.* dotación de personal, esfuerzo humano

mantissa *f.* mantisa

mantle *m.* manto || ***earth's mantle*** *m.* manto terrestre || ***lower mantle*** *m.* manto interno || ***upper mantle*** *m.* manto externo

manual *m.* & *a.* manual

manufacture *f.* manufactura, fabricación

manufacture *vb.* manufacturar, fabricar

manufacturer *m.* fabricante, manufacturero

map *m.* mapa, plano, plan; *f.* carta, planta || ***airphoto map*** *m.* mapa aerofotográfico || ***base map*** *m.* mapa básico, mapa base || ***contour map*** *m.* mapa contornado, mapa con curvas de nivel || ***exploration permit map*** *m.* tablero de comprobación || ***geologic map*** *m.* mapa geológico || ***geophysical map*** *m.* mapa geofísico || ***index map*** *m.* mapa índice, plano de localización || ***isopach map*** *m.* mapa isopáquico || ***lithologic map*** *m.* mapa litológico || ***location map*** *m.* plano de localización || ***map archive*** *f.* mapoteca || ***map sheet*** *f.* hoja || ***map view*** *m.* en mapa || ***quadrangle map*** *m.* cuadrángulo || ***reconnaissance map*** *m.* mapa de reconocimiento || ***seismic map*** *m.* mapa sísmico || ***structure map*** *m.* mapa estructural || ***thematic map*** *m.* mapa temático || ***topographic map*** *m.* mapa topográfico

map *vb.* cartografiar, mapear

mapping *m.* mapeo, mapeamiento; levantamiento de planos; *f.* planimetría; cartografía || ***aerial mapping*** *f.* cartografía aérea || ***subsurface mapping*** *m.* mapeo de subsuelo || ***surface mapping*** *m.* mapeo de superficie

marble *m.* mármol

marcasite *f.* marcasita, marquesita, pirita blanca, pirita de cresta de gallo, pirita prismática, [Bol.] argirita

margarite *f.* margarita

margin *m.* borde, margen || ***continental margin*** *m.* margen continental || ***margin of safety*** *m.* margen de seguridad

marginal *a.* marginal

margins *m.* (com.) márgenes

marialite *f.* marialita

marine *a.* marino, oceánico || ***marine cut terrace*** *f.* terraza de erosión marina || ***open marine*** *a.* mar abierta

maritime *a.* marítimo

mark *f.* marca, señal, impronta || ***gouge mark*** *f.* marca lunada || ***ripple mark*** *f.* marca de oleaje || ***scour mark*** *f.* marca de arrastre || ***tool marks*** *m.* marcadores útiles || ***trade mark*** *f.* marca de fábrica

mark *vb.* (survey, levan.) acotar, ahitar, alindar, señalar, amojonar, [Arg., Ch.] alinderar

marked *a.* señalado

marker *f.* (survey, levan.) acotación, baliza; *m.* coto, mojón || ***marker bed*** *m.* horizonte guía, nivel guía; estrato guía; *f.* capa guía, capa de referencia || ***marker pin*** *f.* aguja

market *m.* mercado || ***export market*** *m.* mercado de exportación

marketable *a.* comerciable

marketing *f.* comercialización, venta

marl, marlstone *f.* marga, greda, greta, [Arg.] marna; *m.* barro, buro ‖ ***marl, marlstone pit*** *m.* marguero ‖ ***red marl, marlstone*** *f.* aga ‖ ***refractory marl, marlstone*** *f.* acendrada

marly *a.* margoso, gredoso

marmatite *f.* christophita

Mars *m.* Marte

marsh *m.* marjal, almarjal, pantano, aguazal, bañado, lodazal, fangal, [PR] cienegal, llamargo, llamazar; *f.* ciénaga, llama, ñanga ‖ ***marsh gas*** *m.* gas de pantano, metano

marshy *a.* marismeño

marsupial *m.* marsupio; *a.* marsupial

martite *f.* martita

mask *f.* máscara ‖ ***dust mask*** *f.* máscara contra el polvo ‖ ***gas mask*** *f.* máscara antigás

mask *vb.* enmascarar, ocultar

mass *f.* masa; *m.* macizo ‖ ***mass balance*** *m.* balance de materia ‖ ***mass spectrometer*** *m.* espectrómetro colectivo (o global, o masa) ‖ ***mass wasting*** *m.* desgaste, desgastando, desmoronamiento ‖ ***rock mass*** *m.* cuerpo

massif *m.* macizo

massive *a.* masivo, macizo, potente ‖ ***massive bedding*** *f.* estratificación masiva

mast *m.* mástil, árbol; *f.* asta

master *m.* & *a.* maestro ‖ ***master joint*** *f.* grieta (o fractura) principal ‖ ***master stream*** *f.* corriente principal

material *m.* material; *f.* materia ‖ ***material balance*** *m.* balance de materiales ‖ ***raw material*** *f.* materia prima; *m.* material en bruto

matildite *f.* plenargirita, schapbachita

matlockite *f.* matloquita

matrix *f.* matriz, ganga, guija, [Méx.] borra, pasta, pasta mineral; *m.* aglomerante; base vidrioso, quijo, soroque

matte *a.* mate

matter *f.* materia ‖ ***organic matter*** *f.* materia orgánica

mattock *m.* zapapico; *f.* [Méx.] talacha

mattox *f.* piqueta

mature *a.* maduro

mature *vb.* madurar

maturity *f.* madurez

maucherite *f.* temiskamita

maximum *a.* máximo

Maxwell *m.* maxvelio

MCF *m.* mil piés cúbicos

meadow *f.* pradera

meadowlands *f.* pradería, pradera

mean *m.* medio; *f.* media

meander *m.* meandro; *m.* vuelta ‖ ***incised meander*** *m.* meandro excavado ‖ ***meander bar*** *f.* barra de meandro ‖ ***meander loop*** *m.* lazo del meandro

meandering *f.* divagancia

meaning *m.* sentido

measure *f.* medida, dimensión

measure *vb.* medir, aforar, tantear

measurement *f.* medición, medida

measuring *m.* aforo ‖ ***measuring instrument*** *m.* medidor

mechanic *m.* mecánico

mechanical *a.* mecánico ‖ ***mechanical analysis*** *m.* análisis mecánico ‖ ***mechanical erosion*** *f.* erosión mecánica

mechanics *f.* mecánica ‖ ***rock mechanics*** *f.* mecánica de rocas

mechanism *m.* mecanismo

medial *a.* intermedia

median *a.* mediano

Mediterranean *a.* mediterráneo

medium *a.* medio, mediano ‖ ***medium grain*** *m.* grano mediano

medusa *f.* medusa

meerschaum *f.* sepiolita, meereschaum, espuma

megacanthopore *m.* megacantóporo

megafold *f.* undación

megafossil *m.* macrofósil, megafósil

megaripples *f.* mega-ondulas

megascopic *a.* megascópico

megaspore *f.* megaspora, macrospora

megatherium *m.* megaterio

megathermal *a.* megatermal

meionite *f.* meionita

melaconite *f.* melaconita

melange *m. & f.* melange; *f.* mezcla

melanite *f.* melanita, pireneita

melanocratic *a.* melanocrático

melaphyre *m.* meláfero

melanterite *f.* melanterita, caparrosa, copaquira

melaphyre *m.* meláfido, meláfiro, meláfero

melilite *f.* melilita, zurlia

mellite *f.* melita

melt *f.* materia fundida, colada, hornada, vaciada || ***melt water*** *f.* agua de derretimiento || ***residual melt*** *f.* fusión residual; *m.* jugo lapídeo

melt *vb.* fundir, fundirse, derretir, deshelarse

meltable *a.* fundible

melted *a.* fundido, derretido

melting *f.* fusión, fundición; *m.* derretimiento; *a.* fundente || ***melting point*** *m.* punto de fusión

member *m.* miembro

membrane *f.* membrana

memory *f.* (comp.) capacidad de almacenamiento, memoría, unidad de memoria || ***memory dump*** *m.* vuelco de la memoria

menaccanite *f.* ilmenita, menaccanita

meniscus *m.* menisco

menu *m.* (comp.) menú || ***pull-down menu*** *m.* menú desplegable

mercaptan *m.* mercaptano

Mercator projection *f.* proyección Mercator

merchandise *f.* mercancías

Mercury, mercury *m.* Mercurio, mercurio, azogue; *f.* (ref.) yapa || ***mercury ore reduction*** *f.* [Sp.] brasa || ***mercury poisening*** *m.* azogado, azogamiento || ***mercury storehouse*** azoguería

mercuric *a.* mercúrico

mercurous *a.* mercuroso

meridian *m.* meridiano

merge *vb.* fusionar, refundir; unir, unirse, integrar

merger *f.* fusión corporativa

merging *a.* integración

meridian *m.* meridiano

meroxene *m.* meroxeno

mesa *f.* mesa, meseta; *m.* otero, tablazo

mesenchyme *f.* mesénquima

mesh *f.* malla, zaranda, criba (o tela) metálica; *m.* retículo, tamiz

mesoclast *m.* mesoclasto

mesocratic *a.* mesocrático

mesohaline *a.* mesohalino

mesolite *f.* zeolita radiada

mesolithic *a.* mesolítico

mesophyte *f.* mesofita

mesophytic vegetation *f.* vegetación mesofítica

mesopore *m.* mesóporo

mesosome *f.* mesosoma

mesosphere *f.* mesosfera

mesothermal *a.* mesotérmico, mesotermal

mesothorax *m.* mesotórax

mesotidal *a.* mesomarea

mesotype *m.* mesotipo

Mesozoic *m.* Mesozóico, era secundaria

mesozone *f.* mesozona

metabentonite *f.* metabentonita

metabolic *a.* metabólico

metabolism *m.* metabolismo

metabolize *vb.* metabolizar

metabrushite *f.* zeugira

metacinnabar *f.* saukovita

metaconglomerate *m.* metaconglomerado

metacryst *m.* metacristal

metagenesis *m.* metagénesis

metagraphite *f.* subgrafíta

metaigneous *a.* metaígneo

metal *m.* metal || ***alkali metal*** *m.* metal alcalino || ***metal sulfate*** *m.* acije || ***native metal*** *m.* metal nativo, metal virgen

metallic *a.* metálico

metalliferous *a.* metalífero

metallization *f.* metalización

metallogenesis *f.* metalogénesis

metallogenic, metallogenetic *a.* metalogenético

metallogeny *f.* metalogenía

metallurgic *a.* metalúrgico

metallurgist *m.* metalúrgico

metallurgy *f.* metalurgia

metamict *a.* metamíctico

metamorphic *a.* metamórfico || ***metamorphic core complex*** *m.* complejo nuclear metamórfico || ***metamorphic facies*** *f.* facies metamórficas || ***metamorphic grade*** *m.* grado de metamorfismo || ***metamorphic rock*** *f.* roca metamórfica

metamorphism *m.* metamorfísmo || ***burial metamorphism*** *m.* metamorfísmo de presión || ***contact metamorphism*** *m.* metamorfísmo de contacto, alometamorfísmo, metamorfismo local || ***dynamic metamorphism*** *m.* metamorfismo dinámico, alometamorfísmo, metamorfísmo de dislocación || ***hydrothermal metamorphism*** *m.* metamorfísmo hidrotermico (o hidrotermal) || ***regional metamorphism*** *m.* metamorfísmo regional (o cáustico, o general) || ***retrograde metamorphism*** *m.* retrometamorfismo || ***thermal metamorphism*** *m.* metamorfísmo térmico (o geotérmico)

metamorphose *vb.* metamorfosear

metamorphosed *a.* metamorfizado

metamorphosis *f.* metamorfosis

metaquartzite *f.* metacuarcita

metasediment *m.* metasedimento

metasedimentary *a.* metasedimentario

metasomatic *a.* metasomático

metasomatism *m.* metasomatismo; *f.* metasomatosis, substitución metasomatica

metatype *m.* metatipo

metavolcanic *f.* metavolcanica

metazoan *m.* metazoo

meteor *m.* meteoro; *f.* estrella fugaz

meteoric *a.* meteórico

meteorite *m.* meteorito, bólido, uranolito

meteorological *a.* meteorológico

meteorologist *m.* meteorologista

meteorology *f.* meteorología

meter *m.* metro; contador, medidor (3,28 pies) || ***electric meter*** *m.* contador eléctrico || ***flow meter*** *m.* contador de flujo, medidor de caudal || ***gas meter*** *m.* contador de gas || ***oil meter*** *m.* contador de petróleo || ***square meter*** *m.* centiárea, metro cuadrado || ***water meter*** *m.* contador de agua, medidor de agua

meter *vb.* medir por contador

metered *a.* medido

metering *f.* medición

methane *m.* metano, gas natural

methanol *m.* metanol

method *m.* método, procedimiento; *f.* práctica

methodology *f.* metodología

methyl *m. & a.* metilo

metric *a.* métrico || ***metric ton*** *f.* tonelada métrica = 1000 kg

Mexican onyx *m.* tecali, alabastro, alabastrino

mho *m.* unidad de conductividad = 1/ohm

miargyrite *f.* kenngotita, miargirita

miarolitic *a.* miarolítico

miascite, miaskite *f.* miascita

mica *f.* mica; *m.* espejuelo || ***mica schist*** *f.* micasita, micacita, mica-pizarra, mica esquisto

micaceous *a.* micáceo

micelle *f.* micela

micrinite *f.* micrinita

micrite *f.* micrita

micritic *a.* micrítico

microbreccia *f.* microbrecha

microclast *m.* microclasto

microclastic *a.* microclástico

microcline *m.* microclino; *f.* microclina

microcrystalline *a.* microcristalino

microfauna *f.* microfauna

microfold *f.* micropliegue

microfolding *m.* plisamiento

microfossil *m.* microfósil

micro-granite *m.* microgranito

microgranitic *a.* micrograníticο

micro-granulite *f.* microgranulita

micrographic *a.* micrográfico

microlite *f.* microlita, triquita; *m.* globolito

microlith *m.* microlitón

microlog *m.* microregistro; *f.* microperfil
micrometer *m.* micrómetro
micron *m.* micrón, micromilímetro; *f.* micra
micropaleobotany *f.* micropaleobotánica
micropaleontologic *f.* micropaleontológico
micropaleontologist *f.* micropaleontólogo
micropaleontology *f.* micropaleontología
micropegmatite *f.* micropegmatita
micropegmatitic *a.* micropegmatítico
microperthite *f.* micropertita
microphone *m.* micrófono
microphotography *f.* microfotografía
microscope *m.* microscopio || ***binocular microscope*** *m.* microscopio binocular || ***microscope stage*** *f.* plataforma, platina || ***monocular microscope*** *m.* microscopio monocular || ***polarizing microscope*** *m.* microscopio polarizante || ***stereoscopic microscope*** *m.* microscopio estereoscópico
microscopic *a.* microscópico
microscopy *f.* microscopía
microseism *m.* microsismo
microseismic *a.* microsísmico
microsporangium *m.* microsporangio
microspore *f.* microspora
microstructure *f.* microestructura
microtectonics *f.* microtectónica
microtexture *f.* estructura microscópica
microwave *f.* micro-onda
mid *a.* medio
middle *a.* medio || ***middle shoreface*** *m.* frente de playa medio
midland *m.* mediterráneo
mid-ocean ridge *m.* cordón interoceánico; *f.* dorsal oceánica
migmatite *f.* migmatita
migmatization *f.* migmatización
migrate *vb.* migrar
migrated *a.* migrado
migration *f.* migración || ***depth migration*** *f.* migración en profundidad || ***depth migration before stacking*** *f.* migración de profundidad antes de apilamiento || ***migration horizon*** *f.* formación conductor || ***migration paths*** *m.* caminos de migración || ***time migration*** *f.* migración en tiempo || ***time migration before stacking*** *f.* migración de tiempo antes de apilamiento
mild *a.* suave, moderado || ***mild steel*** *m.* acero dulce
mile *f.* milla = 1.609 metros
mileage *m.* millaje, kilometraje, recorrido en millas
milepost, milestone *m.* mojón, jalón
mill *m.* molino; taller, moledor, planteaje; *f.* fábrica, (steel) acería, labor de beneficio, moledora || ***mill operator*** *m.* beneficiador, ingenio
mill *vb.* fresar, desbastar, pulir, talar
millable *a.* beneficiable
milled *a.* fresado, rectificado, triturado
miller *m.* beneficiador
millerite *f.* millerita, pirita capilar
milliammeter *m.* miliamperímetro
millibar *m.* milibar
milllidarcy *m.* milidarcy
milligal *m.* miligal
miligram *m.* miligramo
milliliter *m.* mililitro
millimeter *m.* milímetro || ***millimeter of mercury*** *m.* milímetros de mercurio
milling *m.* moliendo, trituración || ***milling machine, milling tool*** *f.* fresadora, fresa
millivolt *m.* milivoltio
minable *a.* explotable
mimetism *m.* mimetismo
mimetite *f.* mimetita, mimatesa, mimetesita
mine *f.* mina; *m.* almadén || ***abandoned mine*** *f.* castaña || ***mine cage*** *f.* caja de extracción, jaula || ***mine car*** *m.* carro || ***mine drainage*** *m.* drenaje de mina || ***mine drillers*** *m.* [Pe.] auquis || ***mine dump*** *m.* vaciadero de mina; *f.* [Pe., Ch.] cancha || ***mine entrance*** *f.* bartolina, bocamina; *m.* emboquillado || ***mine owner*** *m.* minero || ***mine shaft*** *m.* pozo de mina || ***mine workings*** *f.* labor || ***open pit mine*** *f.* mina a cielo abierto || ***strip mine*** *f.* mina en fajas || ***unprofitable mine*** *f.* borrasca
mine *vb.* excavar, labrar, minar || ***mine hard coal*** *vb.* hornaguear

miner *f.* labor; *m.* minador, minero, piquero || ***contract miner*** *m.* [Bol., Ch.] pirquinero

mineral *m.* mineral; *a.* minero || ***accessory mineral*** *m.* mineral accesorio || ***gangue mineral*** *m.* mineral filoniano, mineral de ganga || ***heavy mineral*** *m.* mineral pesado || ***mineral assemblage*** *f.* asociación de minerales || ***mineral belt*** *f.* faja mineralizada || ***mineral deposit*** *m.* almadén, yacimiento mineral, criadero, depósito de mineral, yacimiento || ***mineral oil*** *m.* aceite mineral || ***mineral rights*** *m.* derechos mineros || ***mineral water*** *f.* agua mineral || ***replacement minerals*** *m.* minerales secundarios

mineralization *f.* mineralización

mineralize *vb.* mineralizar

mineralized *a.* mineralizado || ***mineralized area*** *m.* asiento de minas, área mineralizada

mineralized region *m.* asiento mineral; *f.* región mineralizada

mineralized vug or cavity *f.* soyota, soyote

mineralizing *a.* mineralizador

mineralogical *a.* mineralógico

mineralogist *m. & f.* mineralogista

mineralogy *f.* mineralogía

minette *f.* mineta

minimum *a.* minimo, minimum

mining *f.* minería; *m.* laboreo, mineraje || ***contract mining*** *m.* [Bol., Ch.] pirquín || ***mining claim*** *f.* propiedad minera; *m.* fundo minero

minor *a.* menor, secundario

mint *f.* ceca, casa de moneda, [Pe.] callana

minute *m.* minuto || ***minute folding*** *m.* arrugamiento

Miocene *m. & a.* Mioceno, Miocénico

miogeosyncline *f.* miogeosinclinal

mirror *m.* espejo || ***concave mirror*** *m.* espejo concavo || ***oscillating mirror*** *m.* espejo oscilante

miscible *a.* mezclable, miscible

Mississippian *m. & a.* Misisípico, Misisipiano, Misisipiense

Missourian *m. & a.* Misuriano, Misuriense

misourite *f.* misurita

mist *f.* niebla, neblina

mistake *m.* error

mitosis *f.* mitosis

mix (ore for smelting) *vb.* mezclar, casar metales, [Pe.] chacurruscar

mixed *a.* mixto, mezclado || ***mixed base crude*** *m.* aceite crudo de base mixta || ***mixed phase process*** *m.* proceso de fase mixta || ***mixed solvent extraction*** *f.* extracción por disolvente mixto, extracción por solvente

mixer *f.* mezcladora; *m.* mezclador || ***mud mixer*** *m.* mezclador de lodo

mixing circuit *m.* circuito de mezcladura

mixture *f.* mezcla, mixtura

MM *m.* millón (abreviatura)

MMCF *m.* millón de pies cúbicos

mnemonic symbol *m.* símbolo mnemotécnico

moat *m.* foso

mobile *a.* móvil, transportable

mobility *a.* movilidad || ***mobility ratio*** *f.* proporción de movilidad

mobilization *f.* movilización

modal *a.* modal

mode *f.* moda; *m.* modo || ***mode of occurrence*** *f.* manera de presentarse

model *m.* modelo, molde; patrón; *f.* plantilla, muestra || ***numerical model*** *m.* modelo numérico || ***stochastic model*** *m.* modelo estocástico

model *vb.* modelar

modeling *m.* modelaje || ***forward modeling*** *m.* modelaje iterativo || ***reservoir modeling*** *m.* modelaje del yacimiento

modem *m.* módem

moderate *a.* moderado

moderator *m.* regulador, moderador

module *m.* módulo, coeficiente

modulus *m.* módulo || ***bulk modulus*** *m.* módulo de volumen || ***elastic modulus*** *m.* módulo de elasticidad || ***Young's modulus*** *m.* módulo de Young

mofette *f.* mofeta

Mohorovicic discontinuity (Moho) *f.* discontinuidad de Mohorovicic (Moho)

Mohs' scale *f.* escala de Mohs

moist *a.* húmedo

moisture *f.* humedad

moisten *vb.* humedecer

mol *m.* mole, moléculagramo || ***mol percent*** *m.* porcentaje molecular

molar *f.* (chem., quím.) molar, muela; *a.* molar || ***molar solution*** *f.* solución molar || ***molar volume*** *m.* volumen molar || ***molar weight*** *m.* peso molar

molarity *f.* molaridad

molasse *f.* molasa

mold *m.* molde, moho, modelo; *f.* impronta

mold *vb.* moldear

molded *a.* moldeado

moldic *a.* moldico

molding *m.* moldeo

mole *m.* (chem., quím.) mole || ***mole percent*** *m.* porcentaje mole

molecular *a.* molécular || ***molecular sieve*** *m.* tamíz molecular || ***molecular weight*** *m.* peso molecular

molecule *f.* molécula

molluscan *a.* del molusco

mollusk, mollusc *m.* molusco

molten *a.* fundido, derretido

molybdate *m.* molibdato

molybdenite *f.* molibdenita

molybdenum *m.* molibdeno

molybdite *f.* molibdita

moment *m.* momento || ***magnetic moment*** *m.* momento magnético || ***moment of (hydrocarbon) generation*** *m.* momento clave || ***moment of inertia*** *m.* momento de inercia || ***torcion moment*** *m.* momento de torción

momentum *m.* momento, ímpetu; *f.* impulsión, fuerza impulsiva

monadnock *m.* monte reliquia, monte testigo, monadnock, mosor, monte isla, monte residual

monatomic *a.* monoatómico

monaxon *a.* monaxón

monazite *f.* monacita

moncheite *f.* chengbolita

monetize *vb.* monetizar

money *m.* dinero, moneda || ***money order*** *m.* giro postal

moniliform *a.* moniliforme

monitor *m.* (comp.) monitor

monitoring *m.* seguimiento, monitoreo; *f.* supervisión

monkey board *m.* puesto del torrero

monochromatic *a.* monocromático

monoclinal *a.* monoclinal

monocline *m.* monoclinal

monoclinic *a.* monoclínico

monocyclic *a.* monocíclico

monograph *f.* monografía

monolith *m.* monolito

monolithic *a.* monolítico

monomer *m.* mónomero

monomyarian *m.* monomiario

monothalamous *a* monotalámico

monovalent *a.* monovalente

monsoon *m.* monzón

Montian *m. & a.* Montiense, Montiano

monticellite *f.* monticelita

montmorillonite *f.* montmorilonita

monument (survey, levan.) *m.* mojón, hito, monumento; *f.* acotación, guarda raya, mojonera

monument *vb.* ahitar, mojonar

monzonite *f.* monzonita

monzonitic *a.* monzonítico

moon *f.* luna || ***crescent moon*** *f.* luna creciente || ***full moon*** *f.* luna llena || ***new moon*** *f.* luna nueva

moonstone *f.* piedra de luna

moor *m.* marjal, secadal

mooring *m.* fondeo, atraque, anclaje

morrainal *a.* morrénico

moraine *f.* morena; *m.* chancal, canchal || ***end moraine*** *f.* morena terminal || ***frontal moraine*** *f.* morena frontal || ***ground moraine*** *f.* morena de fondo || ***lateral moraine*** *f.* morena lateral || ***recessional moraine*** *f.* morena de retroceso || ***terminal moraine*** *f.* morena terminal, morena frontal

morass *f.* ciénaga; *m.* marjal

morinite *f.* jezekita

morphogenesis *m.* morfogénesis

morphologic *a.* morfológico

morphology *f.* morfología

morphosis *f.* morfosis

morphostructural *a.* morfoestructural

morphotectonics *f.* morfotectónica

mortar *m.* mortero, almirez

mosaic *m.* mosáico || ***airphoto mosaic*** *m.* mosáico aerofotográfico.

mosandrite *f.* johnstrupita, yohnstrupita, khibinita, rinkita, rinkolita

mosasaur *m.* mosasauro

Moscovian *m. & a.* Moscoviano, Moscoviense

moss *m.* moho, musgo

mother liquor *f.* (min.) agua de madre

mother-of-pearl *f.* nácar

motion *m.* movimiento; *f.* moción || ***harmonic motion*** *m.* movimiento armónico || ***periodic motion*** *m.* movimiento periodico

motor *m.* motor || ***motor oil*** *m.* aceite para motor

mottle *vb.* jaspear

mottled *a.* moteado, abigarrado, jaspeado

mottramite *f.* psitacinita

mould *f.* plantilla; *m.* molde

mound *m.* montículo; montón, terraplén; malecón; ribazo

mounded *a.* monticulado

mount *m.* monte || ***sea mount*** *m.* monte marino

mount *vb.* montar

mountain *f.* montaña; *m.* yébel || ***mountain building*** *f.* orogenia; *m.* orogénesis || ***mountain chain*** *f.* cordillera, sierra, cadena de montañas || ***mountain range*** *f.* cordillera, sierra, serranía; *m.* [Arg., Ch.] cordón montañoso || ***table mountain*** *f.* mesa

mountainous *a.* montañoso, alpestre || ***mountainous country*** *f.* serranía

mouse *m.* (comp.) ratón

mouth *f.* boca || ***river mouth*** *f.* boca del río, desembocadura

movable *a.* movible, móvil

movement *m.* movimiento || ***eustatic movement*** *m.* movimiento eustático

moveout *m.* moveout, movimiento hacia afuera; *f.* diferencia en tiempos de llegada entre trazas adyacentes || ***non-hyperbolic moveout*** *m.* moveout no hiperbólico || ***normal (NMO) moveout*** *f.* corrección dinámica

muck *m.* cascote, fango, escombros

muck *vb.* (min.) afrontonar, desescombrar, desalojar escombros

mucker *m.* zafrero

mucking *m.* desescombro

mucronate *a.* mucronato, puntiagudo

mud *m.* lodo, fango, barro, cieno, légamo, limo; *f.* lama, [Arg.] inyección || ***black mud*** *m.* gumbo || ***drilling mud*** *m.* lodo de perforación, bagazo, lodo de circulación || ***gas-cut mud*** *f.* inyección cargada de gas; *m.* lodo gasificado || ***mud bit*** *m.* trépano para formación blanda || ***mud cake*** *f.* costra, cáscara; *m.* revoque || ***mud circuit*** *m.* circuito de inyección || ***mud cracks*** *f.* grietas de desecación; *m.* clizoglifos || ***mud conditioner*** *m.* acondicionador de lodo || ***mud flat*** *f.* llanura de lodo, planicie de barro; *m.* bajío de barro || ***mud flow*** *m.* flujo de fango, torrente de lodo || ***mud gun*** *m.* inyector de lodo || ***mud hole*** *m.* [Ec.] cangilón || ***mud log*** *m.* perfíl de inyección, registro de lodo || ***mud lumps*** *m.* promontorios fangosos, diapiros arcillosos || ***mud off*** *vb.* revocar con inyección, ahogar con inyección || ***mud particles*** *m.* partículas de fango || ***mud pit*** *f.* fosa de lodo || ***mud pump*** *f.* bomba de lodo || ***mud screen*** *f.* zaranda para inyección || ***mud shaker*** *m.* agitador de lodo || ***mud stone*** *f.* lutita, lodolita, arcillita, calcilodita lutolita; *m.* cieno endurecido || ***mud thinnner*** *m.* adelgazador del lodo || ***mud volcano*** *m.* volcán de lodo; *f.* salsa || ***oil-base mud*** *f.* lodo (o inyección) a base de petróleo || ***water-base mud*** *f.* inyección a base de agua

muddy *a.* fangoso, limoso, arcilloso, barroso, lodoso, zurrapiento, zurraposo

mullion structure *f.* estructura de muñón; *m.* mainel

mullite *f.* mullita

multicellular *a.* multicelular

multicolored *a.* policromo

multicostate *a.* multicostillado

multichambered *a.* multilocular, plurilocular

multidisciplinary *a.* multidisciplinario

multiform *a.* multiforme

multinucleate *a.* multinucleado

multiple *a.* múltiple; *m.* (seis.) múltiplo

multiplexing, muxing *f.* multiplexación; *m.* multiplaje

multiplexor *m.* multiplexor

multipurpose *a.* de fines múltiples, propositos multiples

multiserial *a.* multiseriado, multiserial

multispectral *a.* multiespectral

multispiral *a.* multiespiral

multistage *a.* multietapa

multitemporal *a.* multitemporal

multituberculata *m.* multituberculado

multivincular *a.* multivincular

multizone *f. & a.* zona multiple || ***multizone completion*** *f.* terminación en multiples zonas

mummified *a.* momificado

mural *a.* mural || ***mural deposit*** *m.* depósito mural || ***mural pore*** *m.* poro mural

muriatic *a.* muriato || ***muriatic acid*** *m.* ácido muriático

muscle *m.* músculo

muscovite *f.* moscovita, muscovita, mica blanca

mutant *a.* mutante

mutation *f.* mutación

mylonite *f.* milonita

mylonitic *a.* milonítico

mylonitization *f.* milonitización

myophore *m.* mióforo

myriapod *m.* miriápodo

myrmekite *f.* mirmequita, mirmekita

N

nacre *f.* nácar

nacrite *f.* nacrita

nadir *m.* nadir

nadorite *f.* ocrolita

nagyagite *f.* nagyagita, eslamosa

naked eye *a.* a simple vista, a ojo descubierto

Namurian *m. & a.* Namuriense, Namuriano

nanofossil *f.* nanofósil

nanoplankton *m.* nanoplancton

nanosecond *m.* nanosegundo

nantokite *f.* nantoquita

naphtha *f.* nafta || ***naphtha scrubber*** *f.* desnaftadora

naphthalene *m.* naftaleno

naphthene *m.* nafteno || ***naphthene-base crude*** *m.* petróleo crudo de base nafténica

naphthenic *a.* nafténico

nappe *f.* capa cabalgante, capa cobijante, hoja, hoja de arrastre; *m.* cabalgador, nappe, manto

narrow *a.* angosto, estrecho, ahorcado

narrow *vb.* angostarse

narrowing (of a vein, pipe, etc.) *f.* estrangulación; *m.* estrangulamiento

narrowness *f.* angostura, estrechez; *m.* estrechón

narrows *f.* angostura, estrechura; *m.* estrecho, hocino, desfiladero, garganta

national *a.* nacional

nationalization *f.* estatalización, nacionalización

native *a.* nativo, puro, virgen

natrolite *f.* natrolita

natrona *f.* sosa

natrophilite *f.* natrofilita

natural *a.* natural || ***natural bridge*** *m.* puente natural || ***natural gas*** *m.* gas natural, metano || ***natural gas liquids*** *m.* líquidos de gas natural || ***natural history*** *f.* historia natural || ***natural logarithm*** *m.* logaritmo natural || ***natural science*** *f.* ciencia natural || ***natural selection*** *f.* selección natural

nature *f.* naturaleza

naumannite *f.* tascina

nautical *a.* náutico || ***nautical mile*** *f.* milla marina, equivalente a 1,852 km

nautiloid *a.* nautiloideo

nautilus *m.* nautilo

navigation *f.* navegación

neap *a.* marea baja || ***neap tide*** *f.* marea muerta

near, nearby *a.* próximo, cercano

nearby *a.* cercano

nearly *a.* casi

nearshore *a.* costero, cercano a la costa, costeño, costanero

Nebraskan *m. & a.* Nebraskiano, Nebraskiense

nebula *f.* nebulosa

neck *m.* cuello, paso, portezuelo || ***volcanic neck*** *f.* chimenea volcánica; *m.* cuello volcánico, [Méx.] bernal

needle *f.* aguja; (geog.) torre || ***compass needle*** *f.* aguja de brújula

negative *a.* negativo

negotiate *vb.* negociar

negotiation *f.* negociación

nekton *m.* necton

nelsonite *f.* nelsonita

nema *f.* nema

nematoblastic *a.* nematoblástico

nematocyst *m.* nematocisto

Nemausian *m. & a.* Nemausense

neoautochthon *m.* neoautóctono

Neocomian *m. & a.* Neocomiense, Neocomiano, Neócico

neocryst *m. & a.* neocrístico, neocrísto

neodymium *m.* neodimio

neogastropod *m.* neogasterópodo

Neogene *m.* Neógeno, Neoceno

neogenesis *f.* neoformación

neoglaciation *f.* neoglaciación

neohexane *m.* neohexano

neolithic *a.* neolítico

neon *m.* neón

neontologist *m.* neontólogo

neontology *f.* neontología

neopentane *m.* neopentano

neoprene *m.* neopreno, neoprene

neosome *m.* neosomo

neotectonic, neotectonics *m. & a.* neotectónico

neotype *m.* neotipo

neovolcanic *a.* neovolcánico

Neozoic *m. & a.* Neozóico

nepheline *f.* nefelina, eleolita, nefelinita, nefelita; *m.* nefelino || ***nepheline seyenite*** *f.* miascita

nephelinite *f.* nefelinita

nephrite *f.* nefrita; *m.* iu

nepionic *a.* nepiónico

Neptune *m.* Neptuno

neptunian dike *m.* dique clástico

Neptunism *m.* wernerismo, neptunismo

neptunite *f.* neptunita

neptunium *m.* neptunio

neritic *a.* nerítico, sublitoral

nerve *m.* nervio

Nervian *m. & a.* Nerviense, Nerviano

nesosilicate *m.* nesosilicato

nest *vb.* enchufar, anidar

nested *a.* anidado

net *f.* red; *m.* (com.) neto || ***net earnings*** *f.* ganancia neta || ***net-like*** *a.* reticular || ***net pay*** *f.* producción neta || ***net thickness*** *m.* espesor neto || ***net weight*** *m.* peso neto || ***stereographic net*** *f.* red estereograma || ***triangulation net*** *f.* (survey, levan.) red de triángulos

netlike *a.* reticular

network *f.* malla, red, red alviolar; *m.* enmallamiento || ***fracture network*** *f.* red de fracturas

Neudeckian *m. & a.* Neudeckense, Neudeckiano

neutral *a.* neutral || ***neutral axis*** *m.* eje neutro || ***neutral fiber*** *f.* fibra neutra

neutralize *vb.* neutralizar

neutralization *f.* neutralización

neutralized *a.* neutralizado

neutralizer *m.* neutralizador

neutralizing clay *f.* arcilla neutralizador
neutrino *m.* neutrino
neutron *m.* neutrón || ***neutron log*** *m.* registro de neutrón
Nevadan *m. & a.* Nevadiense, Nevadiano
nevadite *f.* nevadita
new *a.* nuevo || ***new (business) ventures*** *f.* empresas nuevas
next *a.* próximo, subsecuente, siguiente
NGLs (natural gas liquids) *m.* liquidos de gas natural
Niagaran *m. & a.* Niagarense, Niagarano, Niagariense
niccolite *f.* nicolita, ñiquelina, pirita roja
niche *m.* nicho
nicholsonite *m.* zinkaragonito
nickel *m.* níquel, níkel || ***nickel-bearing*** *a.* niquelífero || ***nickel pyrite*** *f.* whartonita || ***nickel sulfide*** *f.* copaquira
nickel bloom *f.* annabergita
nickeline *f.* niquelina
nicol *m.* nicol || ***crossed nicols*** *m.* nicoles cruzados || ***nicol prism*** *f.* prisma de Nicol
night *f.* noche; *a.* nocturno
nimbostratus *m.* nimboestrato
nimbus *m.* nimbo
niobium *m.* niobio
niobophyllite *f.* niobofilita
niter *m.* nitro, salitre || ***soda niter*** *f.* nitratina, nitratita
nitrate *m.* nitrato, nitro, azoato; *f.* [Bol., Pe] collpa || ***ammonium nitrate*** *m.* nitrato de amonio || ***nitrate deposit*** *m.* nitral || ***silver nitrate*** *m.* nitrato de plata
nitric *a.* nítrico, azótico || ***nitric acid*** *m.* ácido nítrico, agua fuerte
nitride *m.* nitruro
nitrite *m.* nitrito
nitrocellulose *f.* nitrocelulosa
nitrogen *m.* nitrógeno, ázoe || ***nitrogen dioxide*** *m.* dióxido de nitrógeno
nitroglycerin *f.* nitroglicerina
nitrous *a.* nitroso
nivation *f.* nivación
nobelium *m.* nobelio
noble *a.* noble || ***noble gas*** *m.* gas noble
nodal *a.* nodal
node *m.* nodo
nodular *a.* nodular, arriñonado, noduloso
nodule *m.* nódulo, papa; *f.* haba
noise *m.* ruido, zumbido; *f.* bulla || ***noise level*** *m.* nivel de ruidos
noisy *a.* ruidoso
nomenclature *f.* nomenclatura
nominal *a.* teórica
nomogram, nomograph *m.* nomograma, ábaco, nomogramo
nonadecane *m.* nonadecano
nonane *m.* nonano
noncommercial *a.* no comercial
nonconformable *a.* inconforme, discordante
nonconformity *f.* no concordancia, discordancia
noncorrosive *a.* no corrosivo
nondestructive *a.* no destructivo
nonferrous *a.* no ferroso
nonmetallic *a.* no metálico
non-Newtonian behavior *m.* comportamiento no newtoniano
nonorogenic *a.* anorogenético, anorogénico
non-polar *a.* no polar
nonporous *a.* no poroso
non-producing *a.* improductivo
nonrenewable *a.* no renovable
nonselective *a.* no selectivo
nonseptate *a.* sin septos, sin tabiques
nontoxic *a.* no tóxico
nontronite *f.* nontronita
non-unique *a.* no único
nonvolatile *a.* no volátil
nook *m.* rincón
norbergite *f.* norbergita
nordmarkite *f.* nordmarkita
Norfolk *m. & a.* Norfolkense, Norfolkiano
Norian *m. & a.* Noriense, Noriano
norilskite *f.* norilsquita
norite (diorite) *f.* norita, klausenita
norm *f.* norma, pauta

normal *a.* normal, (angulo) perpendicular

normal *f.* perpendicular, normal || ***normal fault*** *f.* falla normal || ***normal incidence*** *f.* incidencia normal || ***normal moveout (NMO)*** *f.* corrección dinamica || ***normal stress*** *m.* esfuerzo perpendicular || ***optic normal*** *f.* normal óptica

normally *adv.* normalmente

normalization *f.* normalización

normalize *vb.* normalizar

normalizing *f.* corrección al nivel de referencia

normative *a.* normativo

nontronite *f.* pinguita, zamboninita

north *m.* norte || ***magnetic north*** *m.* norte magnético || ***true north*** *m.* norte geográfico, norte verdadero

North America *f.* América del Norte

northern *a.* septentrional

nosean, noselite *f.* nosena, noselita

noseanolith *m.* noseanolíto

nose *f.* nariz; *m.* espolón || ***anticlinal nose*** *f.* nariz anticlinal || ***nose of a fold*** *f.* nariz del pliegue || ***plunging nose*** *f.* nariz buzante

notary public *m.* notario

notation *f.* notación

notch *f.* muesca, ranura, hendedura, mella, terraja; *m.* corte, encaje

note *f.* nota, anotación || ***survey notes*** *f.* minutas, libreta de campo

notebook *f.* libreta; *m.* cuaderno

notice *m.* anuncio, aviso

not indurated *a.* no indurado

notochord *m.* notocordio; *f.* notocorda

novaculite *f.* novaculita

novaculitic *a.* novaculítico

nubbin *m.* tocón, mugrón

nuclear *a.* nuclear

nucleoconch *f.* nucleoconcha

nucleogenesis *f.* nucleogénesis

nucleus *m.* núcleo

nuclide *m.* nuclideo

nuée ardente *f.* nube ardiente

Nuevoleonian *m. & a.* Nuevoleoniense, Nuevoleoniano

nugget *f.* pepita, pepa; *m.* [Vz.] cochano || ***gold or silver nugget*** *f.* roela

null *m. & a.* nulo, cero

number *m.* número || ***atomic number*** *m.* número atómico || ***digital number*** *m.* número digital || ***random number*** *m.* número al azar

numerical *a.* numérico

nummulite *f.* numulita

Nummulitic *m. & a.* Numulítico

nunatak *m.* nunatak

nutrient *m.* nutrimento, nutriente

nutrition *f.* nutrición

nylon *m.* nylon

nymph *f.* ninfa

O

oasis *m.* oasis

obduction *f.* obducción

object *m.* objeto

objective *m. & a.* objetivo, objectivo, propósito; *f.* expectativa || ***objective lens*** *m.* lente objetivo

oblate *a.* oblado

obligation *m.* gravamen; *f.* obligación

oblique *a.* oblicuo

obliqueness *m.* sesgo

oblong *a.* oblongo, alargado

obscure *a.* oscuro

obscure *vb.* obscurecer, enmascarar

obsequent *a.* obsecuente || ***obsequent stream*** *m.* arroyo obsecuente

observation *f.* observación

observatory *m.* observatorio

observe *vb.* observar

observed *a.* observado

observer *m.* observador

obsidian *f.* obsidiana, marecanita; *m.* vidrio volcánico, [Méx.] itztle, itztli, tzinapu

obsolete *a.* anticuado

obstacle *m.* obstáculo, impedimento; *f.* valla

obstruct *vb.* obstruir

obstructed *a.* obstruido

obstruction *f.* obstrucción

obtuse *a.* obtuso

occasional *a.* ocasional, esporádico

occidental *a.* occidental

occluded *a.* ocluído || ***occluded gas*** *m.* gas absorbido

occupational *a.* laboral || ***occupational disease*** *f.* enfermedad laboral

occur *vb.* ocurir, armar, disponerse en, presentarse, pasarse

occurrence *f.* existencia, ocurrencia, presencia

ocean *m.* océano, mar || ***Antarctic Ocean*** *m.* océano Glacial Antártico || ***Arctic Ocean*** *m.* océano Glacial Artico || ***Atlantic Ocean*** *m.* océano Atlántico || ***Indian Ocean*** *m.* océano Indico || ***Pacific Ocean*** *m.* océano Pacífico || ***ocean floor*** *m.* fondo (o piso) oceánico || ***ocean trench*** *f.* fossa oceánica, fosa abisal

oceanic *a.* oceánico

oceanographic *a.* oceánico, oceanográfico

oceanography *f.* oceanografía

ocher, ochre *m.* & *a.* ocre || ***yellow ocher, ochre*** *m.* sil

Ochoan *m.* & *a.* Ochoense, Ochoano

octagon *m.* octágono

octagonal *a.* octágono

octahedrite *f.* octaedrita, anatasa

octahedron *m.* octaedro

octane *m.* & *a.* octano

octaphyllite *f.* octafilíta

ocular *m.* & *a.* ocular

odd *m.* & *a.* impar

odometer *m.* odómetro, cuentapasos

odor *m.* olor

oersted *m.* oersted, oerstedio = gauss = 100 gammas

offer *f.* oferta, propuesta

office *f.* oficina; *m.* despacho || ***branch office*** *f.* sucursal

offlap *m.* traslapo regresivo

off-line *a.* fuera de línea

off of *a.* fuera de

offset *m.* & *a.* excéntrico, sobresaliente, rechazo, desplazamiento, resalte, resalto; salto horizontal; *f.* derivación || ***offset well*** *m.* pozo contrarestante, pozo compensador, pozo adyacente

offshore *f.* costafuera, costa afuera; *a.* marino || ***offshore bar*** *m.* cordón litoral

off-structure *f.* fuera de la estructura

ohm *m.* ohmio

ohmmeter *m.* ohmímetro

oikocryst *m.* oikocrist

oil *m.* aceite, petróleo; *a.* petrolero || ***asphalt-based oil*** *m.* aceite crudo de base asfáltica, crudo nafténico || ***bunker oil*** *m.* petróleo para calderas || ***coal oil*** *m.* aceite de carbón || ***crankcase oil*** *m.* aceite de cárter || ***crude oil*** *m.* aceite bruto, aceite crudo; *f.* nafta || ***cutting oil*** *m.* aceite para corte; aceite para fresar || ***dead oil*** *m.* petróleo muerto; aceite muerto; aceite de creosota || ***drilling oil*** *m.* aceite de taladrar || ***fuel oil*** *m.* aceite combustible, mazut, aciete de horno, combustoleo || ***heating oil*** *m.* petróleo para calefacción || ***heavy oil*** *m.* aceite pesado || ***insulating oil*** *m.* aceite aislante || ***light oil*** *m.* petróleo ligero || ***lubricating oil*** *m.* aceite lubricante || ***mineral oil*** *m.* aceite mineral, petróleo || ***motor oil*** *m.* aceite para motor || ***oil accumulation*** *f.* acumulación de petróleo, acumulación petrolífera || ***oil base mud*** *m.* lodo a base de petróleo || ***oil-bearing*** *a.* petrolífero || ***oil bloom*** *f.* fluorescencia del petróleo || ***oil boom*** *f.* fiebre del petróleo || ***oil camp*** *m.* campamento petrolero || ***oil can*** *f.* aceitera || ***oil cut*** *m.* porcentaje de petróleo || ***oil emulsion*** *f.* emulsión de petróleo || ***oil field*** *m.* campo petrolero || ***oil flushing*** *m.* aceite de enjuagar || ***oil/gas contact*** *m.* contacto gas petróleo || ***oil geologist*** *m.* geólogo de petróleo || ***oil leg*** *f.* zona productora de petróleo || ***oil pipeline*** *m.* oleoducto || ***oil pool*** *m.* criadero de petróleo, yacimiento productor, campo petrolífero || ***oil sand*** *f.* arena petrolífera || ***oil seep*** *f.* manifestación, filtración de petróleo, exudación de petróleo, [Méx.] chapopotera; *m.* zumadero de petróleo, resumidero || ***oil shale*** *f.* lutita petrolífera, kerogenita, turbonita || ***oil show*** *m.* rastro de petróleo || ***oil sump*** *f.* batea de aceite || ***oil tar*** *m.* alquitrán de petróleo, aceite de alquitrán || ***oil/water contact*** *m.* contacto petróleo-agua || ***oil well*** *pozo* de petróleo || ***oil window*** *f.* ventana de generación de petróleo || ***original oil in place (OOIP)*** *m.* petróleo original in-situ || ***parafinic oil*** *m.* aceite parafinico, petróleo lampante || ***raw oil*** *m.* aceite bruto || ***reclaimed or recycled oil*** *m.* aceite recuperado (o reciclado) || ***residual oil*** *m.* aceite residual || ***shale oil*** *m.* aceite de esquisto, aceite de lutita || ***viscous oil*** *m.* aceite viscoso || ***volatile oil*** *m.* aceite volátil || ***wax-free oil*** *m.* aceite sin parafina || ***waxy oil*** *m.* aceite parafínico || ***weathered oil*** *m.* petróleo intemperizado || ***wet oil*** *m.* petróleo húmedo

oily *a.* aceitoso, oleaginoso

old *a.* viejo, antiguo, vetusto, anciano || ***old age*** *f.* senectud, vejez

oleaginous *a.* oleaginoso, oleoso

olefin *m.* olefina

olefinic *a.* olefínico

Olenidian *m. & a.* Johannense, Johanniano

Olenkian *m. & a.* Olenkiense, Olenkiano

Oligocene *m. & a.* Oligoceno, Oligocénico || ***middle Oligocene*** *m. & a.* Jovarense

oligoclase *f.* oligoclasa

oligomictic *a.* oligomíctico

oligotrophic *a.* oligotrófico

oligotrophy *f.* oligotrofía

olistoglyph *m.* olisoglifo

olistostrome *f.* olistostroma

olivine *f.* olivina; *m.* peridoto, olivino

olivinite *f.* olivinita

ombrogenous *a.* ombrogénuo

omission *f.* omisión

omphacite *f.* omfacita

on *a.* sobre

oncolite *m.* oncolito

onlap *m.* traslape; *f.* superposición transgresiva

on-line *a.* en línea, en conexión directa, en corriente, en funcionamiento, en operación

Onondagan *m. & a.* Onondagiense, Onondagiano

onshore *a.* en tierra, terrestre

on-site *a.* in situ, en sitio

on-stream *a.* en funcionamiento, en operación, en corriente

on-structure *a.* en la estructura

Ontarian *m. & a.* Ontariano, Ontariense

ontogenesis *m.* ontogénesis; *f.* ontogenía

onyx *m.* ónice, ónix

ooid *m.* ooide

oolite *f.* oolita

oolith *m.* oolito

oolitic *a.* oolítico

oolitization *f.* oolitización

ooze *m.* fango, limo, légamo, lodo || ***calcareous ooze*** *m.* lodo calcáreo || ***diatomaceous ooze*** *m.* lodo de diatomeas || ***radiolarian ooze*** *m.* lodo de radiolarios || ***siliceous ooze*** *m.* lodo silíceo

ooze *vb.* resumar

opacite *f.* opacita

opacity *f.* opacidad

opal *m.* ópalo, [Méx.] mezcuitlatl; *f.* cachalonga || ***fire opal*** *m.* girasol; *f.* pirofana

opalescence *f.* opalescencia

opalescent, opaline *a.* opalescente, opalino

opalized *a.* opalado, opalizado || ***opalized wood*** *m.* xilópalo

opaque *a.* opaco, oscuro, obscuro

open *a.* abierto || ***open acreage*** *m.* área disponible || ***open cut*** *f.* excavación a cielo abierto, trinchera; *m.* [Pe.] destaje, destajo || ***open hole*** *m.* pozo franco, pozo sin entubar || ***open flow*** *m.* a pleno flujo; *f.* surgencia libre, surgenica sin restricción || ***open flow test*** *m.* ensayo de surgencia libre || ***open fold*** *m.* pliegue abierto || ***open hole pressure*** *f.* presión a pozo abierto || ***open pit*** *f.* excavación a cielo abierto; tajo abierto || ***open sight alidade*** *f.* alidada de pínula o mira abierta || ***open space*** *m.* espacio libre, aire libre || ***open terrain*** *f.* anchura || ***open vein*** *f.* abra

open *vb.* abrir
opening *f.* abertura, apertura, cavidad; abra; (min.) abridura, cholla; *m.* vano, abrimiento, orificio, portillo, tajo
operate *vb.* operar, accionar
operated acreage *m.* área en desarrollo, área en operación
operating *a.* en operación || ***operating condition*** *f.* condición de operación || ***operating cost*** *m.* costo de operación
operating cycle *m.* ciclo de funcionamiento, período de funcionamiento
operating pressure *f.* presión efectiva, presión de trabajo
operation *f.* operación; *m.* funcionamiento
operational *a.* operacional
operator *m.* operario, operador; dueño, administrador, [Bol.] azoguero; *f.* compañía operadora
ophicalcite *f.* oficalcita
ophiolite *f.* ofiolita
ophiolitic *a.* ofiolítico
ophite *f.* ofita
ophitic *a.* ofítico
Opoka *f.* Opoca
opportune *a.* enderezado, oportuno
opportunity *f.* oportunidad
opposite *a.* opuesto
optic, optical *a.* óptico || ***optic angle*** *m.* ángulo óptico || ***optic axis*** *m.* eje óptico || ***optic normal*** *f.* normal óptica || ***optic plane*** *m.* plano óptico || ***optical scanner*** *m.* lector óptico || ***optical sight*** *m.* visor || ***optic sign*** *m.* signo óptico
optics *f.* óptica
optimal *a.* óptimo
optimization *f.* optimización
optimum *a.* óptimo
option *f.* opción
orbicule *m.* orbículo
orbit *f.* órbita
order *m.* pedido, orden, mando, mandato
order *vb.* pedir, exigir
orderly *a.* regular
ordinary *a.* regular
ordonezite *f.* ordoñezita
Ordovician *m. & a.* Ordoviciano, Ordovicense, Ordovícico
ore *m.* mineral, clavo, fruto, metal, metal en bruto; *f.* mena || ***coarse ore*** *m.* cuesco || ***cobbed ore*** *m.* limpio; *f.* pepena || ***crude ore*** *m.* quimbalete || ***disseminated ore*** *m.* mena diseminada || ***dressed ore*** *m.* [Sp.] pepenado || ***fine ore*** *f.* gandinga || ***high-grade ore*** *m.* mineral de alta ley, sábado || ***iron ore*** *m.* mineral de hierro || ***low-grade ore*** *m.* mineral de baja ley, desecho, ordinarios, [Pe.] brocado; *f.* limpia, [Bol.] broza || ***lump ore*** *f.* [Ch.] colpas || ***nodular ore*** *m.* riñón || ***ordinary ore*** *m.* metal común || ***ore amenable to amalgamation*** *m.* azogue || ***ore bag*** *f.* bota || ***ore bin*** *f.* cancha, tolva || ***ore body*** *m.* cuerpo || ***ore bucket*** *f.* [Méx.] calesa, caleza; *m.* chicuite || ***ore buyer*** *m.* rescatador, [Pe.] maquilero || ***ore car*** *m.* vagón || ***ore carrier*** *m.* apiri || ***ore chamber*** *m.* salón || ***ore chunks, fragments*** *m.* gabarro; *f.* [Méx.] granza || ***ore chute*** *f.* bonanza; cámara, chorrera, [Méx.] caíman; *m.* clavo, coladero, chorreadero, resbaladero, rumbadero || ***ore concentrate*** *f.* cabezuela; *f.* mena concentrada || ***ore deposit*** *f.* galera, [Pe.] acomodana; *m.* yacimiento mineral, acumulador de mineral, criadero || ***ore dust*** *m.* finos || ***ore fines*** *m.* ripios || ***ore grade*** *f.* ley || ***ore hopper*** *m.* chifladero; *f.* tolva || ***ore layer*** *f.* capa mineralífera || ***ore nodule*** *m.* boleo || ***ore pile*** *f.* trinchera; *m.* [Méx.] millón || ***ore pipe*** *f.* columna || ***ore pocket*** *f.* ampolla, bolsa, [Pe.] boya; *m.* nido, remolino, tope, [Pe.] bollo, [Col.] buche || ***ore pod*** *m.* ojo || ***ore sack*** *f.* costalera || ***ore seam*** *f.* hijuela || ***ore sorter*** *m.* limpiador, palliri, partidor, quebrador, [Bol., Pe.] pallador; *f.* khareadora || ***ore streak*** *f.* cinta || ***ore thief*** *m.* [Pe.] juquero || ***ore thread*** *m.* hilo || ***ore way*** *m.* way || ***oxide ores*** *m.* colorados || ***pay ore*** *m.* mineral beneficiable || ***placer ore*** *m.* metal corrido || ***purchased ore*** *m.* rescate || ***residual or remaining ore*** *f.* rezaga; *m.* rezago || ***rich ore*** *f.* bolsada, bolsilla, bolsón, bonanza [Bol.] achura; *m.* clavo || ***silver ore*** *f.* tacana; *m.* tacano || ***smelting ore*** *m.* metal de fuego, metal de fundición; pepena || ***sorted ore*** *m.* limpio; *f.* pepena || ***tailings ore*** *m.* polvillo, [Bol., Pe.] pallaco || ***undressed ore*** *f.* grena

Orellan *m. & a.* Orelliense, Orelliano

organic *a.* orgánico, zoógeno || ***organic sediment*** *m.* biolito

organism *m.* organismo

organization *f.* organización

organogenic *a.* organógeno

organosedimentary *a.* organosedimentario

orient *vb.* orientar

oriental *a.* oriental

orientation *f.* orientación || ***crystallographic orientation*** *f.* orientación cristalográfica || ***preferred orientation*** *f.* orientación preferida

oriented *a.* orientado || ***oriented core*** *m.* núcleo orientado

orifice *m.* orificio

origin *m.* origen

Oriskanian *m. & a.* Oriskaniense, Oriskaniano

orocline *m.* oroclino

orocratic *a.* orocrático

orogen *m.* orógeno

orogenesis *m.* orogénesis; *f.* orogénia

orogenetic *a.* orogenético

orogenic *a.* orogénico || ***orogenic belt*** *m.* orogen, orógeno, cinturón orogénico

orogeny *f.* orogénia, levantada, revolución; *m.* orogénesis

orographic *a.* orográfico

orography *f.* orografía

orpheite *f.* orfeíta

orpiment *f.* arsenoblenda amarilla, blenda arsenical; *m.* orpimente, azarnefe, sublimado amarillo

orthite *f.* ortita, allanita

orthoandesite *f.* ortoandesita; *m.* sanukitoide

orthoaxis *m.* orto-eje

orthochlorite *f.* ortoclorita

orthoclase *f.* ortoclasa, ortosa, pegmatolita

orthoclasite *f.* ortoclasita

orthoclastic *a.* ortoclástico

orthoconglomerate *m.* ortoconglomerado

orthogneiss *m.* ortogneis, ortoneis

orthogonal *a.* ortogonal

orthographic *a.* ortográfico || ***orthographic projection*** *f.* proyección ortográfica

orthomagmatic *a.* ortomagmático

orthometric *a.* ortométrico

orthophyre *m.* ortófiro; *f.* ortofelsita

orthophyric *a.* ortofírico

orthopyroxene *m.* ortopiroxeno

ortopyroxenite *f.* ortopiroxenita

orthoquartzite *f.* ortocuarcita

orthoquartzitic *a.* ortocuarcítico

orthorhombic *a.* ortorómbico

orthosilicate *m.* ortosilicato

orthosite *f.* ortosita

orthostratigraphy *f.* ortoestratigrafía

oscillate *vb.* oscilar, vibrar, fluctuar

oscillation *f.* oscilación; *m.* movimiento bascular

oscillator *m.* oscilador

oscilloscope *m.* osciloscopio

osmiridium *m.* osmiridio

osmium *m.* osmio

osmosis *f.* ósmosis

osmotic *a.* osmótico

ossification *f.* osificación

ossify *vb.* osificar || ***to become ossified*** *vb.* osificarse

ostracod *m* . ostrácodo, ostracoide

ounce *f.* onza = 28,3 gramos

out *adv. & p.* fuera, afuera || ***out of operation*** *a.* fuera de operación || ***out of order*** *a.* descompuesto || ***out of phase*** *a.* desfasado, fuera de fase || ***out of sequence*** *a.* fuera de secuencia

outcrop *m.* afloramiento, crestón, asomo, reventón, reventadero, quemazón, [Cu.] brote; *f.* montera, salida, [Méx.] cabeza, [Pe.] eflorescencia || ***small outcrop*** *m.* apuntamiento

outcrop *vb.* aflorar, arrumbar, brotear

outer *a.* exterior, externo

outfit *m.* equipo, apero, apresto; *f.* habilitación

outfit *vb.* equipar, habilitar

outfitting *f.* habilitación

outflow *m.* efluente, flujo; *f.* surgencia, salida

outgas *vb.* desgasificar

outlet *m.* escape, orificio, emisario; conectador; jaguadero; *f.* salida, boca, toma de corriente, caja de salida || ***glacial outlet*** *m.* portón de glaciar || ***river outlet*** *f.* desembocadura, boca del río

outlier *m.* testigo de erosión; *f.* roca apartada, roca extraña

outline *m.* bosquejo

outline *vb.* bosquejar

outlook *f.* perspectiva

outlying *a.* distante de, lejos de

out of sequence *a.* fuera de secuencia

output *m.* abasto (min.); rendimiento, volumen de producción, productos (comp.) datos de salida || ***effective output*** *f.* producción neta

outside *a.* exterior, externo; *adv.* afuera, fuera

outside of *p.* fuera de

outstanding *a.* saliente, sobresaliente

outward *a.* hacia fuera, por fuera, exterior, externo

outwash *f.* deslava; *m.* aluvión, acarreo || ***glacial outwash*** *m.* aluvión fluvio-glacial || ***outwash plain*** *f.* llanura de acarreo, llanura aluvial

oven *m.* horno

over *p.* sobre

overburden *f.* sobrecapa, cubierta, cobertura, cubertura, roca sobre carga; *m.* desmonte, escombro, relleno exterior, manto

overcharge *f.* sobrecarga; *m.* recargo de precio

overcharge *vb.* sobrecargar, recargar el precio

overflow *f.* inundación, corredura, derivación, riada; *m.* derrame, desbordamiento, exceso; aliviadero, rebosadero, rebosamiento

overflow *vb.* rebosar, desbordar(se), derramar(se), llenar

overflowing *m.* desbordamiento, desborde; *a.* desbocado

overgrowth *f.* precipitación de minerales en el sistema de porosidad; *m.* sobrecrecimiento crecimiento excesivo

overhang *f.* protuberancia, comisa; *m.* alero

overhang *vb.* sobresalir

overhanging *a.* sobresaliente, colgante, volado, voladizo

overhaul *m.* reajuste

overhaul *vb.* rehabilitar, reacondicionar, componer, reajustar, repasar, reparar

overhead *a.* hacia arriba, arriba en lo alto; *m.* gastos fijos || ***overhead stope*** *vb.* trabajar de cielo, [Pe.] tajear

overheat *vb.* recalentar

overlap *f.* superposición, sobreposición, imbricación; *m.* traslape, resbalamiento, resbalón, traslapamiento

overlap *vb.* solapar, traslapar, sobreponer, cabalgar, imbricar

overlapping *a.* transgresivo, solapamiento, solapado, imbricado, en echelon, resbalado, solapante, sobrepuesto, superpuesto

overlay *f.* hoja superpuesta, mica; *m.* mapa superpuesto, acetate

overlay *vb.* sobreponer

overlie *vb.* suprayacer, descansar sobre

overload *f.* sobrecarga

overload *vb.* sobrecargar

overlying *a.* sobreyacente, suprayacente

overpressure *f.* superpresión, sobrepresión

overprint *f.* sobreimpresión

overprint *vb.* imprimir sobre, sobreimprimir

overproduced *a.* explotada excesivamente

overproduction *f.* sobreproducción

oversize *a.* sobremedido

oversteepened *a.* sobreempinado

overthrust *m.* sobre-corrimiento, corrimiento, manto de sobreescurrimiento, acabalgamiento, cabalgamiento, sobreempuje; *f.* cabalgadura, cobijadura

overtime *f.* horas extras; *m.* tiempo extra

overturn *vb.* acostar, voltear, volcar

overturned *a.* tumbado, invertido, volcado

overturning *m.* vuelco

overweight *a.* sobrepeso

ovoid *a.* ovoide, ovoideo, oviforme

owner *m.* dueño, propietario

ownership *f.* propiedad

owyheeite *f.* warrenita

oxalic *a.* oxálico

oxammite *f.* oxamita

oxbow lake *f.* vega, resaca, vuelta; *m.* brazo muerto, lago semilunar, lago en herradura

Oxfordian *m. & a.* Oxfordiano, Oxfordiense; Neuvizense

oxidant *m.* oxidante

oxidate *m.* oxidato

oxidation *f.* oxidación

oxide *m.* óxido || ***ferric oxide*** *m.* óxido férrico || ***iron oxide*** *m.* óxido de hierro || ***manganese oxide*** *m.* óxido de manganeso || ***oxide ore*** *m.* colorados, óxidos

oxidize *vb.* oxidar

oxidized *a.* oxidado

oxidizing *a.* oxidante

oxyacetylene *m. & a.* oxiacetileno

oxygen *m.* oxígeno || ***oxygen breathing apparatus*** *m.* aparato respirador de oxígeno

oxygenated *a.* oxigenado

oxygenation *f.* oxigenación

oxyhornblende *f.* oxihornblenda

oxymesostasis *f.* oximesostasis

oyamalite *f.* yamagutilita

ozone *m.* ozono

P

pace *m.* paso

pachydont *m. & a.* paquidonto

pack, backback *f.* mochila

pack *vb.* empaquetar, cargar, envasar

package of rock *m.* paquete de roca, bulto; conjunto

packer *m.* tapón, aguijón; obturador de empaque || ***casing packer*** *m.* obturador del tubo revestidor || ***dual completion packer*** *m.* obturador para pozo a dos zonas || ***tubing packer*** *m.* obturador para tubería || ***wall packer*** *m.* obturador de pared

pack-ice *f.* banquisa; *m.* pack

packing *m.* embalaje, envase, encajonamiento; empaque, espaciador; *f.* empaquetadura || ***crystal lattice packing*** *m.* apilamiento

packstone *f.* caliza lodogranular

pad *m.* relleno; cojín; mango; *f.* almohadilla

paedogenesis *f.* pedogénesis

pahoehoe lava *f.* lava pahoehoe

palagonite *f.* palagonita

palagonization *f.* palagonización

palasome *m.* palasomo

Palatinian *m. & a.* Palatiense, Palatiniano

pale *a.* pálido

paleo *a.* antiguo, viejo

paleoaquifer *m.* paleoacuífero

paleoautochthon *m.* paleoautóctono, paleoautocton

paleobathymetric *a.* paleobatimétrico

paleobiogeography *f.* paleobiogeografía

paleobiology *f.* paleobiología

paleobotany *f.* paleobotánica

paleobotanist *m.* paleobotánico

paleobotanical *a.* paleobotánico

Paleocene *m. & a.* Paleoceno, Paleocénico

paleochannel *m.* paleocanal

paleoclimate *m.* paleoclima

paleoclimatic *a.* paleoclimático

paleoclimatology *f.* paleoclimatología

paleodrainage *m.* paleodrenaje

paleoecology *f.* paleoecología

paleoecologist *m.* paleoecólogo

paleoecological *a.* paleoecológico

paleoequator *m.* paleoecuador

Paleogene *m. & a.* Paleógeno

paleogeographic *a.* paleogeográfico

paleogeography *f.* paleogeografía

paleogeologic *a.* paleogeológico

paleogeomorphology *f.* paleogeomorfología

paleohydrology *f.* paleohidrología

paleokarst *m.* paleocarst

paleolithic *a.* paleolítico

paleomagnetic *a.* paleomagnético
paleomagnetism *m.* paleomagnetismo
paleontologic *a.* paleontológico
paleontologist *m. & f.* paleontólogo
paleontology *f.* paleontología
paleophytology *f.* paleofitología
paleorelief *m.* paleorelieve
paleosol *m.* paleosol
paleotectonic *a.* paleotectónico
paleotemperature *f.* paleotemperatura
paleothermal *a.* paleotermal
paleothermometry *f.* paleotermometría
paleotopography *f.* paleotopografía
paleowind *m.* paleoviento
Paleozoic *m. & a.* Paleozóico, Eozóico, era primaria
paleozoology *f.* paleozoología
palimpsest *m.* palimpsesto
palingenesis *m.* palingénesis
palinspastic *a.* palinspástico || ***palinspastic map*** *m.* mapa palinspático
palintrope *m.* palíntropo
palisade *f.* palizada
palladium *m.* paladio
pallasite *f.* palasita
pallet *f.* paleta
palm *f.* palma
palmate *a.* palmado
paludal *a.* palustre, palúdico, cenagoso, pantanoso, paludal
palus *m.* mástil
palygorskite *f.* paligorskita
palynological *a.* palinológico
palynologist *m.* palinólogo
palynology *f.* palinología
palynomorph *m.* palinomorfo
pampas *f.* pampa
pan *m.* solar;*f.* playa || ***gold pan*** *f.* [Arg., Ch.] poruña
pan *vb.* (min.) aflojar, lavar
panel *m.* panel, tablero || ***display panel*** *m.* panel explicativo || ***instrument panel*** *m.* tablero de instrumentos || ***solar panel*** *m.* panel solar
panethite *f.* panetita
panidiomorphic *a.* panidiomorfico
Pannonian *m. & a.* Pannoniense, Pannoniano
panorama *f.* panorama
pantodont *m. & a.* pantodonto
papil, papilla *f.* papila
papillate, papillose *a.* papilar; papiloso, verrugoso
paraconformity *f.* paraconcordancia
parabituminous *a.* parabituminoso
parabolic *a.* parabólico
Paracelsian *m. & a.* Paracelsiano
parachronology *f.* paracronología
parachrysotile *f.* paracrisotil
paraclase *f.* paraclasa, falla
paraconformable *a.* paraconcordante
paraconformity *f.* paraconcordancia
paradiagenetic *a.* paradiagenético
paradigm *m.* paradigma
paraffin *f.* parafina || ***native paraffin*** *f.* ozoquerita
paraffinic *a.* parafínico
paragenesis *m.* paragénesis
paragenetic *a.* paragenético
parageosyncline *m.* parageosinclinal
paragneiss *m.* paragneiss, paraneiss
paragon *m.* paragón
paragonite *f.* paragonita
paralic *a.* parálico, marginal
parallax *f.* paralaje
parallel *a.* paralelo || ***parallel extinction*** *f.* extinción paralela || ***parallel flow*** *m.* flujo laminar || ***parallel fold*** *m.* pliegue paralelo
parallelogram *m.* paralelogramo
parallochthon *m.* paralóctono
paramagnetic *a.* paramagnético
paramagnetism *m.* paramagnetismo
parameter *m.* parámetro
parametric *a.* paramétrico
paramorph *m.* paramorfo
paramorphism *m.* paramorfismo
parapodia *m.* parapodios

parapsid *m. & a.* parápsido

para-ripple *f.* para-rizadura, para-óndula

parasequence *f.* parasecuencia

parasite *m.* parásito

parasitic *a.* parasítico

parasitism *m.* parasitismo

parastratigraphic *a.* paraestratigráfico

parastratigraphy *f.* paraestratigrafía

paratectonics, paratectonic *m. & a.* paratectónico

paratellurite *f.* paratelurita

paratheca *f.* parateca

paratype *m.* paratipo

paraunconformity *f.* paradiscordancia

parautochthon, parautochthonous *m. & a.* parautóctono

paravane *m.* paravan

parawollastonite *f.* paravolastonita

parcel *m.* bulto; *f.* parcela

parchettite *f.* parchetita

parent rock *f.* roca madre

paroxysm *m.* paroxismo

paroxysmal *a.* paroxismo, paroxismal

pargasite *f.* pargasita

parietal *a.* parietal

park *m.* parque

part *f.* parte, pieza

partial *a.* parcial || ***partial pressure*** *f.* presión parcial

particle *f.* partícula || ***alpha particle*** *f.* partícula alfa || ***mud particle*** *f.* partícula de fango

parting *f.* división, separación, fractura, ruptura, hendimiento

partition *m.* tabique

partly *adv.* en parte, parcialmente

partner *m.* socio, compañero || ***mining partner*** *m.* parcionero, socio minero

partnership *f.* asociación, sociedad || ***limited partnership*** *f.* asociación limitada, compañía en comandita

party *f.* brigada, [Col.] comisión || ***contracting party*** *f.* partes contratantes

Pasadenan *m. & a.* Pasadeniano, Pasadeniense

pass *m.* paso, congosto, desfiladero, collado, puerto, [Ch.] atravieso; *f.* angostura, caluma, estrechura, estrechez, quebrada, [Pe.] apacheta || ***pass through*** *vb.* atravesar

passage *m.* pasaje, conducto, paso, pasillo; *f.* galería, vía de acceso

password *f.* contraseña

paste *f.* pasta

patch *m. & a.* parche || ***patch reef*** *m.* arrecife parche

patent *f. & a.* patente; patentado

patent *vb.* patentar

patented *a.* patentado

paternoster lakes *m.* lagos en rosario

path *f.* trayectoria; vereda, ruta, vía; *m.* recorrido, sendero, camino, trayecto, trazado

patina *f.* pátina

patronite *f.* patronita

pattern *f.* configuración, disposición, pauta, plantilla, regla; *m.* patrón, modelo, esquema, dechado

paurocrystalline *a.* paurocristalino

pavement *m.* pavimento, empedrado

pavilion (of a gem) *m.* pabellón

pay *f.* zona productiva || ***day's pay*** *m.* pago; *f.* paga, [Méx.] raya || ***payday*** *m.* día de pago || ***pay sand*** *f.* arena productiva || ***pay streak*** *m.* clavo

paymaster *m.* pagador

payment *m.* pago, desempeño

payroll *f.* nómina, memoría, planilla

peacock copper *f.* bornita

peak *m.* picacho, pico, cerro, cucurucho; *f.* cumbre, cima, aguja, cúspide, picota, vértice

peak *a.* máximo || ***peak load*** *f.* carga máxima

pearceite *f.* pearceíta

pearl *f.* perla

pearlite *f.* perlita

pearly *a.* nacarado, perlado

pear-shaped *a.* piriforme

peat *f.* turba || ***peat bog*** *f.* turbera, turbal; *m.* pantano turboso || ***peat marsh*** *f.* turbera

pebble *m.* guijarro, canto, canto rodado, cantal, cantillo, chinarro; *f.* guija, piedrecita, china, lleira grava, [Sp.] pudinga || ***quartz pebble*** *f.* peladilla

pebbly *a.* guijarroso, guijoso

pectolite *f.* pectolita, walkerita, yuksporita

pedalfer *m.* pedálfer

pedestal *m.* pedestal, base, pie

pedicle *m.* pedúnculo

pedicular *a.* pedunculado, peduncular

pediment *m. & a.* pedimento, frontón

pedimentation *f.* pedimentación

pedion *m.* pedión

pediplane *f.* llanura de pedimentos

pedogenesis *f.* pedogénesis

pedology *f.* pedología, edafología

pedomorphism *f.* pedomorfosis; *m.* pedomorfismo

peduncle *m.* pedúnculo

peduncular *a.* pedunculado, peduncular

Peelean *m. & a.* Peleano, Pelense

pegmatite *f.* pegmatita

pegmatitic *a.* pegmatítico

pegmatization *f.* pegmatización

pegmatoid *m.* pegmatoide

pelagic *a.* pelágico

pelecypod *m.* lamelibranquio, pelecípodo

peletoid, peletoidal *m. & a.* peletoide, peletoidal

pelite *f.* pelita

pelitic *a.* pelítico

pelitomorphic *a.* pelitomórfico

pellet *f.* pella, pelotilla, partícula, bolita; *m.* nódulo || ***fecal pellet*** *f.* partícula fecal || ***iron pellet*** *m.* perdigón

pellodite, pelodite *f.* pelodita

peloid *m.* peloide

pencil cleavage *m.* clivaje de lápiz

pendant *f.* pendiente

pendulum *m.* péndulo

peneconcordant *a.* peneconcordante

penecontemporaneous *a.* penecontemporáneo, semicontemporáneo

peneplain *m.* peniplano, tronco de montaña; *f.* penillanura, peniplanicie, llanura troncal, planicie troncal

peneplanation *f.* peneplanación

penesaline *a.* penesalino

penetrate *vb.* atravesar, penetrar

penetrating *a.* penetrante

penetration *f.* penetración

peninsula *f.* península

penninite *f.* penninita, japanita

pennite *f.* pennina

Pennsylvanian *m. & a.* Pensilvánico, Pensylvánico, Pensylvaniano, Pensilvaniense, Pensilvaniano

Penokean *m. & a.* Penokeiense, Penokeiano

penstock *m.* tubo de entrada; *f.* tubería de carga

pentacyclic *a.* pentacíclico

pentagon *m.* pentágono, pentagonal

pentagonal *a.* pentagonal

pentane *m.* pentano

pentamerous *a.* pentámero

pentlandite *f.* pentlandita, nicopirita, pirita ferroniquelífera

penumbra *f.* penumbra

Penutian *m. & a.* Penutiense, Penutiano

per *a.* por || ***per-acre yield*** *f.* rendimiento por acre || ***per annum*** *a.* anual, por año || ***per diem*** *a.* por día

peralkaline *a.* peralcalino

peraluminous *a.* peraluminoso

percent *m.* por ciento, porcentaje

percentile *f.* centila

percentage *m.* porcentaje || ***mole percentage*** *f.* porcentaje de mol

perched groundwater, perched water table *f.* agua subterránea endicada

percolate *vb.* percolar, filtrar, colarse, resumar

percolating *f.* acción de percolar

percolation *f.* colación, filtración

percolator *m.* filtro, colador

percrystalline *a.* percristalino

percussion *f.* percusión || ***percussion drilling*** *f.* perforación por percusión

perenial *a.* perenne

perfect *a.* perfecto

perforate *vb.* perforar, agujerar, punzonar

perforated *a.* perforado
perforating gun *f.* pistola perforadora; *m.* punzonador
perforation *f.* perforación
perforated liner *m.* caño filtro, caño punzonado, filtro
perforator *m.* aparato perforador, cañón perforador, punzonador || ***casing perforator*** *f.* perforadora de tubería de revestimiento
performance *m.* rendimiento, desempeño, funcionamiento; *f.* eficacia, ejecución
per hour *a.* por hora
perhyaline *f.* herhialina
pericardium *m.* pericardio
periclase *f.* periclasa
pericline *m. & a.* periclinal; *m.* domo
peridot *m.* peridoto
peridotite *f.* peridotita
periembryonic *a.* periembriónico
perigee *m.* perigeo
perigenic *a.* perigénico
periglacial *a.* periglacial
perihelion *m.* perihelio
perimeter *m.* perímetro, ámbito; *f.* circunvalación
period *m.* período || ***amortization period*** *m.* período de amortización || ***glacial period*** *m.* período glacial
periodic motion *m.* movimiento periódico
periodic table *m.* cuadro periódico; *f.* tabla periódica
periostracum *m.* perióstraco, periostraco
peripheral *a.* periférico; *m.* (comp.) auxiliar
periphery *f.* periferia
peristome *f.* peristoma
peritectic *a.* peritéctico
perithecа *f.* periteca
perlite *f.* perlita
perlitic *a.* perlítico
perloffite *f.* perlofita
permafrost *m.* pergelisol, permagel
permanent *a.* permanente
permeability *f.* permeabilidad || ***absolute permeability*** *f.* permeabilidad absoluta
permeable *a.* permeable
permeameter *m.* permeámetro
permeate *vb.* penetrar, impregnar, calar, saturar
permeation *f.* permeación
Permian *m. & a.* Pérmico, Permiano
permineralization *f.* permineralización
permissible *a.* admisible
permit *m.* permiso; *f.* patente
permit *vb.* licenciar
Permo-Carboniferous *m. & a.* Permocarbonífero
permutation *f.* permutación
perovskite *f.* perofskita, perovskita, perowskita
peroxide *m.* peróxido
perpendicular *a.* perpendicular
personnel *m.* personal
perspective *f.* perspectiva
perthite *f.* pertita
perthitic *a.* pertítico
perthosite *f.* pertosita
perturbation *f.* perturbación
pervasive *a.* penetrante
pervious *a.* permeable
pestle *f.* maja; *m.* majadero
petalite *m.* castor
petcock *f.* llave de purga; *m.* purgador
peter out *vb.* desaparecer, empobrecer, acuñar(se)
petition for a mining claim *f.* denuncia
petrification *f.* petrificación
petrified *a.* petrificado
petrify *vb.* petrificar
petrochemical *a.* petroquímico
petrochemistry *f.* petroquímica
petrofabric *f.* petrofábrica
petrogenesis *f.* petrogenia; *m.* petrogénesis
petrogeny *m.* petrogénesis
petrographer *m.* petrógrafo
petrographic *a.* petrográfico
petrography *f.* petrografía
petrol *f.* gasolina
petrolatum *m.* petrolato
petrolene *m.* petroleno

petroleum *m.* petróleo, aceite mineral; *a.* petrolero || ***petroleum system*** *m.* systema de petroleo
petroliferous *a.* petrolífero
petrologic *a.* petrológico
petrologist *m.* petrólogo
petrology *f.* petrología
petrophysical data *m.* datos petrofísicos
petrophysicist *m.* petrofísico
petrophysics *f.* petrofísica
petzite *f.* petzita
Pfalzian *a.* Palatiense, Palatiano; Pfalziense, Pfalziano
phacolite, phacolith *m.* facolito
phaneritic *a.* fanerítico, fanerocristalino, granular, granudo, faneròmero
phantom horizon *m.* horizonte imaginario
pharmacolite *f.* farmacolita
phase *f.* fase, etapa || ***liquid phase*** *f.* fase líquida || ***mixed phase*** *f.* fase mixta || ***phase correction*** *f.* corrección de fase || ***phase diagram*** *m.* diagrama de fases || ***phase difference*** *m.* desplazamiento de fase || ***phase rule*** *f.* regla de fases || ***phase shift*** *m.* cambio de fase
phenakite *f.* fenacita, fenaquita
phenoclast *m.* fenoclasto
phenoclastic *a.* fenoclástico
phenocryst *m.* fenocristal
phenol *m.* fenol
phenology *f.* fenología
phenolphthalein *f.* fenolftaleína
phenomenon *m.* fenómeno
phenylamine *f.* fenilamina
phenylbenzene *m.* fenilbenceno
phialine *a.* fialino
phillipsite *f.* phillipsita
phlogopite *f.* flogopita, mica parda
phoenicochroite *f.* scheibeita
phonolite *f.* fonolita; *m.* fonolito || ***olivine phonolite*** *f.* polenita
phosgenite *f.* fosgenita
phosphate *m.* fosfato, fosfeno
phosphatic *a.* fosfático
phosphite *m.* fosfeno
phosphoresce *vb.* fosforecer
phosphorescence *f.* fosforescencia
phosphorescent *a.* fosforescente
phosphoric *a.* fosforado, fosfórico || ***phosphoric acid*** *m.* ácido fosfórico
phosphorite *f.* fosforita
phosphorous *m.* fósforo
photic *a.* fótico
photocopy *vb.* fotocopiar
photoelectric *a.* fotoeléctrico || ***photoelectric cell*** *f.* celda fotoeléctrica
photogeology *f.* fotogeología
photogrammetry *f.* fotogrametría
photograph *f.* fotografía, foto || ***aerial photograph*** *f.* fotografía aérea, aerofoto
photograph *vb.* fotografiar, retratar
photographic *a.* fotográfico
photography *f.* fotografía
photomap *m.* fotomapa
photometer *m.* fotómetro
photomicrograph *f.* foto del microscopio
photomicrography *f.* fotomicrografía
photon *m.* fotón
photosensitive *a.* fotosensible
photosynthesis *f.* fotosíntesis
phragmacone *m.* fragmacono
phreatic *a.* freático
phrenite *f.* frenita
phthalic *a.* ftálico
phyletic *a.* filético
phyllite *f.* filita || ***phyllite-mylonite*** *f.* filonita
phyllotheca *f.* filoteca
phylogenetic *a.* filogenético
phylogeny *m.* filogénesis; *f.* filogenia
phylum *m.* tipo, filo
physical *a.* físico || ***physical constant*** *m.* factor físico constante
physicist *m.* físico
physics *f.* física
physiochemical *a.* fisicoquímico
physiographic *a.* fisiográfico
physiography *f.* fisiografía
phytogenic *a.* fitógeno
phytoplankton *m.* fitoplancton

phytosaur *m.* fitosaurio

Piacenzian *m. & a.* Piacentino

pick (tool) *m.* pico, despuntador, picaporro, [Col.] barretón; *f.* piqueta, picota, escota, alcotana, piocha, [Méx.] talacha; (seis.) cima, selección, superficie || ***small pick*** *f.* punterola || ***velocity pick*** *f.* recogida de la velocidad

pick *vb.* picar; (seis.) seleccionar

pickaxe *m.* guíngaro, pico, zapapico; *f.* piqueta, [Méx.] talacha

picking of seismic traces *m.* trazado de reflecciones

picotite *f.* picotita

picrite *f.* picrita

picritic *a.* picrítico

picrochromite *f.* picrocromita

picromerite *f.* schoenita

picture *f.* figura, ilustración, imagen; *m.* dibujo || ***picture element*** *m.* pixel

piedmont *m.* piamonte; pie de monte

piedmontite *f.* piamontita

piece *m.* pedazo; *f.* pieza || ***piece of ground*** *m.* tramo

pier *m.* macho, muelle

pierce *vb.* atravesar, agujerar

piercing *a.* taladrador, penetrante

piercement *f.* penetración || ***passive piercement*** *f.* penetración pasiva

piezocrystallization *f.* piezocristalización

piezoelectric *a.* piezoeléctrico

piezoelectricity *f.* piezoelectricidad

peizometric *a.* piezométrico || ***peizometric surface*** *f.* superficie peizométrica; *m.* nivel hidrostático

pig *m.* taco de limpiar, diablo, chanchito, raspatubos, limpiatubos; *f.* plancha || ***pig iron*** *m.* hierro en lingotes || ***pig of copper, lead, or tin*** *m.* galápago

pig (to clean a pipeline) *vb.* limpiar con taco, limpiar con diablo

pigeonite *f.* pigeonita

piggyback basin *f.* cuenca acuestas, cuenca transportada

pile *f.* pila, tonga; *m.* montón; pilote, rimero, [Car.] volcán || ***pile driver*** *m.* martinete || ***waste pile*** *f.* trincha

pile, pile up *vb.* amontonar

piled *a.* colmado

piling *m.* amontonamiento, hincado de pilotes

pilite *f.* pilita

pillar *m.* pilar, pináculo, torre; *f.* aguja

pillow *f.* almohada || ***pillow lava*** *f.* lava en almohada, lava cordada, almohadilla || ***salt pillow*** *f.* almohada de sal

pilot *m.* piloto, guía || ***pilot plant*** *m.* sitio de pruebas; *f.* planta piloto

pilotaxic *a.* pilotaxítico

pin *f.* aguja

pinacoid *m. & a.* pinacoide

pin line *f.* linea alfiler, linea fijada

pinched out *a.* agotado

pinching *m.* estrechamiento

pinchout *m.* acuñamiento, adelgazamiento, remate

pinch out *vb.* acuñarse, adelgazarse, cerrarse, emborrascarse, perderse

pingo *m.* hidrolacolito, palse

pinite *f.* pinita

pinnacle *m.* pináculo, pilar de tierra, torre; *f.* aguja || ***pinnacle-shaped*** *a.* acumbrado

pinnate *a.* pinado

pint *f.* (meas.) pinta (473 cc)

pipe *f.* chimenea, tubería, cañería, pipa; *m.* conducto, tubo, tapón, caño || ***coated pipe*** *f.* tubería revestida || ***drill pipe*** *m.* caño (o tubo) de perforación; *f.* tubería vástago || ***flanged pipe*** *m.* caño bridado || ***flow pipe*** *f.* tubería de disparo || ***intake pipe*** *m.* tubo de admisión || ***perforated pipe*** *m.* tubo colador, caño perforado, caño filtro || ***pipe bender*** *m.* curvatubos || ***pipe clay*** *m.* blanquizal || ***pipe joint*** *f.* unión de cañería || ***pipe straightener*** *f.* enderezadora || ***pipe stock*** *f.* tarraja, terraja || ***seamless pipe*** *m.* tubo sin costura (o sin soldadura) || ***surface pipe*** *f.* columna conductora

piped *a.* enviado por cañería

pipeline *m.* oleoducto, ducto; *f.* tubería de transporte || ***pipeline rates*** *f.* tarifa de oleoductos

pipette *f.* pipeta; *m.* gotero

piracy *f.* captura, captación || ***river piracy*** *f.* captura del río

pirobituminous *a.* pirobituminoso

pisolite *f.* pisolita

pisolitic *a.* pisolítico

pit *m.* hoyo, hoyuelo, foso, apique, pozo; *f.* cantera, cata, cava, fosa, hoya, sima || ***corrosion pit*** *f.* picadura de corrosión || ***mud pit*** *m.* foso de lodo, foso de ripio, foso de desperdicios, depósito de lodos || ***prospect pit*** *f.* zanja || ***sand pit*** *f.* cantera de arena

pitch *m.* (petrol.) pez, brea, asfalto, betún, alquitrán; (geol.) hundimiento; *f.* inclinación || ***pitch coal*** *f.* picurita

pitch *f.* pendiente, pitch; *m.* ángulo de inclinación, hundimiento, ángulo de cabeceo, buzamiento

pitchblende *f.* pechblenda, pecblenda, uranopicco

pitchstone *f.* piedra pez, retinita, cantalita

pitted *a.* cacarañado || ***pitted outwash plain*** *f.* llanura glacifluvial hoyosa

pitticite *f.* piticita

pitting *f.* cacaraña, picadura

pivot *m.* pivote, muñón, gorrón

pivot *vb.* pivotar

pixel *m.* elemento matricial, pixel

place *f.* ubicación, localización; *m.* lugar

place *vb.* colocar, ubicar, localizar

placental *a.* placentario

placer *m.* placer, aluvión metalifero, aventadero, [Col.] batir || ***placer miner*** *m.* barranquero, gambusino || ***placer pit*** *f.* cata

placer *vb.* [Col.] barequear

placoderm *m.* placodermo

plagioclase *f.* plagioclasa

plagionite *f.* plagionita

plain *m.* (geog.) llano, [Bol., Par.] chaco; *f.* llanura, planicie, llaneza, llana, llanada, sábana; *a.* simple, sencillo, común || ***abyssal plain*** *f.* planicie abisal, llanura abisal || ***alluvial plain*** *f.* planicie alluvial; *m.* llano aluvial || ***coastal plain*** *f.* llanura costera, llanura costanera || ***constructional plain*** *f.* llanura de adición || ***delta plain*** *f.* llanura intradelta, planicie deltáica || ***denudation plain*** *f.* planicie de abrasión || ***destructional plain*** *f.* llanura de erosión || ***flood plain*** *m.* lecho de creciente; *f.* planicie de inundacion, llanura aluvial, vega || ***high mountain plain*** *m.* altiplano || ***open plain*** *f.* vega || ***outwash plain*** *m.* valle aluvial, llano aluvio-glaciar

Plaisancian *m. & a.* Plaisanciense, Plaisanciano

plan *m.* plan, trazo; *f.* heliografía, proyección horizontal, planta || ***plan view*** *f.* vista en planta

plan *vb.* planear

planar *a.* planar, plano || ***planar flow structure*** *f.* estructura de flujo laminar || ***planar structure*** *f.* estructura laminar

planation *f.* planación

plane *m.* plano, cepillo; *a.* plano, llano || ***axial plane*** *m.* plano axial || ***bedding plane*** *f.* comisura estratigráfica; *m.* plano de estratificación || ***cleavage plane*** *m.* plano de clivaje, plano de crucero || ***datum plane*** *m.* plano de referencia, plano de cota zero || ***fault plane*** *m.* plano de falla || ***fracture plane*** *m.* plano de fractura || ***joint plane*** *m.* plano de fractura (o de junta) || ***optic plane*** *m.* plano óptico || ***plane table*** *f.* plancheta; [Sp.] goniógrafo

plane *vb.* acepillar, cepillar, alisar, nivelar

planet *m.* planeta

planetary *a.* planetario

planetesimal *f.* planetesimal

planetoid *m.* planetoide, asteroide

planimeter *m.* planímetro

planimetric *a.* planimétrico

planimetry *f.* planimetría

planispiral *a.* planispiral

plankton *m.* planctón

planning *m.* planeamiento; *f.* programación, planificación

planoconvex *a.* planoconvexo

plankton *m.* plancton

planktonic *a.* planctónico

plant *f.* planta, instalación, fábrica, estación; *m.* ingenio, planteaje, plantel, planteamiento || ***acid recovery plant*** *f.* planta de recuperación de ácidos || ***extraction plant*** *f.* planta de extracción || ***gas plant*** *f.* fábrica de gas, planta de gas || ***pilot plant*** *f.* planta piloto || ***plant debris*** *m.* residuos de plantas || ***recycling plant*** *f.* planta de recirculación

plant *a.* vegetal

plasma *f.* plasma

plaster *m.* yeso, revoque; *f.* argamasa || ***plaster of Paris*** *m.* yeso de París, yeso mate

plasterer *m.* yesero

plastic *m. & a.* plástico

plasticity *f.* plasticidad

plastotype *m.* plastotipo

plat *m.* mapa administrativo

plate *f.* placa; plancha, lámina || ***lithospheric plate*** *f.* placa litosférica || ***plate tectonics*** *a.* tectónica de placas || ***tectonic plate*** *f.* placa tectónica

plateau *f.* meseta, altiplanicie, altillanura, alcarría; *m.* altiplano, [Col.] altillano

platform *f.* plataforma || ***airborne platform*** *f.* plataforma aeroportada || ***continental platform*** *m.* zócalo || ***drilling platform*** *f.* plataforma de perforación || ***external platform*** *f.* plataforma externa || ***floating platform*** *f.* plataforma flotante || ***interior platform*** *f.* plataforma interna || ***space-based platform*** *f.* plataforma espacial || ***tension-leg platform*** *f.* plataforma de cables en tensión || ***wave-cut platform*** *f.* plataforma de abrasión

platinum *m.* platino, [Bol.] chumpi; *f.* platina

platy *a.* laminado, escamiforme, escamoso

platyhelminthes *m.* platelmintos

play (exploration) *m.* "play," juego, prospecto; *f.* localización

playa *f.* playa; *m.* [Ch.] solar

Pleistocene *m. & a.* Pleistoceno, Pleistocénico

pleochroic *a.* pleocróico, pleocroisto

pleochroism *m.* pleocroísmo, pleocromatismo

pleonaste *f.* pleonasta

pleosponge *f.* pleoesponja

plesiosaur *m.* plesiosauro

pleural *a.* pleural

pleurite *m.* pleurito

pleuron *f.* pleura

pleystenite *f.* kreuzbergita

pliable *a.* ductil, flexible, plegable, plegadizo

pliability *f.* ductilidad, flexibilidad

plicate *a.* plegado

plication *m.* plegamiento; *f.* plicación

plica *m.* pliegue

Pliensbachian *m. & a.* Pliensbaquense, Pliensbaquiano

Plinian *m. & a.* Pliniano, Pliniense

Pliocene *m. & a.* Plioceno, Pliocénico

plot *m.* solar; terreno, terruño; gráfico; *f.* parcela; impresión || ***crossplot*** *m.* gráfico de ____ contra ____ || ***log–log plot*** *f.* escala logarítmica || ***semi-log plot*** *f.* escala semilogarítmica

plot *vb.* trazar, delinear; imprimir, graficar; ubicar

plotted *a.* graficado, restituído, impreso

plotter *m.* impresor, graficador, restituidor; *f.* trazadora de gráficos

plotting *m.* graficado; *f.* impresión, restitución

pluck *vb.* coger, recoger

plucking *f.* extracción; *m.* extrayendo

plug *m.* tapón, botón, obturador, taco || ***casing plug*** *m.* tapón de tubería de revestimiento || ***cement plug*** *m.* tapón de cementación || ***plug back*** *f.* cementación de fondo || ***volcanic plug*** *m.* tapón volcánico

plug (a formation) *vb.* obturar (la formación), taponar, tapar, obstruir || ***back plug*** *vb.* retrotaponar, taponar el fondo de un pozo

plugged *a.* taponado, tapado || ***plugged bit*** *f.* barrena taponada || ***plugged well*** *m.* pozo taponado

plugging (a well) *m.* puente || ***back plugging*** *m.* retrotaponamiento

plumb, plumb bob, plumb line *f.* plomada; *m.* pendulo

plumb-line deviation *f.* desviación de la vertical

plume *f.* (geol.) pluma, intrusión; *m.* penacho

plumose *a.* plumoso

plunge *m.* buzamiento, hundimiento

plunge *vb.* buzar, sumergir, invertir, zambullir, hundirse || ***plunge of a structure*** *f.* inclinación de la estructura

plunger *m.* émbolo buzo, émbolo macizo; chupón

plunging *a.* buzante || ***plunging fold*** *m.* pliegue buzante

Pluto *m.* Plutón

pluton *m.* plutón; *f.* intrusión

plutonic *a.* plutónico, intrusivo || ***plutonic rock*** *m.* agalísico

plutonism *m.* plutonismo

plutonium *m.* plutonio

pluvial *a.* pluvial, lluvial

ply *f.* hoja, capa, tela

pneumatic *a.* neumático

pneumatolytic *a.* pneumatolítico, neumatolítico, neumatológeno, neumatógeno

pocket *m.* bolsillo; *f.* cavidad, bolsa || ***pocket knife*** *f.* cortaplumas, cuchilla, navaja de bolsillo || ***pocket of ore*** *m.* remolino

pod-like *a.* vainiforme

podsol *m.* podsol

poikilitic *a.* poiquilítico, xenoíquico

poikiloblastic *a.* poiquiloblástico

poikilothermic *a.* poiquilotermo, exotermo

poikilotopic *a.* poikilotópico

point *m.* cabo, punto; *f.* punta || ***boiling point*** *m.* punto de ebullición || ***branch point*** *m.* punto de bifurcación || ***break-even point*** *m.* punto de muerto || ***bubble point*** *m.* punto de burbuja, punto de burbujeo || ***cloud point*** *m.* punto de enturbamiento, punto de opacidad, punto de neblina || ***common depth point (CDP)*** *m.* punto común profundo || ***condensation point*** *m.* punto de condensación || ***control point*** *m.* punto de control, punto de comprobación || ***datum point*** *m.* punto de referencia || ***depth point*** *m.* punto de profundidad || ***dew point*** *m.* punto de rocío, punto de niebla || ***end point*** *m.* punto final || ***flash point*** *m.* punto de inflamación || ***focal point*** *m.* punto de origen || ***freezing point*** *m.* punto de congelación || ***melting point*** *m.* punto de fusión, punto de escurrimiento || ***point bar*** *f.* barra de punta; *m.* albardón semilunar || ***point bar deposit*** *m.* albardón semilunar || ***point count*** *m.* conteo de puntos || ***point diagram*** *m.* diagrama de puntos || ***pour point*** *m.* punto de fluidez, punto mínimo de fluidez, punto de escurrimiento || ***reference point*** *m.* punto de referencia, testigo || ***saturation point*** *m.* punto de saturación || ***shot point*** *m.* punto tiro, punto de la explosión || ***spill point*** *m.* punto de derrame || ***stress concentration point*** *m.* punto de concentración del esfuerzo || ***survey point*** *m.* punto de marca || ***vapor point*** *m.* punto de vaporización || ***view point*** *m.* punto de vista || ***yield point*** *m.* punto cedente, punto de deformación

pointed *a.* puntiagudo, puntoagudo

pointy *a.* puntual

poise *m.* poise; medida de viscosidad

poison *m.* veneno

poisonous *a.* venenoso, letal

Poisson's ratio *m.* coeficiente de Poisson

polar *a.* polar || ***polar coordinates*** *f.* coordenadas polares || ***polar wandering*** *m.* desplazamiento polar; *f.* deriva polar

polarimeter *m.* polarímetro

polarimetry *f.* polarimetría

polarity *f.* polaridad

polarization *f.* polarización

polarized *a.* polarizado

polarizer *m.* polarizador

polarizing *a.* polarizador || ***polarizing microscope*** *m.* microscopio de polarización

pole *m.* polo, poste, jalón, mástil; *f.* asta, vara, varilla || ***geographic pole*** *m.* polo geográfico || ***magnetic pole*** *m.* polo magnético

policy *f.* política, póliza || ***insurance policy*** *f.* póliza de seguro

poligonal *a.* poligonal

polimerize *vb.* polimerizar

polish *m.* pulimento, lustre
polish *vb.* pulir, lustar
polished *a.* pulido, fino || ***polished section*** *m.* lado pulido
polisher *m.* pulidor
polishing *f.* pulimentación
polje *m.* polie
pollen *m.* polen
pollenite *f.* polenita
pollute *vb.* contaminar
polluted *a.* contaminado
pollution *f.* contaminación
polonium *m.* polonio
polucite *f.* polucita
polyargyrite *f.* poliargirita
polyaxon *a.* poliaxona, poliaxónida, poliaxónica
polybasite *f.* polibasita, plata agria, plata azul, polvorilla; *m.* acercado, teposteles
polyconic *a.* policónico
polycrase *m.* policraso
polycrystal *m.* policristal
polychrome *a.* policromo
polycyclic *a.* policíclico, poligénico
polydymite *f.* polidimita
polyethylene *m.* polietileno
polygenetic *a.* polígeno, poligénico
polygon *m.* polígono
polygonal *a.* poligonal
polyhaline *a.* polihalino
polyhalite *f.* polihalita
polykinematic *a.* policinemático
polymer *m.* polímero
polymerization *f.* polimerización
polymerized *a.* polimerizado
polymetallic *a.* polimetálico
polymetamorphism *m.* polimetamorfismo
polymictic *a.* polimítico
polymineralic *a.* polimineràlico
polymodal *a.* polimodal
polymorph *m.* polimorfo
polymorphic *a.* polimorfo, multiforme
polymorphism *m.* polimorfismo; *f.* polimorfía
polymorphous *a.* polimorfo, multiforme
polyp *m.* pólipo
polyphase *a.* polifásico
polyphyletic *a.* polifilético
polypide *m.* polípido
polypropylene *m.* polipropileno
polyquartz *m.* policuarzo
polyserial *a.* poliseriado
polysynthetic *a.* polisintético || ***polysynthetic twinning*** *f.* maclación polisintética
polythalamous *a.* politalámico, multilocular
polytype *m.* politipo
polytypism *m.* politipismo
polyvalence *f.* plurivalencia, polivalencia
pond *f.* alberca, charca, charcal, albufera, alcubilla, [Bol., Ec.] pond; *m.* estanque, chafariz, chortal, charco, ojo, [Sp.] chabarco, [Vz.] estero || ***salty evaporite pond*** *f.* era; *m.* charco evaporítico salino || ***tailings pond*** *m.* estanque para colas
Pontian *m. & a.* Pontiense, Pontiano || ***lower Pontian*** *m. & a.* Josarense
Pontic *m. & a.* Pontico
pontoon *m.* pontón
pony rod *f.* varilla corta
pool *m.* criadero, yacimiento, depósito; aljibe, regato, remanso, [Ec.] aguatal; *f.* alberca, poceta, tinaja
pop safety valve *f.* válvula de seguridad de resorte (o de disparo)
poppet valve *f.* válvula de disco con movimiento vertical
population *f.* población; (estadística) populación || ***reference population*** *f.* población de referencia || ***type population*** *f.* población tipo
porcelain *f.* porcelana
porcellaneous *a.* porcelanado
porcellanite *f.* porcelainita, porcelanita
pore *m.* poro; *f.* oquedad || ***pore space*** *m.* espacio poral || ***pore throat*** *f.* la garganta del poro || ***pore throat radius*** *m.* radio de la garganta del poro || ***pore volume*** *m.* volumen de poro
porifera *m.* poríferos, espongiarios

porosity *f.* porosidad || ***absolute porosity*** *f.* porosidad total || ***effective porosity*** *f.* porosidad efectiva || ***primary porosity*** *f.* porosidad primaria || ***secondary porosity*** *f.* porosidad secundaria

porosimeter *m.* porosímetro

porous *a.* poroso, ampolloso, filtrante

porphyrin *f.* porfirina

porphyritic *a.* porfídico, porfiroide, porfirítico

porphyroblast *m.* porfidoblasto

porphyroclasitc *a.* porfiroclástico

porphyry *m.* pórfido, porfirio; *f.* porfidita, porfirita || ***decomposed porphyry*** *f.* tosca

port *m.* puerto

portable *a.* portátil, movible, móvil

portal *m.* acceso; *f.* entrada, bocamina

portfolio *f.* cartera

portion *m.* lote

Portlandian *m. & a.* Portlandiense, Portlandiano

position *f.* posición; *m.* posicionamiento

positive *a.* positivo

post *m.* pilar, poste; *a.* poste

post (on the log; on the map) *vb.* markar (en el registro; en el mapa)

posterior *a.* posterior

post-glacial *a.* posglacial, post glacial

post-orogenic *a.* posorogénico

post-stack *a.* despues de apilamiento

post-tectonic *a.* postectónico

potamic *a.* potamógeno

potash *f. & a.* potasa; potásico || ***potash feldspar*** *m.* feldespato potásico || ***potash mica*** *f.* muscovita, mica potásica

potassic *a.* potásico

potassium *m.* potasio || ***potassium–argon*** *m.* K–Ar, potasio–argón

potency *f.* potencia, fuerza

potential *m. & a.* potencial || ***drawdown potential*** *f.* producción potencial || ***potential fields*** *m.* campos potenciales, metodos potenciales || ***rated potential*** *m.* potencial nominal || ***spontaneous potential*** *m.* potencial espontaneo

potentiometer *m.* potenciómetro

potentiometric *a.* potenciométrico

pothole *f.* marmita, olla de gigante, olla de remolineo, pila, [Col.] centenaja; *m.* bache, molino, nido

Potsdam, Potsdamian *m. & a.* Potsdamiense, Potsdamiano; Olenense, Olenidense

potter's clay *f.* arcilla para cerámica, caolín

pound *f.* libra (0,4534 kg)

pound *vb.* golpear

pounding *m.* golpeo, golpeteo

pour *vb.* vertir || ***pour out*** *vb.* derramar || ***pour point*** *m.* punto de fluidez, punto de escurrimiento || ***pour test*** *f.* prueba de fluidez a baja temperatura

powder *m.* polvo; *f.* pólvora || ***blasting powder*** *f.* pólvora para voladura || ***powder magazine*** *m.* polvorín

powellite *f.* powellita

power *f.* potencia, fuerza motriz; energía, fuerza; *m.* poder || ***nuclear power*** *f.* potencia nuclear || ***power house*** *f.* casa de motores; central eléctrica, planta de energía || ***power of attorney*** *f.* (legal) carta poder, poder legal || ***power plant*** *f.* central de energía, usina eléctrica || ***power take-off*** *f.* toma de fuerza, tomafuerza

powerful *a.* potente, poderoso

pozzolan *a.* pozolano

pozzolana *m.* puzol

practical *a.* práctico

practice *f.* práctica

practice *vb.* practicar

Praetiglian *m. & a.* Praetigliense, Praetigliano

prairie *f.* pradera, pampa

praseodymium *m.* praseodimio

pre-Alpine *a.* prealpino

Precambrian *m. & a.* Precámbrico, Precambriano, Antecambriano

precious *a.* precioso, valioso || ***precious stones*** *f.* pedrería, piedras preciosas

precipice *m.* asomo, derrumbadero, derrumbo, jorfe, precipicio, desbarrancadero, despeñadero

precipitant *m.* precipitante

precipitate *m.* precipitado; *f.* gotera

precipitate *vb.* precipitar
precipitated *a.* concrecionado, precipitado
precipitation *f.* precipitación
precipitator *m.* precipitador
precipitous *a.* acantilado, despeñadero
precise *a.* exacto, preciso
precision *f.* precisión, exactitud
preclude *vb.* evitar, impedir
precursor *m.* precursor
predator *m.* depredador
predazzite *f.* predazita
pre-diagenesis *f.* prediagénesis
predict *vb.* predecir, pronosticar
predictable *a.* previsible
prediction *f.* predicción; *m.* pronóstico
prefabricated *a.* prefabricado
preface *m.* prefacio, prologo
preferential absorption *f.* absorción preferencial
preferred *a.* preferido || ***preferred orientation*** *f.* orientación preferida
prehensile *a.* prensil
prehistoric *a.* prehistórico
prehnite *f.* prehnita, chiltonita, jacksonita, zeolita del Cabo
preliminary *a.* preliminar
pre-orogenic *a.* preorogénico
prepare *vb.* preparar, confeccionar
preplanning *m.* anteproyecto
pre-print *f.* prueba
present *vb.* presentar
present-day *a.* actual, hoy día, hoy en día
preservation *f.* preservación
preservative *m.* preservativo, conservador de_____
preserve *vb.* preservar
press *f.* prensa || ***press conference*** *f.* conferencia de prensa
press *vb.* prensar, apretar
pressed steel *m.* acero prensado
pressure *f.* presión, compresión; *m.* empuje || ***abnormal pressure*** *f.* alta presión, presión anormal || ***absolute pressure*** *f.* presión absoluta || ***atmospheric pressure*** *f.* presión atmosférica || ***back pressure*** *f.* contrapresión || ***bottom-hole pressure*** *f.* presión de fondo de pozo || ***bubble-point pressure*** *f.* presión de burbuja || ***capillary pressure*** *f.* presión capilar || ***casing pressure*** *f.* presión en el interior de la tubería || ***critical pressure*** *f.* presión crítica || ***differential pressure*** *f.* presión diferencial || ***directed pressure*** *f.* presión dirigida || ***flowing pressure*** *f.* presión de flujo || ***flowing bottom-hole pressure*** *f.* presión de fondo de pozo fluyendo || ***fluid pressure*** *f.* presión de flujo || ***formation pressure*** *f.* presión de formación || ***high pressure*** *f.* alta presión, presión elevada || ***hydrostatic pressure*** *f.* presión hidrostática || ***open hole pressure*** *f.* presión a pozo abierto || ***partial pressure*** *f.* presión parcial || ***pore fluid pressure*** *f.* presión de fluídos en los poros || ***pressure buildup*** *m.* fortalecimento de la presión || ***pressure buildup test (PBUT)*** *f.* prueba de incremento de presión || ***pressure depletion*** *m.* agotamiento de presión || ***pressure distillation*** *f.* destilación a presión || ***pressure gauge*** *m.* manómetro || ***pressure gradient*** *m.* gradiente de presión || ***pressure maintenance*** *f.* conservación de la presión; *m.* mantenimiento de presión || ***pressure regulator*** *m.* regulador de presión || ***pressure relief valve*** *f.* válvula de alivio de presión || ***pressure ridge*** *f.* colina de presión || ***pressure solution*** *f.* disolución por presión || ***pressure transient test*** *f.* prueba de presión momentánea || ***pressure vessel*** *m.* recipiente a presión || ***pressure wave*** *f.* onda primaria || ***shut-in pressure*** *f.* presión cerrada, presión encerrada, presión hermética || ***solution pressure*** *f.* presión de solución || ***static pressure*** *f.* presión estática || ***steam pressure*** *f.* presión de vapor (PV) || ***tubing pressure*** *f.* presión en la tubería || ***vapor pressure*** *f.* presión de vapor, tensión de vapor || ***working pressure*** *f.* presión efectiva, presión de trabajo
pre-stack *a.* pre-apilamiento
prestressing *m.* preestiramiento, pretensado
pre-tectonic *a.* pretectónico
pretreatment *m.* pretratamiento

Priabonian *m. & a.* Priaboniense, Priaboniano

price *m.* precio, costo, coste

priceite *f.* pandermita

primacord *m.* cordón explosivo

primary *a.* primario

prime *a.* principal || ***prime meridian*** *m.* primer meridiano || ***prime mover*** *m.* motor primario, motor primordial

prime *vb.* cebar, preparar

primer *m.* imprimador, aprestador, tapaporos

priming *f.* preparación

primitive *a.* primitivo

primordial *a.* primordial

print *f.* impresión || ***blue print*** *f.* impresión azul, copia heliográfica

print *vb.* imprimir

printer *m.* impresor; *f.* impresora

printout *f.* salida impresa

prioritize *vb.* jerarquizar, ordenar, priorizar

priority *f.* prioridad

prism *f.* prisma || ***Nicol prism*** *f.* prisma de Nicol || ***quartz prism*** *f.* prisma de cuarzo

prismatic *a.* prismático

prismoidal *a.* prismoide

private *a.* privado, particular, propio, secreto || ***private enterprise*** *f.* empresa particular || ***private property*** *f.* propiedad particular, propiedad privada

probability *f.* probabilidad

probable *a.* probable

probe *m.* electrodo explorador; *f.* tienta

probertite *f.* kramerita

problem *m.* problema; (mech.) accidente || ***three-point problem*** *m.* problema de tres puntos

procedure *m.* procedimiento, método

proceeding *f.* acta

proceeds *f.* ganancias

process *m.* proceso, procedimiento, método

process *vb.* elaborar, procesar

processed *a.* beneficiado

processing *m.* proceso || ***batch processing*** *m.* proceso por lotes || ***data processing*** *m.* proceso de datos || ***digital processing*** *m.* tratamiento digital || ***processing engineer*** *m.* ingeniero refinador || ***processing equipment*** *m.* equipo para fabricación, tratamiento, beneficio, elaboración, o manufactura || ***real-time processing*** *m.* proceso en tiempo real

procurement *f.* provisión

prodelta *f.* parte más externa del delta

produce *vb.* producir, rendir, rentar

producer *m.* productor || ***producer gas*** *m.* gas pobre, gas de aire

producible *a.* recuperable, extraible

producing *a.* productivo || ***producing gas–oil ratio*** *f.* relación gas–petróleo de producción || ***producing horizon*** *m.* horizonte productivo || ***producing sand*** *f.* arena productiva || ***producing well*** *m.* pozo productor

product *m.* producto, (min.) fruto || ***waste product*** *m.* producto de desperdicio, producto sin valor

productid *m.* prodúctido

production *f.* producción, extracción; *m.* rendimiento || ***cumulative production*** *f.* producción acumulada || ***gross production*** *f.* producción en bruto || ***initial production*** *f.* producción inicial || ***primary production*** *f.* producción primaria || ***production ceiling*** *m.* tope de producción || ***production cost*** *m.* costo de producción || ***production curve*** *f.* curva de producción || ***production engineer*** *m.* ingeniero de producción || ***production equipment*** *m.* equipo de producción || ***production secondary*** *f.* producción secundaria || ***shared production*** *f.* producción compartida || ***ultimate production*** *f.* producción última, producción final

productive *a.* productivo || ***productive interval*** *f.* zona productora, zona productiva

productivity *f.* productividad || ***productivity index*** *m.* coeficiente de productividad

professional *m. & a.* profesionista

profile *m.* perfil || ***deep seismic reflection profile*** *m.* perfil sísmico de reflexión profunda || ***graded profile*** *m.* perfil de equilibrio || ***profile shooting*** *m.* disparo con detectores en línea recta || ***sample profile*** *m.* perfil según muestras || ***soil profile*** *m.* perfil de suelo, perfil edáfico || ***vertical velocity profile*** *m.* perfil de velocidad vertical

profiling *m.* perfilaje || ***continuous profiling*** *m.* perfilaje continuo

profit *m.* provecho, beneficio, producto; *f.* utilidad, ganancia, ventaja || ***estimated profit*** *f.* utilidad probable

profitability *f.* rentabilidad

profitable *a.* rentable, costeable

progeny *f.* progenie, prole

prognosis *f.* prognosis

progradation *f.* progradación

prograde *vb.* progradar

prograding *a.* progradante || ***prograding shoreline*** *f.* línea de ribera en avance || ***prograding wedge*** *f.* cuña progradante

program *m.* levantamiento, programa, plan || ***computer program*** *m.* programa de computación || ***geologic program*** *m.* programa geológico || ***pilot program*** *m.* programa piloto || ***seismic program*** *m.* levantamiento sísmico

programmer *m.* (comp.) programador

programming *f.* programación

progress *m.* progreso, avance

project *m.* proyecto

project *vb.* proyectar, sobresalir, emerger

projectile *m.* proyectil

projecting *a.* sobresaliente, emergente

projection *f.* proyección; *m.* saliente, vuelo, resalte || ***crystal projection*** *f.* proyección cristalina || ***down-plunge projection*** *f.* proyección descendente, proyección busamiento hacia abajo || ***lower hemisphere projection*** *f.* proyección en el hemísfero inferior || ***Mercator projection*** *f.* proyección Mercator || ***polyconic projection*** *f.* proyección policónica || ***stereographic projection*** *f.* proyección estereográfica

prolate *a.* prolado

prolific *a.* prolífico

promontory *m.* promontorio, resalte, saliente

promote *vb.* promover, fomentar

proof *f.* prueba

pro-ostracum *m.* proóstraco

prop *m.* (min.) adema, ademe, fajado; *f.* puntal

propagate *vb.* propagar, propagarse

propagation *f.* propagación || ***propagation time*** *m.* tiempo de propagación || ***propagation velocity*** *f.* velocidad de propagación

propane *m.* propano

proparian *m.* & *a.* propario

property *f.* propiedad; finca; característica || ***mining property*** *m.* predio minero || ***private property*** *m.* propiedad particular (o privada)

proportion *f.* proporción, relación, (chem., quím.) dosis

proportional *a.* proporcional || ***proportional dividers*** *m.* compás de proporción

proprietary *a.* propietario, patentado

propulsion *f.* propulsión

propylene *m.* propileno, propeno

propylite *f.* propilita

propylitization *f.* propilitización

prorate *vb.* proratear

prorated *a.* prorateado || ***prorated production*** *f.* producción prorateada

proration *m.* prorrateo; *f.* cuota

prorationing *m.* reparto de la demanda

prosiphonate *a.* prosifonado

prosobranch *m.* prosobranquio

prosocline *m.* & *a.* prosoclino

prosodetic *a.* prosodético

prosogyre *m.* prosogiro

prosoma *f.* prosoma

prosopon *m.* ornamento

prospect *m.* prospecto; *f.* perspectiva || ***prospect pit*** *f.* cata, [Sp.] cala

prospect *vb.* explorar, catar, catear, buscar, reconocer; tirar un labor

prospecting *f.* prospección, exploración, búsqueda; *m.* cateo, reconociendo, reconocimiento

prospective *a.* con posibilidades, con indicios

prospector *m.* explorador, buscador, buscón, busconero, cateador, gambusino, prospector

prospectus *m.* prospecto

protect *vb.* proteger

protective coating *f.* capa protectora

protegulum *m.* protégulo
proterogenesis *m.* proterogénesis
Proterozoic *m. & a.* Proterozóico, Agnostozóico
protheca *f.* proteca
protista *f.* protista
protobranch *a.* protobranquiado
protochordate *m.* protocordado
protoclastic *a.* protoclástico
protoconch *f.* protoconcha
protoecium *m.* protecio
protogastropod *m.* protogasterópodo
protogenic *a.* protogénico
protomesentery *m.* protomesenterio
protomylonite *f.* protomilonita
proton *m.* protón
protoplasm *f.* protoplasma
protopodite *m.* protopodito
protopore *m.* protoporo
protore *f.* protomena
protoquartzite *f.* protocuarzita
protoseptum *m.* protosepto
prototheca *f.* prototeca
protozoan *m.* protozoario, protozoo
protractor *m.* transferidor, transportador
protuberance *f.* protuberancia
proustite *f.* proustita, plata roja clara; *m.* cochiso, cochizo, rosicler
prove *vb.* comprobar, probar
proved, proven *a.* comprobado, probado
provenance *m.* origen; *f.* procedencia, fuente de origen
province *f.* provincia || ***geologic province*** *f.* provincia geológica || ***metalogenic province*** *f.* provincia metalogénica, provincia metalogenética
provisioning *m.* abasto
provitrian *m.* provitriano
proximal *a.* próximo
pry *vb.* bronquear
psammite *f.* sammita, psammita
psephite *f.* sefita, psefita
pseudobreccia *f.* suedobrecha
pseudoceratite *f.* seudoceratita
pseudochlorite *f.* seudoclorita
pseudocrystal *m.* seudocristal
pseudofossil *m.* seudofósil
pseudo-igneous *a.* seudoígneo
pseudokarst *m.* seudocarst
pseudomatrix *f.* seudomatriz
pseudolamellibranchiate *a.* seudolamelibranquiado
pseudoleucite *f.* seudoleucita
pseudomalachite *f.* tagilita
pseudomorph *m.* seudomorfo
pseudomorphic, pseudomorphous *a.* seudomórfico
pseudomorphism *m.* seudomorfismo
pseudoplanktonic *a.* seudoplanctónico
pseudoporphyritic *a.* seudoporfírico
pseudoporphyroblastic *a.* seudoporfirítico
pseudospondilium *m.* seudoespóndilio
pseudo-steady state *m.* seudorégimen estacionario
pseudostratification *f.* seudoestratificación
pseudo-tachylite *f.* seudotaquilita
pseudovitrinite *f.* seudovitrinita
psilomelane *f.* silomelana, psilomelana
psilophytes *f.* psilofita
psychrometer *m.* sicrómetro || ***sling psychrometer*** *m.* sicrómetro giratorio
pterobranch *m.* pterobranquio
Pterocerian *m. & a.* Pteroceriense, Pterocериano
pterodactyl *m.* pterodáctilo
pteropod *m.* pteropodo
ptygmatic folding *m.* plegamiento ptigmático
Puercan *m. & a.* Puerciano, Puerciense
pull *f.* tracción; *m.* tiro
pull *vb.* jalar, tirar, arrastrar, arrancar, sacar || ***pull apart*** *vb.* separar
puller *m.* extractor || ***core puller*** *m.* extractor de testigo
pulley *f.* polea, roldana, garrucha
pulmonate *a.* pulmonado
pulsar *m.* púlsar
pulsate *vb.* pulsar

pulsating *a.* pulsatorio

pulsation *f.* pulsación

pulse *m.* pulso

pulverization *f.* pulverización

pulverize *vb.* pulverizar, atomizar

pulverizer *m.* pulverizador, atomizador

pumaceous *a.* pomáceo

pumice *f.* pumita, piedra pómez; *m.* púmice

pump *f.* bomba || ***booster pump*** *f.* bomba reforzadora || ***bottomhole pump*** *f.* bomba del fondo de pozo || ***cementing pump*** *f.* bomba de cementación || ***circulating pump*** *f.* bomba de circulación || ***gasoline pump*** *m.* surtidor de gasolina || ***lifter pump*** *f.* bomba aspirante || ***mud pump*** *f.* bomba para lodos || ***pump jack*** *m.* caballete de bombeo, gato de bombeo || ***submersible pump*** *f.* bomba eléctrica sumergible || ***suction pump*** *f.* bomba aspirante || ***sump pump*** *f.* bomba recolectora || ***volcanic pump*** *f.* bomba volcánica

pumpellyite *f.* pumpeliita

pumper *m.* bombeador, bombero

pumping *m.* bombeo

punch *m.* punzón

punctate *a.* punteado, punctuado

Punjabian *m. & a.* Penjabiano, Penyabiano, Punjabiano, Purbeckiano

Purbeckian *m. & a.* Purbeckiense, Purbeckiano

purchase *f.* compra; aparejo; *m.* palanca

purchase *vb.* comprar

purchaser *m.* comprador

purchasing department *m.* departamento de compras

pure *a.* puro, fino, obrizo, nativo, virgen || ***pure shear*** *f.* cizalla pura

purge *vb.* purgar

purging *f.* purga, purgación

purification *f.* purificación

purified *a.* purificado; acrisolado

purifier *m.* purificador, depurador

purify *vb.* purificar, acrisolar, apurar, depurar

purifying *a.* depurador

purity *f.* pureza

purpose *m.* propósito; *f.* intención

push *m.* empuje || ***push rod*** *f.* varilla de empuje, levantaválvula

push *vb.* empujar

put out *vb.* apagar

putrefaction *f.* putrefacción

putty *f.* masilla, mastique

puzolana *f.* puzzólana

pycnite *f.* picnita

pygidium *m.* pigidio

pyralmandite *f.* piralmandita

pyralspite *f.* piralspita

pyramid *f.* pirámide

pyramidal *a.* piramidal

pyrargyrite *f.* pirargirita, argirosa, blenda de plata, plata roja oscura; *m.* cochiso, cochizo

pyrex *m.* pirex || ***pyrex glass*** *m.* vidrio pirex, vidrio refractario

pyrite *f.* pirita, xantopirita, [Pe.] carhua; *m.* acerado, bronce, chino || ***argentiferous pyrite*** *m.* [Méx.] banco platero || ***iron pyrite*** *f.* pirita de azufre, pirita de fierro, pirita de hierro, pirita marcial || ***pyrite arsenopyrite mix*** *f.* [Ch.] acontita || ***pyrite-bearing*** *a.* piritoso || ***pyrite cubes*** *m.* cubos de pirita, [Col.] bauleo

pyritic *a.* piritoso, pirítico

pyritization *f.* piritización

pyritize *vb.* piritizar

pyritohedron *m.* piritoedro

pyroaurite *f.* piroaurita

pyrobelonite *f.* pirobelonita

pyrobitumen *m.* pirobitumen || ***asphaltic pyrobitumen*** *f.* wurtzilita

pyrochlore *m.* pirocloro; *f.* koppita || ***manganese pyrochlore*** *f.* yodostonolita

pyroclastic *a.* piroclástico

pyrogenesis *f.* pirogénesis

pyrogenic *a.* pirogénico, pirógeno

pyrolisis *m.* pirólisis

pyrolite *f.* pirolita

pyrolitic *a.* pirolítico

pyrolusite *f.* pirolusita, jabón de vidrieros, manganesa, manganesia; *m.* óxido de manganeso

pyrolysis *f.* pirólisis

pyromeride *f.* piromerida

pyrometamorphism *m.* pirometamorfismo

pyrometasomatism *m.* pirometasomatismo

pyrometry *f.* pirometría

pyrometer *m.* pirómetro

pyromorphite *f.* piromorfita; *m.* plomo abigarrado, plomo pardo, plomo verde

pyromorphous *a.* piromorfo

pyronaptha *f.* pironafta

pyrope *m.* piropo || ***manganese pyrope*** *f.* yohnstonolita

pyrophane *f.* pirofana

pyrophyllite *f.* pirofilita

pyropissite *f.* piropisita

pyroretinite *f.* piroretinita

pyrosmalite *f.* pirosmalita

pyrosphere *m.* pirosfero

pyrostilpnite *f.* pirostilpnita

pyroxene *f.* piroxena; *m.* piroxeno

pyroxenite *f.* piroxenita

pyroxenoid *m.* piroxenoide

pyrrhotite *f.* pirrotina, pirita magnética

Q

quadrangle *m.* cuadrángulo

quadrant *m.* cuadrante

quadratic equation *f.* ecuación cuadrática

quadrivalent *a.* tetravalente

quagmire *m.* cenagal

quake *m.* temblor

qualified *a.* capacitado

qualitative *a.* cualitativo

quality *f.* calidad, cualidad || ***high quality*** *a.* de alta calidad || ***poor quality*** *a.* de baja calidad

quantification *f.* cuantificación

quantify *vb.* cuantificar

quantitative *a.* cuantitativo

quantity *f.* cantidad

quantum *a.* cuanto

quarantine *f.* cuarentena

quark *m.* quark

quarry *f.* abra, cantera, cantería, llera, pedrera; *m.* lanchar, losero, pedrejal, pedragal || ***quarry waste*** *m.* cascajo

quarrying *f.* explotación de canteras

quarryman *m.* canterista

quarterly *a.* trimestral, por trimestre

quartz *m.* cuarzo || ***green quartz*** *m.* prasio || ***honeycombed quartz*** *f.* carrascal || ***hyaline quartz*** *m.* iris || ***quartz-bearing*** *a.* cuarcífero, soporte de cuarzo || ***quartz crystal*** *m.* cristal de cuarzo || ***quartz diorite*** *f.* cuarzodiorita || ***quartz monzonite*** *f.* cuarzomonzonita || ***quartz pebbles*** *m.* clavaque || ***quartz sand*** *f.* arena cuarzosa || ***quartz wedge*** *f.* cuña de cuarzo, prisma de cuarzo

quartziferous *a.* cuarcífero

quartzite *f.* cuarcita

quartzitic *a.* cuarcítico

quartzose *a.* cuarzoso

quasar *m.* quásar, cuasar

Quaternary *m. & a.* Cuaternario, Neozoico

quay *m.* muelle

quench *vb.* apagar, templar, enfriar

quenched *a.* templado

questionable *a.* cuestionable, problemático

queue *f.* cola; fila de espera

quick *a.* rápido

quicklime *m.* óxido de calcio

quicksand *f.* arena movediza

quicksilver *m.* mercurio, azogue

quiescence *f.* quietud

quitclaim *m.* finiquito

quota *m.* contingente; *f.* cuota, prorata

quotation *f.* cotización; cita textual

R

rack *f.* casillero; *f.* tarima; cremallera, percha || ***pipe rack*** *f.* tarima para tuberías

radar *m.* radar

radial *a.* radial || ***radial drainage*** *m.* drenaje radial; desagüe, avenamiento radial || ***radial flow*** *f.* corriente radial

radian *m.* radián

radiance *f.* radiancia

radiant *a.* radiante || ***radiant energy*** *f.* energía radiante

radiate *vb.* radiar

radiating *a.* radiado

radiation *f.* radiación || ***electromagnetic radiation*** *f.* radiación electromagnética || ***radiation belt*** *m.* cinturón de radiación

radiator *m.* radiador

radical *m. & a.* radical

radio *m.* radio || ***radio frequency*** *f.* radiofrecuencia || ***radio receiver*** *m.* receptor de radio || ***radio transmitter*** *m.* transmisor de radio

radioactive *a.* radioactivo || ***radioactive material*** *f.* sustancia radioactiva, material radioactiva

radioactivity *f.* radioactividad || ***radioactivity detector*** *m.* radiodetector

radiochemistry *f.* radioquímica

radiogenic *a.* radiogénico

radiogram *m.* radiograma; inalámbrico

radioisotope *m.* radioisótopo

radiolaria *m.* radiolarios

radiolarian *a.* radiolario

radiolarite *f.* radiolarita

radiole *f.* radiola

radioluminescence *f.* radioluminiscencia

radiolysis *f.* radiólisis

radiometer *m.* radiómetro

radiometric *a.* radiométrico

radiosonde *f.* radiosonda

radium *m.* radio

radius *m.* radio, rayo || ***drainage radius*** *m.* radio de drene

radix *f.* raíz

radon *m.* radón

radula *f.* rádula

radular *a.* radular

raffinate *a.* refinado

raft *f.* almadía

raft *vb.* apilar, llevar, transportar

rafted *a.* llevado, apilado, transportado || ***rafted ice*** *m.* hielo apilado

rafting *m.* transporte

rail *m.* carril, ferrocarril

railroad *m.* ferrocarril || ***railroad freight*** *m.* flete férreo

rain *f.* lluvia || ***rain forest*** *m.* bosque lluvioso; *f.* selva || ***rain shower*** *m.* chubasco || ***rainwater*** *f.* agualluvia

rainy *a.* lluvioso

raise *f.* chimenea; *m.* ascenso; tiro de alcancía; aumento de salario

raise *vb.* levantar, alzar, aumentar

raised beach *f.* playa levantada

raised working face *m.* realce

rake *m.* rastrillo, rastro; rascacenizas; (geol.) buzamiento; *f.* inclinación, raspadura

rake *vb.* rastrillar; inclinarse

ram *m.* ariete empaquetador; pisón || ***ram type blowout preventer*** *m.* impiderreventones de ariete

ram *vb.* apisonar, pisonear

ramification *f.* ramificación

ramp *f.* rampa, rampla || ***frontal ramp*** *f.* rampa frontal || ***hanging wall ramp*** *f.* rampa colgada (o colgante) || ***lateral ramp*** *f.* rampa lateral

rampart *m.* terraplén

ramsayite *f.* ramsayita

Rancholabrean *m. & a.* Rancholabreiense, Rancholabreano

randanite *f.* zeyssatita

random *a.* aleatorio || ***random access*** *m.* (comp.) acceso al azar, acceso indirecto || ***random variable*** *m.* álea numérico

randomness *m.* álea

range *m.* rango, alcance; límite, amplitud, tramo; *f.* extensión, gama; (geog.) serranía || ***dynamic range*** *f.* gama dinámica || ***mountain range*** *f.* cordillera, sierra, serranía || ***range finder*** *m.* telémetro

rank *a.* grosero, aspero, basto || ***rank wildcat (well)*** *m.* pozo exploratorio

rank *vb.* jerarquizar, ordenar, poner en fila

rapakivi granite *f.* wiborgita

rape oil *m.* aceite de nabo, aceite de colza

rapid *a.* rápido

rapids *m.* rápidos, cabezón, rabiones, recial, raudales; *f.* cachuela, raudas

rare *a.* raro, escaso || ***rare earths*** *f.* tierras raras

rasorite *f.* rasorita

rate *f.* tasa, proporción, razón; tarifa, rapidez, relación; *m.* caudal; índice, régimen; tipo, ritmo || ***discount rate*** *m.* tipo de descuento || ***flow rate*** *m.* régimen (o tasa) de flujo, caudal, real de fontanero || ***freight rate*** *f.* tarifa de fletes || ***growth rate*** *f.* ritmo de crecimiento || ***input rate*** *m.* régimen de entrada || ***penetration rate*** *f.* velocidad de penetración || ***pipeline rate*** *f.* tarifa de oleoductos || ***replacement rate*** *f.* tasa de reposición || ***sampling rate*** *f.* tasa de sondeo (o muestreo) || ***utilization rate*** *f.* tasa de utilización

rated *a.* valuado, calculado, asignado || ***rated capacity*** *f.* capacidad nominal, capacidad estimada, capacidad de clasificación; *m.* potencial nominal

rathite *f.* wiltshireíta

rathole *f.* ratonera; *m.* hueco de descanso, barreno de fondo de pozo, pozo piloto || ***rathole pipe*** *m.* tubo revestidor de la ratonera

rating *f.* clasificación, capacidad normal; tasación

ratio *f.* proporción, relación, razón; *m.* cociente || ***carbon–hydrogen ratio*** *f.* relación carbono–hidrógeno || ***mobility ratio*** *f.* proporción de movilidad || ***oil–gas ratio*** *f.* relación gas–petróleo || ***Poisson's ratio*** *m.* coeficiente de Poisson

Rauracian *m. & a.* Rauraciense, Rauraciano

ravine *f.* quebrada, barranca, hondonada, cañada, quebradura, quebrantadura, cárcava, hoz, rambla, torrentera; *m.* cauce, barranco, torrente

raw *a.* crudo || ***raw material*** *f.* materia prima || ***raw oil*** *m.* petróleo crudo || ***raw ore*** *m.* mineral bruto

ray *m.* rayo || ***cathode ray*** *m.* rayo catódico || ***cosmic ray*** *m.* rayo cósmico || ***electromagnetic ray*** *m.* rayo electromagnético || ***gamma ray*** *m.* rayo gamma || ***ray path*** *f.* trayectoria de la onda sísmica || ***seismic ray*** *m.* rayo sísmico || ***X-ray*** *m.* rayo equis, rayo x

Rayleigh wave *f.* onda Rayleigh

reabsorb *vb.* resorber

reach of river *m.* tramo; *f.* parte recta de un río

react *vb.* reaccionar

reactance *f.* reactancia

reactant *m.* reactante

reaction *f.* reacción || ***chain reaction*** *f.* reacción en cadena || ***chemical reaction*** *f.* reacción química

rectivate *vb.* reactivar

reactivation *f.* reactivación

reactive *a.* reactivo

reactivity *f.* tendencia reactiva

reactor *m.* reactor; *f.* cámara de reacción

reader (tape drive) *m.* (comp.) lector; *f.* lectora || ***card reader*** *f.* lectora de cartas

reading *f.* lectura

readout *f.* salida impresa

reagent *m.* reactivo, agente químico

real *a.* real, verdadero || ***real estate*** *m.* bienes raíces, bienes inmuebles || ***real time*** *a.* en tiempo real || ***real-time processing*** *m.* proceso en tiempo real

realgar *f.* rejalgar; *m.* sublimado rojo, sulfuro rojo

ream *vb.* escariar, ensanchar

reamer *m.* ensanchador, escariador, rectificador

rear *a.* trasero, posterior, atrás

rearrange *vb.* arreglar de nuevo

reassemble *vb.* rearmar

rebate *m.* descuento; *f.* rebaja

rebate *vb.* rebajar, descontar

rebore *vb.* rectificar

rebound *m.* rebote, resalte, resalto ‖ ***isostatic rebound*** *m.* rebote isostático

rebound *vb.* rebotear

recapitulation *f.* recapitulación

receipt *m.* recibo; *f.* nota, factura, carta de pago

receiver *m.* depositario, receptor

Recent *m. & a.* Reciente, Holoceno

recess *m.* rebajo; *f.* caja, cajuela, rebajada

recession *m.* retroceso

recessional *a.* retroceso ‖ ***recessional moraine*** *f.* morena regresiva

recharge *m.* incremento; *f.* recarga

recharge *vb.* volver a cargar, recargar, reponerse

reciprocal *a.* recíproco, alterno, alternativo

reciprocating pump *f.* bomba alternativa, bomba recíproca

recirculated *a.* recirculado

recirculation *f.* recirculación

reclaim *vb.* aprovechar

reclaimable *a.* recuperable

reclaimed *a.* recuperado

reclaimer *m.* recuperador

reclamation *f.* regeneración, recuperación, reclamación; *m.* aprovechamiento

recognition *m.* reconocimiento

recognize *vb.* reconocer

recompletion *f.* reconstrucción (de un pozo)

recompression *f.* recompresión

recondition *vb.* reacondicionar, rehabilitar, repasar

reconditioning *m.* reacondicionamiento

reconnaissance *m.* reconocimiento; *f.* exploración

reconnoiter *vb.* reconocer

reconstruction *f.* reconstrucción

reconstructive *a.* reconstructivo

record *m.* registro, diario; *f.* mención, anotación, ficha, memoría ‖ ***header record*** *m.* registro de cabecera

record *vb.* registrar, gravar

recorder *m.* registrador, gravador, (seis.) recopilador; *f.* registradora

recording *m.* registro

records *m.* fichero

recover *vb.* desempeñar, recuperar, recobrar, extraer

recoverable *a.* recuperable, extraible

recovered *a.* recuperado

recovery *f.* recuperación; *m.* recobro, recobramiento, rescate ‖ ***enhanced recovery*** *f.* recuperación mejorada, extracción vigorizada, recuperación asistida ‖ ***primary recovery*** *f.* recuperación primaria ‖ ***recovery factor*** *m.* factor de recuperación ‖ ***secondary recovery*** *f.* recuperación secundaria ‖ ***tertiary recovery*** *f.* recuperación terciaria ‖ ***ultimate recovery*** *f.* recuperación última, recuperación final, producción total

recrystallization *f.* recristalización

rectangle *m.* rectángulo ‖ ***rectangle drainage*** *m.* avenamiento rectangular

rectification *f.* fraccionación, rectificación

rectify *vb.* rectificar

rectilinear *a.* rectilíneo

recumbent *a.* acostado, recostado, recumbente, yacente, tumbado, volcado

recuperate *vb.* recuperar

recurrent *a.* recurrente, se repite

recycle *vb.* repasar, reciclar

recycling *f.* recirculación; *m.* reciclaje, repaso, reprocesamiento, reciclo

redbed *m.* lecho rojo, banco rojo

redeem *vb.* desempeñar

redeemable *a.* amortizables

redemption *m.* desempeño

redeposit *vb.* redepositar

redeposited *a.* redepositado

redeposition *m.* redepósito; *f.* redeposición

redistillation *f.* redestilación; *m.* desnaftado

redistilled *a.* redestilado

red ocher *m.* ocre rojo; *f.* hematita terrosa

Redonian *m. & a.* Redoniense, Redoniano

redrill *vb.* reperforar

reduce *vb.* reducir, rebajar

reduced crude *m.* crudo reducido

reducer *m.* reductor

reducing *f. & a.* reducción, destilación; reductor || ***reducing agent*** *m.* reductor

reduction *f.* reducción

reef *m.* arrecife, escollo, banco de coral, bajío cabezo, rompiente; (min.) abrojos || ***atoll reef*** *m.* corolígeno || ***back reef*** *a.* postarrecife || ***barrier reef*** *m.* arrecife de barrera o de barra || ***coral reef*** *m.* arrecife coralino o coralígeno || ***fringing reef*** *m.* arrecife marginal, arrecife litoral, arrecife costero, arrecife franjeante || ***patch reef*** *m.* arrecife parche || ***reef complex*** *m.* complejo arrecifal || ***reef flat*** *f.* llanura arrecifal || ***reef talus*** *m.* talud arrecifal

reefal *a.* arrecifal

reel *m.* tambor

reentrant *m. & a.* reentrante; *m.* rincón

reface bits *vb.* reparar trépanos

reference *f.* referencia || ***reference level*** *m.* nivel de referencia || ***reference point*** *m.* punto de referencia; testigo || ***sounding datum reference*** *f.* referencia cero de sondeo

refine *vb.* refinar, destilar; (metal) acendrar, afinar, acrisolar, beneficiar, refinar

refined *a.* refinado; acrisolado, obrizo, puro

refiner *m.* refinador, afinador, destilador

refinery *f.* refinería, destilería, [Pe.] callana; *m.* afino, ingenio, refino || ***refinery run*** *f.* cochada

refining *f.* refinación, afinación, afinadura, refinadura; *m.* afinado, afino, refinamiento, refino || ***acid refining process*** *f.* refinación ácida

reflect *vb.* reflejar

reflectance *f.* reflectancia; *m.* reflejo

reflected *a.* reflejado, reflejo || ***reflected light*** *f.* luz reflejada || ***reflected wave*** *f.* onda reflejada

reflecting layer *f.* capa de reflexión

reflection *f.* reflexión, reflección || ***reflection coefficient*** *f.* coeficiente de reflexión

reflectivity *f.* reflectividad

reflector *m.* reflector || ***slow reflector*** *m.* reflector lento

reflux *m.* reflujo

refolding *m.* replegamiento

refoliation *f.* refoliación

reform *vb.* reformar

reformation (petrol. products) *f.* (ref.) reformación

reforming *f.* reformación

refract *vb.* refractar, refringir

refracted *a.* refractada

refraction *f.* refracción; *m.* refractario || ***atmospheric refraction*** *f.* refracción atmosférica || ***refraction seismograph*** *m.* sismógrafo de refracción

refractive *a.* refractivo || ***refractive index*** *m.* índice de refracción

refractivity *f.* refractividad, refrigencia

refractometer *m.* refractómetro

refractometry *f.* refractometría

refractory *a.* refractario, ignífugo

refringence *f.* refringencia

refrigerant *m.* refrigerante

refrigerate *vb.* refrigerar

refrigeration *f.* refrigeración

Refugian *m. & a.* Refugiense, Refugiano

regenerate *vb.* regenerar

regenerated *a.* regenerado

regeneration *f.* regeneración

region *f.* región, comarca, zona; *m.* asiento

regional *a.* regional || ***regional dip*** *m.* buzamiento regional, echado regional || ***regional geology*** *f.* geología regional || ***regional gradient*** *m.* gradiente regional || ***regional elevation*** *m.* nivel regional || ***regional metamorphism*** *m.* metamorfismo regional || ***regional unconformity*** *f.* discordancia regional

register *m.* registro || ***claim register*** *m.* padrón || ***land register*** *m.* catastro

register *vb.* registrar, asentar

registration *m.* registro; *f.* registración

regmatic *a.* regmático

regolith *m.* regolito; *f.* regolita

regradation *f.* regradación

regression *f.* regresión; *m.* movimiento retrógrado || ***linear regression*** *f.* regresión lineal

regressive *a.* regresivo

regular *a.* regular

regulate *vb.* regular, tasar

regulation *f.* regulación

regulator *m.* regulador || ***pressure regulator*** *m.* regulador de presión

rehabilitation *f.* rehabilitación

reimbursement *m.* reembolso

reinforced concrete *m.* hormigón armado

reinforcement *m.* refuerzo

rejuvenation *f.* rejuvenecimiento

related *a.* relacionado

relation *f.* relación

relational database *m.* base relacional

relative *a.* relativo || ***relative humidity*** *f.* humedad relativa || ***relative sea level change*** *m.* cambio relativo del nivel del mar

relaxation *f.* relajación

relay *m.* disyuntor

release *vb.* soltar, aflojar, desembragar, desacoplar

reliability *f.* regularidad funcional, seguridad de funcionamiento, confiabilidad

relic *f. & a.* reliquia; antigüedad; *m.* resto, vestigio, remanente

relict *a.* relicto, reliquia, residual, remanente; *m.* indicio, remanento

relief *m.* alivio; (geog.) relieve || ***relief map*** *m.* mapa de relieve, mapa topográfico || ***relief valve*** *f.* válvula de seguridad || ***relief well*** *m.* pozo de auxilio

relinquish *vb.* renunciar

relinquishment *f.* renuncia (de una concesión)

reluctance *f.* reluctancia || ***magnetic reluctance*** *f.* reluctancia magnética

remainder *m. & a.* remanente; *m.* residuo

remaining *a.* remanente || ***remaining reserves*** *f.* reserva remanente

remains *m.* restos, residuos

remanent *a.* remanente, residual || ***remanent magnetism*** *m.* magnetismo remanente

remark *f.* observación

remedial *a.* correctivo

remediation *f.* reparación

remittance *f.* remesa; *m.* giro

remnant *m.* remanente || ***erosional remnant*** *m.* isleo cobijante

remote *a.* a distante, remoto || ***remote access*** *m.* acceso a distancia || ***remote control*** *f.* regulación a distancia || ***remote recorder*** *m.* teleregistrador || ***remote sensing*** *f.* teledetección; *m.* sensores remotos

removable *a.* reemplazable, removible

removal *f.* remoción

remove *vb.* sacar, remover, quitar

remover *m.* disolvente

renamed *m. & a.* redenominado

renew *vb.* renovar, remendar, rehacer; extender, prorogar

renewable *a.* renovable

renewal *f.* renovación, prórroga; *m.* renuevo

reniform *a.* reniforme

rent *f.* renta

rent *vb.* arrendar, alquilar, rentar

reopen *vb.* reabrir

reorganization *f.* reconversión

reorganize *vb.* reformar

repack *f.* reempaquetadura

repair *f.* reparación; *m.* reparo

repair *vb.* reparar, componer

reparation *f.* reparación

repeat *vb.* repetir

repeatability, repetition *f.* repetabilidad, repetición

Repettian *m. & a.* Repettiense, Repettiano

replace *vb.* reemplazar, reponer

replacement *m.* reemplazo, reemplazamiento; *f.* substitución, reposición, restitución, renovación

replica *f.* réplica, copia

repopulation *f.* repoblación

report *m.* informe, [C.A., Méx.] reporte; *f.* memoría, relación || ***driller's report*** *m.* informe de perforación || ***financial report*** *m.* informe de estado de cuentas

report *vb.* presentar, reportar, informar

repose *m.* reposo || ***angle of repose*** *m.* ángulo de reposo

reposition *f.* reposición

represent *vb.* representar

representation *f.* representación

representative *m. & f.* representante

repressuring *f.* recompresión

reprint *f.* separata, tirada aparte, reimpresión

reprocess *vb.* reprocesar

reproducibility *f.* reproducibilidad

reproduction *f.* reproducción

request *m.* pedido; *f.* solicitación, solicitud

request *vb.* solicitar

rerun *f.* repetición, nueva pasada

rerunning *f.* redistilación

resample *vb.* remuestrear

research *f.* investigación

research *vb.* investigar

resedimentation *f.* resedimentación

resedimented *a.* resedimentado

resequent *a.* resecuente, reconsequente || ***resequent stream*** *m.* arroyo resecuente

reserve *vb.* reservar

reserves *f.* reservas || ***estimated proved reserves*** *f.* reservas comprobadas estimadas || ***proven reserves*** *f.* reservas probadas || ***recoverable reserves*** *f.* reservas recuperables

reservoir *m.* almacén, depósito, yacimiento, vaso, embalse, zafareche, tanque, [Arg.] reservorio; *f.* cisterna, roca almacenadora; (lago) presa, represa || ***reservoir pressure*** *f.* presión de yacimiento || ***reservoir rock*** *f.* almacenadora, roca almacenadora, roca reservora, roca productiva, roca recipiente, roca reservoro; *m.* [Arg.] reservorio, estrato productivo

reset *vb.* poner a cero

residual *a.* residual || ***residual magnetism*** *m.* magnetismo remanente || ***residual volcanic soil*** *m.* [Méx.] tepetate || ***residual weathering*** *m.* meteorización residual

residue, residuum *m.* resíduo, restos

residues (from washing ore) *m.* relaves

resilience *f.* elasticidad; *m.* resalto, rebote

resiliency *f.* elasticidad

resilium *m.* resilio

resin *f.* resina

resinous *a.* resinoso

resistance *f.* resistencia

resistant *a.* resistente

resistivity *f.* resistividad || ***electric resistivity*** *f.* resistividad eléctrica || ***soil resistivity*** *f.* resistividad de la tierra

resolution *f.* resolución || ***resolution of forces*** *f.* descomposición de fuerzas

resolving power *m.* poder de resolución

resonant *a.* resonante

resorption *f.* reabsorción, resorción

resource *m.* recurso || ***nonrenewable resource*** *m.* recurso no renovable || ***proven resources*** *m.* recursos comprobados

respiration *f.* respiración

respirator *m.* respirador

response *f.* respuesta, reacción

responsibility *f.* responsibilidad

rest on *vb.* asentar, yacer sobre

restart *f.* reanudación

restite *f.* restita

restitution *f.* restitución

restoration *f.* restauración, reparación; restitución; *m.* reparo

restore *vb.* restaurar, reparar, restablecer, restaurarse

restored *a.* restaurado, regenerado, restituído || ***restored section*** *f.* sección restaurado

restricted *a.* restringido

resultant force *f.* fuerza resultante

resupinate *a.* resupinada

resurgence *f.* resurgencia, surgencia; *m.* resurgimiento

resurgent *a.* resurgente

resurrected stream *m.* cauce reactivado

retail *a.* al por menor, al detalle || ***retail dealer*** *m. & f.* detallista, minorista

retainer *m.* (com.) fiador, seguro, retenedor; *m. & f.* minorista

retaining wall *m.* muro de sostén, muro de contención

retardation *m.* retraso, retardo

retention *f.* retención

reticle, reticule *m.* retículo

reticular, reticulated *a.* reticular

reticulated *a.* reticulado

reticulum *m.* retículo

retinite *f.* retinita

retire *vb.* retirar, jubilar

retirements *m.* retiros

retort *f.* retorta, cucúrbita, desazogadera; *m.* quemadero || ***mercury retort*** *m.* [Pe.] cántaro

retort *vb.* quemar, refogar

retorting *f.* quema; *m.* queme

retractable *a.* retractable

retractile *a.* retráctile

retrievable *a.* recuperable

retrieve *vb.* recuperar, recobrar

retroactive *a.* retroactivo

retrodeformable *a.* retrodeformable

retrogradation *f.* retrogradación

retrograde *a.* retrógrado, retrogradante || ***retrograde metamorphism*** *m.* retrometamorfismo

retrogressive *a.* regresivo

retrosiphonate *a.* retrosifonado

return (on investment) *f.* (com.) rentabilidad, renta; *m.* producto, rendimiento || ***return on capital employed*** *f.* utilidad del capital empleado || ***return on shareholder equity*** *f.* utilidad del capital propio de los accionistas, utilidad del activo de los accionistas

returns *f.* ganancias

Retzian *m. & a.* Retziano, Retziense

reveal *vb.* manifestar, revelar, desplegar, demostrar

revenue *f.* rentas; *m.* ingresos, rédito, cobros, rendimiento

reverberatory *a.* reverberante || ***reverberatory furnace*** *m.* horno de revebero, galeme

reversal *f.* inversión, reversión

reverse *a.* atrás, retroceso, inverso, reverso || ***reverse circulation*** *f.* circulación inversa || ***reverse fault*** *f.* falla inversa

reversible *a.* reversible

review *m.* repaso

review *vb.* repasar, revisar

revise *vb.* revisar

revised *a.* revisado, corregido

revision *m.* repaso

revolution *f.* revolución, vuelta, (geol.) orogenia; *m.* giro; (geol.) diastrofismo

revolutions per minute (RPM) *f.* revoluciones por minuto

revolve *vb.* girar

revolving *a.* giratorio || ***revolving letter of credit*** *f.* carta de crédito permanente

rewinding *a.* rebobinado

reward *f.* recompensa

rework *vb.* reparar, rehacer, retrabajar || ***rework an abandoned mine*** *vb.* rascar

reworked *m.* redepositados, retrabajados

Rhaetian *m. & a.* Retiense, Retiano

rheidity *m.* reodid

rhenium *m.* renio

rheoglyph *m.* reoglifo

rheologic *a.* reológico

rheology *f.* reología

rheomorphic *a.* reomórfico

rheostat *m.* reostato

rheotropism *m.* reotropismo

rhizopoda *m.* rizópodos

Rhodanian *m. & a.* Rhodaniense, Rhodaniano

rhodesite *f.* rodesita

rhodite *f.* rodita

rhodochrosite *f.* rodocrosita, rodocrusita; *m.* espato de manganeso

rhodolite *f.* rodolita

rhodonite *f.* rodonita; [Méx.] alajite; *m.* espato de manganeso || ***altered rhodonite*** *f.* wittingita

rhombic *a.* rombal, rómbico, rombiforme, romboidal, rombo

rhombohedral *a.* romboédrico

rhombohedron *m.* romboedro

rhomboid *m.* romboide; *a.* romboideo

rhomboidal *a.* romboidal

rhyodacite *f.* riodacita

rhyolite *f.* riolita, liparita, llanita || ***nodular rhyolite*** *f.* piromerida

rhythmic *a.* rítmico

rhythmically *adv.* rítmicamente

rhythmite *f.* rítmita

ria *f.* ría

rib *f.* costilla, pestaña; *m.* nervio

ribbed *a.* acostillado

rich *a.* rico

riches *f.* riqueza

richness *f.* (min.) nobleza, riqueza

ridge *f.* cresta, serranía, cumbrera, sierra, crestería, cuchilla, loma; *m.* lomo, filo, caballón || ***accretionary ridge*** *f.* cresta acrecionada || ***pressure ridge*** *f.* colina de presión

riebeckite *f.* riebeckita

riffles *m.* listones, rabiones, rifle

rift *f.* hendidura, grieta, cuarteadura, fosa, apertura, trinchera || ***rift valley*** *m.* valle hendido

rift *vb.* rajarse, agrietarse, cuartearse, hundirse, abrir

rig *f.* perforadora, plataforma de perforación; *m.* taladro, [Vz.] sancocho

rig *vb.* instalar

rigging *f.* cabria, perforadora || ***rigging-up*** *m.* mantaje

right *m. & a.* derecho || ***right angle*** *m.* ángulo recto || ***right lateral*** *a.* lateral derecho, dextral || ***right lateral fault*** *f.* falla lateral derecho (o dextral)

right-of-capture *m.* derecho de apropiación

right-of-way *m.* derecho a la vía; *f.* servidumbre de paso

rigid *a.* rígido

rigidity *a.* rigidez

rill *m.* surco de escurrimiento, cavas || ***rill mark*** *f.* marca de escurrimiento; huellas de riachuelos, marca de cava

rim *f.* corona, llanta, orilla; *m.* reborde, borde, margen || ***rim syncline*** *m.* sinclinal marginal

rimmed shelf *f.* plataforma bordeada

rimpylite *f.* rimpilita

rimrock *m.* filo de roca

rimstone *m.* depósito de borde

ring *m.* anillo, aro; *f.* argolla, virola || ***ring dike*** *m.* dique anular

ringing *m.* zumbido || ***earth ringing*** *m.* zumbido terrestre

rip current *f.* corriente de resaca

rip *vb.* rasgar

ripidolite *f.* proclorita

ripple *f.* rizadura, rejilla, ondulita, [Arg.] óndula, [Sp.] ondulación || ***laminated ripple*** *a.* laminación de corriente || ***ripple mark*** *f.* rizadura, señal de olas, marca de oleaje, [Vz.] encrespadura, [Arg.] óndulita, [Sp.] ondulación || ***ripple scour*** *f.* rizadura, ondulación u ondulita de cauce

rip-up *a.* rotura, rasgadura || ***rip-up clast*** *m.* clasto rasgado

rise *f.* cuesta, loma, elevación, altura; *m.* alto

rise (of sea level) *a.* subida, ascenso, incremento (del nivel del mar)

riser *m.* tubo de subida, tubo ascendente, tubo vertical, tubo subiente

risk *m.* riesgo, álea

river *m.* río || ***river bank*** *f.* ribera, playa; *m.* banco, borde, margen, [Arg.] barranco || ***river bed*** *m.* lecho, cauce, álveo, fondo; *f.* [Ch.] caja || ***river bend*** *m.* recodo || ***river capture*** *f.* captura de río || ***river channel*** *m.* cauce, álveo || ***river mouth*** *f.* disembocadera, desembocadera, boca del río, [Col.] bocana || ***river outlet*** *m.* jaguadero, bocadero || ***river piracy*** *f.* captura de río; piratería de río || ***river system*** *m.* sistema fluvial || ***river terrace*** *f.* terraza fluvial || ***river valley*** *f.* arribera; *m.* valle

rivulet *m.* riachuelo, arroyuelo

road *m.* camino, sendero; *f.* carretera || ***road bed*** *m.* terraplén || ***road gravel*** *m.* recebo || ***road metal*** *m.* ripio

roadbuilding *m.* ruteo

roast *vb.* tostar, quemar, calcinar, (min.) cocer metal, reverberar, oxidar

roaster *m.* calcinador

roasting *f.* cocción, cochura, quema, tostación, tostadura, reverberación; *m.* queme, tueste || ***roasting furnace*** *m.* tostador

rob (pillars) *vb.* despilar (pilares)

robbed (pillars) *a.* (pilares) robado

robbing (pillars) *m.* despilamiento

robust *a.* robusto

roches moutonees *f.* rocas aborregadas

rock *f.* roca, piedra, peña; *m.* jejo, roquedo || ***basement rock*** *f.* roca de basamento; *m.* sótano, basamento, zócalo || ***cap rock*** *f.* cresta de domo salino, roca que cubre; *m.* capuchón, estrato de cobertura || ***clastic rock*** *f.* roca clástica || ***country rock*** *f.* rocas vecinas, roca de espaldos, roca del lugar, roca madre || ***cover rock*** *f.* roca de cubierta || ***crushed rock*** *f.* roca picada || ***crystalline rock*** *f.* roca cristalina || ***extrusive rock*** *f.* roca efusiva || ***host rock*** *f.* roca huesped, roca encajonante || ***igneous rock*** *f.* roca ígnea || ***intrusive rock*** *f.* roca intrusiva || ***loose rock*** *m.* [Méx.] bronco || ***massive barren rock*** *f.* puertas || ***metamorphic rock*** *f.* roca metamórfica || ***plutonic rock*** *f.* roca plutónica || ***rock burst*** *m.* estallido por presión || ***rock fall*** *m.* deslave, desplome de montaña; *f.* caída de roca, alud de roca || ***rock flour*** *f.* harina de roca || ***rock glacier*** *m.* glaciar de roca || ***rock mass*** *m.* mole, peñol, peñon || ***rock mechanics*** *f.* mecánica de rocas || ***rock pile*** *f.* trinchera || ***rock salt*** *f.* sal gema, sal de roca, halita || ***rock shelf*** *f.* cama de roca || ***rock slide*** *f.* alud de rocas; *m.* chiflón || ***rock unit*** *f.* unidad litoestratigráfica || ***rock waste*** *m.* desperdicio rocoso || ***sedimentary rock*** *f.* roca sedimentária || ***source rock*** *f.* roca madre, roca fuente, roca generadora, roca petrolífera || ***submerged rock*** *m.* cabezo || ***volcanic rock*** *f.* roca volcánica || ***wall rock*** *f.* roca de caja, roca vecina

rocker arm *m.* balancín, balancín de válvulas

rocking a well *f.* iniciación de surgencia por balanceo de presión

rocky *a.* rocoso; fragoso, escabroso, pedregoso, pedrizo, pétreo, rocalloso, roqueño || ***rocky ground*** *m.* chancal, canchal, llerón, pedragal, pedriscal, peñascal; *f.* llera || ***rocky place*** *f.* roqueda; *m.* roquedal

rod *f.* varilla, barra, baliza, mira, vara; vareta, (meas.) pértiga; *m.* vástago || ***divining rod*** *f.* vara mágica, vareta divinatoria, varilla mágica || ***level rod*** *m.* estadal, jalón de mira, mira de corredera; *f.* estadia || ***leveling rod rod*** *f.* estadia, mira de nivelación || ***lightening rod*** *m.* pararrayos || ***measuring rod*** *f.* vara de medir, vara de agrimensor || ***rod man*** *m.* jalonero, portamira, estadalero, [Méx.] balicero, portabandera || ***rod shaped*** *a.* bastoniforme, bacular || ***rod target*** *f.* mira || ***stadia rod*** *f.* mira taquimétrica, varilla estadimétrica || ***sucker rod*** *f.* varilla de bombeo, varilla de succión || ***survey rod*** *m.* jalón, reglón; *f.* regla, vara de topografía

roepperite *f.* roeperita

roesslerite, rösslerite *f.* roeslerita

roggianite *f.* rogianita

roll *f.* onda, ola || ***ground roll*** *f.* onda superficial

roll *vb.* laminar, arrollar, rodar

roller *m.* rodillo || ***salt roller*** *m.* rodillo de sal || ***shale roller*** *m.* rodillo de arcilla

rolling mill *m.* taller de laminación

rollover *m.* cierre contra la falla

römerite *f.* weslienita

roof *m.* cielo, techo; (min.) pendiente; *f.* cubierta, tapa || ***roof pendant*** *m.* colgajo

room *m.* (min.) anchurón; *f.* cámara, sala || ***conference room*** *f.* sala de conferencias || ***control room*** *f.* sala de control, sala de gobierno || ***drafting room*** *f.* sala de proyectos || ***pump room*** *f.* sala de bombas || ***room and pillar*** *a.* cámaras y pilares || ***room temperature*** *f.* temperatura ambiente || ***work room*** *f.* sala de trabajo

root *f.* raiz || ***square root*** *f.* raiz cuadrada || ***root zone*** *f.* zona de raíz

rope *f.* cuerda, piola; *m.* cable

ropy lava *f.* lava acordonada

roscoelite *f.* roscoelita

rose (compass) *f.* rosa de los vientos

rose diagram *m.* diagrama roseta

rosembuschite *m.* zircón-jacinto, zircón-pectolita

rose diagram *m.* diagrama roseta

rose quartz *m.* cuarzo rosada, rubí de bohemia

rosette *f.* escarapela

rostrum *m.* rostro; *f.* guarda

rotary *a.* rotatorio, giratorio, rotativo || ***rotary bit*** *f.* broca para equipo rotatorio || ***rotary drill*** *f.* barrena rotatorio, taladro rotatorio || ***rotary drive*** *m.* mando rotatorio || ***rotary table*** *f.* mesa rotatoria

rotate *vb.* girar

rotation *f.* rotación; *m.* giro

rotational *a.* rotatorio, rotativo, rotacional || ***rotational fault*** *f.* falla rotativa, falla en pivote

rotifer *m.* rotífero

rough *a.* áspero, tosco, rugoso, (geog.) arisco, accidentado, cerril, quebrado, brusco || ***rough ground*** *m.* reventadero

roughneck *m.* ayudante de perforador, peón de perforación, [Méx., Col.] chango

roughness *m.* accidente; *f.* aspereza

round *a.* redondo || ***round trip*** *m.* (perf.) viaje de ida y vuelta; *f.* entrada y salida de la tubería, carrera completa

rounded *a.* redondeado, rodado

roundness *f.* redondez

roustabout *m.* obrero en trabajos de producción, mozo, operario, peón, peón de cuadrilla, peón suelto

route *f.* ruta, trayectoria; *m.* recorrido, trayecto

row *f.* fila, hilera; *m.* renglón

royalty *f.* regalía; *m.* derecho industrial, (com.) canón || ***fee royalty*** *m.* canón del propietario || ***fixed royalty*** *m.* canón fijo || ***gross royalty*** *m.* canón bruto || ***landowner royalty*** *f.* parte del propietario || ***non-participating royalty*** *m.* canón exento de cargos || ***ore royalty*** *m.* tequio || ***overriding royalty*** *m.* canón de adición || ***royalty interest*** *m.* canón de participación

RPM (revolutions per minute) *f.* revoluciones por minuto

rubber *m.* caucho, [Méx.] hule; *f.* goma

rubble *m.* cascote, desmonte, ripio; *f.* rocalla

rubellite *f.* rubelita

rubidium *m.* rubidio

ruby *m.* rubí || ***rough or unpolished ruby*** *m.* cabuchón, cabujón, rubí áspero || ***ruby copper*** *f.* cuprita || ***ruby silver*** *f.* pirargirita, proustita, pétanque, petlanque; *m.* cardenillo, rosicler || ***star ruby*** *m.* rubí estrellado

rudimentary *a.* rudimentario

rudistid *f.* rudista; *m.* rudístido

rudite *f.* rudita, psefita

ruga *f.* arruga

rugged *a.* abrupto, áspero

rugose *a.* rugoso

rugosity *f.* rugosidad

rule *f.* regla; *m.* reglamento || ***phase rule*** *f.* regla de las fases || ***rule of thumb*** *f.* regla práctica || ***rule of v's*** *f.* regla de uves || ***slide rule*** *f.* regla de cálculo

ruler *f.* regla, medida

run *f.* tanda, jornada, cochura, cochada; *m.* ciclo de operación || ***test run*** *f.* operación de ensayo, operación de prueba

run *vb.* correr || ***runoff*** *vb.* derramarse, escurrirse || ***run out*** *vb.* agotarse, terminarse, vaciar || ***run pipe*** *vb.* (perf.) bajar cañería

running *a.* corriente || ***running water*** *f.* agua corriente || ***running wild*** *a.* fuera de control

runoff *m.* escurrimiento superficial, aflujo, aporte, derramiento, derrame; *f.* afluencia, vertientes, aportación

Rupelian *m. & a.* Rupeliense, Rupeliano

rupture *f.* ruptura, quebradura, fractura, quebrantadura; *m.* rompimiento, quebrantamiento || ***rupture strength*** *f.* resistencia a la ruptura; *m.* límite de ruptura

Ruscinian *m. & a.* Rusciniense, Rusciniano

rust *m.* orín, moho; *f.* herrumbre, corrosión || ***rust remover*** *f.* quitaherrumbre

rusty *a.* mohoso, oxidado

rut *m.* surco, bache

ruthenium *m.* rutenio

rutilated *a.* rutilado, sagenítico

rutile *m.* rutilo

Ryazanian *m. & a.* Riazaniense, Riazaniano

S

S wave *f.* onda S, onda secundario

Saalian *m. & a.* Saaliense, Saaliano

Sabinian *m. & a.* Sabiniense, Sabiniano, Wilcoxiano, Wilcoxiense

saccharoidal *a.* sacaróideo

sack *m.* saco

saddle *f.* silla, silleta, caballete, depresión; *m.* collado

safe *a.* seguro ǁ ***safe load*** *f.* carga admisible

safeguard *vb.* proteger

safety *f.* seguridad ǁ ***safety factor*** *m.* factor de seguridad ǁ ***safety valve*** *f.* válvula de seguridad

sag *f.* comba; *m.* doblegamiento, pandeo ǁ ***sag pond*** *f.* laguna de hundimiento, laguna de desplome

sag *vb.* combar

sagittate *a.* sagital

Sahelian *m. & a.* Saheliense, Saheliano

Sakmarian *m. & a.* Sakmariense, Sakmariano

sal ammoniac *m.* amoniaco

salar *m.* saladar

salary *m.* sueldo, salario

sale(s) *f.* venta(s) ǁ ***bulk sale(s)*** *f.* venta a granel, venta al por mayor ǁ ***cash sale(s)*** *f.* venta a contado ǁ ***retail sale(s)*** *f.* venta al por menor ǁ ***whole sale(s)*** *f.* venta al por mayor

salesman *m.* agente vendedor, dependiente, vendedor

salient *m.* saliente; resalte, resalto; *a.* saliente

Salinan *m. & a.* Saliniense, Saliniano

saline *a.* salado, salino, salobral ǁ ***saline aquifer*** *m.* salacuifer

salinity *f.* salinidad, salsedumbre

salinization *f.* salinización

Salopian *m. & a.* Salopiano, Salopiense

salt *f.* sal; *m.* cloruro de sodio, [Pe.] cachí ǁ ***impure salt*** *f.* sal de barco, saltierra ǁ ***pure salt*** *f.* sal de corazón ǁ ***rock salt*** *f.* sal de roca, gema, sal piedra, sal pedres, sal común ǁ ***salt and pepper*** *a.* blanco y negro ǁ ***salt deposit*** *m.* salitral ǁ ***salt dome*** *m.* domo salino, diapiro; *f.* cúpula salina ǁ ***salt flats*** *f.* salada; *m.* salado, salar ǁ ***salt lake*** *f.* marisma ǁ ***salt lick*** *m.* lamedero, salegar ǁ ***salt marsh*** *f.* albina, salina; *m.* saladar, salar ǁ ***salt mine*** *m.* salar, salero; *f.* salina, mina de sal ǁ ***salt pan*** *f.* salina; *m.* salar ǁ ***saltpeter*** *m.* nitrato de chile, nitro, salitre ǁ ***saltpeter deposit*** *m.* salitral; *f.* salitrera ǁ ***salt pool*** *f.* albina ǁ ***salt tectonics*** *f.* halocínesis ǁ ***saltwater*** *f.* agua salada, agua salobre ǁ ***salt works*** *f.* salinas

salt *vb.* falsificar el mineral

saltation *f.* saltación

saltiness *f.* salobridad

saltpeter *m.* salitre, nitro, caliche

saltpetrous *a.* salitral, salitroso

salty *a.* salado, salino, salinero, salobre, salobreño

salvage *m.* salvamento

samarium *m.* samario

samarskite *f.* itroilmenita, ytroilmenita, nuevita, uranotantalita, yttrocolombita, yttroniobita

sample *f.* muestra, ejemplar; *m.* testigo ǁ ***channel sample*** *m.* muestreo por canaleta ǁ ***chip sample*** *m.* muestreo por picado ǁ ***sample catcher*** *f.* recogemuestras

sample *vb.* muestrear, probar, catear, catar

sampler *m.* probador, extractor de muestras, cateador, cojemuestras, muestrador, sacador; *f.* recogemuestras, sacamuestras ǁ ***bottom sampler*** *m.* probador de fondo ǁ ***sidewall sampler*** *f.* sacamuestras de pared

sampling *m.* muestreo, cateo, desmuestre

sand *f.* arena ǁ ***coarse sand*** *f.* arena gruesa, arena gorda, arena recia; *m.* sablón ǁ ***fine sand*** *f.* arena fina, arenilla, jable ǁ ***gas sand*** *f.* arena gasífera ǁ ***gold-bearing sand*** *f.* arena aurífera ǁ ***loamy sand*** *f.* arena franca ǁ ***oil sand*** *f.* arena petrolífera ǁ ***pay sand*** *f.* arena productiva ǁ ***quicksand*** *f.* arena movediza ǁ ***ratty sand*** *f.* arena sucia ǁ ***sandbank*** *m.* banco de arena; *f.* barra, restinga ǁ ***sandbar*** *f.* barra arenosa, barra litoral, barra, restinga; *m.* médano, secano, tómbolo arenal, banco de arena, alfaque || ***sandblast*** *vb.* sopletear con arena || ***sand drift*** *f.* acumulación de arena || ***sand dune*** *f.* duna de arena, algaida || ***sand gall*** *f.* torca || ***sand percent map*** *m.* mapa de porcentaje de arena || ***sand pit*** *f.* arena de cava, arena de fosa, arena de guija, arena de hoyo, arena de tajo || ***sand/shale map*** *m.* mapa de relacion arena/pelita (o arena/lutita) || ***sand sheet*** *m.* manto de arena || ***sandstorm*** *f.* tormenta de arena || ***sand trap*** *m.* colector de arena, separador de arena || ***sand up*** *vb.* tapar(se) con arena, enarenar(se) || ***shoestring sand*** *f.* arena cordoniforme, cordones estratigráficos || ***tar sand*** *f.* arena asfáltica || ***volcanic sand*** *f.* peperita

salty *a.* salobre

sandstone *f.* arenisca, [Bol.] amoladera; *m.* asperón, gres || ***feldspar-rich sandstone*** *f.* arenisca rica en feldespatos; *m.* maciño || ***ratty sandstone*** *f.* arenisca sucia

sandstorm *f.* tormenta de arena

sandy *a.* arenoso, arenáceo, arenisco, sabuloso

Sangamonian *m. & a.* Sangamoniense, Sangamoniano

sanidine *f.* sanidina, ortosa sódica, riacolita; *m.* feldespato transparente, feldespato vítreo

Sannoisian *m. & a.* Sannoisiense, Sannoisiano

Sannonian *m. & a.* Sannoniano, Sannoniense

Santonian *m. & a.* Santoniense, Santoniano

sap *f.* savia

sapanthracite *f.* sapantracita

saponification *f.* saponificación

saponify *vb.* saponificar

saponite *f.* saponita, piotina

sapperite *f.* saperita

sapphire *m.* zafiro

sapping *f.* zapa, socavación

saprolite *f.* saprolita

saprolitic *a.* saprolítico

sapropel *f.* sapropelita; *m.* sapropel || ***sapropel clay*** *f.* arcilla sapropélica

saprophyte *f.* saprofita

saprozoic *a.* saprozóico

sard *m.* sardio, sardo; *f.* sardónica

Sardic *m. & a.* Sardiense, Sardiano

sardonyx *m.* sardónice, sardonio, sardónique, sardónix

sarkinite *f.* xantoarsenita

Sarmatian *m. & a.* Sarmatiense, Sarmatiano

sassolite *f.* sassolina

satellite *m.* satélite

satin spar *m.* espato satinado, yeso fibroso

saturate *vb.* saturar, aguachinar

saturated *a.* saturado, aguachento || ***under saturated*** *a.* bajo saturado || ***supersaturated*** *a.* super saturado

saturation *f.* saturación, aguada || ***irreducible water saturation*** *f.* saturación irreducible del agua

Saturn *m.* Saturno

Saucesian *m. & a.* Saucesiense, Saucesiano

saussuritization *f.* saussuritización

savannah *f.* sábana

save *vb.* (comp.) guardar, salvaguardar

savings *m.* ahorros || ***annualized pre-tax savings*** *m.* ahorros anuales brutos

Saxonian *m. & a.* Sajoniense, Sajoniano, Sajónico, Saxoniense, Saxoniano

saxonite *f.* sajonita

S–C fabric *f.* fábrica S–C, textura S–C

scabland *m.* terreno costroso; *f.* roca pelada

scacchite *f.* scachita

scale *f.* escala; incrustación, escama, costra, regla; báscula, balanza, (rust) laminilla; *m.* sarro || ***bar scale*** *f.* escala gráfica || ***scale model*** *m.* modelo a escala || ***scale remover*** *f.* quitacostra || ***time scale*** *f.* escala de tiempo

scale-forming *a.* incrustante

scalenohedral *a.* escalenoédrico || ***scalenohedral calcite*** *m.* dientes de perro

scalenohedron *m.* escalenoedro

scaly *a.* escamoso, escamiforme, sarroso

scale-like *a.* escamiforme

scalenohedral *a.* escalenoédrico

scan *vb.* escandir, escanear

scanner *m.* haz explorador; explorador, explorador de barrido, escáner

scanning *a.* barrido

scaphopod *m.* escafópodo

scapolite *f.* escapolita, parranthina

scar *f.* impresión, huella, cicatriz; *m.* surco || ***avalanche scar*** *m.* surco de alud

scarbroite *f.* tucanita

scarce *a.* escaso

scarp *f.* escarpa; *m.* acantilado, risco, resbalo, talud, pendiente || ***fault scarp*** *f.* escarpa de falla o de fractura; *m.* escalón de falla, escalón de fractura || ***fault line scarp*** *f.* escarpa de línea de falla

scatter *vb.* difundir, dispersar

scattered *a.* disperso, diseminado

scattering *f.* dispersión, difusión

scavenger *m.* basurero, carroñero, necrófago

scavenging *m.* barrido

scfh (standard cubic feet per hour) *m.* pies cúbicos estándard por hora

schedule *m.* programa, horario, itinerario

scheelite *f.* escheelita, scheelita

schematic *a.* esquemático

science *f.* ciencia

scientific *a.* científico

scientist *m. & f.* científico

schist *m.* esquisto || ***amphibolite schist*** *m.* esquisto anfibolítico || ***chlorite schist*** *m.* esquisto clorítico || ***green schist*** *m.* esquisto clorítico, esquisto verde || ***mica schist*** *m.* esquisto micáceo || ***weathered argillaceous schist*** *f.* láguena

schistoid *m.* esquistóideo

schistose *a.* esquistoso

schistocity *f.* esquistosidad, foliación

schizodont *m. & a.* esquizodonto

schlieren *f.* faja; *m.* schlieren

schollendome *m.* túmulo; *f.* ampolla de lava

schorl *m.* chorlo

schroekingerite *f.* schroekingerita

scintillometer *m.* centellómetro, centellador, cintilómetro, contador de centelleo, escintilómetro

scissors fault *f.* falla pivotal (o giratoria, o tijera)

sclerenchyme *f.* esclerénquima

sclerite *f.* esclerita

scleroblast *m.* escleroblasto

scolecite *f.* escolecita; *m.* weissian

scolecodont *m.* escolecodonto

scoop mark *f.* marca de cavidad

scorch *vb.* chamuscar, abrasar

scope *m.* alcance, rango

scoria *f.* escoria, horrura, ramosita; *m.* cagafierro || ***iron scoria*** *m.* moco

scorodite *f.* jorgyanita

scour *f.* abrasión, erosión || ***scour and fill*** *a.* erosión y relleno de cauce || ***scour cast*** *m.* turboglifo || ***scour mark*** *f.* marca de arrastre

scour *vb.* desoxidar, desengrasar, limpiar, lavar; arrastrar

scoured *a.* erosionado

scouring *f.* limadura, erosión, socavación

scout *vb.* explorar, reconocer

scout *m.* explorador, informador; *f.* exploradora, exploracionista

scrap *f.* chatarra; *m.* rezago || ***scrap iron*** *f.* chatarra

scrape *vb.* raer, raspar

scraper (casing) *m.* raspatubos

scraping *f.* raedura

scrapings *f.* raspadura

scratch *f.* rasgadura, marca, estría; *m.* tachón || ***scratch mark*** *f.* raya

scree *m.* derrubios

screen *f.* criba, tela metálica, zaranda, malla; (comp.) pantalla; *m.* colador, cedazo, cernedor, cernidor, garbillo || ***computer screen*** *f.* pantalla || ***CRT screen*** *f.* pantalla de tubo de rayos catódicos || ***mud screen*** *f.* criba de lodo, zaranda, tela metálica de colador sacudidor; *m.* colador de lodo || ***sand screen*** *m.* filtro de pedregullo, tubo colador || ***screen analysis*** *m.* análisis por tamizado || ***video screen*** *f.* pantalla

screen *vb.* cernir, cerner, cribar, tamizar, acribar, garbillar, zarandar, zarandear

screened *a.* zarandado

screening *f.* cribadura; *m.* zarandeo

screenings *f.* cerniduras, zaborras; *m.* recebo

screw *m.* tornillo || ***levelling screw*** *m.* tornillo nivelador (o de nivelación)

scrub *vb.* depurar, lavar, limpiar

scrubber *m.* depurador || ***gas scrubber*** *m.* depurador de gas, lavador de gas; depurador de aire || ***steam scrubber*** *m.* separador de agua

scud *f.* pirita de carbón

sculpture, sculpt *vb.* esculpir, labrar

sculptured *a.* labrado, esculpido

scyphozoan *m.* escifozoos

Scythian *m. & a.* Scytiense, Scytiano

sea *m. & f.* mar || ***Bering Sea*** *m.* mar de Bering || ***Black Sea*** *m.* mar Negro || ***Caribbean Sea*** *m.* mar Caribe || ***Caspian Sea*** *m.* mar Caspio || ***China Sea*** *m.* mar de China || ***deep sea*** *a.* marino profundo || ***epeiric sea*** *m.* mar epicontinental || ***Mediterranean Sea*** *m.* mar Mediterráneo || ***North Sea*** *m.* mar del Norte || ***Red Sea*** *m.* mar Rojo || ***sea coast*** *f.* costa marina || ***sea floor*** *m.* fondo del mar; lecho, piso, o fondo marino || ***sea floor spreading*** *f.* expansión del fondo oceanico || ***sea ice*** *m.* hielo marino || ***sea level*** *m.* nivel del mar || ***sea mount*** *m.* promontorio submarino, guyot || ***seashore*** *f.* ribera marina, costa marina || ***sea wall*** *m.* malecón

seal *m.* sello, roca sello; cierre; hermeticidad

seal *vb.* sellar, tapar, cerrar

sealed basin *f.* cuenca cerrada

sealing fault *f.* falla sellante

seam *m.* horizonte, criadero, jazido, venero; *f.* hendedura, grieta, fractura, fisura, capa, vena, veta || ***ore seam*** *f.* hijuela

seamed *a.* avetado

seamless *a.* sin costura

search *f.* búsqueda, investigación, busca, indagación, reconocimiento

search *vb.* buscar, reconocer, explorar

seashore *f.* ribera, playa marina

season *f.* estación || ***wind and snow season*** *m.* paramar

seasonal *a.* estacional

sea star *m.* asteroideo; *f.* estrella de mar

sea urchin *m.* equinoideo, erizo de mar

seaward stepping *m.* avance hacia el mar

seaweed *f.* alga marina

second *m.* segundo; *a.* segundo

secondary *a.* secundario || ***secondary deposit*** *m.* depósito secundario || ***secondary enrichment*** *m.* enriquecimiento secundario || ***secondary growth*** *m.* crecimiento secundario || ***secondary porosity*** *f.* porosidad secundaria || ***secondary recovery*** *f.* producción secundaria, segunda recuperación; *m.* recobro secundario

second of arc *m.* segundo de arco

secretion *f.* secreción

section *f.* sección; *m.* perfil, sector || ***balanced section*** *f.* sección balanceada; *m.* transecto balanceado || ***columnar section*** *m.* perfil estratigráfico || ***condensed section*** *f.* sección condensada || ***cross section*** *m.* transecto, corte transversal; *f.* sección transversal, sección || ***expanded section*** *f.* sección aumentada || ***geologic section*** *f.* sección geológica || ***lithologic section*** *m.* perfil litológico || ***missing section*** *f.* sección perdida || ***polished section*** *f.* sección pulida, sección pulimentada || ***restored section*** *f.* sección restaurada || ***stratigraphic section*** *f.* sección geológica || ***thin section*** *f.* lámina delgada, sección delgada || ***type section*** *f.* localidad tipo

sector *m.* sector

secure *a.* firme, seguro

securities *m.* valores || ***marketable securities*** *m.* valores vendibles (o cotizables)

security *f.* seguridad

sedentary *a.* sedentario

sediment *m.* sedimento, depósito, asiento; *f.* zurrapa || ***aeolian sediments*** *m.* sedimentos eólicos || ***clastic sediments*** *m.* sedimentos clásticos || ***consolidated sediments*** *m.* sedimentos consolidados

sedimentary *a.* sedimentario || ***sedimentary breccia*** *f.* brecha sedimentaria || ***sedimentary iron ore*** *m.* mineral de hierro sedimentario || ***sedimentary petrography*** *f.* petrografía sedimentaria || ***sedimentary structure*** *f.* estructura sedimentaria

sedimentation *f.* sedimentación

sedimentology *f.* sedimentología

seep *f.* (geol.) manifestación, emanación, exudación, filtración, percolación, salida || ***gas seep*** *f.* fuente de gas || ***oil seep*** *f.* manadero de petróleo, rezumadero, [Méx.] chapopotera || ***water seep*** *m.* brotadero

seep *vb.* colar, colarse, pasar, brotar, rezumarse, filtrarse, rezumar

seepage *f.* coladura, filtración, percolación, sorbencia, [Méx.] chapopotera; *m.* manadero

segment *m.* segmento

segmentation *f.* segmentación

segregate *vb.* segregar

segmented *a.* segmentado

segregation *f.* segregación, separación

seif ridge *f.* cresta seif

seism *m.* sismo

seismic *f.* sísmica; *a.* sísmico || ***2D seismic*** *f.* sísmica de dos dimensiones || ***3D seismic*** *f.* sísmica de tres dimensiones || ***seismic line*** *f.* linea sísmica, sísmica || ***seismic prospecting*** *f.* prospección sísmica || ***seismic party*** *f.* comisión sísmica, brigada sísmica || ***seismic record*** *m.* sismograma || ***seismic reflection*** *f.* reflexión sísmica || ***seismic refraction*** *f.* refracción sísmica || ***seismic signal*** *f.* señal sísmica || ***seismic waves*** *f.* ondas sísmicas

seismicity *f.* sismicidad

seismite *f.* sismita

seismoscope *m.* sismescopo

seismogram *m.* sismograma

seismograph *m.* sismógrafo || ***reflection seismograph*** *m.* sismógrafo de reflexión || ***refraction seismograph*** *m.* sismógrafo de refracción || ***seismograph crew*** *f.* comisión de sismólogos, cuadrilla de sismólogos || ***seismograph party*** *f.* expedición de sismólogos, cuadrilla de sismólogos || ***seismograph recorder*** *m.* observador || ***seismograph survey*** *m.* estudio sismográfico; *f.* investigación sismográfica || ***synthetic seismograph*** *m.* sismógrafo sintético

seismographic *a.* sismográfico

seismologic *a.* sismológico || ***seismologic observatory*** *m.* observatorio sismológico

seismologist *m.* sismólogo

seismology *f.* sismología

seismometer *m.* sismómetro

seismometry *f.* sismometría

seize *vb.* embargar

seizure *m.* embargo

selbite *f.* plata azul de catorce

select *vb.* seleccionar, escoger, elegir

selected *a.* seleccionado

selection *f.* selección || ***natural selection*** *f.* selección natural

selective *a.* selectivo

selectivity *f.* selectividad

selector *m.* selector || ***range selector*** *m.* selector de campo

selenite *f.* selenita; *m.* yeso espático, yeso especular || ***crystal selenite*** *m.* padre de sal

selenium *m.* selenio

self *a.* por sí mismo, auto || ***self-potential*** *a.* autopotencial, potencial espontáneo

seller *m.* vendedor

selling *f.* venta

selva *f.* selva

selvage *m.* orillo; salbanda

semiarid *a.* semiárido

semicrystalline *a.* semicristalino

semisubmersible *a.* semisumergible, semisumersible || ***semisubmersible rig*** *f.* plataforma semisumergible

semivariogram *m.* semivariograma

senescence *m.* envejecimiento

senile *a.* senil, geróntico

senility *m.* senectud

Senonian *m. & a.* Senoniense, Senoniano

sense *m.* sentido

sensitive *a.* sensible

sensitivity *f.* sensibilidad

sensor *m.* captor, sensor || ***electromagnetic sensor*** *m.* sensor electromagnético || ***pressure sensor*** *m.* sensor de presión || ***sun sensor*** *m.* sensor solar

separate *vb.* separar, despegar, desprender, disgregar

separation *f.* separación, disociación || ***normal horizontal separation*** *f.* distancia horizontal de afloramientos || ***vertical separation*** *f.* distancia vertical

separator *m.* separador, deflegmador || ***cyclone separator*** *m.* separador ciclonal

sepia *f.* sepia

sepiolite *f.* sepiolita; espuma de mar

septal *a.* septal

septate *a.* con septos, con tabiques

septibranch *a.* septibranquiado

septic tank *m.* tanque séptico

septotheca *f.* septoteca

septum *m.* septo, tabique; *f.* septa

Sequanian *m. & a.* Secuaniense, Secuaniano

sequence *f.* secuencia; *m.* orden de sucesión || ***deformation sequence*** *f.* secuencia de deformación || ***out of sequence*** *f.* fuera de secuencia || ***sequence boundary*** *m.* límite de la secuencia || ***sequence of deformation*** *f.* secuencia de deformación || ***sequence stratigraphy*** *f.* estratigrafía de secuencias, estratigrafía secuencial || ***stratigraphic sequence*** *f.* secuencia estratigráfica

sericite *f.* sericita

serial *a.* serie

series *f.* serie, sarta, sucesión, tanda; *m.* paquete || ***homologous series*** *f.* serie homóloga

serpentine *f.* serpentina, zöblitzita || ***fibrous serpentine*** *f.* webskyita, satelita; *m.* xylotilo

serpentinite *f.* serpentinita

serrate, serrated *a.* serrado, aserrado, dentado, endentado

server *m.* (comp.) servidor

service *m.* servicio; *f.* reparaciones || ***service station*** *f.* estación de servicio, estación de gasolina; *m.* surtidor de gasolina

service *vb.* reparar, hacer ajustes, rehabilitar, mantener

sessile *a.* sésil

set *m.* juego, conjunto || ***fracture sets*** *m.* juegos de fracturas || ***setup*** *f.* instalación, montaje, arreglo

set *vb.* (survey, levan.) establecer, amojonar || ***set pipe on bottom*** *vb.* asentar o clavar la entubación en el fondo del pozo || ***set up*** *vb.* instalar

setting *m.* marco; fraguado || ***setting pipe*** *f.* (perf.) entubación || ***tectonic setting*** *m.* marco tectónico

settle *vb.* asentarse, sedimentarse

settling *m.* asentamiento; *f.* sedimentación, deposición || ***differential settling*** *m.* asentamiento diferencial || ***settling basin*** *m.* estanque de sedimentación || ***settling period*** *m.* período de asentamiento || ***settling pond*** *m.* depósito de asentamiento

severance pay *f.* indemnización, cesantías

sewer *f.* cloaca, acequia; *m.* desagüe, albañal

sextant *m.* sextante

seyenite *f.* ortoclasita, pertosita

shade, shadow *f.* sombra

shade *vb.* sombrar

shaded *a.* sombreado

shading *m.* sombreado

shadow *f.* sombra || ***shadow zone*** *f.* zona de sombra

shaft *m.* pozo; (min.) ahonde, chiflón; eje, arbol, [Col.] apique, [Bol., Ch., Pe.] pique; *f.* chimenea, escotilla, lumbrera, sima || ***auxiliary shaft*** *f.* bocamejora || ***inclined shaft*** *f.* tiro de recueste, [Pe.] media barreta || ***mine shaft*** *m.* pozo principal, tiro de mina || ***shaft collar*** *f.* bocamina || ***shaft slope*** *m.* arrastre || ***stoping shaft*** *m.* zurdo || ***ventilation shaft*** *f.* lumbrera

shake *vb.* sacudir, agitar, zarandar

shaker *f.* sacudidora; *m.* batidor, agitador

shaking *f.* sacudida

shale *f.* lutita, pelita, arcilla, [Vz.] arcilla laminar; pizarra sedimentaria, [Sp.] pizarra arcillosa || ***bituminous shale*** *f.* lutita bituminosa || ***carbonaceous shale*** *f.* lutita carbonosa, lutita carbonífera || ***oil shale*** *f.* turbonita, lutita petrolífera, kerogenita, lutita bituminosa || ***shale above or below coal*** *m.* abrigo del carbon || ***shale break*** *f.* lutita intrusa; *m.* estrato intruso de lutita, interyacente de lutita || ***shale out*** *vb.* transformarse en arcilla, volverse en lutita || ***shale parting in coal*** *f.* soleta || ***shale shaker*** *m.* agitador de lodo, colador de lodo, colador vibratorio, tamiz vibratorio para lodos; *f.* zaranda vibratoria, tela metálica de colador sacudidor

shallow *a.* somero, poco profundo, capotero, pando || ***shallow water*** *f.* agualita, agua somera

shaly *a.* arcillosa, lutítico

shape *f.* forma, figura, configuración

shape *vb.* formar

shaped *a.* en forma de

shapeless *a.* amorfo

shard *m.* fragmento de vidrio volcánico; *f.* triza

share *f.* (com.) acción, parte, participación, prorata; *m.* (porción) contingente, lote || ***limited liability share*** *f.* barra aviada || ***market share*** *f.* participación en el mercado || ***share paying assessment*** *f.* barra aviadora

share *vb.* proratear; compartir

shareholder *m. & f.* accionista, socio accionista

sharp *a.* pronunciado, bien definido; agudo; fuerte

sharpener *f.* afiladora

shatter *vb.* romper, romper en pedazos

shattering *m.* quebrantamiento

shear *f.* cizalla; *m.* cizallamiento, corte, [P.R.] cortante || ***angular shear*** *f.* cizalla angular || ***flexural shear*** *f.* cizalla flexural || ***inclined shear*** *f.* cizalla inclinada || ***pure shear*** *f.* cizalla pura || ***Reidel shear*** *f.* cizalla de Riedel || ***simple shear*** *f.* cizalla simple || ***shear cleavage*** *f.* foliación de cizalla || ***shear joint*** *f.* diaclasa de cizalla, diaclasa de cortante || ***shear strain*** *f.* deformación interna por cizalla || ***shear strength*** *f.* resistencia al corte, resistencia al cizallamiento || ***shear stress*** *m.* esfuerzo cortante, esfuerzo cizallante, cortadura, cizallamiento || ***shear wave*** *f.* onda cizallante, onda cortante

shear *vb.* cizallar, recortar

shearing *m.* deslizamiento, cizallamiento

sheath fold *m.* pliegue de cubierta

sheer *a.* acantilado

sheet *f.* (geog.) hoja; (geol.) capa, placa, lámina, laminilla, sábana; *m.* estrato || ***metal sheet*** *f.* plancha || ***sheet flood*** *f.* inundación de avenida || ***sheet flow*** *m.* flujo paleal || ***sheet sand*** *m.* manto de arena || ***sheet wash*** *m.* lavaje en mantos, lavaje laminar || ***thrust sheet*** *m.* bloque cabalgante

shelf *f.* plataforma, cama de roca, prominencia; *m.* anaquel, cantil, banco || ***continental shelf*** *f.* plataforma continental, pendiente continental; *m.* zócalo || ***continental break shelf*** *m.* borde continental

shelfal *a.* relativo a plataforma

shell *f.* concha; [Arg.] conchilla; *m.* carapacho, caparazón, casco, casquete || ***shell hash*** *f.* brecha conchífera

shelly *a.* conchudo

shield *m.* escudo, nucleo, kratógeno; *f.* pantalla || ***face shield*** *f.* careta de mano || ***Precambrian shield*** *m.* escudo precambrico || ***shield volcano*** *m.* volcán en escudo

shielded *a.* blindado, protegido

shift *m.* desplazamiento, salto horizontal, salto, rechazo, (laboral) cuarto, turno

shifting sand *a.* acarreadizo

shining *a.* relumbroso

shiny *a.* lustroso

shingle *f.* grava marina

shingled *a.* escalonado, imbricado

ship *m.* buque, barco || ***drill ship*** *m.* buque de perforación

ship *vb.* mandar, enviar, transmitir, embarcar, transportar

shipment *m.* embarque, despacho, cargamento; *f.* partida, remisión

shoal *m.* bajío, escollo, banco de arena, alfaque, bajo, cayo, rompiente; *f.* restinga, barra de río

shoaling *a.* somerizando

shock *m.* choque, golpe, temblor; *f.* sacudida

shock *vb.* choquear, golpear

shoestring sand *f.* arena cordoniforme, cintas estratigráficas de arena; *m.* cordones estratigráficos

shonkinite *f.* shonkinita

shoot *vb.* explotar, dinamitar

shooting *m.* torpedeamiento; *f.* dinamitación || ***shooting a well*** *f.* dinamitación de un pozo || ***shooting star*** *f.* estrella fugaz

shop *m.* taller; *f.* tienda

shore *f.* orilla, costa, playa, marina; *m.* (min.) ademe, adema; (geog.) litoral || ***back shore*** *f.* playa alta || ***fore shore*** *f.* playa baja || ***shore face*** *f.* anteplaya, zona de inframarea; *m.* frente de ribera

shore *vb.* acodalar, acodar, ademar, apear

shoring *m.* acodalamiento, acodamiento

shoreline *f.* orilla, línea de litoral, línea costanera

short *a.* corto || ***short term*** *a.* corto plazo

shorten *vb.* acortar

shortening (geol.) *m.* acortamiento; *f.* contracción

shot *m.* tiro, barreno, disparo; *f.* voladura || ***patterned shot*** *m.* tiro en pauta || ***shot hole*** *f.* hoya de la explosión, perforación de disparo || ***shot line*** *f.* línea de fuentes, línea de tiros || ***shot pattern*** *m.* dispositivo de tiro || ***shot point*** *m.* punto tiro, punto de tiro, punto de disparo || ***shot point grouping, shotpoint spacing*** *m.* espaciamientos sísmicos

shovel *f.* pala || ***small shovel*** *f.* paleta

shovel *vb.* palear

shoveller *m.* paleador, palero

shovelling *m.* paleaduro, paleaje

show of oil or gas *f.* manifestación, impregnación; *m.* vestigio, rastro, indicio

show *vb.* presentar

shower *m.* chubasco, chaparrón; *f.* lluvia

shredder *f.* trituradora

shrimp *m.* camarón

shrink *vb.* contraerse, acortarse

shrinkage *m.* encogimiento; *f.* contracción, reducción || ***shrinkage crack*** *f.* grieta de desecación

shrinking *f.* contracción

shrub *m.* arbusto; *f.* mata

shungite *f.* chungita

shunt *f.* derivación

shut *a.* cerrado, parado || ***shutdown*** *m.* cierre || ***shut in*** *a.* tapado, cerrado || ***shut in pressure*** *f.* presión encerrada, presión hermética || ***shutoff valve*** *f.* válvula de paso

shut-in *vb.* cerrar

shutter lens *m.* obturador

sial *m.* sial

sialic *a.* siálico

sialite *f.* sialita

Sicilian *m. & a.* Siciliano, Siciliense

side *m.* lado, costado, flanco; *f.* margen; *a.* ladero || ***sidetrack*** *vb.* desviar || ***side view*** *f.* vista lateral || ***side wall*** *m.* [Arg.] hastial || ***side wall core*** *m.* testigo lateral, testigo de pared || ***side wall sampler*** *f.* sacamuestras de pared

sideswipe *m.* barrido lateral

sidetrack *f.* desviación

sidetracking *f.* desviación del hoyo

siderite *f.* siderita, siderosa, calibita, chalibita, calamita, junkerita; *m.* hierro espático, carbonato de hierro, espato de hierro, espato ferrífero, [Bol.] cachí, [Sp.] campanil

Siegenian *m. & a.* Siegeniense, Siegeniano

sierra *f.* sierra

sierranite *f.* sierranita

sieve *m.* cedazo, tamiz, cernedor, cernidor; *f.* criba, zaranda || ***sieve texture*** *f.* textura cribosa

sieve *vb.* ahechar, cerner, cernir, cribar, zarandar, zarandear, tamizar

sieved *a.* zarandado

sieving *m.* zarandeo

sieze *vb.* embargar, acribar, confiscar

siezure *m.* embargo; *f.* confiscación

siferna *f.* siferna

sift *vb.* acribar, ahechar, cribar, tamizar

sifter *m.* tamiz

sifting *f.* cribadura

sight *f.* mira, (elect.) pínula, visual

sigmoid *a.* sigmoideo

sign *f.* señal, seña || ***optic sign*** *m.* signo óptico

signal *f.* señal || ***electromagnetic signal*** *f.* señal electromagnético

signatory *m.* signatorio; *m. & f.* firmante

significance *f.* significación; *m.* significado, realce

significant *a.* significante

signpost *m.* señalizador

silcrete *f.* silcreta

Silesian *m. & a.* Silesiense, Silesiano

silex *m.* sílex

silexite *f.* silexita, paracidita

silica *m.* sílice, sílex

silicate *m.* silicato; *a.* silicado || ***alumino silicate*** *m.* silicato de aluminio || ***cyclosilicate*** *m.* silicato en anillo || ***inosilicate*** *m.* silicato en cadena || ***nesosilicate*** *m.* silicato aislado || ***phyllosilicate*** *m.* silicato en hoja || ***tectosilicate*** *m.* silicato en andamiaje

siliceous *a.* silícico, silíceo, silicoso

silicic *a.* silícico

siliciclastic *a.* siliciclástico, silicoclastico

silicification *f.* silicificación, silificación

silicified *a.* silicificado

silicify *vb.* silicificar

silicoflagellate *m.* silicoflagelado

silicon *m.* silicio

silicosis *m.* mal de choco, silicosis

silky *a.* sedoso

sill *m.* silo, manto interno, dique concordante, surco, dique estrato, filón; *f.* capa intrusiva, plancha intrusiva

silled basin *f.* cuenca con umbral

sillimanite *f.* silimanita

silt *m.* limo, cieno, arrastres, azolve, barro; *f.* lama

silt *vb.* aterrar, azolvarse

silting up *m.* aluvionamiento, aterramiento, atierre

siltstone *f.* limolita; *m.* limolito

silty *a.* limoso, barroso, fangoso

Silurian *m. & a.* Silúrico, Siluriano; Gotlandiense

silver *f.* plata; *m.* [Bol., Pe.] colqui || ***massive native silver*** *m.* reventón || ***native silver*** *f.* plata bruto, plata blanca, plata nativa, plata natural; *m.* [Bol.] anco, [Pe.] cincel, machacado || ***refined silver*** *f.* secas || ***ruby silver*** *f.* pétanque, petlanque || ***silver amalgam*** *f.* plata pella, [Ch.] arquería || ***silver-bearing*** *a.* argentífero, argentoso || ***silver-bearing carbonate ore*** *m.* azulillo || ***silver-bearing pyrite*** *m.* [Méx.] azurronado || ***silver bullion*** *f.* plata en barra || ***silver chloride*** *f.* plata ceniza; *m.* metal cenizo, [Méx.] bodoque || ***silver ingot weighing 75 lb*** *f.* [Méx.] barra de plata || ***silver mill*** *m.* arrastre || ***silver nitrate*** *m.* nitrato de plata || ***silver ore*** *m.* tacano, [Méx.] azogue, azulaque, [Méx.] calichal, [Ch., Pe.] anco; *f.* tacana, [Sp.] anguena; [Pe.] anquería, colpa || ***silver payments*** *m.* situados || ***silver salts*** *m.* alimonados || ***silver shipments*** *m.* situados || ***silver sulfides*** *m.* alimonados || ***wrought silver*** *f.* plata labrada

silvery *a.* argentado, argénteo, argentino

sima *f.* sima, siferna

simatic *a.* simático

similar *a.* semejante, similar, conforme || ***similar fold*** *m.* pliegue semejante

simple *a.* simple || ***simple harmonic motion*** *m.* movimiento armónico simple || ***simple shear*** *f.* cizalla simple

simulate *vb.* simular, fingir

simulated *a.* simulado

simulation *f.* simulación

simulator *m.* simulador

simultaneous *a.* simultáneo

simultaneously *adv.* simultáneamente, sincronizadamente

Sinemurian *m. & a.* Sinemuriense, Sinemuriano

singenite *f.* kaluszita

single-chambered *a.* unilocular

Sinian *m. & a.* Siniense, Siniano, Siniánico

sinistral *a.* sinestral, siniestro, levógiro, izquierdo

sink *f.* dolina || ***sink hole*** *f.* fosa, dolina, torca; *m.* sumidero, hueco, [Mex.] cenote

sink *vb.* hundir, ahondar, bajar, zahondar || ***sink a hole*** *vb.* perforar un pozo

sinkhole *f.* dolina, torca; fosa de hundimiento, [Cu.] casimba; *m.* sumidero, karst, embudo, [Sp.] avenc, [Méx.] cenote

sinter *f.* toba; incrustación, sinter; *m.* [Méx.] bataque || ***siliceous sinter*** *f.* toba silícea

sinter *vb.* concrecionar

sinuous *a.* sinuoso, tortuoso

sinuosity *f.* sinuosidad

sinus *m.* seno

Siphean *m. & a.* Siniense, Siniano

siphon *m.* sifón
siphonate *a.* sifonado
siphon off *vb.* drenar
siphuncle *m.* sifúnculo
sismondine *f.* sismondina
site *f.* localización, ubicación, situación; *m.* local, sitio; solar, terreno, terruño
site *vb.* ubicar
size *m.* tamaño; *f.* dimensión || ***grain size*** *m.* tamaño de grano
skarn *m.* criadero de contacto; *f.* tactita
skein *m.* ovillo
sketch *m.* dibujo, bosquejo, trazo, croquis, diseño; *f.* esquema || ***sketch book*** *f.* libreta de croquis
sketch *vb.* bosquejar, dibujar
skeletal *a.* esqueletal
skeleton *m.* esqueleto
sketch *m.* dibujo; *f.* esquema
skew *m.* & *a.* sesgo
skewed *a.* oblicuo, sesgado, torcido
skewness *f.* asimetría
skid *vb.* arrastrar, deslizar || ***skid the derrick*** *vb.* trasladar la torre arrastrándola
Skiddavian *m.* & *a.* Skiddaviense, Skiddaviano
skill *f.* pericia, destreza
skim *vb.* desnatar
skimmer *m.* desnatador
skimming *f.* desnatación; *m.* desnatado
skin *f.* piel || ***skin effect*** *m.* efecto superficial, efecto pelicular
sklodowskite *f.* chinkolobwita
skutterudite *f.* skutterudita
sky *m.* cielo || ***skyline*** *m.* horizonte
Skythian *m.* & *a.* Skytiense, Skytiano
slab *f.* laja, hoja, lámina, lancha, lastra, losa, placa, zamarra, tabla
slab *vb.* cortar longitudinalmente
slabby *a.* losáceo
slack *a.* flojedad; *f.* carbonilla || ***slack water, slack tide*** *f.* agua represada, agua tranquila
slake *vb.* desleir, remojar, apagar
slacked lime *f.* cal apagada
slag *f.* escoria, ceniza, costra, chatarra, grasa; *m.* cagafierro, moco || ***slag heap*** *f.* escorial; *m.* grasero
slakings *m.* deleznamiento
SLAR (Side Looking Airborne Radar) *m.* radar aerotransportado de barrido lateral
slate *f.* pizarra; *m.* esquisto || ***slate quarry*** *f.* pizarrería; *m.* pizarral
slaty *a.* pizarroso, pizarreño, esquistoso, apizarrado, pizarrosidad || ***slaty cleavage*** *m.* clivaje esquistoso, clivaje pizarroso, apizarramiento; *f.* pizarrosidad
sledge hammer *f.* almádena, mandarria, acotilla, porra, [Pe.] comba; *m.* macho, marro, mazo
sleet *f.* aguanieve, cellisca
slice *f.* tajada, longa, (geol.) escama, (piedra) rebanada
slice *vb.* cortar, tajar || ***slice off*** *vb.* cercenar
slicing *m.* rebanado, tajado
slickenside *m.* espejo de falla, espejo de fricción, deslizamiento brilloso, liso, plano de resbalamiento (o de fricción), [Col.] jabón; *f.* estrías, relices de fallas, salbanda, superficie de deslizamiento
slickensided *a.* pulido, estriado
slide *m.* deslizamiento, derrumbe || ***slide plane*** *m.* plano de deslizamiento || ***slide rule*** *f.* regla de cálculos || ***slide zone*** *f.* zona de desplazamiento
slide *vb.* deslizar(se), derrumbar(se), resbalar
sliding *m.* deslizamiento
slightly *adv.* ligeramente
slime *m.* fango, légamo; *f.* lama
slimy *a.* lamoso, legamoso
slimhole rig *m.* equipo para perforación de diámetro reducido
sling psychrometer *m.* psicrómetro giratorio
slip *m.* deslizamiento, desplazamiento; derrumbe, desprendimiento, resbalón || ***apparent slip*** *m.* desplazamiento aparente || ***bedding plane slip*** *m.* deslizamiento capa-a-capa, deslizamiento a lo largo de planos de estratificación || ***flexural slip*** *m.* deslizamiento flexural, resbalón flexible || ***slip cleavage*** *m.* clivaje de transposición || ***slip face*** *f.* cara de deslizamiento || ***slip plane*** *m.* plano de deslizamiento

slip *vb.* resbalar, patinar, deslizarse, desprenderse, escurrirse

slippage *m.* resbalamiento, deslizamiento

slippery *a.* resbaloso

slipstream *f.* corriente retrógrada

slit *f.* escotadura, ranura, raja, hendidura

sliver *f.* escama, astilla

slope *f.* pendiente, ladera, falda, costanera, cuesta, declividad, escarpa, inclinación, rampa, [Ch., Col.] cuelga; *m.* declive, talud, abajadero, derrame, escarpe, recuesto, repecho, vertiente, zopetero || ***back slope*** *m.* talud de corte || ***continental slope*** *f.* pendiente continental || ***dip slope*** *f.* pendiente estructural || ***fore slope*** *m.* talud interior || ***natural slope*** *m.* talud de reposo || ***slope fan*** *m.* abanico de talud || ***slope wash*** *f.* lavaduras de falda, lavaduras de pendiente || ***talus slope*** *f.* ladera de derrubios || ***unstable slope*** *f.* pendiente inestable

slope *vb.* taludar, inclinar, inclinarse, sesgar

sloping *a.* descendiendo, inclinado, desnivelado, [Ch.] achiflonado

sloppy *a.* desprolijo || ***sloppy work*** *m.* trabajo desprolijo

slot *f.* ranura, hendedura

slotted pipe *f.* cañería filtro ranurada

slough *m.* derrumbe, desprendimiento, cenegal

slough *vb.* derrumbarse, desprenderse, deslizarse

slow *a.* lento

sludge *m.* cieno, barro, lodo, fango mineral, sedimento; *f.* heces

slug *m.* golpe; *f.* cantidad || ***slug of water*** *m.* golpe de agua

sluice *f.* esclusa, limpiadora; *m.* ábaco, canalón, tragante || ***sluice box*** *f.* caja de la esclusa || ***sluice gate*** *f.* trampa || ***sluice way*** *m.* vaciadero

sluice *vb.* mover con corriente de agua, aflojar

slump *m.* derrumbe, depósito por deslizamiento, derrumbamiento; *f.* avalancha || ***slump mark*** *f.* marca a sotavento || ***slump scar*** *f.* grieta de deslizamiento

slump *vb.* derrumbar(se), echar abajo, desgastar, resbalar

slumped *a.* resbalado

slumping *m.* desgastando, desgaste

slurry *f.* mezcla aguada; lecheda, emulsión, papilla

slush *m.* barro, lodo || ***slush pit*** *m.* foso del lodo de perforación || ***slush pump*** *f.* bomba para barro de circulación

small *a.* pequeño

smaltite *f.* esmaltita, esmaltina

smear *f.* mancha

smear *vb.* manchar

smeared *a.* manchado

smectite *m.* jabón de soldado, jabón mineral; *f.* esmectita, arcilla expansible

smell *m.* olor

smelt *vb.* fundir, beneficiar, cebar

smelter *m.* apartador, fundidor; *f.* fusalina, hacienda

smelting *f.* fundición, fusión, colicuación, conflación; *m.* apartado, beneficio

Smithian *m. & a.* Smithiense, Smithiano

smithsonite *f.* smithsonita, calamina, esmitsonita; *m.* espato de cinc, zinc carbonatado

smog *f.* contaminacion del aire

smoke *m.* humo || ***smokestack*** *f.* chimenea

smoky *a.* ahumado

smooth *a.* suave, liso, llano

smooth *vb.* homogenizar, suavizar

smoothed *a.* suavizado

snail *m.* caracol

snow *f.* nieve || ***snow capped peak*** *m.* ventisquero || ***snow-covered*** *a.* nevado || ***snow drift*** *f.* nieve a la deriva; *m.* montón de derive || ***snow field*** *m.* nevero || ***snow slide*** *m.* deslizamiento de nieve

snowline *f.* línea de nieves perpetuas

snowstorm *f.* nevada, nevasca, nevisca

snub in *vb.* entubar a presión

snubber *m.* insertador de tubería

soak *vb.* empapar

soapstone *m.* talco, jabón de sastre; *f.* esteatita

socket *m.* alvéolo; *f.* foseta

soda *f.* sosa, soda; *m.* bicarbonato, carbonato de sodio || ***caustic soda*** *f.* soda cáustica || ***soda ash*** *f.* ceniza de soda, sosa; *m.* carbonato de sodio calcinado || ***soda niter*** *m.* nitro sódico, nitro de Chile

sodalite *f.* sodalita

sodic *a.* sódico

sodium *m.* sodio || ***sodium carbonate*** *m.* carbonato de sodio || ***sodium chloride*** *m.* cloruro de sodio || ***sodium hydroxide*** *m.* hidrato de sodio; *f.* soda cáustica || ***sodium nitrate*** *m.* nitrato de sodio

soft *a.* suave, blando, maleable, dúctil, flexible || ***soft coal*** *f.* hulla grasa || ***soft earth*** *f.* blandura || ***soft water*** *f.* agua blanda

softener *m.* suavizador, ablandador || ***water softener*** *m.* suavizador de agua

softening *m.* ablandamiento

softness *f.* blandura, suavedad

software *m.* soporte lógico, soporte de programación, componentes lógicos, software; *f.* dotación lógica

soil *m.* suelo, terreno; *f.* tierra || ***soil analysis*** *m.* análisis del suelo || ***soil gas analysis*** *m.* análisis del gas del suelo || ***soil horizon*** *m.* astrago, horizonte de suelo || ***soil mechanics*** *f.* mecánica de suelos || ***soil moisture*** *f.* humedad del suelo || ***soil profile*** *m.* perfil del suelo || ***soil sample*** *f.* muestra de suelos

solar *a.* solar || ***solar system*** *f.* sistema solar

sole *m.* subestrato || ***sole mark*** *f.* marca de base

solfatara *f.* solfatara; *m.* azufral

solfataric *a.* solfatáricc

solid *m.* sólido, macizo; *a.* sólido, compacto || ***solid/liquid equilibrium*** *m.* equilibrio entre sólido y líquido || ***solid solution*** *f.* solución en sólidos

solidification *f.* solidificación

solidify *vb.* solidificar

solidity *a.* solidez

solidus *m.* sólidus

solifluction *f.* solifluxión, soliflucción

solitary *a.* solitario

solstice *m.* solsticio || ***summer solstice*** *m.* solsticio de verano || ***winter solstice*** *m.* solsticio de invierno

solubility *f.* solubilidad, disolubilidad

soluble *a.* soluble, disoluble

solute *m.* soluto

solution *f.* solución, disolución || ***colloidal solution*** *f.* solución coloidal || ***hydrothermal solution*** *f.* solución hidrotermal || ***molar solution*** *f.* solución molar || ***non-unique solution*** *f.* solución no única || ***solution gas*** *m.* gas en solución || ***solid solution*** *f.* solución sólida

solvation *f.* solvación

solvent *m.* solvente, disolvente || ***solvent extraction*** *f.* extracción por disolvente, refinación a base de disolventes || ***solvent treatment*** *m.* tratamiento con solvente

somasteroid *m.* somasteroideo

somite *m.* somito

sonar *m.* sonar, asdic

sonde *f.* sonda

sonic *a.* sónico || ***sonic delay time*** *f.* línea de retardo sónico || ***sonic log*** *m.* perfil sónico

sonograph *m.* sonógrafo

soot *m.* hollín

sorbent *m.* & *a.* absorbente

sorption *f.* adsorción

sorosilicate *m.* sorosilicato

sort *vb.* seleccionar, escoger, (min.) apartar, clasificar, estriar, ordenar, arreglar || ***hand sort*** *vb.* escoger a mano || ***sort ore*** *vb.* pallaquear, pallar, pepenar

sorted *a.* seleccionado, surtido, [Col.] calibrado || ***poorly sorted*** *adv.* pobremente seleccionado || ***well sorted*** *a.* bién seleccionado

sorter *f.* clasificadora; *m.* pepenador

sorting *m.* seleccionado, surtido, (min.) apartado; *f.* chanca, calibración || ***grain size sorting*** *f.* calibración del grano || ***hand sorting*** *m.* apartado a mano

sound *f.* (geog.) boca, sonda; *m.* boquerón, estrecho o brazo de mar; (phys., física) sonido, ruido || ***sound damper*** *f.* sordina; *m.* silenciador, amortiguador de ruido || ***sound wave*** *f.* onda sonora, onda acústica

sounder *f.* sonda, sondaleza || ***echo sounder*** *m.* sondador

sounding *m.* sondeo, sondaje; *f.* sonda

sounding *m.* sondaje

soundng *vb.* sondar, sondear

sour *a.* agrio, ácido, corrosivo || ***sour crude*** *m.* crudo sulfuroso, crudo agrio || ***sour gas*** *m.* gas sulfuroso

source *f.* fuente, mana; *m.* brotadero, manantial, origen || ***calibration source*** *f.* fuente de calibración || ***seismic source*** *f.* fuente de emisión, fuente sísmica

source rock *f.* roca madre, roca generadora, roca fuente, roca petrolífera

south *m.* & *a.* sur, meridional

South America *f.* América del Sur, Sudamérica

southern *a.* del sur, meridional

spa *m.* balneario

space *m.* espacio

space *vb.* espaciar

spacer *m.* espaciador, separador

spacing *m.* espaciado, espaciamiento; *f.* separación

spacious *a.* dilatado, espacioso, vasto, amplio, extenso

spade *f.* pala, garlancha; azada

spall *f.* laja, astilla de piedra, lasca, escalla, astilladura; *m.* cantal

spall *vb.* salvar, franquear, descostrarse, lascar, desconcharse, fisurarse, machar, quebrantar

spalled rock *m.* ripio

spalling *f.* fisuración || ***spalling off*** *m.* astillamiento

spalt *m.* espalto

span *m.* tramo

span *vb.* atravesar

spar *m.* espato || ***Iceland spar*** *m.* espato de Islandia || ***satin spar*** *m.* espato lustroso

sparry *a.* espático

sparse *a.* ralo, escaso, disperso

spark *f.* chispa; *m.* brillo || ***spark arrester*** *m.* arrestador de chispas, chispero

Sparnacian *m.* & *a.* Esparnaciense, Esparnaciano, Sparnaciense, Sparnaciano

Spathian *m.* & *a.* Spatiense, Spatiano, Espatiense, Espatiano

spathic *a.* espático

spatter cone *m.* cono acandilado

species *f.* especies

specific *a.* específico || ***specific gravity*** *f.* gravedad específica, densidad relativa; *m.* peso específico || ***specific heat*** *m.* calor específico

specification *f.* especificación

specimen *f.* muestra; *m.* ejemplar, espécimen, testigo || ***rich specimens*** *m.* gallos

speck *f.* manchita, motita

speckled *a.* moteado

spectrometer *m.* espectrómetro || ***infrared spectrometer*** *m.* espectrómetro de rayos infrarrojos || ***mass spectrometer*** *m.* espectrómetro colectivo, espectrómetro global, espectrómetro de masa

spectroscopic analysis *m.* análisis espectroscópico

spectrum *m.* espectro

specular *a.* especular || ***specular hematite*** *f.* hematita especular

specularite *f.* especularita, dionisia; *m.* hierro brillante

speed *f.* velocidad

spelaean *a.* cavernoso

speleology *f.* espeleología

spelter *m.* peltre

spelunking *f.* espeleología

spend *vb.* gastar

spent *a.* usado, gastado

sperm *f.* esperma

sperrylite *f.* sperrylita

spessartite *f.* espessartina, espesartita

sphalerite *f.* esfalerita, blenda, blenda de zinc; *m.* copalillo, esmanil, metal negro, [Bol., Pe.] chumbe || ***disseminated sphalerite*** *m.* [Méx.] acanelado

sphene *m.* esfeno; *f.* esfena, titanita

sphenolith *m.* esfenolito

sphere *f.* esfera; *m.* globo

spherical *a.* esferoide, esférico

sphericity *f.* esfericidad

spheroid *f.* esferoide

spheroidal *a.* esferoidal

spherulite *f.* esferulita; *m.* esferolito

spherulitic *a.* esferolítico

spicule *f.* espícula

spider *f.* araña

spike *m.* (min.) clavo, perno; *f.* (elect.) pulsación

spike *vb.* clavar, enclavar

spilite *f.* espilita

spilitic *a.* espilítico

spill *m.* derrame, rebosamiento; *f.* clavija, broca || ***spill point*** *m.* punto de derrame, punto de fuga

spill *vb.* derramar

spillage, spilling *m.* derramiento

spillway *m.* aliviadero, vertedero

spin *vb.* girar || ***spin the bit*** *vb.* gira el trépano

spindle-shaped *a.* ahusado

spine *f.* espina

spinel *f.* espinela || ***red spinel*** *m.* balaje

spinose *a.* espinoso

spiracle *m.* espiráculo

spiral *a.* espiral, caracol

spirallium *m.* brazo espiralado

spire *f.* espira, aguja; *m.* pilar, pináculo, torre

spirifer *m.* espirífero

spiriferid *m.* espiriférido

spirit level *m.* nivel de burbuja

spirotheca *f.* espiroteca

spit *f.* lengua de arena unida a la costa, flecha, flecha litoral; *m.* espetón

splay *m.* derrame aluvial; *f.* bifurcación

splice *m.* empalme, acoplamiento; *f.* unión

splice *vb.* empalmar, juntar, unir

spline *f.* ranura, estría

splinter *f.* astilla, astilladura

split *f.* hendedura, cuarteadura, raja, escisión, grieta, ruptura, quebraja; *m.* cisma, resquebrajo; *a.* desquebrajado, rajado || ***split spread*** *m.* dispositivo simétrico

split *vb.* hender, rajar, cuartear, resquebrar, desquebrajar, henderse, quebrajar, quebrantar(se), ramalearse, partir, dividir

splitter (casing) *m.* rajatubos, separador

splitting *m.* hendimiento; *f.* descomposición

spodumene *f.* espodumena, zeolita de Sudermania, zeolita de Suecia; *m.* espodumen

spoil *m.* desperdicio, despojo || ***spoil bank*** *m.* caballero, vaciadero

sponge *f.* esponja

spontaneous *a.* espontáneo || ***spontaneous combustion*** *f.* combustión espontánea || ***spontaneous potential*** *m.* potencial espontáneo

sporadic *a.* esporádico

spore *f.* espora

sporinite *f.* esporinita

sporomorph *m.* esporomorfo

sporonin *f.* esporonina

spot *m.* sitio, lugar || ***bright spot*** *m.* punto brillante

spot *vb.* ubicar

spotted *a.* manchado, moteado, jaspeado, mosqueado

spotter *m.* observador

spotty *a.* esporádico

spray *f.* rociada; *m.* rocío

spray *vb.* rociar, regar

spread *f.* distribución, extensión, propagación, diseminación, difusión; *m.* [Arg.] despliegue || ***cross spread*** *m.* despliegue en cruz || ***detector spread*** *m.* despliegue de detectores || ***seismic spread*** *f.* distribución

spread *vb.* desparramar, desplegar, extender

spreading *f.* expansión || ***seafloor spreading*** *f.* expansión del fondo oceanico

spreadsheet *f.* hoja electrónica, hoja de cálculo

spring *f.* fuente, manantial, mana, alfaguara, azanca, venera, casimba, tinaja, [Sp.] alzaguera, [Ch.] aguada, [Arg.] cachimba; (estación) primavera; *m.* brotadero, ojo de agua, chorro, nacimiento, venero, [Col.] aljibe, amagamiento, [Ch., Pe.] puquio || ***artesian spring*** *f.* fuente artesiana; *m.* manantial artesiano, manantial surtidor || ***boiling spring*** *f. & m.* hervidero || ***hot spring*** *f.* manantial termal, burga, fuente termal, fuente caliente, terma; *m.* balneario || ***mineral spring*** *f.* fuente mineral || ***spring tide*** *f.* marea viva || ***spring water*** *f.* agua de manantial

sprout *m.* brote, retoño

sprout *vb.* echar, hacerse, brotar, retoñar

spud a well *vb.* iniciar la perforación

spudder *f.* perforadora inicial, barrena principiadora, perforadora a balancín

spudding *f.* perforación inicial

spur *f.* trepadera, estribación; *m.* escalador, contrafuerte, picacho

squad *f.* brigada

squall *f.* turbonada

squamous *a.* escamoso

square *m. & a.* cuadro; cuadrado

squeeze *m.* estrujón; *f.* presión, estrujadura || ***squeeze cementing*** *f.* cementación a presión

squeeze-up (volcanic) *f.* extumescencia

squib *m.* detonador, detonante

stability *f.* estabilidad

stabilization *f.* estabilización

stabilize *vb.* estabilizar

stabilizer *m.* estabilizador; *a.* estabilizante

stable *a.* estable, firme

stack *f.* pila, columna; chimenea; *m.* montón, peñasco, farallón, pilar de erosión; (seis.) apilamiento || ***stack gas*** *m.* gas de chimenea

stack *vb.* apilar, hacinar, amontonar

stacking *m.* apilamiento || ***stacking pattern*** *m.* patrón de apilamiento || ***stacking velocity*** *f.* velocidad de apilamiento

stadia *f.* estadia; *m.* taquímetro || ***stadia arc*** *m.* arco taquimétrico || ***stadia cross-hairs*** *m.* hilos de estadia || ***stadia rod*** *f.* mira taquimétrica || ***stadia survey*** *f.* taquimetría || ***stadia tables*** *f.* tablas taquimétricas

Staffordian *m. & a.* Estaffordiense, Estaffordiano, Staffordiano, Staffordiense

stage *f.* etapa; *m.* piso, estado || ***geomorphic stage*** *m.* estado geomórfico || ***glacial stage*** *m.* tardiglacial || ***microscope stage*** *f.* plataforma, platina

stagger *vb.* escalonar

staggered *a.* al tresbolillo, en zigzag, salteado, en echelon, escalonado

staggering *a.* alternado

stagnant *a.* estancado, dormido

stagnation *m.* estacionamiento

stain *f.* mancha, tinta, tintura; *m.* color, colorante

stain *vb.* teñir, manchar || ***copper stain*** *m.* verdiones

stained *a.* manchado, teñido

stainless steel *m.* acero inoxidable

stake (survey, levan.) *m.* jalón, piquete; *f.* estaca

stake out *vb.* jalonar, ubicar

stalactite *f.* estalactita

stalagmite *f.* estalagmita

stall *vb.* parar, ahogar, atascar, pararse

Stampian *m. & a.* Estampiense, Estampiano, Stampiense, Stampiano

stamp mill *m.* bocarte

stand *m.* soporte, sostén || ***stand by*** *vb.* tener en reserva; esperar

standard(s) *m. & a.* de norma, estándard, estándar, standar(d); patrón; *f.* especificaciones, normas || ***standard deviation*** *f.* desviación estándar || ***standard gauge*** *f. & a.* plantilla normal || ***standard silver for assays*** *m.* parragón

standardization *f.* estandardización, normalización, unificación, estandarización

standardize *vb.* uniformar, estandarizar, estandardizar

standing water *f.* agua estancada

standpipe *m.* tubo vertical

standstill *m.* paro (laboral); *f.* quietud

stannekite *f.* estanekita

stannite *f.* estannina, estannita, estañina, pirita de estaño

stannous *a.* estanoso, estañífero, estañoso

star *f.* estrella

starboard *a.* estribor

starfish *m.* asteroideo; *f.* estrella de mar

star-shaped *a.* esteliforme, estrellado

start *vb.* arrancar, empezar, iniciar

starved basin *f.* cuenca empobrecida

state *m.* estado || ***natural state*** *m.* estado natural, estado virgen || ***original state*** *m.* estado original

state-of-the-art *a.* estado de arte

static *f.* estática; *a.* estático

statics *f.* estática || ***refraction statics*** *f.* estáticas de refracción

station *f.* estación; ventanilla; *m.* despacho || ***base station*** *f.* estación de base || ***booster station*** *f.* estación auxiliar || ***control station*** *f.* estación de control || ***filling station*** *m.* surtidor || ***fire station*** *f.* central de bomberos || ***gas station*** *f.* estación de gasolina || ***gathering station*** *f.* estación recolectora || ***power station*** *f.* central hidroeléctrica, estación generadora || ***pump station*** *f.* estación de bombas || ***service station*** *f.* estación de servicio, estación de gasolina || ***survey station*** *m.* punto de marca

stationary *a.* estacionario, fijo

statistical *a.* estadístico

statistics *f.* estadística

status *m.* status

statute mile *f.* milla terrestre ordinaria (1.609 m)

staurolite *f.* estaurolita

steady *a.* firme || ***steady state*** *m.* régimen estacionario, estado estacionario

steam *m.* vapor, vaho, vapor vivo || ***steam condenser*** *m.* condensador de vapor || ***steam cracker*** *m.* desintegrador de vapor || ***steam cracking*** *m.* craqueo con vapor || ***steam distillation*** *f.* destilación a vapor || ***steam scrubber*** *m.* interceptor de agua || ***steam stripping*** *f.* destilación al vapor || ***steam turbine*** *f.* turbina de vapor || ***steam valve*** *f.* válvula de vapor

stearate *m.* estearato

steatite *f.* & *m.* talco, jabón de sastre; *f.* esteatita

steinkern *m.* molde interno

steel *m.* acero || ***acid steel*** *m.* acero ácido || ***carbon steel*** *m.* acero al carbono || ***corrosion-resistant steel*** *m.* acero resistente a la corrosión || ***heat-resistant steel*** *m.* acero resistente al calor || ***nickel steel*** *m.* aceroníquel || ***pertaining to iron and steel*** *a.* siderúrgico || ***stainless steel*** *m.* acero inoxidable || ***steel and iron industry*** *f.* siderurgia || ***structural steel*** *m.* acero estructural, acero de construcción || ***tungsten steel*** *m.* acero tungsteno

steelworks *f.* acería

steep *a.* empinado, abrupto, acantilado, despeñadero, escarpado, fuerte

stellar *a.* estelar, astral

stem *m.* tallo

stenobathic *a.* estenobático

stenobiontic *a.* estenobióntico

stenohaline *a.* estenohalino

stenothermal *a.* estenotermo

stenotopic *a.* estenotópico

step *m.* paso; escalón, grado

step-out *a.* de avanzada

Stephanian *m.* & *a.* Estefaniense, Estefaniano, Stefaniano, Stefaniense

stephanite *f.* estefanita, plata agria, plata prismática, psaturosa

steppe *f.* estepa, tundra

steppe-like *a.* estepario

stepped *a.* escalonado

stepping *m.* escalonamiento

stereogram *m.* estereograma

stereographic *a.* estereográfico

stereo pair *m.* par estereoscópico

stereoscope *m.* estereoscopio

stereoscopic *a.* estereoscópico

stereoscopy *f.* estereoscopía

sterile *a.* estéril, infecundo, [Ch.] bobo

sternite *m.* esternito

stibnite *f.* estibina, estibnita, antimonita

sticky *a.* pegajoso

stilbite *f.* estilbita, zeolita nacarada

still *m.* alambique, destilador; *f.* destilería, destiladora || ***standstill*** *f.* tranquilidad, quietud

stimulation *f.* estimulación

stinger *m.* aguijón

stipulation *f.* estipulación

stir *vb.* agitar

stishovite *f.* stipoverita

stock *m.* (geol.) tronco, trunco, pivote, stock; *f.* (com.) acciones; (materiales) existencias, materia prima || ***stock market*** *f.* bolsa de valores, bolsa || ***stock tank*** *m.* tanque de almacenamiento

stockwork *m.* & *a.* entreverado; *f.* esponja, red filoniana

stolon *m.* estolón

stolzite *f.* estolzita

stomach *m.* estómago || ***stomach stone*** *m.* gastrolíto

stone *f.* piedra, china; *m.* cantal, cantillo, canto, chinazo, guijarro, jejo || ***ashlar stone*** *f.* sillería || ***building stone*** *f.* cantera, cantería || ***crushed stone*** *m.* piedrín; *f.* [Arg.] granza || ***paving stone*** *f.* zaborra || ***rough stone*** *m.* pedrusco || ***stone cutter*** *m.* pedrero, picapedrero || ***stone trough*** *m.* pilón

stonecutting tool *f.* azada

stonemason *m.* cantero

stony *a.* pétreo, pedregoso, guijarroso, guijoso, pedrizo, rocalloso, rocoso || ***stony ground*** *m.* cantizal, cantorral, pedragal, pedriscal; *f.* guijarra || ***stony meteorite*** *m.* aerolito

stopcock *f.* llave de cierre, llave de paso

stop, labor stoppage *m.* paro, alto; *f.* huelga

stop *vb.* parar, detener

stope *f.* excavación en escalones, bancada, grada, labor escalonada; *m.* casarón, escalón, rajo || ***overhand stope*** *m.* testero, [Méx.] alevante || ***underhand stope*** *m.* desborde, zanjeo, rebaje; *f.* labor de bancos || ***vaulted stope*** *f.* bóveda

stope *vb.* desbordar(se) || ***underhand stope*** *vb.* rebajar

stoping *m.* realce || ***overhand stoping*** *m.* realce lateral

stoppage (labor) *m.* paro (laboral); *f.* huelga

storage *m.* almacenamiento

store *f.* tienda; *m.* mercado || ***company store*** *f.* pulpería

storeroom *m.* almacen, depósito; *f.* bodega, despensa

storm *f.* tormenta; temporal

straight *a.* recto, derecho, en línea

straighten *vb.* enderezar, rectificar

strain *f.* deformación, reacción al esfuerzo aplicado; *m.* esfuerzo interno || ***bulk strain*** *f.* deformacion global || ***infinitesimal strain*** *f.* deformación infinitesimal || ***penetrative strain*** *f.* deformación penetrante || ***plane strain*** *f.* deformación en plano (o plana) || ***shear strain*** *f.* deformación interna por cizalla || ***strain ellipse*** *f.* elipsoide de deformación || ***strain gage*** *m.* medidor de deformación, indicador de deformación || ***strain rate*** *f.* velocidad de deformación

strained *a.* deformado

strainer *m.* filtro, colador, tamizador

straining *f.* coladura

strait *m.* estrecho; *f.* sonda

strand *f.* playa, ribera || ***strand line*** *m.* límite de playa

strata *m.* estrato, lecho; *f.* capa, cama, camada

stratabound deposit *m.* manto

stratal *a.* estratal || ***stratal patterns*** *m.* patrones estratales

strategic *a.* estratégico

strategy *f.* estrategia

stratified *a.* estratificado

stratiform *a.* estratiforme

stratification *f.* estratificación, stratificación

stratify *vb.* estratificar

stratigrapher *m.* estratígrafo

stratigraphic *a.* estratigráfico || ***stratigraphic column*** *f.* columna estratigráfica || ***stratigraphic interval*** *m.* intervalo estratigráfico || ***stratigraphic range*** *f.* extensión estratigráfica; *m.* rango estratigráfico || ***stratigraphic sequence*** *m.* orden estratigráfico || ***stratigraphic test*** *f.* prueba estratigráfica; *m.* ensayo estratigráfico || ***stratigraphic throw*** *f.* separación estratigráfica || ***stratigraphic trap*** *f.* trampa estratigráfica

stratigraphy *f.* estratigrafía || ***sequence stratigraphy*** *f.* estratigrafía de secuencias, estratigrafía secuencial

stratocumulus *m.* estratocúmulo

stratosphere *f.* estratósfera

strato-volcano *m.* estratovolcán

stratum *m.* lecho, estrato, horizonte; *f.* capa, hebra

stratus clouds *m.* nubes estrato

streak *f.* raspadura, traza, cinteada, raya, banda

streaky *a.* fajeado

stream *m.* curso, arroyo, caño, río, flujo, regato, torrente; *f.* corriente, corriente fluvial, quebrada, torrentera, acequia || ***adjusted stream*** *f.* corriente ajustada || ***antecedent stream*** *m.* curso antecedente || ***braided stream*** *m.* curso ramificado (o trenzado) || ***consequent stream*** *m.* curso consequente || ***dry stream*** *f.* rambla, yasa, quebrada seca || ***Gulf Stream*** *f.* corriente del Golfo || ***offset stream*** *m.* curso desplazado || ***stream bed*** *m.* lecho de un arroyo || ***stream capture*** *f.* captación de río || ***stream gauge*** *f.* mira || ***stream pattern*** *m.* diseño de avenamiento || ***stream terrace*** *f.* terraza fluvial || ***subsequent stream*** *m.* curso subsequente || ***superposed stream*** *m.* curso sobrepuesto

streaming cut *f.* efusión de petróleo bajo de la influencia de un solvente

streamlined *a.* aerodinámico, perfilado, correntilíneo, corriente de línea, fuselado, hidrodinámico

strength *f.* resistencia; fortaleza, fuerza, potencia; solidez || ***cohesive strength*** *f.* resistencia cohesiva || ***crushing strength*** *f.* resistencia a la compresión || ***shear strength*** *f.* resistencia al cizallamiento, resistencia al corte || ***tensile strength*** *f.* resistencia a la tensión, resistencia tensora

stress *m.* esfuerzo, stress; *f.* fatiga, tensión, fuerza, solicitación || ***bending stress*** *m.* esfuerzo de flexión || ***compressive stress*** *m.* esfuerzo por compresión || ***effective normal stress*** *m.* esfuerzo normal efectivo || ***residual stress*** *m.* esfuerzo restante || ***shear stress*** *m.* esfuerzo cizallante || ***soil stress*** *f.* fuerza de adherencia || ***stress corrosion*** *f.* corrosión por esfuerzo || ***stress ellipsoid*** *m.* elipsoide de esfuerzo || ***tensile stress*** *m.* esfuerzo de tensión || ***torque stress*** *m.* esfuerzo de torsión || ***ultimate stress*** *m.* esfuerzo de rotura, fuerza de rotura || ***yield stress*** *f.* fuerza de rotura; *m.* esfuerzo de rotura, punto de cesión

stretch *vb.* estirar, extender, alargar

stretching *m.* estiramiento; *f.* expansión || ***histogram stretching*** *f.* expansión del histograma

striae, stria *f.* estría, franja, ranura de deslizamiento || ***glacial striae, stria*** *f.* amoladura glacial

striated *a.* estriado; aborregado

striation *f.* estría, estriación, estriadura

strike *m.* (geol.) rumbo, arrumbamiento, (de una veta) corrido; *f.* dirección, orientación; *m.* (min., perf.) hallazgo, descubrimiento; (laboral) *f.* huelga || ***along strike*** *a.* isorrumbado, a lo largo del rumbo || ***strike slip fault*** *f.* falla lateral, falla transcurrente, falla de rumbo

strike *vb.* (geol.) dirigirse || ***strike oil*** *vb.* encontrar petróleo

string *f.* sarta; *m.* (perf.) juego de herramientas || ***casing string*** *f.* sarta de tubería || ***drill string*** *f.* tubería || ***producing drill string*** *f.* tubería de producción

stringer *m.* larguero, ramito, filamento; *f.* capa discontinua; (min.) aguja, venilla, venita, vetita

strip *f.* (geog.) franja, faja, tira || ***strip mine*** *f.* mina arrasada, mina laborada

strip *vb.* desguarnecer; desencofrar, desmoldar, agotar || ***strip off overburden*** *vb.* desmontar

stripe *f.* lista

striped *a.* bandeado, listado, rayado

stripper well *m.* pozo marginal, pozo de bajo producción

stripped *a.* despojado || ***stripped gas*** *m.* gas despojado de condensables || ***stripped oil*** *m.* petróleo despojado de fracciones livianas

stripper tower *f.* despojadora, estabilizadora; *m.* rectificador

stromatolite *m.* estromatolito

stromatoporoid *m.* estromatoporoideo

stromeyerite *f.* stromeyerita

strong *a.* fuerte, resistente, potente

strontianite *f.* estroncianita

strontium *m.* estroncio

structural *a.* estructural ‖ ***imbricate structural*** *f.* estructura imbricada ‖ ***structural closure*** *m.* cierre estructural ‖ ***structural cross section*** *m.* corte estructural, transecto, corte transversal; *f.* sección estructural ‖ ***structural geology*** *f.* geología estructural ‖ ***structural gradient*** *m.* gradiente estructural ‖ ***structural high*** *m.* alto estructural ‖ ***structural map*** *m.* mapa estructural ‖ ***structural relief*** *m.* relieve estructural ‖ ***structural terrace*** *f.* terraza estructural ‖ ***structural trap*** *f.* trampa estructural

structure *f.* estructura, contextura, estructuración, trama ‖ ***atomic structure*** *f.* red atómica ‖ ***crystal structure*** *f.* red ‖ ***down structure*** *a.* estructura abajo ‖ ***sedimentary structure*** *f.* estructura sedimentaria ‖ ***(salt) turtle structure*** *f.* estructura tortuga (de sal) ‖ ***structure contour*** *m.* contorno estructural; *f.* curva de nivel de la estructura ‖ ***up structure*** *a.* estructura arriba

structuring *f.* estructuración

Strunian *m. & a.* Estruniense, Estruniano, Struniano, Struniense

strut *f.* puntal

stuck *a.* adherido, fijo, pegado ‖ ***stuck pipe*** *f.* cañería aprisionada

study *m.* estudio

style *m.* estilo ‖ ***tectonic style*** *m.* estilo tectónico

stylolite *f.* estilolita; *m.* estilolito

stylonodular *a.* estilonodular

Styrian *m. & a.* Styriano, Styriense

subaerial *a.* al aire, subaéreo

subalkalic *a.* subalcálico

subalkaline *a.* subalcalino

suballuvial *a.* subaluvial

subalpine *a.* subalpino

subaluminous *a.* subaluminoso

subangular *a.* semi-angular, subangular, subanguloso

subarctic *a.* subártico

subaqueous *a.* bajo el agua, subágueo

subautomorphic *a.* subidiomórfico

subbasin *f.* subcuenca

sub-bituminous *a.* sub-bituminoso

subclass *f.* subclase

subcontinent *m.* subcontinente

subcrop *m.* subafloramiento

subcrustal *a.* subcortical

subdivision *f.* subdivisión

subduction *f.* subducción ‖ ***subduction zone*** *f.* zona de subducción

subfamily *f.* subfamilia

subgenus *m.* subgénero

subgreywacke *m.* subgrauwacke

subgroup *m.* subgrupo

subhedral *a.* subhedral

Subhercinian *m. & a.* Subherciniano, Subherciniense

subincline *f.* subrampa

subjacent *a.* subyacente

subject *m.* sujeto

subject *vb.* subjetar

(to be) subjected to *vb.* sujeta a

sublease *m.* subarriendo

sublimate *m.* sublimado

sublimate, sublime *vb.* sublimar

sublimation *f.* sublimación

sublitharenite *f.* sublitarenita

sublittoral *a.* sublitoral

submarine *a.* submarino ‖ ***submarine bar*** *f.* barra submarina, restinga ‖ ***submarine canyon*** *m.* cañon submarino ‖ ***submarine earthquake*** *m.* maremoto ‖ ***submarine fan*** *m.* abanico submarino ‖ ***submarine ridge*** *f.* dorsal submarina ‖ ***submarine slope*** *f.* pendiente submarina

submarginal *a.* submarginal

submature *a.* submaduro, submaturado

submerge *vb.* ahogarse, ahogar, sumergir, inundar, sumir, hundir

submerged *a.* sumergido, ahogado, hundido

submergence *f.* sumersión; *m.* sumergimiento

submersible *a.* sumergible ‖ ***submersible electric pump*** *f.* bomba eléctrica sumergible ‖ ***submersible rig*** *f.* plataforma sumergible

submersion *f.* sumersión

submetallic *a.* submetálico

subophitic *a.* subofítico

suborder *m.* suborden

subperiod *m.* subperíodo

subpolar *a.* subpolar

subsampling *m.* submuestreo

subsea *a.* submarino, bajo el nivel del mar || ***subsea completion*** *f.* terminación submarina

subsequent *a.* subsecuente || ***subsequent stream*** *m.* arroyo subsecuente || ***subsequent valley*** *m.* valle subsecuente

subside *vb.* hundirse, abajar, bajar, hundir

subsidence *f.* subsidencia, rafa; *m.* hundimiento, abajamiento, descenso, revenimiento, socavón

subsidiary *a.* suplementario, subsidiario

subsidize *vb.* subsidiar

subsidy *m.* subsidio

subsoil *m.* subsuelo

subspecies *f.* subespecie

substage *f.* subetapa

substance *f.* substancia, sustancia

substitute *m.* substituto

substitution *f.* substitución

substrate, substratum *m.* subestrato, subhorizonte; *f.* capa inferior

substructure *f.* subestructura

subsurface *m.* subsuelo, subterráneo, bajo la superficie; *a.* freático

subterranean *a.* soterraño, soterrado, subterráneo

subtidal *a.* baja la influencia de las mareas, submarea

subvolcanic *a.* subvolcánico

subzero *a.* bajo cero

succession *f.* sucesión || ***faunal succession*** *f.* sucesión faúnica o faunística

sucker rod *f.* varilla de bombeo, varilla de succión

suction *f.* aspiración, succión

Sudetic *m.* & *a.* Sudético

suite *m.* juego, conjunto; *f.* serie

sulfate, sulphate *m.* sulfato || ***copper sulfate, sulphate*** *m.* vitriolo azul || ***zinc sulfate, sulphate*** *m.* vitriolo blanco

sulfatite, sulphatite *f.* sulfatita

sulfide, sulphide *m.* sulfuro; *f.* pirita || ***hydrogen sulfide*** *m.* sulfuro de hidrógeno || ***massive sulfides*** *m.* sulfuros en masa

sulfite *m.* sulfito

sulfonate *m.* sulfonato

sulfosalt *m.* sulfosal

sulfur, sulphur *m.* azufre, sulfuro || ***sulfur dioxide*** *m.* dióxido de sulfuro; anhídrido sulfuroso

sulfuration *f.* sulfuración

sulfuretted *a.* sulfurado

sulfuric *a.* sulfúrico || ***sulfuric acid*** *m.* ácido sulfúrico

sulfurization *f.* sulfidización, sulfurización

sulfurous, sulphurous *a.* sulfuroso, azufrado

sum *f.* suma; *m.* total

summary *m.* resumen, sumario; *f.* recopilación

summer *m.* verano

summit *f.* cumbre, cima, altura, cabecera, culminación, [Pe.] apacheta; *m.* vértice, cerro, picacho, colmo, cucurucho

sump *m.* sumidero, resumidero, foso de lodo; *f.* poceta, pileta

sun *m.* sol

sunken *a.* hundido

sunspot *f.* mancha solar

superanthracite *f.* superantracita

superficial *a.* superficial

supergene *a.* secundario, supergénico, supergeno

supergroup *m.* supergrupo

superheated *a.* recalentado, sobrecalentado

superimpose *vb.* sobreponer

superimposed *a.* sobrepuesto, superpuesto

superintendent *m.* encargado, superintendente

superior *a.* superior, mejor

supernova *f.* supernova

superperiodicity *f.* superperiodicidad

superplasticity *f.* superplasticidad

superpose *vb.* sobreponer

superposed stream *m.* arroyo superpuesto

superposing, superimposition *f.* sobreimposición

superposition *f.* superposición, yuxtaposición

supersaturate *vb.* sobresaturar, supersaturar

supersaturated *a.* sobresaturado, supersaturado

supersaturation *f.* sobresaturación, supersaturación

superseries *f.* superserie

superstage *f.* superetapa

supersystem *m.* supersistema

supervisor *m.* sobrestante, superintendente

supplementary *a.* suplementario

supplies *m.* (min.) aperos

supply, supplying *m.* monto, suministro, abastecimiento, abasto || ***supply and demand*** *f.* oferta y demanda

supply *vb.* abastar, abastecer, suministrar

support *m.* apoyo, sustento, soporte, sostén

support *vb.* mantener, apoyar

supportive covering *m.* revestimiento

supracrustal *a.* supracortical

supratenuous fold *m.* pliegue supratenue

supratidal *a.* supralitoral

surcharge *f.* sobrecarga

surety *m.* fiador; *f.* fianza

surf *m.* rompiente, mar de fondo, mar de leva, oleaje; *f.* resaca

surface *f.* superficie; *m.* firme || ***bedding surface*** *f.* superficie de estratificación || ***earth's surface*** *f.* superficie terrestre || ***equipotential surface*** *f.* superficie equipotencial || ***erosion surface*** *f.* superficie de erosión || ***flooding surface*** *f.* superficie de inundacion || ***fracture surface*** *f.* superficie de fracturación || ***isostatic equilibrium surface*** *f.* superficie de equilibrio || ***maximum flooding surface*** *f.* superficie de inundación máxima || ***piezometric surface*** *f.* superficie piezométrica || ***surface mapping*** *m.* mapeo de superficie || ***surface pipe*** *f.* columna conductora, cañería guía || ***surface tension*** *f.* tensión superficial || ***surface texture*** *m.* relieve superficial || ***surface wave*** *f.* onda superficial, onda Rayleigh

surfactant *m.* agente tensoactivo

surficial *a.* superficial

surge *f.* oleada; *m.* oleaje || ***surge tank*** *m.* tanque compensador

surge *vb.* fluctuar, surgir

surging *f.* fluctuación

surplus *m.* superávit, superavitario

surround *vb.* rodear, envolver

surrounding *a.* adyacentes

survey *f.* prospección, agrimensura, encuesta, campaña, mensura, planimetría; *m.* estudio; apeo, levantamiento, reconocimiento, examen || ***aerial survey*** *m.* levantamiento aerofotogramétrico || ***cadastral survey*** *m.* estudio catastral || ***geologic survey*** *m.* levantamiento geológico || ***land survey*** *f.* mojona || ***plane table survey*** *m.* levantamiento a plancheta || ***seismic survey*** *m.* levantamiento sísmico || ***survey corner*** *f.* vértice || ***survey line*** *f.* picada || ***survey marker*** *m.* balate || ***survey party*** *f.* brigada, comisión || ***topographic survey*** *f.* altimetría; *m.* relevamiento

survey *vb.* levantar, cartografiar, estudiar, apear, examinar, mensurar, reconocer, recorrer

surveyed *a.* mensurado

surveying *f.* agrimensura, topografía, apeadura; *m.* levantamiento

surveyor *m.* topógrafo, agrimensor, examinador, apeador, deslindador

survive *vb.* sobrevivir

susceptibility *f.* susceptibilidad

suspended *a.* colgado, colgante; suspendido || ***suspended matter*** *m.* arrastres

suspension *f.* suspensión

sussexite *f.* susexita

suture *f.* sutura, línea lobal

sutured *a.* suturado

swab, swabber *m.* chupador, pistón de achique

swab *vb.* excitar, activar, limpiar, suavear

swabbing *f.* limpieza con escobillón; *m.* achique, achique de activación, achique de excitación, suaveando

swale *m.* terreno pantanoso, légamo, bajío; *f.* hondonada

swamp *f.* ciénaga, llama, marisma, ñanga, zubía; *m.* pantano, aguazal, chapatal, llamargo, llamazar, chamarcal, atascadero, atolladero, bañado, suampo, [Arg., Ch.] agualotal, [Ec.] aguatal, cenagal, [Arg., Méx.] estero

swampy *a.* cenagoso, pantanoso, guadaloso, palúdico, palustre

swarm *m.* enjambre || ***dike swarm*** *m.* enjambre de diques

swash *m.* vaivén || ***swash line*** *f.* línea de oleaje || ***swash mark*** *f.* marca de oleaje

swath *f.* guadaña, ringlera

swaying *a.* bamboleo

sweep *m.* barrido

sweep *vb.* barrer

sweet crude *m.* crudo dulce

sweet gas *m.* gas dulce

sweeten oil *vb.* desazufrar el petróleo, destufar, desulfurar, endulzar

swell *f.* (naut.) onda, marejada, oleada, (geog.) elevación suave, protuberancia; *m.* umbral continental, (naut.) oleaje

swell *vb.* engrosarse, hincharse

swelling *m.* engrosamiento, intumescencia

swinging *a.* bamboleo

switch *m.* interruptor || ***on–off switch*** *m.* interruptor de encendio y apagado || ***reset switch*** *m.* interruptor de reinicialización

switch *vb.* cambiar, desviar

swollen *a.* hinchado; abombado

syenite *f.* sienita

syenodiorite *f.* sienodiorita

sylvanite *f.* silvanita; *m.* metal gráfico

sylvite *f.* silvita, silvina

symbiosis *f.* simbiosis

symbol *m.* símbolo

symbolize *vb.* representar

symmetric *a.* simétrico

symmetry *f.* simetría

symplectite *f.* simplectita

synantectic *a.* sinantéctico

synapsid *m. & a.* sinápsido

synchroneity *f.* sincroneidad, sincronía

synchronism *m.* sincronismo

synchronization *f.* sincronización

synchronize *vb.* sincronizar

synchronous *a.* sincrónico, síncrono

synchysite *f.* sinchisita

synclinal *a.* sinclinal

syncline *m.* sinclinal || ***rim syncline*** *m.* sinclinal marginal

synclinorium *m.* sinclinorio, sinclinal compuesto

synclise *f.* sineclisa

syndeposition *f.* sindeposición

syndepositional *a.* sindeposicional

synecological *a.* sinecológico

synecology *f.* sinecología

syneresis *f.* sinéresis || ***syneresis cracks*** *f.* grietas de desecación

synform *f.* sinforma

synformal *a.* sinformal

syngenesis *m.* singénesis

syngenetic *a.* singenético, idiógeno

synkinematic *a.* sinkinemático

synoptic *a.* sinóptico

synorogenic *a.* sinorogénico

synsedimentary *a.* sinsedimentario

syntectonic *a.* sintectónico

synthesis *f.* síntesis

synthesize *vb.* sintetizar

synthetic *a.* sintético || ***synthetic fibers*** *f.* fibras sintéticas

syphon *m.* sifón

system *m.* sistema || ***circulation system*** *m.* sistema de circulación || ***crystal system*** *f.* singonia cristalográfica; *m.* sistema cristalino || ***fault system*** *m.* sistema de fallas || ***geographic information system (GIS)*** *m.* sistema de información geográfico || ***global positioning system (GPS)*** *m.* sistema de posicionamiento global || ***loop gathering system*** *m.* sistema de oleoductos recolectores || ***petroleum system*** *m.* sistema petrolero || ***river system*** *m.* sistema hidrográfico || ***system tract*** *m.* intervalo del sistema || ***UTM coordinate system*** *m.* sistema MTU

systematic *a.* sistemático

systematics *f.* sistemática

T

Tabianian *m. & a.* Tabianinao, Tabianiense

table *m.* cuadro; *f.* tabla; mesa || ***light table*** *f.* mesa de luz; *m.* [Méx.] tormento || ***periodic table*** *m.* cuadro periódico || ***rotary table*** *m.* mesa rotaria || ***table mountain*** *f.* mesa, meseta || ***table land*** *m.* ajarafe, meseta || ***water table*** *m.* nivel freático

tabular *a.* tabular, laminado

tabulate *vb.* tabular; *a.* tabular

tabulate coral *m.* coral tabulado

tachyhydrite *f.* taquihidrita

tachylyte *f.* taquilita, sordawalita; *m.* jaspoide

tachymetry *f.* taquimetría

tacky *a.* pegajoso

Taconic *m. & a.* Tacónico

Taconian *m. & a.* Taconiano, Taconiense

tactite *f.* tactita

tag *f.* etiqueta

Taghanican *m. & a.* Taghanicano, Taghanicense

tailings *m.* desechos, residuos, chicharrones, deslave, acopio, polvillo, relaves, vaciadero, [Bol. Pe.] pallaco; *f.* colas, colillas, cangalla, lama, [Méx.] arenilla || ***rich tailings*** *m.* polvillos, [Méx.] jales, jalsontles

tailrace *m.* socaz

talc *m.* talco || ***talc ore*** *m.* [Méx.] jabón || ***talc schist*** *f.* talquita || ***talc seam*** *f.* tosca

talcite *f.* talcita

talcose, talcous *a.* talcoso

talcum *m. & a.* talco

talus *m.* talud, eluvión, talus, argayo, derrubios; [Arg.] pedregullo de ladera; *f.* escarpa, rocalla, [Bol.] llampera; *m.* llampo || ***reef talus*** *m.* talud recifal || ***talus fan*** *m.* cono de deyección || ***talus slope*** *f.* pendiente de escombros de talud

tangent *f. & a.* tangente

tangential *a.* tangencial

tank *m.* tanque, depósito; *f.* cisterna || ***blowdown tank*** *m.* tanque colector de descarga || ***floating roof tank*** *m.* tanque de techo pontón || ***fuel tank*** *m.* tanque de combustible || ***mud return tank*** *m.* tanque de lodo devuelto || ***pressure tank*** *m.* tanque a presión, tanque de presión || ***septic tank*** *m.* tanque séptico || ***settling tank*** *m.* tanque asentador, reposador || ***storage tank*** *m.* tanque de almacenamiento || ***surge tank*** *m.* tanque compensador || ***tank car*** *m.* vagón cisterna, vagón tanque || ***tank farm*** *m.* patio de tanques || ***tank trailer*** *m.* tanque remolque || ***tank truck*** *m.* camión tanque || ***water tank*** *m.* estanque

tanker *m.* barco petrolero, barco tanquero, buque tanques, petrolero

tantalite *f.* tantalita

tantalum *m.* tantalio, tántalo

tape *f.* cinta || ***magnetic tape*** *f.* cinta magnética || ***measuring tape*** *f.* cinta, lienza, [Bol., Ch.] huincha || ***tape drive*** *m.* lector

taper *m.* ahusamiento

taper *vb.* ahusar

tapered *a.* ahusado

tapering *a.* aguzado

taphonomy *f.* tafonomía

taphrogenic *a.* tafrogénico

taphrogenesis *f.* tafrogénesis

taphrogeny *f.* tafrogenía

taphrogeosyncline *m.* tafrogeosinclinal

tar *f.* brea, pez; *m.* alquitrán; [Méx.] chapopote, [Arg., Cu.] chapuro

tar *vb.* embrear, alquitranar

tarbuttite *f.* salmoita

tare weight *f.* tara

target *m.* objetivo, prospecto

tarnish *vb.* deslustrar

tarriff *f.* tarifa; *m.* arancel, derecho de aduana

tarn *m.* lago de circo glacial, ibón

Tartarian *m. & a.* Tartariense, Tartariano

task *f.* faena, tarea, labor, tanda

taut line *m.* cable tieso

tax *m.* impuesto, gravamen, (com.) canón; *f.* contribución, tasa || ***annual mine tax*** *f.* [Bol., Ch.] patente || ***crude oil excise tax (COET)*** *m.* impuesto al consumo del petróleo crudo || ***excise tax*** *m.* impuesto indirecto, impuesto interior || *ore tax m.* tequio || *property tax f.* contribución predial || *severance tax m.* impuesto de producción

tax *vb.* tasar

taxites *f.* taxitas

taxodont *m. & a.* taxodonto

taxonomic *a.* taxonómico

taxonomy *f.* taxonomía

Tayloran *m. & a.* Tayloriano, Tayloriense, Taylorense

TCF *m.* trillón de pies cúbicos, terapiés cúbicos, gigapiés cúbicos

teach *vb.* enseñar

teacher *m.* profesor, preceptor, maestro

team *m.* equipo

tear *m.* rasgón, rasgo, desgarrón, desgarre, desgarramiento || *tear fault m.* desgarre, desgarramiento

tear *vb.* rasgar, desgarrar, lacerar, romper

technetium *m.* tecnecio

technical *a.* técnico || *technical advisor m.* asesor técnico || *technical drawing m.* dibujo técnico

technician *m.* técnico

technique *f.* técnica

technology *f.* tecnología || *high technology f.* tecnología más avanzada || *information technology f.* tecnología informática

tectite *m.* uranolito

tectogene *m.* tectógeno

tectonic *a.* tectónico || *tectonic front m.* frente tectónico

tectonics *f.* tectónica, geotectónica || *alpino-type tectonics m.* ortotectonismo || *gravity tectonics f.* tectónica de gravidad || *plate tectonics f.* tectónica de placas || *salt tectonics f.* tectónica salina

tectonite *f.* tectonita

tectonization *f.* tectonización

tectonophysics *f.* tectonofísica

tectonosphere *f.* tectonósfera

tectosphere *f.* tectósfera

tectosilicate *m.* tectosilicato

tectum *m.* tecto

tegilla *f.* tegilla

tegmentum *m.* tegmento

tektite *f.* tectita

tektonite *f.* tectonita

telehemic *a.* silicotélico

telemeter *m.* telémetro

telemeter *vb.* telemedir

telemetering *f.* medición a distancia

telescope *m.* telescópio, anteojo || *reflecting telescope m.* telescopio reflector || *refracting telescope m.* telescopio refractor

telescope *vb.* enchufar

telescopic, telescoping *a.* telescópico

telethermal *a.* teletermal

tellurate *m.* telurato

telluric *a.* telúrico

telluride *f.* telúrido

tellurite *f.* telurita

tellurium *m.* telurio, teluro

telogenetic *a.* telogenético

telson *m.* telsón

temper, tempering *m.* temple

temper *vb.* templar

temperate *a.* templado

temperature *f.* temperatura || *absolute temperature f.* temperatura absoluta || *critical temperature f.* temperatura crítica || *temperature gradient m.* gradiente térmico

tempered *a.* recocido, templado

tempest *f.* tempestad

template *f.* plantilla, cercha; *m.* gálibo, patrón

temporary *a.* provisional

temporate *a.* templado

tendency *f.* tendencia

tender *f.* (com.) oferta, propuesta

tengerite *f.* ytterita

tennantite *f.* tennantita, julianita

tenorite *f.* tenorita, melaconisa; *m.* óxido cúprico

tensile *a.* tensil, extensible || *tensile stress m.* esfuerzo tensil || *tensile strength f.* resistencia tensora, resistencia a la tensión

tension *f.* tensión, tracción || ***interfacial tension*** *f.* tensión interfacial || ***surface tension*** *f.* tensión superficial || ***tension crack*** *f.* grieta de tracción || ***tension failure*** *f.* rotura causada por la tensión || ***tension joint*** *f.* diaclasa de tracción, fractura de tensión || ***tension leg platform*** *f.* plataforma de cables en tensión || ***tension stress*** *m.* esfuerzo de tensión

tent *f.* tienda

tentacle *m.* tentáculo

tephra *f.* tefra

tephrite *f.* tefrita

tephrochronology *f.* tefrocronología

tephroite *f.* tefroíta

terephthalic acid *m.* ácido tereftálico

tergite *m.* tergito

tergopore *m.* tergóporo

tergum *m.* tergo

term (geol.) *m.* plazo || ***long-term*** *m.* largo plazo || ***short-term*** *m.* corto plazo

terminal *a.* terminal

terminal *f.* estación; (comp.) terminal; *a.* terminal || ***terminal moraine*** *f.* morena frontal, morena terminal

terms *f.* condiciones

ternary *a.* ternario || ***ternary axis of rotation*** *m.* trigiro || ***ternary diagram*** *m.* diagrama ternario

terrace *f.* terraza; *m.* terraplén, balate, [Bol., Pe.] andén || ***aggradational terrace*** *f.* terraza de deposición || ***alluvial terrace*** *f.* terraza aluvial || ***constructional terrace*** *f.* terraza de acumulación || ***erosional terrace*** *f.* terraza de erosión || ***low-tide terrace*** *f.* terraza de marea baja || ***marine terrace*** *f.* terraza marina || ***river terrace*** *f.* terraza fluvial || ***stream terrace*** *f.* terraza fluvial || ***structural terrace*** *f.* terraza estructural || ***wave-built terrace*** *f.* terraza de marea

terraced *a.* abancalado

Terracian *m. & a.* Terraciano, Terraciense

terra cotta *f.* terracota

terrain, terrane *m.* terreno; *f.* topografía || ***broken, rough terrain*** *a.* barrancoso, cerril || ***terrain correction*** *f.* corrección topográfica

terra rossa *f.* terra rossa

terreplain *m.* terraplén; *f.* valla

terrestrial *a.* terrestre

terrigenous *a.* terrígeno

territory *m.* territorio; *f.* comarca

Tertiary *m. & a.* Terciario

teschenite *f.* teschenita

test *m.* ensayo, ensaye, tanteo; *f.* prueba, determinación; *f.* (paleo., biol.) [Sp., Cu., Vz.] concha, testa [Méx.] testa, [Arg.] concilla, cónchula || ***chi-squared test*** *f.* prueba de ji cuadrado || ***drill stem test*** *m.* ensayo de capas || ***likelihood ratio test*** *f.* prueba de razón versimilitud || ***nondestructive test*** *f.* prueba no destructiva || ***significance test*** *f.* prueba de significación || ***stratigraphic test*** *m.* ensayo de pozo estratigrafico || ***student test*** *f.* prueba de estudiante || ***test bore*** *f.* cala de prueba || ***test of independence*** *f.* prueba de independencia || ***test pit*** *f.* [Arg.] calicata || ***test site*** *m.* sitio de pruebas || ***test tube*** *f.* probeta; *m.* tubo de ensayo || ***test well*** *m.* pozo de ensayo; *f.* calicata || ***unbiased test*** *f.* prueba sin sesgo

test *vb.* ensayar; probar, tantear, verificar

tester *m.* probador, ensayador, verificador || ***drill stem tester*** *m.* ensayor de la formación por la tubería vástago

Tethys *m.* Tetis

tetrabranchiate *m. & a.* tetrabranquial

tetracyclic *a.* tetracíclico

tetrad *f.* tétrada

tetradymite *f.* tetradimita

tetraethyl *a.* tetraetilo || ***tetraethyl lead*** *m.* plomo tetraetílico, plomo tetraetilo

tetragonal *a.* tetragonal

tetrahedrite *f.* tetraedrita, pétanque, petlanque; *m.* abronzado, cobre gris, metal negro argentífero, [Pe.] pavonado

tetrahedron *m.* tetraedro

tetrahexahedron *m.* tetraexaedro

tetravalence *f.* tetravalencia

tetravalent *a.* tetravalente

textural *a.* textural

texture *f.* textura, contextura

thallium *m.* talio

thalweg *f.* vaguada; *m.* talweg

thanatocenosis *m. & f.* tanatocenosis

Thanetian *m. & a.* Thanetiense, Taneciense, Thanetiano, Taneciano

thaw *m.* deshielo, deshiele

thaw *vb.* derretir, deshelarse

thawing *m.* derretimiento

theca *f.* teca

thecal *a.* tecal

thecarium *m.* tecario

thematic *a.* temático

theodolite *m.* teodolito

theorem *m.* teorema

theory *f.* teoría, tesis

theralite (olivine) *f.* kilita, teralita

therapsid *m.* terápsido

terbium *m.* terbio

therm *f.* terma; *m.* termío

thermal *a.* térmico, termal || ***thermal conductivity*** *f.* conductividad térmica || ***thermal cracking*** *f.* disociación térmica || ***thermal diffusion*** *f.* difusión térmica || ***thermal diffusivity*** *f.* difusividad térmica || ***thermal expansion*** *f.* expansión termal || ***thermal gradient*** *m.* gradiente térmico || ***thermal recovery*** *f.* recuperación térmica || ***thermal spring*** *f.* fuente termal, burga; *m.* balneario || ***thermal stability*** *f.* estabilidad térmica || ***thermal unit*** *f.* unidad térmica

thermistor *m.* termistor; *f.* resistencia térmica

thermocline *f.* termoclina; *m.* termoclinal, termoclino

thermocouple *m.* par térmico; *f.* termocopla, pila termoeléctrica, termopar

thermodynamic *a.* termodinámico

thermodynamics *f.* termodinámica

thermoelastic *a.* termoelástico

thermoelectric *a.* termoeléctrico

thermogene *m.* termógeno

thermogenesis *f.* termogénesis

thermogenic, thermogenetic *a.* termogénico, termogenético

thermogram *m.* termograma

thermography *f.* termografía

thermokarst *m.* termocarst

thermoluminescence *f.* termoluminiscencia

thermometer *m.* termómetro

thermometry *f.* termometría || ***fluid inclusion thermometry*** *f.* termometría de inclusión de flúidos

thermonatrite *f.* termonatrita

thermo-osmosis *f.* termoósmosis

thermophile *m.* termófilo

thermoplastic *a.* termoplástico

thermosphere *f.* termósfera

thermostat *m.* termostato

thernadite *f.* thernadita

thesis *f.* tesis

thick *a.* grueso, espeso, potente || ***thick bed(ded)*** *f.* estratificación gruesa; *adv.* estratificado gruesamente

thicken *vb.* espesar(se), engrosarse

thickener *m.* espesador, concentrador

thickening *m.* engrosamiento, espesamiento

thicket *f.* maleza

thickness *m.* espesor, grueso, grosor; *f.* potencia, espesura

thief *m.* ladrón, ratero, pillo || ***ore thief*** *m.* [Pe.] juquero

thin *a.* delgado, ralo || ***thin bed(ded)*** *a.* en bancos delgados; *adv.* estratificado finamente (o delgadamente) || ***thin section*** *f.* lámina delgada, sección delgada; *m.* corte de roca, corte delgado

thin *vb.* adelgarse, acuñarse, angostarse, estrecharse, cerrarse, [Pe.] estrellarse (la veta, etc.)

thinned out *a.* agotado

thinning *a.* adelgazamiento, estrechamiento

thinolite *f.* tinolita

thin out *vb.* acuñarse, adelgazarse

thiophane *m.* tiofano

thixotropy *f.* tixotropía

tholeiitic *a.* tholeiítico

thompsonite *f.* thompsonita

thorax *m.* tórax

thorium *m.* torio || ***thorium thalenite*** *f.* yttrialita

thorutite *f.* smirnovita

thread-like *a.* filiforme

three-dimensional (3D) seismic *f.* sismica de tres dimensiones

three-point problem *m.* problema de tres puntos

threshold *m.* umbral, nivel crítico

throughput *m.* volumen de materia tratada, rendimento específico; *f.* productividad

throw *m.* desplazamento vertical; salto vertical, rechazo vertical; *f.* dislocación

thrust *m.* empuje; cabalgamiento, corrimiento, [Méx.] coceo || ***thrust belt*** *m.* cinturón corrido || ***thrust fault*** *f.* cabalgadura, falla de corrimiento, falla de sobrecorrimiento; *m.* acabalgamiento, corrimiento || ***thrust sheet*** *m.* manto; *f.* hoja

thrust *vb.* empujar, cabalgar

thulite *f.* thulita

thulium *m.* tulio

Thuringian *m. & a.* Turingiense, Turingiano, Thuringiense, Thuringiano

thuringite *f.* turingita

tidal *a.* de la marea, maremotar, maremotriz || ***tidal bar*** *f.* barra de marea || ***tidal bore*** *f.* ola de marea; *m.* pororoca || ***tidal channel*** *m.* canal de marea || ***tidal flat*** *m.* llano de marea || ***tidal inlet*** *f.* bocana de flujo de mareas; *m.* canal de marea (o de ingreso) || ***tidal marsh*** *f.* marisma || ***tidal wave*** *m.* tsunami

tide *f.* marea || ***ebb tide*** *m.* reflujo; *f.* menguante || ***flood tide*** *f.* pleamar, llenante; *m.* influjo || ***high tide*** *f.* marea alta, pleamar; *m.* creciente || ***low tide*** *f.* marea baja, bajamar || ***red tide*** *f.* marea roja || ***rising tide*** *m.* flujo || ***tide lands*** *f.* tierras inundables

tie *f.* ligadura, atadura; *m.* vínculo, lazo, enlace

tie *vb.* atar, amarrar, ligar, enlazar

Tiffanian *m. & a.* Tiffaniano, Tiffaniense

tiger's eye *m.* ojo de tigre

tight *a.* sin porosidad, de baja porosidad; apretado; tieso, teso; hermético, estanco, ahorcado || ***tight formation*** *f.* formación compacta || ***tight hole*** *m.* pozo secreto, pozo propietario || ***tight sand*** *f.* arena compacta

tightness *f.* tensión, estrechez

Tiglian *m. & a.* Tigliano, Tigliense

till *f.* arcilla pedregosa, til; *m.* acarreo glacial || ***till plain*** *f.* llanura de till

tillite *f.* tilita, tillita

tilt *f.* inclinación; *m.* ladeo, basculamiento

tilt *vb.* inclinarse, ladearse; volcar, bascular

tilted *a.* inclinado, volcado || ***tilted block*** *m.* bloque dislocado, bloque inclinado || ***tilted plane*** *m.* plano inclinado

tilting *f.* volcadura; *m.* vuelco, canteamiento, basculamiento

timber *f.* madera; *m.* arboses, bosque, madero, cuartón, (min.) ademado, ademe, (min.) zarzo || ***timberline*** *m.* límite del bosque

timber *vb.* (min.) ademar, componer con madera

timberman *m.* ademador

time *m.* tiempo || ***absolute time*** *m.* tiempo absoluto || ***access time*** *m.* tiempo acceso || ***arrival time*** *m.* tiempo de llegada || ***delay time*** *m.* tiempo de retardo || ***down, idle time*** *m.* tiempo pasivo || ***intercept time*** *m.* tiempo de intercepción || ***overtime*** *m.* tiempo extra || ***propagation time*** *m.* tiempo de propagación || ***reaction time*** *m.* tiempo de reacción || ***relative time*** *m.* tiempo relativo || ***time limit*** *m.* plazo; límite de tiempo || ***time line*** *f.* línea de tiempo || ***time migration*** *f.* migración en tiempo || ***time rock unit*** *f.* unidad cronoestratigráfica || ***time scale*** *f.* escala de tiempo || ***time section*** (seis.) *f.* línea sismica en tiempo || ***time slice*** (seis.) *f.* sección horizontal en tiempo, corte horizontal en tiempo || ***time stratigraphic*** *a.* cronoestratigráfico || ***time zone*** *m.* huso horario || ***travel time*** *m.* tiempo de propagación || ***two-way travel time*** *m.* tiempo doble || ***up time*** *m.* tiempo activo || ***uphole time*** *m.* tiempo vertical en el punto de disparo, tiempo de propagación ascendente

timer *m.* sincronizador, cronómetro || ***timer driven*** *adv.* accionado por reloj

timing *f.* sincronización

tin *m.* estaño || ***tin-bearing*** *a.* estañifero || ***tin ore*** *m.* estaño || ***tin sulfide*** *f.* zinkisa

Tioughniogan *m.* *&* *a.* Tioughniogense, Tioughniogiano

tip *f.* punta, boquilla

tissue *m.* tejido

titanic *a.* titánico

titaniferous *a.* titanífero

titanite *f.* titanita; *m.* esfeno

titanium *m.* titanio || ***titanium dioxide*** *m.* óxido de titanio

Tithonian *m.* *&* *a.* Titoniano, Titoniense

title *m.* (survey, levan.) título (de propiedad) || ***title block*** *f.* leyenda

titration *m.* análisis volumétrico, análisis por titración; *f.* titración, dosificación

Toarcian *m.* *&* *a.* Toarciense, Toarciano

toe *m.* pie, base; *f.* punta

toeset *m.* pie de capa frontal

tolerance *f.* tolerancia

toluene *m.* tolueno

tombolo *m.* tómbolo

ton *f.* tonelada, salma || ***gross ton*** *f.* tonelada bruta || ***long ton*** *f.* tonelada larga (1015 kg) || ***metric ton*** *f.* tonelada métrica; *m.* tonelámetro (1000 kg) || ***net, short ton,*** *f.* tonelada neta (907 kg)

tonalite *f.* tonalita

Tonawadan *m.* *&* *a.* Tonawadiano, Tonawadense

tone *f.* tonalidad; *m.* tono

Tongrian *m.* *&* *a.* Tongriense, Tongriano

tongue *f.* (geol.) lengua || ***salt tongue*** *f.* lengua de sal

tonnage *f.* capacidad; *m.* tonelaje || ***gross tonnage*** *m.* tonelaje bruto

Tonolawayan *m.* *&* *a.* Tonolawayense, Tonolawayano

tools *f.* herramientas; *m.* implementos, útiles || ***cable tool*** *f.* herramienta de cable || ***fishing tool*** *f.* herramienta de pesca; *m.* útiles de extracción || ***sidetracking tool*** *f.* herramienta desviadora || ***tool carrier*** (min.) *m.* abajador || ***tool extractor*** *m.* extractor de herramientas || ***tool pusher*** *m.* jefe de cuadrilla de perforadores, perforador en jefe

tooth *m.* diente

toothed *a.* denticulado, dentado

top *f.* cima, cumbre; *m.* ápice, tope || ***top water*** *f.* agua superyacente

tops *f.* cimas de las formaciones

topaz *m.* topacio

topographer *m.* topógrafo

topographic *a.* topográfico || ***topographic map*** *m.* mapa topográfico || ***topographic survey*** *f.* altimetría; *m.* relevamiento

topography *f.* topografía

topology *f.* topología

topometry *f.* topometría

topping *f.* destilación primaria

topset bed *f.* capa dorsal; *m.* depósito superior

topsoil *f.* tierra franca, capa vegetal; *m.* [Col.] capote

torbernite *f.* torbernita

torch *f.* mecha, tea

tornado *m.* tornado, ciclón

torque *m.* momento o impulso de rotación, esfuerzo de torsión; *f.* torsión

Torrejonian *m.* *&* *a.* Torrejoniano, Torrejoniense

torrent *m.* torrente, raudal, [And.] volcán

torrential *a.* torrencial

torrid *a.* tórrido

Torridonian *m.* *&* *a.* Torridoniano, Torridoniense

torsion *f.* torsión || ***torsion angle*** *m.* ángulo de torsión || ***torsion balance*** *f.* balanza de torsión

Tortonian *m.* *&* *a.* Tortoniense, Tortoniano

tortuosity *f.* tortuosidad

total *f.* suma; *a.* total, global || ***total depth (TD)*** *m.* fondo del pozo; *f.* profundidad total || ***total dissolved solids*** *m.* total de sólidos disueltos

touchstone *m.* toque

toughness *f.* fortaleza

tourmaline *f.* turmalina

tourmalinization *f.* turmalinización

tourmalite *f.* turmalita

Tournaisian *m.* *&* *a.* Tournaisiense, Tournaisiano

tour of duty *m.* turno

tow *vb.* remolcar

toward the basin *av.* hacia la cuenca

tower *m.* torre, pilar || ***cooling tower*** *m.* torre de enfriamiento || ***distillation tower*** *m.* torre de destilación || ***fractionating tower*** *m.* torre fraccionadora (o de fraccionamiento) || ***scrubbing tower*** *m.* torre depuradora, torre lavadora || ***stripping tower*** *m.* torre de extracción, torre de desprendimiento || ***water tower*** *m.* torre de agua

toxic *a.* tóxico

toxin *f.* toxina

trace *f.* traza (de capa), huella, línea de registro; *m.* vestigio, rastro, color, indicio, trazo || ***trace element*** *m.* elemento traza, oligoelemento, microelemento

trace *vb.* trazar

tracer *m.* rastreador, indicador, trazador

trachyandesite *f.* traquiandesita

trachybasalt *m.* traquibasalto

trachyophitic *a.* traquiofítico

trachyte *f.* traquita; *m.* traquito

trachytic *a.* traquítico

trachytoid *m.* traquitoide

tracing *m.* calco, trazado

track *f.* pisada, huella, traza; vereda; (comp.) pista; *m.* rastro; sendero || ***fission track*** *f.* huella (o traza) de fisión

tract *m.* tramo; intervalo

traction *f.* tracción

tractionite *f.* traccionita

trade *m.* comercio, negocios, intercambio || ***trademark*** *f.* marca de fábrica || ***trade winds*** *m.* alisios

traffic *m.* tráfico

tragacanth *m.* tragacanto

trail *f.* pista, brecha, vereda, senda, huella; *m.* sendero, atajo

trailing *a.* colgado || ***trailing edge*** *m.* borde posterior

train *m.* tren, ferocarril

train *vb.* capacitar, preparar, enseñar

trained *a.* capacitado, entronado

trainee *m.* practicante

training *f.* capacitación; *m.* entrenamiento

trait *m.* rasgo; *f.* característica

trajectory *f.* trayectoria; *m.* trayecto

tram, tramway *f.* aerovía de cable, cablevía, tranvía

tranquillityite *f.* tranquilidadita

transactions *m.* trabajos; *f.* actas

transceiver *m.* emisor

transcurrent *a.* transcurrente || ***transcurrent fault*** *f.* falla de rumbo, falla lateral

transcontinental *a.* transcontinental

transducer *m.* transductor

transect *m.* transecto

transfer *f.* transferencia; *m.* traslado

transfer *vb.* transferir

transfluence *f.* transfluencia

transform *vb.* transformar

transformation *f.* transformación || ***Intensity–Hue–Saturation (IHS) transformation*** *f.* transformación de la intensidad, el tono, y la saturación || ***Laplace transformation*** *f.* transformación de Laplace

transgression *f.* transgresión, trasgresión, inundación; *m.* avance del mar

transgressive *a.* transgresivo

transient *a.* impulso momentáneo, impulso transitorio

transit *m.* teodolito; *m.* & *a.* tránsito; *f.* dioptra || ***transit rod*** *m.* [Méx.] báculo

transition *f.* transición

transitional *a.* de transición, transicional

translate (language) *vb.* traducir

translate (move) *vb.* trasladar

translation *f.* traslación || ***translation gliding*** *m.* deslizamiento por traslación

translational *a.* traslacional

translucent *a.* translúcido, traslúcido, translucente, traslucente, transluciente

transmission *f.* transmisión

transmissivity *f.* transmisividad

transmit *vb.* transmitir

transmitted light *f.* luz transmitida

transmitter *m.* emisor, transmisor

transparant *a.* transparente

transparency *f.* diafanidad, diapositiva, transparencia; *a.* transparente

transpiration *f.* transpiración

transponder *m.* transpondedor

transport *vb.* arrastrar, llevar

transportation *m.* transporte; *f.* transportación, conducción

transported *a.* arrastrado, exótico, alóctono || ***transported soil*** *m.* terreno de acarreo

transporting *m.* acarreo

transpression *f.* transpresión

transpressive, transpressional *a.* transpresivo

transtension *f.* transtensión

transtensional *a.* transtensivo

transverse *a.* transversal, trasversal, transverso || ***transverse dune*** *f.* duna transversal

trap *f.* trampa; *m.* sifón, interceptor; colector, (petrol.) trapp || ***combination trap*** *f.* trampa combinación (o combinada) || ***steam trap*** *m.* separador de agua || ***stratigraphic trap*** *f.* trampa estratigráfica || ***structural trap*** *f.* trampa estructural || ***trap rock*** *f.* dolerita || ***vacuum trap*** *m.* interceptor al vacío

trap *vb.* atrapar

trapezohedron *m.* trapezoedro

travel *m.* viaje, recorrido || ***travel time*** *m.* tiempo de propagación

travel *vb.* viajar

traverse *f.* línea transversal; travesía, carrera; *m.* juego, trazado, travesaño, través, [Arg.] poligonal

traverse *vb.* trazar, trasladar, cruzar, recorrer

traversing *m.* recorrimiento, [Ec.] caminamiento

travertine *f.* travertina, toba calcárea; *m.* travertino

treacherous *a.* falso

treasure *m.* tesoro

treasurer *m.* tesorero

treasury *f.* tesorería || ***national treasury*** *m.* fisco nacional, hacienda nacional

treat *vb.* tratar

treated *a.* tratado

treatment *m.* tratamiento

treaty *m.* tratado || ***North American Free Trade Agreement (NAFTA)*** *m.* Tratado de Libre Comericio (TLC)

tree *m.* (botan.) árbol || ***Christmas tree*** *m.* (perf.) arbol de conexiones, arbol de navidad || ***tree line*** *m.* límite del bosques

trellis *a.* acostillado || ***trellis drainage*** *m.* drenaje acostillado, drenaje enrejado

Tremadocian *m. & a.* Tremadociense, Tremadociano

tremolite *f.* tremolita

tremor *m.* temblor, choque sísmico, tremor; *f.* vibración

Trempealeauan *m. & a.* Trempealeauiano, Trempealeauense

trench *f.* trinchera, zanja, fosa, canal, cuneta, galería, trancada; *m.* foso, zanjón, tajo, apique || ***ocean trench*** *f.* fosa abisal, fosa oceánica

trench *vb.* zanjar, zanjear

trenching machine *f.* cavadora de zanjas, zanjadora

trend *f.* tendencia, dirección, faja; *m.* rumbo, curso, giro, trend, play

trend *vb.* dirigirse

trending from ___ to ___ *a.* dirigido de ___ a ___

Trentonian *m. & a.* Trentoniano, Trentoniense

trespass *f.* invasión

trestle *f.* armazón; *m.* viaducto de caballetes

trial *m.* ensayo; *f.* prueba || ***trial and error*** *f.* a fuerza de ensayo, por experiencia; *m.* métodos de ensayo, tanteo || ***trial run*** *f.* pasada de prueba (o de comprobación)

triangle *m.* triángulo

triangular *a.* triangular

triangulate *vb.* triangular

triangulation *f.* triangulación

Triassic, Trias *m. & a.* Triásico, Trias

triaxial *a.* triaxial

triaxon *m.* triaxón

triboluminescence *f.* triboluminiscencia

tributary *m. & a.* tributario; *a.* confluente; *f.* afluente, rama

trichroism *m.* tricroísmo

triclinic *a.* triclínico

tricostate *a.* de tres costillas

tricyclic *a.* tricíclico

tridymite *f.* tridimita, christensenita || ***tridymite Y*** *f.* spencita

trigonal *a.* trigonal

trihedral *a.* triedro, trihedro

trilobite *f.* trilobita

trilocular *a.* trilocular

trim *vb.* despojar, recortar

trimorphism *m.* trimorfísmo

Trinitian *m. & a.* Trinitiano, Trinitiense

trinominal *a.* trinominal

trip *f.* excursión, jornada; *m.* viaje

triphylite *f.* trifilita

triple junction *f.* triple unión

tripod *m.* trípode

tripping out *f.* sacada de la tubería; *m.* viaje

triserial *a.* triserial

tritium *m.* tritio

trivalent *a.* trivalente

trivium *m.* trivio

trochoceracone *m.* trococono

trochoid *a.* trocoide

trochophore *m.* trocóforo

troctolite *f.* troctolita

troilite *f.* troilita

trona *f.* trona, [Bol.] ckollpa, [Bol., Pe.] collpa

trondjemite *f.* trondjemita

tropic, tropics *m.* trópico || ***Tropic of Cancer*** *m.* trópico de Cáncer || ***Tropic of Capricorn*** *m.* trópico de Capricornio

tropical *a.* tropical

troposphere *f.* tropósfera

trouble *f.* dificultad; *m.* contratiempo, disgusto, inconveniente

troubleshooting *f.* localización de errores o de averías; investigación y reparación de averías

trough *f.* fosa, graben, grabenfosa, depresión, hoya, cubeta, pila, artesa; *m.* seno, hoyo, canal, ábaco, pilón || ***fault trough*** *f.* fosa tectónica, *m.* hoyo de falla || ***wash trough*** *m.* artesón; *f.* batea, pila

trowel *f.* paleta

troy weight *m.* peso de joyería

truck *m.* camión || ***drilling truck*** *m.* camión de taladro || ***recording truck*** *m.* camión de registro || ***tank truck*** *m.* camión tanque

true *a.* en línea recta, verdadero, derecho || ***true amplitude*** *f.* amplitud verdadera

truncate *vb.* acuñarse, truncar(se)

truncated *a.* truncado || ***truncated spur*** *m.* espolón truncado

truncation *m.* truncamiento; *f.* truncación

trunk pipeline *m.* oleoducto troncal; gasoducto troncal

truss *f.* armazón, armadura, cabriada

trustee *m.* curador, depositario

trydimite *f.* tridimita

tsunami *f.* ola de mar producida por un temblor submarino, tsunami

tube *m.* tubo, caño, conducto || ***capillary tube*** *m.* tubo capilar || ***cathode ray tube*** (CRT) *m.* tubo de rayos catódicos || ***condenser tube*** *m.* tubo condensador || ***test tube*** *m.* tubo de ensayo || ***tube cutter*** *m.* cortatubos || ***tube expander*** *m.* abocinador de tubos, mandril de expansión || ***U tube*** *m.* tubo en U

tubing *f.* tubería, cañería, tubería de disparo; *m.* tubo productor || ***pull tubing*** *vb.* sacar tubería || ***run tubing*** *vb.* instalar tubería || ***seamless tubing*** *f.* tubería sin costura || ***tubing head*** *m.* estoperol

tuberose *a.* tuberoso

tubing *f.* tubería; *m.* tubo productor

tuboscope, tubescope *m.* tuboscopio

tubular *a.* tubular

tufa *f.* tufa, toba calcárea; *m.* tras

tuff *f.* toba, toba volcánica (o de cenizas), tufa; *m.* tras || ***tuff quarry*** *m.* tobar || ***welded tuff*** *f.* toba soldada, ignimbrita

tuffaceous *a.* tobáceo, toboso, tufaceo

tuffite *f.* tufita

tug boat *f.* lancha remolcadora

tumescence *f.* inflación

tumid *a.* túmido

tundra *f.* tundra

tune *vb.* sintonizar

tungstate *m.* tungstato

tungsten *m.* tungsteno, wolfram || ***tungsten-bearing*** *a.* wolframífero

tunnel *m.* túnel, socavón; *f.* galería || ***tunnel driver*** *m.* socavonero

Tuorian *m. & a.* Tuoriano, Tuoriense

turbation *f.* turbación

turbid *a.* turbio, túrbido, zurrapiento, zurraposo

turbidite *f.* turbidita, corriente de densidad

turbidity *f.* turbiedad, turbidez, turbieza || ***turbidity current*** *f.* corriente turbidítica o de densidad

turbine *f.* turbina || ***steam turbine*** *f.* turbina de vapor || ***turbine pump*** *f.* bomba turbina

turbulence *f.* turbulencia

turbulent *a.* turbulento || ***turbulent flow*** *m.* flujo turbulento

turf *f.* turba

turgid *a.* túrgido, abultado, turgente

turgite *f.* turgita

turn *f.* vuelta, revolución; *m.* giro

turn *vb.* girar, revolver, doblar

turnaround time *m.* tiempo de ida y vuelta

turning radius *m.* radio de viraje

turnover of inventories *f.* rotación de inventarios

Turolian *m. & a.* Turoliano, Turoliense

Turonian *m. & a.* Turoniano, Turoniense

turpentine *f.* trementina, aguarrás

turquoise *f.* turquesa; *m.* [Méx.] chalchihuitl, xihuitl

tusk *m.* diente

twin *f.* macla; *m.* gemelo || ***twin crystal*** *f.* macla; *m.* cristal geminado || ***twin gliding*** *m.* deslizamiento por maclado

twinned *a.* maclado

twinning *f.* geminación, macla

twist *f.* torsión, torcedura || ***twist off*** *f.* rotura por torsión

twist *vb.* retorcer, girar

twisted *a.* torcido, retorcido, desviado

two-dimensional *a.* de dos dimensiones

two-way radio *m.* radioteléfono en dos direcciones

two-way travel time *m.* tiempo doble

type *m. & a.* tipo || ***type fossil*** *m.* fósil típico || ***type location*** *f.* localidad tipo || ***type section*** *f.* sección tipo

typhoon *m.* tifón

typical *a.* típico, característico

typomorphic *a.* tipomórfico || ***typomorphic mineral*** *m.* mineral tipomorfo

tyrolite *f.* tirolita

Tyrrhenian *m. & a.* Tyrrheniano, Tyrrheniense

tyuyamunite *f.* tyuyamunita

U

U-shaped *a.* en forma de U

Uintan *m. & a.* Uintense

Ulatsian *m. & a.* Ulatsiense, Ulatsiano

ulexite *f.* ulexita, natroborocalcita; *m.* borato

Ulsterian *m. & a.* Ulsteriense, Ulsteriano

ultimate *a.* último || ***ultimate production*** *f.* producción final || ***ultimate recovery*** *f.* recuperación última || ***ultimate stress*** *m.* esfuerzo de ruptura

ultrabasic *a.* ultrabásico, ultramáfico

ultrabasite *f.* quevellita

ultramafic *a.* ultramáfico, ultrabásico

ultramafite *f.* perknita

ultramarine *m.* zafiro

ultrametamorphism *m.* ultrametamorfismo

ultramylonite *f.* ultramilonita

ultrasonic *a.* ultrasónico

ultraviolet *a.* ultravioleto || ***ultraviolet rays*** *m.* rayos ultravioletos

Ultravulcanian *m. & a.* Ultravulcaniense, Ultravulcaniano

ulvöspinel *f.* ulvita

umbilical *a.* umbilical

umbilicus *m.* ombligo

umbo *m.* umbo, umbón, corchete

umbonal *a.* del umbo

umbra *f.* umbra

unattached *a.* no fijo, no adherido

unbalanced *a.* no balanceada

unbroken *a.* intacto, continuo

uncased *a.* sin entubar

uncertainty *f.* incertidumbre; *m.* incierto

uncoiled *a.* desenrollado

uncompacted *a.* inconsistente, sin compactar

uncompahgrite *f.* uncompahgrita

unconformable *a.* discordante

unconformity *f.* disconformidad, discordancia || ***angular unconformity*** *f.* discordancia angular || ***parallel unconformity*** *f.* discordancia paralela || ***regional unconformity*** *f.* discordancia regional

unconsolidated *a.* suelto, sin consolidar, no consolidado, incoherente

uncover *vb.* descubrir, descapotar

uncovered *a.* descubierto, descapotado

undeformed *a.* indeformado

under *adv.* abajo, debajo; *p.* debajo de

underbalanced *a.* con insuficiente contrapeso

underclay *f.* arcilla debajo de una capa de carbón, sotoarcilla

undercut, undercutting *f.* regadura, regola, roza, rozadura; *m.* reguero, socavón || ***undercut, undercutting slope*** *f.* pendiente socavada

undercut *vb.* socavar, rozar

underflow *m.* subdesbordamiento

undergo *vb.* sufrir, sujeta a

underground *m.* subterrano; *a.* subterráneo, soterraño, soterrado

underhand stope *f.* labor de banco; *m.* rebaje

underlain *vb.* tener un subyacente

underlie *vb.* subyacer

underlying *a.* subyacente, infrayacente, infrapuesto

undermine *vb.* socavar, carcomar, minar, zapar

undermining *m.* socavamiento; *f.* regadura, zapa, socavación, socava

underpressure *f.* subpresión

underproduction *f.* producción subnormal, producción insuficiente

undersaturate *vb.* subsaturar

undersaturated *a.* subsaturado

undersaturation *f.* saturación insuficiente, subsaturación

undersigned *m.* & *f.* firmante; *m.* infraescrito

understanding *m.* entendimiento

underthrust *a.* bajoescurrimiento, bajocorrimiento; *f.* falla de bajo escurrimiento

underwater *a.* sumergido, subacuático, subácueo

underwriter *m.* asegurador

undisturbed *a.* no perturbado, imperturbado, intacto

undulate, undulated *a.* ondulado

undulating *a.* alomado, ondulante

undulation *f.* ondulación

uneroded *a.* joven

uneven *a.* desigual, irregular, accidentado, impar, (geog.) arisco, cerril, quebrado, ondulado

uneveness (geog.) *m.* accidente, desnivel; *f.* aspereza

unexploitable *a.* inexplotable

unexploited *a.* inexplotado

unexplored *a.* inexplorado

unfolded *a.* desdoblado

unfoliated *a.* no foliado

uniaxial *a.* uniaxial, uniáxico || ***uniaxial interference figure*** *f.* figura de interferencia uniaxial

unicellular *a.* unicelular

uniclinal *a.* monoclinal

unicostate *a.* con una costilla

uniform *a.* uniforme, homogéneo, constante

uniformitarianism *m.* uniformitarismo

uniformity *f.* uniformidad

unify *vb.* unificar

unilocular *a.* unilocular

unindurated *a.* suelto, no consolidado, no endurecido

uninhabited *a.* yermo

union *f.* unión, junta; *m.* empalme || ***labor, trade union*** *m.* sindicato, gremio; sindicatura; *a.* sindical

unionization *f.* sindicación

unionize *vb.* sindicalizar, sindicar
unionism *m.* sindicalismo
uniserial *a.* uniserial, uniseriado
unisexual *a.* unisexual
unit *f.* unidad, formación || ***biostratigraphic unit*** *f.* unidad bioestratigráfica || ***British Thermal Unit (BTU)*** *f.* unidad térmica británica, unidad térmica inglesa || ***central processing unit (CPU)*** *f.* unidad central de procesamiento || ***chronostratigraphic unit*** *f.* unidad cronoestratigráfica || ***processing unit*** *f.* (comp.) unidad de proceso || ***solvent extraction unit*** *f.* unidad de extracción de solventes || ***stratigraphic unit*** *f.* unidad estratigráfica || ***unit operation*** *f.* administración unificada || ***vacuum distillation unit*** *f.* unidad de destilación al vacío
unite *vb.* unir, unirse, juntar, conjuntar
united *a.* unido, conjuntado, juntado
unitization *f.* unificación
unitize *vb.* unificar, combinar, consolidar
univalve *m. & a.* univalvo
universe *m.* universo
universal *a.* universal
unknown *m. & a.* incognito; *f.* incognita
unleaded *a.* sin plomo
unlevelled *a.* desnivelado, desigual
unlimited *a.* ilimitado
unload *vb.* descargar
unmanned *a.* no tripulado
unmineralized *a.* pobre, estéril
unpolished *a.* mate
unrefined *a.* no refinado
unrolled *a.* desenrollado
unsaturated *a.* no saturado
unsaturates *f.* substancias no saturadas
unstable *a.* inestable, inconsistente || ***unstable slope*** *f.* pendiente inestable
unstratified *a.* no estratificado
unsymmetrical *a.* asimétrico
unweathered *a.* inalterado, no meteorizado
unworked *a.* no explotado, sano
up *a.* arriba
uparching *m.* abovedamiento
upbuilding *f.* intrusión, construcción hacia arriba
update *vb.* actualizar, poner al día
updating *f.* actualización
updip *m.* echado arriba, buzamiento arriba; *f.* inclinación arriba
updraft *m.* tiro hacia arriba, tiro ascendente
upgrade *a.* cuesta arriba
upgrade *vb.* modernizar, potenciar, mejorar
upheaval *m.* levantamiento, movimiento ascensional
uphill *a.* cuesta arriba
uphole time *m.* tiempo vertical sísmico en el punto de disparo, tiempo vertical
upkeep *f.* conservación; *m.* mantenimiento
upland *f.* tierra alta, tierra arriba
uplift *m.* levantamiento, ascenso, erguimiento, solevamiento, solevantamiento; *f.* sublevación
uplift *vb.* levantar, altearse
uplifted *a.* levantado || ***uplifted block*** *m.* pilar tectónico, horst;f. gleba tectónica
upper *a.* superior || ***upper limit*** *m.* límite superior || ***uppermost*** *a.* superior extremo || ***upper part*** *f.* parte superior || ***upper shoreface*** *m.* frente de playa alto
upright *a.* ramplón || ***upright fold*** *m.* pliegue parado
upstream *a.* corriente arriba, agua(s) arriba, rio arriba; (com.) exploración y producción
upstructure *a.* estructura arriba
upthrown *a.* levantado || ***upthrown block*** *m.* bloque alto, labio alto, bloque levantado || ***upthrown side*** *m.* labio superior
upthrust *f.* falla de alto empuje, empuje ascendente, empuje hacia arriba
uptime *m.* tiempo activo
upwards *a.* hacia arriba
upwarp *m.* abovadamiento
upwelling *f.* surgencia, corriente surgente, aguas de afloramiento, llamada de fondo
Uralian *m. & a.* Uraliano, Uraliense
uralite *f.* uralita
uralitization *f.* uralitización
uranic *a.* uránico

uraniferous *a.* uranífero

uraninite *f.* urnaninita, pechblenda; *m.* óxido de urano, uranopicco || ***altered uraninite*** *f.* yttrogummita

uranite *f.* uranita, kalkuranita

uranium *m.* uranio || ***uranium oxide*** *m.* urano

uranophane *m.* uranofano

uranopyrochlore *f.* yttrohatchettolita

uranosphaerite *f.* uranosfaerita

uranotile *m.* uranotilo

Uranus *m.* Urano

ureyite *m.* kosmocloro

Uriconian *m. & a.* Uriconiense, Uriconiano

useful *a.* útil

utilities *m.* servicios auxiliares

utility *f.* utilidad

utilized capacity *f.* capacidad aprovechada

uvarovite *f.* uvarovita, uwarowita; *m.* granate crómico

V

V-shaped *a.* en forma de V

vacuole *f.* vacuola

vacuum *m. & a.* vacío, al vacío, en vacío, vacuo || ***vacuum distillation*** *f.* destilación al vacío || ***vacuum filter*** *m.* filtro al vacío || ***vacuum gauge*** *m.* vacuómetro || ***vacuum pump*** *f.* bomba de vacío || ***vacuum relief valve*** *f.* ventosa al vacío || ***vacuum tube*** *m.* tubo al vacío

vadose *a.* vadosiana || ***vadose water*** *f.* agua meteórica

Valanginian *m. & a.* Valanginiense, Valanginiano; Nemausense

vale *f.* cañada; *m.* valle

valence *f.* válencia

Valendisian *m. & a.* Valendisiano, Valendisiense

Valentian *m. & a.* Valentiense, Valentiano

valentinite *f.* exitela

valid *a.* válido

validate *vb.* dar validez a

validity *f.* validez

valley *m.* valle; *f.* cañada, vega || ***alluvial valley*** *m.* valle aluvial || ***antecedent valley*** *m.* valle epigenético || ***hanging valley*** *m.* valle colgante || ***humid valleys east of the Andes*** *f.* [And.] yungas || ***rift valley*** *m.* valle de hundimiento || ***river valley*** *f.* arribera || ***submerged valley*** *m.* valle sumergido || ***subsequent valley*** *m.* valle subsequente || ***superimposed valley*** *m.* valle epigenético || ***valley***fill *m.* relleno del valle || ***valley***floor *m.* piso del valle || ***valley***glacier *m.* glaciar de valle

Valmeyeran *m. & a.* Valmeyerense, Valmeyeriano

valuable *a.* valioso, estimable

valuation *f.* valía, tasa, apreciación, tasación; *m.* avalúo, aforamiento, aforo, aprecio

value *m.* valor; *f.* valía || ***book value*** *m.* valor contable || ***caloric value*** *m.* valor calórico, valor térmico || ***calorific value*** *m.* BTU

value *vb.* apreciar, avaluar, tasar, estimar, valorar, evaluar

valve *f.* valva, válvula || ***check valve*** *f.* válvula de retención || ***flow valve*** *f.* válvula reguladora de flujo || ***hydraulic valve*** *f.* válvula hidráulica || ***master gate valve*** *f.* válvula maestra || ***pressure relief valve*** *f.* válvula de alivio, válvula de seguridad || ***safety valve*** *f.* válvula de seguridad || ***standing valve*** *f.* válvula fija

vanadate *m.* vanadato

vanadiferous *a.* vanadífero

vanadinite *f.* vanadinita, johnstonita; *m.* vanadato

vanadium *m.* vanadio

vane (weather) *f.* veleta

vapor *m.* vapor, vaho || ***vapor barrier*** *f.* barrera de vapor || ***vapor line*** *f.* cañería de vapor || ***vapor lock*** *f.* traba de vapor, obstrucción por vapor || ***vapor pressure*** *f.* presión (o tensión) de vapor || ***vapor phase*** *m.* fase vapor || ***vapor phase cracking*** *m.* craqueo por el método de fase de vapor || ***vapor pressure*** *f.* tensión de vapor

vaporization *f.* vaporización || ***vaporization temperature*** *f.* temperatura de vaporización

vaporizer *m.* vaporizador

vara *f.* vara (medición antigua que varia entre 0,8–1,1 metros, o aprox. 33 pulgadas)

variable *a.* variable

variance *f.* variancia, varianza

variation *f.* variación

varicolored *a.* abigarrado, multicolor

varied *a.* variado

variegated *a.* abigarrado, jaspeado, multicolor

varient *f. & a.* variante

variety *f.* variedad, gama

variolite *f.* variolita

variolitic *a.* variolítico

variometer *m.* variómetro

Variscan *m. & a.* Varíscico, Varisco

variscite *f.* variscita, utahlita

varnish *m.* barniz || ***desert varnish*** *m.* barniz del desierto; *f.* pátina desértica

varve *m.* estrato anual, depósito anual; *f.* varva

vary *vb.* variar

varying *a.* cambiante, diverso

vascular *a.* vascular

vat *f.* cuba

Vauclusian *m. & a.* Vauclusiano, Vauclusiense

vaulted *a.* abovado

vector *m.* vector

vegetable *a.* vegetal

vegetation *f.* vegetación

vein *f.* veta, vena, fibra, hebra, [Col.] aguja, [Ch.] sirca; *m.* filón, ganga, mena, panino, [Pe.] farallón || ***antitaxial fibrous vein*** *f.* venas fibrosas antitaxiales || ***barren vein*** *f.* veta en borra || ***branch vein*** *f.* ramita de filón, veta ramal || ***companion vein*** *f.* veta socia || ***large vein*** *m.* vetarrón || ***ore-bearing vein*** *f.* veta en frutos || ***small vein, veinlet*** *m.* hilo; *f.* venilla, venita || ***syntaxial fibrous vein*** *f.* venas fibrosas sintaxiales || ***variable striking vein*** *f.* veta serpenteada || ***vein-filling clay*** *f.* tosca || ***vein wall*** *m.* reliz, respaldo

veined *a.* avetado, venoso, veteado

veining *f.* filonitización

veinlet *f.* vetica, veticilla, vetilla, venilla; *m.* filoncillo

vein-like *a.* filoniano

velocity *f.* velocidad || ***apparent velocity*** *f.* velocidad aparente || ***average velocity*** *f.* velocidad promedia || ***instantaneous velocity*** *f.* velocidad instantánea || ***interval velocity*** *f.* velocidad del intervalo || ***stacking velocity*** *f.* velocidad de apilamiento || ***velocity pick*** *f.* recogida de la velocidad

veneer *f.* pelicula, cubierta

vent *m.* conducto, respiradero, acis; *f.* chimenea

ventifact *m.* ventifacto; *f.* gliptolita

ventilation *f.* ventilación

ventilator *m.* ventilador

venting *f.* purga

ventral *a.* ventral

ventricle *m.* ventrículo

Venturian *m. & a.* Venturiano, Venturiense

Venturi meter *m.* venturímetro

Venus *f.* Venus

vergence *f.* vergencia

verification *f.* verificación

verify *vb.* comprobar, verificar, probar

vermiculite *f.* vermiculita

verrucose *a.* verrugoso

Versilian *m. & a.* Versiliano, Versiliense

vertebra *f.* vertebra

vertebrate *m.* vertebrado

vertex *f.* vértice

vertical *a.* vertical, [Col.] clavado || ***vertical hole*** *m.* pozo vertical || ***vertical interval*** *m.* intervalo vertical || ***vertical magnetic intensity*** *f.* intensidad magnética vertical || ***vertical section*** *f.* sección vertical || ***vertical seismic survey*** *m.* perfil de velocidad vertical || ***vertical separation*** *f.* separación vertical || ***vertical shear*** *f.* cizalla vertical

vesicle *f.* vesícula; *m.* vesículo

vesicular *a.* vesicular

vessel *f.* vasija; *m.* recipiente, vaso || ***pressure vessel*** *m.* recipiente a presión

vestibule *m.* vestíbulo
vestige *m.* vestigio
vestigial *a.* atrofiado
Vesulian *m. & a.* Vesuliense, Vesuliano
vesuvianite *f.* vesuvianita, vesubianita, vesubiana, jewreinowita
viable *a.* viable
viborgite *f.* wiborgita
vibracula *f.* vibracula
vibrate *vb.* vibrar
vibrating *a.* acción de vibrar
vibration *f.* vibración, oscilación; *m.* balance || ***vibration dampener*** *m.* amortiguador de vibraciones
vibrator *m.* vibrador
vibrofrac *m.* torpedeamiento por vibración
vibroseis *f.* sísmica de vibradoras, vibroseis, vibrosísmica
Vicksburgian *m. & a.* Vicksburgiano, Vicksburgiense
view *f.* vista, visión || ***synoptic view*** *f.* visión sinóptica
view *vb.* mirar, avistar
viewfinder *m.* anteojo buscador
viewgraph *m.* acetato
viewpoint *m.* punto de vista, punto de observación
vigilence *f.* vigilancia
Villafranchian *m. & a.* Villafranquense, Villafranquiano
Vindobonian *m. & a.* Vindobonense, Vindoboniano
vinyl *m.* vinilo || ***vinyl chloride*** *m.* cloruro de vinilo
virgalia *m.* virgalio
virgation *f.* virgación
virgella *f.* virgella
Virgilian *m. & a.* Virgiliano, Virgiliense
Virglorian *m. & a.* Virgloriense, Virgloriano
Virgulian *m. & a.* Virguliense, Virguliano
viridine *f.* viridina
visceral *a.* visceral
viscometer *m.* viscómetro
viscosimeter *m.* viscosímetro
viscosity *f.* viscosidad
viscous *a.* viscoso
Visean *m. & a.* Viseense, Viseano
visibility *f.* visibilidad
vision *f.* visión
visor *f.* visera
visual *a.* visual
vital *a.* vital
vitreous *a.* vítreo, vidrioso
vitrian *f.* vitrita
vitric *a.* vítrico, vitreo
vitrification *f.* vitrificación
vitrified *a.* vitrificado
vitrify *vb.* vitrificar
vitrinite *f.* vitrinita
vitriol *m.* acije, vitriolo
vitrite *f.* vitrita
vitroclarian *f.* vitroclarita
vitrophyre *m.* vitrófiro, vitriófido, vitrófido
vitrophyric *a.* vitrofídico, vitrófido
vivianite *f.* vivianita; *m.* hierro azul
void *m.* vacío, hueco, poro, intersticio
volatile *a.* volátil
volatility *f.* volatilidad
volatilization *f.* volatilización
volatiles *m.* ligeros
volatilize (mercury) *vb.* desazogar
volborthite *f.* uzbekita
volcanic *a.* volcánico, eruptivo || ***volcanic agglomerate*** *m.* aglomerado volcánico || ***volcanic ash*** *f.* ceniza volcánica, toba de cenizas; *m.* espodito || ***volcanic bomb*** *f.* bomba volcánica; *m.* proyectil volcánico || ***volcanic breccia*** *f.* brecha volcánica || ***volcanic cone*** *m.* cono volcánico || ***volcanic crater*** *f.* caldera || ***volcanic dome*** *m.* domo volcánico || ***volcanic neck*** *f.* chimenea volcánica; *m.* cuello volcánico || ***volcanic pipe*** *f.* chimenea volcánica || ***volcanic plug*** *m.* tapón volcánico || ***volcanic sand*** *f.* peperita || ***volcanic system*** *m.* aparato volcánico || ***volcanic vent*** *m.* conducto volcánico, acis
volcanism *m.* vulcanismo, volcanismo, plutonismo; *f.* vulcanicidad, volcanicidad
volcanologist *m.* volcanólogo, vulcanólogo, vulcanologista

volcanology *f.* vulcanología

volcanite *f.* volcanita

volcano *m.* volcán || ***extinct volcano*** *m.* volcán apagado || ***mud volcano*** *m.* volcán de fango, hervidero, volcán de lodo; *f.* salsa || ***shield volcano*** *m.* volcán en escudo

volcanogenic *a.* volcanogénico

volcanology *f.* vulcanología

volt *m.* voltio

voltage *m.* voltaje

voltmeter *m.* voltímetro

volume *m.* volumen; caudal, cubaje, cubicaje || ***pore volume*** *m.* volumen poroso

volumetric *a.* volumétrico

volumetrics *f.* volumetría

vonsenite *f.* paigeita

vortex *m.* vórtice

voucher *m.* comprobante, justificativo, documento

Vraconian *m. & a.* Vraconiano, Vraconiense

vug *f.* vacuola, cavidad, ampolla, oquedad; *m.* hueco, poro

vuggy *a.* poroso, ampolloso, gongorado

vugular *a.* cavernoso || ***vugular porosity*** *f.* porosidad cavernosa

vulcanism *m.* vulcanismo, plutonismo

W

Waalian *m. & a.* Waaliano, Waaliense

wachenrodite *f.* wackenrodita

wacke *f.* vacia, waca, wacka

wackestone *f.* caliza lodosa

wad *m.* taco, grafito, tapón; wad; *f.* acerdesa

wadi *m.* wadi, waudi, cañadón seco

wage(s) *m.* jornal, salario, sueldo

wagon *m.* carro

waiting on *a.* en fragüe de

wakefieldite *f.* kusuita

walking beam *m.* balancín

wall *f.* pared, muralla; *m.* muro || ***foot wall*** *m.* muro de base || ***hanging wall*** *m.* pendiente, bloque alto, bloque levantado, yacente, bloque colgante, muro colgante; *f.* tapa || ***lateral wall***of an excavation *m.* hastial || ***retaining wall*** *m.* muro de detención, muro de sostén, muro de contención || ***wall rock*** *f.* caja; *m.* peña, [Bol.] farallón

wallongongite *f.* wallongita

walpurgite, waltherite *f.* walpurgina

Waltonian *m. & a.* Waltoniano, Waltonense, Waltoniense

wandering *m.* desvío

warehouse *m.* almacén, depósito; *f.* bodega

warm *a.* caliente; templado, tibio; cálido, caluroso

warning *m.* aviso

warp *f.* curvatura, comba; *m.* torcimiento, encorvamiento

warp *vb.* combarse

warped *a.* torcido, combado, acombado

warping *m.* combado, doblegado

warwickite *f.* warwicita

Wasatchian *m. & a.* Wasatchiano, Wasatchiense

wash *m.* barranco; *f.* barranca || ***granite wash*** *m.* lavado de granito || ***slope wash*** *f.* lavaduras de falda

wash *vb.* (min.) aflojar, lavar

wash away *vb.* quitar, remover

washing *m.* lavaje, lavado; *f.* remoción, lava, lavada, lavadura

Washitan *m. & a.* Washitano, Washitense

washout *m.* socavamiento, derrumbamiento, derrumbe, arrastre

wash out *vb.* lavar, enjuagar, derrubiar

washover fan *m.* abanico de desborde

waste *m.* desperdicio, desechos, ataques, desmonte, despojo; *f.* estopa, limpia, merma || ***hazardous waste*** *m.* desperdicio perjudicial para la salud || ***solid waste*** *m.* residuos sólidos || ***waste disposal system*** *m.* sistema de eliminación de desperdicios || ***waste gas*** *m.* gas de desecho, gas de desperdicio || ***waste heap*** *m.* acopio; *f.* escombrera || ***waste heat*** *m.* calor perdido || ***wasteland*** *m.* terreno baldío;

f. gándara || ***waste pile*** *f.* trincha || ***waste product*** *m.* producto sin valor, producto de desperdicio || ***waste rock*** *m.* atierres, caballo, caballete, escombro, cascote, desecho, rezago; *f.* rezaga, zafra, [Pe.] bazofía || ***waste rock dump*** *m.* boleo

waste *vb.* desgastar

watch *vb.* vigilar

watchfulness *f.* vigilancia

watchman *m.* centinela, velador, vigía

water *f.* agua, agüilla, aguosidad || ***artesian water*** *f.* agua artesiana, agua surgente || ***bottom water*** *f.* agua de fondo, agua subyacente || ***bound water*** *f.* agua limitada || ***brackish water*** *f.* agua salobre || ***capillary water*** *f.* agua capilar || ***carbonated water*** *f.* agua carbonatada || ***compaction water*** *f.* agua de expulsión, agua de consolidación || ***connate water*** *f.* agua congénita, agua connata, agua intersticial, agua de formación, agua fósil, agua innata, agua singenética || ***crystallization water*** *f.* agua de cristalización || ***deep water*** *f.* agua profunda || ***discharge water*** *f.* agua de descarga || ***distilled water*** *f.* agua destilada || ***drinking water*** *f.* agua potable || ***edge water*** *f.* agua marginal, agua subyacente || ***flood waters*** *f.* agua de inundación, [Méx.] broncas || ***flowing water*** *f.* agaus vivas || ***formation water*** *f.* agua de formación || ***fossil water*** *f.* agua fósil || ***fresh water*** *f.* agua dulce || ***groundwater*** *f.* agua subterránea, agua freática, agua de fondo, cloacales de infiltración, [Arg.] agua telúrica || ***hard water*** *f.* agua calcárea, agua cruda, agua dura, agua gorda || ***juvenile water*** *f.* agua juvenil || ***low water*** *f.* aguas estiales || ***magmatic water*** *f.* agua magmática || ***melt water*** *f.* agua deshielo || ***metamorphic water*** *f.* agua metamórfica || ***meteoric water*** *f.* agua meteórica, agua de gravidad || ***mine water*** *f.* agua de mina || ***mineral water*** *f.* agua mineral || ***oxygenated water*** *f.* agua oxígenada || ***perched ground water*** *f.* agua endicada || ***percolation water*** *f.* agua infiltrada || ***phreatic water*** *f.* agua freática || ***pore water*** *f.* agua de retención || ***potable water*** *f.* agua potable || ***rainwater*** *f.* agua pluvial || ***saltwater*** *f.* agua salada, salmuera, agua salobre || ***seawater*** *f.* agua de mar || ***shallow water*** *f.* agua somera || ***soil water*** *f.* agua del suelo || ***soft water*** *f.* agua blanda, agua delgada || ***spring water*** *f.* agua de manantial, agua natural, agua de pie, agua viva || ***stagnant water*** *f.* agua dormida, agua estancada || ***standing water*** *f.* agua estancada || ***streambed water*** *f.* agua subálvea || ***surface water*** *f.* agua freática, agua superficial, [Arg.] agua somera || ***thermal water*** *f.* agua termal || ***treatment water*** *f.* agua de tratamiento || ***underground water*** *f.* agua subterránea || ***unpotable water*** *f.* agua sucia || ***vadose water*** *f.* agua vadosiana, agua meteórica, agua de gravidad, agua vadosa || ***waste water*** *f.* agua de descarga, agua residual, aguas inmundas || ***water-based mud*** *m.* lodo a base de agua || ***water-bearing*** *a.* acuífera || ***water course*** *m.* arroyado, curso de agua, lecho de río || ***water coning*** *f.* conificación del agua || ***water cut*** *m.* porcentaje de agua || ***water drive*** *m.* empuje hidrostático, empuje de agua, desplazamiento por agua || ***water encroachment*** *f.* intrusión acuífera, avance de aguas marginales || ***waterfall*** *f.* cascada, catarata, caída de agua; *m.* salto, aguacero, [PR] cangilón || ***water fowl*** *f.* aves acuáticas || ***water flood*** *f.* inyección de agua || ***water flooding*** *f.* inundación artificial || ***water gap*** *m.* valle de drenaje; *f.* abra || ***water hammer*** *m.* ariete hidráulico || ***waterhole*** *m.* charco, abrevadero || ***water injection*** *f.* inyección de agua || ***water leaching*** *f.* percolación de agua || ***water leg*** *f.* placa de agua; *m.* hervidero || ***water level*** *m.* nivel de agua || ***waterlogged*** *a.* empapado; anegado, inundado || ***water off*** *vb.* inundar || ***water–oil ratio (WOR)*** *f.* relación agua–petróleo || ***waterproof*** *a.* impermeable || ***water pump*** *f.* bomba de agua || ***water resources*** *f.* aguas aprovechables || ***water rights*** *f.* [Sp.] almoceda || ***water scrubber*** *m.* depurador de agua || ***water shed*** *f.* hoya hidrográfica, divisoria de vertientes, divisoria de aguas, vaguada || ***water softener*** *m.* ablandador de agua || ***water-soluble*** *a.* soluble en agua, acuosoluble || ***water spout, water jet*** *m.* chorro || ***water supply*** *m.* abastecimiento de agua || ***water table*** *m.* nivel freático, nivel hidráulico, nivel de saturación, botaguas; *f.* lámina acuífera, napa freática || ***water tank*** *m.* estanque || ***watertight*** *a.* impermeable, estanco || ***water vapor*** *m.* vapor de agua || ***waterway*** *f.* vaguada, via de agua || ***water well*** *m.* pozo de agua || ***well water*** *f.* agua de pozo, [Méx.] noria

watershed *f.* área alimentadora, área vertiente, cuenca, cuenca tributaria, divisoria de aguas, divisoria de vertientes, hoya hidrográfica, [Col.] área de escurrimiento || ***watershed divide*** *f.* [Vz.] arista hidrográfica

watering out (of a wellbore) *m.* influjo de agua

watery *a.* acuoso, aguoso

watt *m.* vatío

wattmeter *m.* vatímetro

Waucoban *m.* & *a.* Waucobiano, Waucobiense, Waucobense

wave (ocean) *f.* ola || ***internal wave (ocean)*** *f.* ola interna || ***shoaling wave*** *f.* ola de bajio || ***surface wave*** *f.* ola de superficie || ***tidal wave*** *f.* ola de marea; *m.* tsunami || ***wave ripple*** *f.* marca de oleaje

wave (seis., phys., física) *f.* onda || ***acoustic wave*** *f.* onda acústica || ***compressional wave*** *f.* onda compresional, onda de condensación || ***dilational wave*** *f.* onda de dilatación || ***elastic wave*** *f.* onda elástica || ***Love wave*** *f.* onda de Love || ***P wave, primary*** *f.* onda P, onda primaria || ***Rayleigh wave*** *f.* onda Rayleigh || ***S wave*** *f.* onda S || ***seismic wave*** *f.* onda sísmica || ***shear wave*** *f.* onda de cizalla || ***sound wave*** *f.* onda sonora || ***surface wave*** *f.* onda superficial || ***tidal wave*** *f.* onda de marea; *m.* tsunami || ***wave equation*** *f.* ecuación de onda || ***wave erosion*** *f.* erosión de marea || ***wave front*** *m.* frente de ola, frente de onda || ***wave train*** *m.* grupo de ondas, tren de olas

wave-cut platform *f.* plataforma de abrasión

wave-cut terrace *f.* terraza cortada por las olas

wavelength *f.* longitud de onda

wavelet *f.* ondícula sísmica

wavellite *f.* wavelita

wavy *a.* ondulante || ***wavy bedding*** *f.* estratificación ondulada || ***wavy extinction*** *f.* extinción ondulada

wax *f.* parafina, cera || ***wax-free oil*** *m.* aceite sin parafina

waxy *a.* parafínico, ceroso, céreo || ***waxy crude*** *m.* crudo parafínico

way *m.* sendero, camino; *f.* vía

weak *a.* débil

weaken *vb.* debilitar

weakness *f.* debilidad

Wealden *m.* & *a.* Wealdiano, Wealdiense, Wealdense

wealth *m.* caudal; riqueza

wear and tear *m.* desgaste, deterioro

wear away *vb.* desgastar, gastar, raer, consumir

wear down *vb.* desgastar, gastar; agotar, rendir

wear out *vb.* desgastar, gastar, romper

weather *f.* intemperie; *m.* clima, tiempo || ***weather proof*** *a.* a prueba de intemperie, impermeable, estanco

weather *vb.* meteorizar, desgastar

weathered *a.* meteorizado, intemperizado || ***weathered layer*** *m.* capa superficial

weathering *f.* meteorización (de las rocas), denudacion, erosión; pátina del tiempo; *m.* desgaste, intemperismo, rumbón || ***chemical weathering*** *f.* meteorización química || ***differential weathering*** *f.* meteorización diferencial || ***physical weathering*** *f.* meteorización mecánica || ***weathering of ore*** *m.* [Bol.] broseo

wedge *f.* cuña, calza, llave; *m.* calce, calzo || ***critical wedge*** *f.* cuña critica || ***quartz wedge*** *f.* prisma de cuarzo || ***tectonic wedge*** *f.* cuña tectónica

wedge *vb.* calzar, acuñar

wedge-shaped *a.* cuneiforme

wedgeout *m.* adelgazamiento, acuñamiento, remate

wedge out *vb.* adelgazarse, acuñarse

wedging *m.* encuñamiento, acuñamiento

weigh *vb.* pesar

weight *m.* peso; *f.* pesa, carga || ***atomic weight*** *m.* peso atómico || ***avoirdupois weight*** *m.* peso avoirdupois || ***gross weight*** *m.* peso bruto || ***molecular weight*** *m.* peso molecular || ***net weight*** *m.* peso neto || ***troy weight*** *m.* peso joyería

weighted average *m.* promedio ponderado, promedio geométrico

weir *m.* vertedero, rebosadero, dique; *f.* presa

weld *f.* soldadura || ***salt weld*** *f.* soldadura después de sal

weld *vb.* soldar, caldear

welded *a.* soldado || ***welded tuff*** *f.* ignimbrita, toba soldada

well *m.* pozo, aljibe, hoyo, pocillo; *f.* casimba, fuente || ***abandoned well*** *m.* pozo abandonado, pozo muerto || ***artesian well*** *m.* pozo artesiano || ***delineation well*** *m.* pozo de delineación || ***development well*** *m.* pozo de desarrollo, pozo de explotación || ***deviated well*** *m.* pozo desviado || ***directional well*** *m.* pozo direccional, pozo dirigido || ***discovery well*** *m.* pozo descubridor || ***disposal well*** *m.* pozo de descarga || ***dry well*** *m.* pozo seco || ***exploratory well*** *m.* pozo exploratorio || ***flowing well*** *m.* pozo brotante, pozo saltante, pozo surgente || ***gas well*** *m.* pozo de gas || ***horizontal well*** *m.* pozo horizontal || ***infill well*** *m.* pozo de explotación || ***injection well*** *m.* pozo inyector || ***obligation well*** *m.* pozo obligatorio || ***offset well*** *m.* pozo contrarestante, pozo compensador, pozo adyacente || ***oil well*** *m.* pozo petrolífero || ***producing well*** *m.* pozo productivo || ***pumping well*** *m.* pozo a bomba || ***relief well*** *m.* pozo de alivio || ***shallow well*** *f.* [Arg.] cachimba || ***shut-in well*** *m.* pozo cerrado, pozo parado || ***step-out well*** *m.* pozo de avancada || ***stripper well*** *m.* pozo mermado, pozo pobre, pozo casi agotado || ***test well*** *f.* cala de prueba || ***water well*** *m.* pozo de agua || ***wildcat well*** *m.* pozo de cateo, pozo exploratorio || ***well bore*** *m.* pozo || ***well casing*** *f.* tubería de revestimiento || ***well completion*** *f.* terminación de pozo || ***well head*** *m.* cabezal de pozo; *f.* boca de pozo || ***well log*** *m.* registro, perfil de pozo, informe de pozo || ***well logging*** *m.* perfilaje de pozos || ***well mouth*** *m.* agujero || ***well site*** *m.* sitio de pozo || ***well site geology*** *f.* geología de pozo || ***well site geologist*** *m.* geólogo de pozo || ***well workover*** *f.* reparación de un pozo || ***workover well*** *m.* pozo reacondicionado

wellbore skin *m.* efecto superficial en el pozo

wellbore storage coefficient *m.* coeficiente de almacenaje del pozo

welt *m.* ribete; refuerzo, costurón

Wemmelian *m. & a.* Wemmeliano, Wemmeliense, Wemmelense

Wenlockian *m. & a.* Wenlockiense, Wenlockiano

Wentworth grade scale *f.* escala granulométrica de Wentworth

Werfenian *m. & a.* Werfeniense, Werfeniano

wernerite *f.* wernerita

west *m. & a.* oeste, occidente, poniente

westerlies *m.* viento del oeste

western *a.* occidental, del oeste, del poniente

Westphalian *m. & a.* Westfaliense, Westfaliano

wet *a.* húmedo, mojado, impregnado || ***wet gas*** *m.* gas húmedo

wettability *f.* humectabilidad

wetting *a.* humectante || ***wetting fluid*** *m.* flúido húmedo

Weybournian *m. & a.* Weybourniano, Weybourniense

whaleback *m.* lomo de ballena

wharf *m.* embarcadero

wheelbarrow *f.* carretilla

Wheelerian *m. & a.* Wheeleriano, Wheeleriense

whetstone *f.* piedra de afilar

while *p.* durante; mientras

whipstock *m.* desviador; *f.* guíabarrena, guíasondas, cuña desviadora

whirlpool *m.* cabezón, remolino, torbellino, vórtice; *f.* olla, vorágine

Whitbian *m. & a.* Whitbiano, Whitbense

Whitcliffian *m. & a.* Whitcliffiano, Whitcliffiense

whitecap *f.* cresta de espuma

white clay *m.* calocar

white earth *f.* albariza, albica

white hot *a.* albar, albaro, albero

white mica *f.* muscovita, mica blanca

Whitneyan *m. & a.* Whitneyense

whole *a.* todo, entero, total || ***whole rock*** *f. & a.* roca entera

wholesale *a.* al por mayor || ***wholesale dealer*** *m. & f.* mayorista

whorl *f.* vuelta, circunvolución, espira

wichtisite *f.* wichtina

wide *a.* ancho, anchuroso, extenso, vasto, grande, amplio || ***widespread*** *a.* ampliamente separado, ampliamente distribuido

widen an opening *vb.* abocinar, abocarder, ensancharse

widening *m.* abocinamiento, ensanchamiento; *f.* expansion; *a.* ensancho

width *m.* ancho, latitud; *f.* anchura || ***width of a vein*** *m.* abrigo; *f.* anchura

wild *a.* salvaje, fuera de control

wildcat *m.* pozo exploratorio, pozo de cateo

wildcatter *m.* explorador de petróleo

wildcatting *f.* campaña de exploración

willemite *f.* willemita, wilhemita, vilemita

winch *m.* cabrestante, guinche, huinche, winche

wind *m.* viento || ***wind abrasion*** *f.* abrasión eólica || ***windblown*** *a.* soplado por el viento || ***wind erosion*** *f.* erosión eólica || ***wind gap*** *m.* ventisquero

windlass *f.* montacargas

window *f.* (geol.) ventana, ventanilla, fenestra, fenster; peladura || ***oil-generation window*** *f.* ventana de generación de petróleo || ***read–write window*** *f.* ventana de lectura y grabación

windrow ridge *m.* lomo eólico

windward *m.* barlovento

winged headland *m.* promontorio alado

winnow *vb.* despajar

winter *m.* invierno

winze *m.* tiro ciego, coladero, [Col.] apique; *f.* caldera, calderilla, clavada

wireline *m.* cable de alambre, cable de las poleas

Wisconsian *m. & a.* Wisconsiano, Wisconsiense

withdrawal *f.* saca, retirada; *m.* retiro

witherite *f.* witherita, whitherita

withstand *vb.* resistir, soportar

wittichenite *f.* wittichita

wobble *m.* tambaleo, bamboleo

wobble *vb.* bambolear

wobbling *a.* bamboleo

Wolfcampian *m. & a.* Wolfcampiano, Wolfcampiense

wolframite *f.* wolframita, volframita; *m.* volframio, tungsteno, wolfram

wollastonite *f.* wollastonita; *m.* espato en tablas

wollongongite *f.* wollongita

wood *f.* madera || ***fossilized wood*** *f.* xilantita || ***opalized wood*** *m.* xilópalo, xilopal || ***petrified wood*** *f.* xylolitha

Woodbinian *m. & a.* Woodbiniano, Woodbiniense

woodland *m.* monte

woods *f.* selva

woody *a.* leñoso

work *m.* trabajo; *f.* faena, obra || ***day's work*** *m.* jornal || ***dirty work, grunt work*** *m.* trabajo bruto (o sucio) || ***workday*** *f.* jornada || ***work gang*** *m.* peonaje || ***workload*** *f.* carga de trabajo || ***workman*** *m.* jornalero, obrador, obrero || ***work shift*** *f.* tanda || ***workshop*** *m.* taller || ***workstation*** *f.* (comp.) estación de trabajo || ***work trip*** *f.* campaña

work *vb.* beneficiar, escopetar, trabajar || ***work ore*** *vb.* maquilar, explotar, tirar un labor || ***work stone*** *vb.* tallar piedra

workable *a.* beneficiable

worked *a.* labrado

worker, workman *m.* obrador, obrero, jornalero, operario, mozo, peón

working face *f.* rampa

working a mine *m.* beneficio

working beam *m.* balancín

working *m.* laboreo; funcionamiento; *f.* labra; operación || ***working face*** *f.* labor || ***working model*** *m.* modelo de guía || ***working pressure*** *f.* presión efectiva

workings *f.* labores; *m.* trabajos || ***exploratory workings*** *f.* labor de reconocimiento || ***underground workings*** *m.* minado

workmanship *f.* ejecución, hechura, manufactura; *m.* mano de obra

workover *f.* rehabilitación de un pozo, reperforación, limpieza de pozo; *m.* reacondicionamiento

workover *vb.* reacondicionar

workshop *m.* taller

world *m.* mundo

worldwide *a.* mundial

worm-shaped *a.* vermicular, vermiforme

worn *a.* raído, desgastado || ***worn out*** *a.* gastado, usado, estropeado

worth *f.* valía; *m.* valor

wrap *vb.* embozar

wrapping *f.* envoltura

wrench *f.* llave || ***wrench fault*** *f.* falla lateral, falla transcurrente || ***wrench tectonics*** *f.* tectónica de corte, tectónica de torsión

wrinkle *f.* rugosidad, arruga

wrinkled *a.* rugoso, arrugado

write *vb.* escribir || ***write protection*** *f.* portección contra la grabación || ***write-up*** *vb.* confeccionar

writedown *f.* disminución en el valor

wrought *a.* forgado, batido, fundido, labrado || ***wrought iron*** *m.* hierro forjado, hierro dulce, hierro fundido, hierro batido || ***wrought silver*** *f.* chafalonía

wulfenite *f.* wulfenita, vulfenita; *m.* plomo amarillo

wursilite *f.* wursilita

wurtzite *f.* wurtzita

wurzilite *f.* wurzilita, betún

X

xanthiosite *f.* xantiosita

xanthite *f.* xantita

xanthoconite *f.* xantoconita

xanthophyllite *f.* waleuwita

xanthoxenite *m.* xanotoxeno; *f.* xantoxenita

xenoblast *m.* xenoblasto

xenoblastic *a.* xenoblástico

xenocryst *m.* xenoblasto, xenocristal, xenobloque, xenocristo

xenoikic *a.* xenoíquico

xenolith *m.* xenolito, gabarro; *f.* xenolita

xenolithic *a.* xenolítico

xenomorph *m.* xenomorfo

xenomorphic *a.* xenomórfico, xenomorfo, alotriomorfo

xenomorphism *m.* xenomorfismo

xenon *m.* xenón

xenophyre *m.* xenofiro

xenoporphyritic *a.* zenoporfirítico

xenothermal *a.* xenotermal

xenotime *f.* xenótima, wiserina; *m.* espato de itrio

xenotopic *a.* xenotópico

xeric *a.* xérico

xerophyte *f.* xerofita

xerophytic *a.* xerofítico

xerothermic *a.* xerotérmico

xilolite *f.* xilolita

xiphosuran *m.* xifosuro

Xmas tree *m.* arbol de navidad

x-ray *m.* rayo x || ***x-ray diffraction*** *f.* difracción radiográfica

xylanthite *f.* xilantita

xylene *m.* xileno

xylenite, xylinite *f.* xylenita, xylinita

xylinoid *m.* xylinoide

xylite *f.* xylita

xyloid *m.* xiloide, xyloide; *a.* xilito

xyloid lignite *m.* lignito xiloide

xylokryptite *f.* xylocriptita

xylopal *m.* silópalo

xylotile *m.* xilotilo

Y

Yakutsk *m. & a.* Yakutense, Yakutiano

yama *m.* pozo cárstico

yamaskite *f.* yamasquita

yard *f.* yarda (0,9144 metros); *m.* patio

yardage *m.* yardaje

yardang *m.* yardang

Yarmouthian *m. & a.* Yarmouth

yaw *f.* guiñada, oscilación

yeatmanite *f.* yeatmanita

y connection *f.* conexión en Y, conexión de estrella

yellow copper *f.* calcopirita, pirita de cobre

yentnite *f.* yentnita

Yeovilian *m. & a.* Yeoviliano, Yeovilense

yield (on investment) *f.* rentabilidad; producción; rendición; *m.* rendimiento de un producto, máquina, o empresa; (min.) fruto, producto || ***yield point*** *m.* punto cedente; límite elástico aparente, punto de deformación || ***yield strength*** *f.* fuerza de rotura; *m.* esfuerzo de rotura, punto de cesión || ***yield stress*** *m.* límite de cedencia

yield *vb.* rendir, rentar

Ynezian *m. & a.* Ynezіano, Yneziense

yoderite *f.* yoderita

yogoite *f.* yogoita

Yorkian *m. & a.* Yorkiano, Yorkiense

young *a.* joven; reciente

Young's modulus *m.* módulo de Young

youth *m.* juventud

youthful *a.* joven || ***youthful topography*** *f.* topografía juvenil

Ypresian *m. & a.* Ypresiense, Ypresiano

Y-shaped valley *m.* valle en forma de Y

Y-tombolo *m.* tómbolo en Y

ytterbic *a.* itérbico

ytterbium *m.* iterbio, yterbio

yttria *f.* itria

yttrian garnet *m.* ytriogranate

yttrian fluorite *f.* yttrocalcita, yttrofluorita

yttrium *m.* itrio, ytrio || ***yttrium apatite*** *m.* yttriumapatito || ***yttrium titanite*** *f.* yttrotitanita

yttrobetafite *f.* yttrobetafita

yttropyrochlore *m.* yttropirocloro

yuconite *f.* yuconita

Z

Zanclean *m. & a.* Zancleano, Zanclenense

Zemorrian *m. & a.* Zemorriano, Zemorriense

zenith *m.* cenit, zenit

zeolite *f.* ceolita, zeolita

zeolitic *a.* zeolítico

zeolitization *f.* zeolitización

zeophyllite *f.* zeofilita

zero *m.* cero || ***absolute zero*** *m.* cero absoluto || ***zero pour*** *f.* fluidez cero

zinalcite *f.* zinalsita

zinc *m.* cinc, zinc, zinque || ***zinc-bearing*** *a.* cincífero, zincoso || ***zinc blende*** *f.* esfalerita, blenda || ***zinc carbonate*** *f.* galmeíta || ***zinc oxide*** *m.* óxido de cinc; *f.* cincita || ***zinc silicate*** *f.* galmeíta || ***zinc sulfate*** *m.* vitriolo blanco; *f.* zincosita, zinkosita

zincite *f.* zincita, zinkita, cincita; *m.* óxido de cinc

zircon *m.* circón, zircón; *f.* zironita || ***red zircon*** *m.* jacinto || ***yellow gem zircon*** *m.* jargón

zirconium *m.* circonio, zirconio || ***zirconium carbonate*** *f.* zircarbita

zircophyllite *f.* zirkofilita

zircosulfate *m.* zircosulfato

zirkelite *f.* zirconolita

Zoantarian *m. & a.* Zoantariano, Zoantariense

zoisite *f.* zoicita

zonation *f.* zonación

zone *f.* zona, faja, franja || ***accomodation zone*** *f.* zona de acomodación || ***alteration zone*** *f.* zona de alteración || ***Benioff zone*** *f.* zona de Benioff || ***bypass zone*** *f.* zona desviada || ***capillary fringe zone*** *f.* zona de capilaridad || ***contact zone*** *f.* zona de contacto || ***depletion zone*** *f.* zona de agotamiento || ***detachment zone*** *f.* zona de despegue || ***development zone*** *f.* zona de desarollo || ***discharge zone*** *f.* zona de descargo || ***fault zone*** *f.* zona de fallas, zona fallada || ***flushed zone*** *f.* zona lavada, zona barrida || ***fracture zone*** *f.* zona de fracturación, zona de fracturamiento, zona fracturada || ***Fresnel zone*** *f.* zona de Fresnel || ***injection zone*** *f.* zona de inyección || ***invaded zone*** *f.* zona invadida || ***invasion zone*** *f.* zona de invasión || ***oil zone*** *f.* zona petrolífera || ***pay zone*** *f.* zona de desarollo, zona de explotación, zona petrolífera, zona productiva, zona productora || ***porous zone*** *f.* zona porosa || ***producing zone*** *f.* zona de desarollo, zona de explotación, zona productiva, zona productora || ***root zone*** *f.* zona de arranque || ***saturation zone*** *f.* zona de saturación || ***shadow zone*** *f.* zona sombreada || ***shear zone*** *f.* zona de cizallamiento, zona transcurrente || ***subduction zone*** *f.* zona de subducción || ***subsidence zone*** *f.* zona de hundamiento || ***time zone*** *m.* huso horario || ***transfer zone*** *f.* zona de transferencia || ***transition zone*** *f.* zona de transición || ***triangle zone*** *f.* zona triangular || ***untested zone*** *f.* zona no comprobada || ***wrench zone*** *f.* zona transcurrente || ***zone of aeration*** *f.* zona de aereación || ***zone of saturation*** *f.* zona de saturación || ***zone of weakness*** *f.* zona de debilidad || ***zone of weathering*** *f.* zona de meteorización

zonal *a.* zonal

zonation, zoning *f.* zonación

zoned *a.* zonado, zonal

zoning *f.* zonificación, zonalidad; *m.* zonamiento

zonule *f.* zónula; *m.* zonulo

zoarium, zooarium *m.* zoario, zooario

zooecium *m.* zooecio

zoogenic *a.* zoógeno

zoogeography *f.* zoogeografía

zooid *m.* zooide

zoolith *f.* zoolita

zoological *a.* zoológico

zoologist *m.* zoólogo

zoology *f.* zoologia

zonochlorite *f.* zonoclorita

zoophasmic *a.* zoofásmico

zoophyte *f.* zoofita

zooplankton *m.* zooplancton

zoospore *f.* zoospora

zooxanthella *f.* zooxantela

Zuloagan *m.* & *a.* Zuloagense, Zuloagiano

zunyite *f.* zunyita

zygote *m.* cigoto, zigoto

DICCIONARIO DE TERMINOS DE GEOCIENCIAS

ESPAÑOL - INGLÉS

A

aa lava *f.* aa lava
Aalenense, Aaleniense, Aaleniano *m. & a.* Aalenian
aas *m.* type of sand dune
ábaco *m.* trough, sluice; graph, chart, nomogram
abajadero *m.* slope, downgrade
abajador *m.* (min.) toolcarrier
abajamiento *m.* subsidence
abajar *vb.* to lower, let down, fall, descend, subside
abajo *adv. & p.* down, below, under
abajor *m.* depression
abaluartado *a.* rock bulwark
abancalado *a.* terraced
abandonado *a.* abandoned
abandonamiento, abandono *m.* abandonment
abandonar *vb.* to abandon, leave
abandonos *m.* (legal) abandonments
abanico *m.* fan || ***abanico aluvial*** *m.* alluvial fan || ***abanico de desborde*** *m.* washover fan || ***abanico de desembalse*** *m.* crevasse splay || ***abanico de deyección*** *m.* talus fan || ***abanico de piso de cuenca*** *m.* basin floor fan || ***abanico submarino*** *m.* submarine fan
abarrancar *vb.* to excavate, form gullies
abarranco *a.* excavated
abastar *vb.* to supply, furnish
abastecer *vb.* to supply, provide
abastecimiento *m.* supply || ***abastecimiento de agua*** *m.* water supply
abasto *m.* (min.) output, provisioning, supplying, abundance
abataje *m.* breaking ground
abatón *m.* inaccessible location
aberración *f.* aberration
aberrante *a.* aberrant
abertal *a.* cracked by drought
abertura *f.* aperture, opening, fissure, crevice
abierto *a.* open; (geog.) flat, level
abigarrado *a.* variegated, mottled
abiogénesis *f.* abiogenesis
abioglifo *m.* abioglyph
abiótico *a.* abiotic
abisal, abísico, abismal *a.* abyssal
abismo *m.* abyss
abisolito *m.* batholith
ablación *f.* ablation, erosion
ablandamiento *m.* softening || ***ablandamiento de agua*** *m.* water softening
ablución *f.* ablution
abocardar *vb.* to widen an opening
abocardo *m.* drill
abocinador de tubos *m.* tube expander
abocinamiento *m.* widening
abocinar *vb.* to widen an opening
abombado *a.* convex, swollen, arched
abombamiento *m.* dome
aborregado *a.* glaciated, striated by glaciers
aborto de oro *m.* gold nuggets found in potholes
abovedado, abovado *a.* arched, vaulted
abovedamiento *m.* upwarp, arching
abra *f.* quarry, open vein, fissure, cave, opening, gap, cove, clearing, cavity, pass || ***abra de agua*** *f.* water gap || ***abra de viento*** *f.* wind gap

abrasión *f.* abrasion, scour, erosion || ***plataforma de abrasión*** *f.* wave-cut terrace

abrasivo *m.* abrasive

abrazadera *f.* clamp, clip, cleat, clasp, buckle

abrazar *vb.* to enclose, to make a structure to prevent caving or spalling of a mine roof

abrevadero *m.* flooded mine or excavation

abridura *f.* opening or enlargement of a mine gallery

abrigo *m.* width of a vein || ***abrigo el carbon*** *m.* shale above or below coal

abrillantar *vb.* to cut gems

abrimiento *m.* opening

abrir *vb.* to rift; to open a basin, drift, shaft, or well; to bore a tunnel

abrojos *m.* (min.) reef, (naut.) hidden rocks

abronzado *m.* tetrahedrite, [Méx.] copper ores, chalcopyrite

abrupto *a.* abrupt, steep, rugged

absoluto *a.* absolute

absorbedor *m.* absorber

absorbente *a.* absorbent

absorber *vb.* absorb

absorción *f.* absorption || ***absorción atmosferica*** *f.* atmospheric absorption || ***absorción del suelo*** *f.* ground absorption || ***absorción preferencial*** *f.* preferential absorption

abundamiento *m.* abundance

abundancia *f.* abundance

abundante *a.* abundant

abundar *vb.* to be abundant, abound

abundoso *a.* abundant

acabaigamiento *m.* thrust fault, overthrust

acabamiento (de un pozo) *m.* (well) completion

Acadiense, Acadiano, Acadico *m. & a.* Acadian

acamado *a.* horizontal

acampanar *vb.* to clean a placer to paydirt

acampar *vb.* to camp

acanalado *a.* fluted, grooved, corrugated

acanaladura *f.* groove

acanalamiento *m.* grooving

acanalar *vb.* to groove

acanelado *m.* [Méx.] disseminated sphalerite, argentite, and pyrargyrited proustite

acantilado *m. & a.* escarpment, cliff, bluff; steep, sheer, precipitous

acantita *f.* acanthite

acantóporo *m.* acanthopore

acaparrosado *a.* copper colored

acardenillado *a.* copper carbonate-stained

acarreadizo *a.* shifting sand

acarreador *m.* carrier

acarreo *m.* transported material, alluvium, float; transporting, hoisting || ***acarreo de glaciar*** *m.* glacial drift

acarreos *m.* float rock

acavernar *vb.* to make cave-like, excavate

acceso *m.* entrance, access, portal

accesorio *m. & a.* accessory, fitting || ***accesorio de taller*** *m.* shop fittings || ***accesorio para calderas*** *m.* boiler fittings || ***accesorio para tanques*** *m.* tank fittings || ***accesorio para tubería*** *m.* pipe fittings

accidentado *a.* rough, uneven, hilly

accidental *a.* accidental

accidente *m.* (geog.) roughness, uneveness || ***accidentes mecánicos*** *m.* mechanical problems

acción *f.* action; (com.) stock, share || ***acción de gravedad, gravitativa*** *f.* gravitational attraction || ***acción galvánica*** *f.* galvanic action

accionado manualmente *adv.* hand-actuated || ***accionado manualmente por pedal*** *a.* foot activated || ***accionado manualmente por reloj*** *a.* timer driven

accionamiento eléctrico *m.* electric drive

acciones *f.* stocks || ***acciones ordinarias*** *f.* common stock

accionista *m. & f.* shareholder

aceite *m.* oil, petroleum (véase petróleo) || ***aceite aislante*** *m.* insulating oil || ***aceite bruto*** *m.* raw oil, crude oil || ***aceite combustible*** *m.* fuel oil || ***aceite condensado*** *m.* condensate || ***aceite crudo*** *m.* crude oil || ***aceite crudo de base asfáltica*** *m.* asphalt-based oil || ***aceite crudo de base mixta*** *m.* mixed base crude || ***aceite crudo nafténico*** *m.* asphalt-based oil || ***aceite de alquitran** m.* tar oil || ***aceite de alumbrado** m.* kerosene || ***aceite de carbón** m.* coal oil, kerosene || ***aceite de cárter** m.* crankcase oil || ***aceite de creosota** m.* dead oil || ***aceite de enjuagar** m.* oil flushing || ***aceite de esquisto** m.* shale oil || ***aceite de horno** m.* fuel oil || ***aceite de lámpara** m.* [Mex] kerosene || ***aceite de lutita** m.* oil shale || ***aceite de taladrar** m.* drilling oil || ***aceite desparafinado** m.* dewaxed oil || ***aceite lampante** m.* [Sp.] kerosene || ***aceite lubricante** m.* lubricating oil || ***aceite mineral** m.* petroleum || ***aceite muerto** m.* dead oil || ***aceite para corte** m.* cutting oil || ***aceite espeso** m.* heavy oil || ***aceite para corte** m.* cutting oil || ***aceite para fresar** m.* cutting oil || ***aceite para máquinas** m.* engine oil || ***aceite para motor** m.* motor oil || ***aceite parafínico** m.* waxy oil, parafinic oil || ***aceite pesado** m.* heavy oil || ***aceite pesado refinado** m.* bright stock || ***aceite recuperado** m.* reclaimed or recycled oil || ***aceite residual** m.* residual oil || ***aceite sin parafina** m.* wax-free oil || ***aceite virgin** m.* virgin stock || ***aceite viscoso** m.* viscous oil || ***aceite volátil** m.* volatile oil

aceitera *f.* oil can || ***aceitera de goteo** f.* oil driller

aceitón *m.* thick oil, dregs

aceitoso *a.* oily

aceleración *f.* acceleration || ***aceleración gravitacional** f.* gravitational acceleration

acelerar *vb.* to accelerate

acelerómetro *m.* accelerometer

acenaftileno *m.* acenaphthylene

acendrada *f.* refractory marl

acendrar *vb.* to refine metal

acequia *f.* flume, canal, channel, ditch; brook

acerado *m.* [Arg.] silver sulfide-bearing pyrite; [Méx.] gold-bearing pyrite, guadalcazarite, stephanite, cerargyrite; [Bol.] jamesonite

acerado *a.* steely, steel grey color

acerar *vb.* to chill, harden

acerdesa *f.* wad, bog manganese

acería *f.* steel works, mill

acerillo *m.* small, granular, silver-rich galena

acero *m.* steel || ***acero ácido** m.* acid steel || ***acero al carbono** m.* carbon steel || ***acero al cromo** m.* chrome steel || ***acero al molibdeno** m.* moly steel || ***acero al níquel** m.* nickel steel || ***acero anticorrosivo** m.* corrosion-resistant steel || ***acero con tratamiento térmico** m.* heat-treated steel || ***acero de construcción** m.* structural steel || ***acero estructural** m.* structural steel || ***acero forjado** m.* forged steel || ***acero inoxidable** m.* stainless steel || ***acero laminado** m.* rolled steel || ***acero laminado en frio** m.* cold-drawn steel || ***acero resistente a la corrosión** m.* corrosion-resistant steel || ***acero resistente al calor** m.* heat-resistant steel || ***acero tratado al calor** m.* heat-treated steel || ***acero tungsteno** m.* tungsten steel

aceroníquel *m.* nickel steel

acetato *m.* acetate, viewgraph || ***acetato butilico** m.* butyl acetate

acético *a.* acetic

acetileno *m.* acetylene || ***mechero de acetileno** m.* acetylene torch || ***quemador de acetileno** m.* acetylene torch

acetona *f.* acetone

acíclico *a.* acyclic

acicular *a.* acicular

aciculita *f.* aikinite

acidez *f. & a.* acidity

acidificación *f.* acidification

acidificador *m.* acidifier

acidificar *vb.* acidify, to acidize

acidímetro *m.* acid meter

ácido *m.* acid || ***ácido acético** m.* acetic acid || ***ácido bórico** m.* boric acid || ***ácido carbólico** m.* carbolic acid || ***ácido carbónico** m.* carbonic acid || ***ácido ciánico** m.* cyanic acid || ***ácido clorhídrico** m.* hydrochloric acid || ***ácido fénico** m.* carbolic acid || ***ácido fluorhídrico** m.* hydrofluoric acid || ***ácido fórmico** m.* formic acid || ***ácido fosfórico** m.* phosphoric acid || ***ácido ftalico** m.* phthalic acid || ***ácido graso** m.* fatty acid || ***ácido hidroclórico** m.* hydrochloric acid || ***ácido húmico** m.* humic acid || ***ácido isobutírico** m.* isobutyric acid || ***ácido muriático** m.* muriatic or hydrochloric acid || ***ácido nítrico** m.*

nitric acid, aqua fortis || ***ácido orgánico m.*** organic acid || ***ácido oxálico m.*** oxalic acid || ***ácido sulfhídrico m.*** hydrogen sulfide || ***ácido sulfúrico m.*** sulphuric acid || ***ácido tereftálico m.*** terephthalic acid

acídula *f.* carbonic acid-rich mineral water

acidulación *a.* acidizing

acidular *vb.* to acidize

acije *m.* metal sulfate, vitriol || ***acije amarillo m.*** copiapite || ***acije blanco m.*** goslarite || ***acije de hierro m.*** iron vitriol

acimut *m.* azimuth

acimutal *a.* azimuthal

acis *m.* volcanic vent; diapiric fold

aclimatación *f.* acclimatization

aclínico *a.* aclinic

acmita *f.* acmite

acobrado *a.* copper-colored

acocadera *f.* stone cutting chisel

acodalamiento *m.* shoring, bracing

acodalar, acodar *vb.* to shore, brace

acodamiento *m.* shoring, bracing

acombado *a.* bent, warped

acometer *vb.* to branch

acomodación *f.* accommodation

acomodada *a.* accommodated

acomodana *f.* [Pe.] ore deposit

acomodar *vb.* to accommodate

aconamiento de agua *m.* water coning

aconcharse *vb.* [Ch.] to deposit sediment

acondicionador de lodo *m.* mud conditioner

acóndrito *m.* achondrite

acontita *f.* [Ch.] pyrite-arsenopyrite mix

acopio *m.* tailings, waste heap, dump; stockpile

acoplador *m.* coupling || ***acoplador hidráulico m.*** hydraulic coupling

acoplamiento *m.* coupling, joint, splice, connection || ***acoplamiento mútuo m.*** interface

acoplar *vb.* to couple, engage

acordamiento *m.* concordance of strata

acordancia *f.* conformity

acordanza *f.* tectonic-caused conformity

acortamiento *m.* shortening

acortar *vb.* to shorten

acortarse *vb.* to shrink

acostado *a.* recumbent

acostar *vb.* to lay over, lay down, overturn, incline

acostarse *vb.* to change dip or direction, to lay over

acostillado *a.* ribbed

acotación *f.* boundary marker, monument, elevation mark; [Cu.] elevation || ***acotación paralela f.*** parallel contouring

acotamiento *m.* setting boundary markers

acotar *vb.* to delimit, mark boundaries, mark elevations, set monuments

acotilla *f.* sledge hammer

acre *m.* acre (4.047×10^{-3} km^2)

acrecencia, acreción *f.* accretion

acrecentamiento *m.* accretion

acrecentar *vb.* to accrue

acreedor *m.* creditor

acrepié *m.* acre-foot (1233 m^3)

acribar *vb.* to sift, sieve, screen

acrílico *m.* acrylic

acrisolar *vb.* to refine, assay, or purify metal in a crucible

acrisolado *a.* refined, purified in a crucible

acroíta *f.* achroite

acromático *a.* achromatic

acta *f.* proceeding

actinolita, actinota *f.* actinolite

actínoto *m.* actinolite

activación *f.* activation

activado *a.* activated

activador (de gas) *m.* (gas) actifier

activar (el pozo) *vb.* activate (a well)

actividad *f.* (chem.) activity

activo *m.* assets; *a.* live, active || ***activo circulante m.*** liquid assets

Actual *m.* Holocene

actualización *f.* updating

actualizar *vb.* to update

acuático *a.* aquatic

acueducto *m.* aqueduct

ácueo *a.* aqueous || ***ácueo igneo*** *a.* hydroplutonic

acuerdo *m.* agreement

acuicierre *m.* aquiclude

acuiclusa *f.* aquiclude

acuicluso *m.* aquiclude

acuífero *m.* aquifer; *a.* water-bearing || ***acuífero artesiano*** *m.* artesian aquifer

acuitardo *m.* aquitard

acumbrado *a.* pinnacle-shaped

acuminado *a.* conical

acumulación *f.* accumulation, deposition, accretion, deposit || ***acumulación metalífera*** *f.* ore deposit || ***acumulación petrolífera*** *f.* oil pool, oil accumulation || ***por acumulación*** *a.* constructional, by accumulation

acumulado *a.* cumulative, accumulated

acumulador *m.* battery || ***acumulador de mineral*** *m.* ore deposit

acumular *vb.* to aggrade, accumulate

acuñamiento *m.* pinchout, wedgeout

acuñarse *vb.* to pinch out, thin out, truncate

acuoso *a.* aqueous, watery

acuosoluble *a.* water-soluble

acústico *a.* acoustic

acutángulo *a.* acute-angled

achatado *a.* flattened

achatamiento *m.* levelling by erosion

achatar *vb.* to level by erosion

achicador *m.* bailer || ***achicador de arena*** *m.* sand bailer

achicamiento *m.* draining of a mine

achicar *vb.* scoop, bail, drain, dewater

achiflonado *a.* [Ch.] sloping

achique *m.* bailing, swabbing, dewatering || ***achique de activación*** *m.* swabing a well || ***achique de excitación*** *m.* swabing a well

achúa *f.* [Ch.] pan to test the amalgamation of silver ore

achura *f.* [Bol.] rich ore

adamantino *a.* adamantine

adamelitea *f.* adamellite

adamina *f.* adamite

adaptabilidad *f.* adaptability

adaptación *f.* adaptation

adaptar *vb.* to adapt

adaptador *m.* adaptor || ***adaptador de bomba*** *m.* pump adapter || ***adaptador de tubería*** *m.* casing adapter

adarme *m.* half dram; 1/16 Spanish ounce; 1.8 grams

adelantado *a.* advanced

adelantar *vb.* to advance, make headway, move forward

adelanto *m.* advance on wages; lead

adelgazador del lodo *m.* mud thinner

adelgazamiento *m.* wedgeout, pinchout, thinning

adelgazarse *vb.* wedge out, thin out; [Col.] to separate gold from gravel

adema, ademe *m.* shore, prop

ademado *m.* timbering, bracing; [Méx.] well casing || ***ademado de cajon*** *m.* cribwork

ademador *m.* timberman

ademadora *f.* mine ax

ademantino, adiamantado *a.* adamantine, diamond-like

ademar *vb.* (min.) to shore, timber

ademe *m.* mine timbers

adentro *adv. & p.* inside, within

adherido *a.* stuck, attached

adherir *vb.* adhere

adhesión *f.* bond, adhesion

adiabático *a.* adiabatic

adicional, adicionante *m.* admixture

adición por sedimentación *f.* aggradation

adipita *f.* chabazite

aditamentos *m.* fittings

aditivo *m.* additive

adjunto *a.* enclosed

administración unificada *f.* unit operation

admisible *a.* admissible, allowable

admisión *f.* intake, inlet, input

admixtura *f.* admixture, mixture

adobe *m.* clay soil, clay brick

adobera, adorbería *f.* brickyard

adolescente *a.* adolescent
adoquín *m.* desert pavement
adquisición *f.* acquisition
adsorbedor *m.* adsorber
adsorbente *a.* adsorbent
adsorber *vb.* adsorb
adsorción *f.* adsorption
aduana *f.* customs house, customs
aductor *a.* adductor
adularia *f.* adularia
ad valorem *a.* (com.) ad valorem
adyacente *a.* adjacent
aegerina, aegirina *f.* acmite
aeración *f.* aeration
aerador, aereador *m.* aerator
aereado *a.* aeriated, weathered
aerear *vb.* aerate
aéreo *m.* & *a.* aerial
aerífero *a.* copper-bearing
aeróbico, aerobio *a.* aerobic
aerodinámico *a.* aerodynamic, streamlined
aerofotografía *f.* aerial photography
aerofotogrametría *f.* aerial mapping
aerolito *m.* stony meteorite
aeromagnético *a.* aeromagnetic
aeromagnetometría *f.* aeromagnetics
aerómetro *m.* hydrometer
aerosol *m.* aerosol
aerovía de cable *f.* aerial tram
afanesa *f.* aphanite, clinoclase
afanita *f.* amphibolite
afanítico *a.* aphanitic, fine-grained
afilado *a.* tapering, sharp
afiladora *f.* grinder, sharpener
afilar(se) *vb.* to grind, to taper
afinación *f.* refining
afinado *m.* refining
afinador *m.* refiner
afinadura *f.* refining
afinar *vb.* to refine metals
afinidad *f.* affinity
afino *m.* refining, refinery
aflojadero *m.* [Méx.] soft part of a vein
aflojamiento *m.* slacking
aflojar *vb.* remove ore trapped in chute || ***aflojarel canalón*** *vb.* [Col.] to pan or wash material in a sluice
afloramiento *m.* outcrop, exposure
aflorar *vb.* to outcrop, to be exposed
afluencia *f.* influx, abundance, runoff
afluente *a.* abundant, affluent; *f.* tributary
afluir *vb.* to flow in, be tributary to, to open, lead
aflujo *m.* inflow, runoff
aforamiento *m.* valuation, appraisal
aforar *vb.* to gauge, measure, appraise
aforo *m.* measuring, guaging, appraisal, assessment, valuation, river flow, capacity
afótico *a.* aphotic
afrentar un hilo *vb.* [Col.] to cut across a vein to determine thickness, dip, strike
Africa *f.* Africa
afrontonar *vb.* to muck ore
Aftoniano, Aftonense *m.* Aftonian
afuera *adv.* & *p.* out, outside || ***costa afuera*** *a.* offshore
aga *f.* red marl
agalísico *m.* plutonic rock
agarradera *f.* clamp, handle grip
ágata *f.* agate
agencia *f.* agency
agenda *f.* agenda
agente *m.* agent || ***agente catalítico*** *m.* catalytic agent || ***agente de tierras*** *m.* landman || ***agente floculador*** *m.* flocculating agent || ***agente químico*** *m.* reagent || ***agente reductor*** *m.* reducing agent || ***agente tensoactivo*** *m.* surfactant
agitación *f.* agitation
agitado *a.* churned, agitated
agitador *m.* agitator || ***agitador de lodo*** *m.* shale shaker
agitar *vb.* agitate, stir, churn
aglomeración *f.* agglomeration
aglomerado *m.* agglomerate
aglomerante *m.* matrix
aglomerar *vb.* to agglomerate
aglutinación *f.* caking, coalescence

aglutinado *a.* arenaceous

aglutinante *m.* cement

aglutinar *vb.* to cement, aggregate

Agnostozóico *m.* Algonkian, Proterozic

agojía *f.* channel cut to dewater mine

agolpamiento *m.* impounding

agotado *a.* depleted, exhausted, pinched out, thinned out, drained

agotamiento *m.* depletion, exhaustion

agotar(se) *vb.* strip, drain off, exhaust, deplete

agradación *f.* aggradation

agrandamiento *m.* enlargement

a granel *a.* in bulk, bulk

agregado *m.* aggregate; *a.* attached || ***agregado coloidal*** *m.* admixture || ***agregado escalonado*** *m.* graded aggregate || ***agregado graduado*** *m.* graded aggregate || ***agregado tal como sale*** *m.* run-of-mine aggregate

agregar *vb.* to add

agrietado *a.* cracked

agrietadura *f.* crack, fissure

agrietamiento *m.* (véase fracturamiento) fracturing

agrietarse *v.* to fracture

agrimensor *m.* surveyor

agrimensura *f.* surveying

agrio *a.* sour, acid

agronomía *f.* [Méx.] surveying

agrónomo *m.* [Méx.] surveyor

agrupación molecular *f.* molecular arrangement

agua *f.* water || ***agua abajo*** *f.* downstream || ***agua arriba*** *f.* upstream || ***agua artesiana*** *f.* artesian water || ***agua blanda*** *f.* soft water || ***agua calcárea*** *f.* hard water || ***agua capilar*** *f.* capillary water || ***agua carbonatada*** *f.* carbonated water || ***agua congénita*** *f.* connate water || ***agua connata*** *f.* connate water || ***agua contaminada*** *f.* polluted water || ***agua corriente*** *f.* running water || ***agua cruda*** *f.* hard water || ***agua de arriba*** *f.* top water || ***agua de consolidación*** *f.* water of compaction || ***agua de cristalización*** *f.* crystallization water || ***agua de descarga*** *f.* waste water, discharge water || ***agua de deshielo*** *f.* meltwater || ***agua de expulsión*** *f.* compaction water || ***agua de fondo*** *f.* bottom water, ground water || ***agua de formación*** *f.* connate water, formation water || ***agua de gravedad*** *f.* vadose, meteoric water || ***agua de lluvia*** *f.* rainwater || ***agua de madre*** *f.* mother liquor || ***agua de manantial*** *f.* spring water || ***agua de mar*** *f.* sea water || ***agua de mina*** *f.* mine water || ***agua de noria*** *f.* [Méx.] well water || ***agua de pie*** *f.* spring water || ***agua de retencion*** *f.* pore water || ***agua de tratamiento*** *f.* treatment water || ***agua del suelo*** *f.* soil water || ***agua delgada*** *f.* soft water || ***agua destilada*** *f.* distilled water || ***agua dormida*** *f.* stagnant water || ***agua dulce*** *f.* fresh water || ***agua dura*** *f.* hard water || ***agua estancada*** *f.* stagnant or standing water || ***agua fósil*** *f.* connate water || ***agua freática*** *f.* phreatic water, groundwater || ***agua fuerte*** *f.* nitric acid || ***agua gorda*** *f.* hard water || ***agua infiltrada*** *f.* percolation water || ***agua innata*** *f.* connate water || ***agua intersticial*** *f.* connate water || ***agua juvenil*** *f.* juvenile water || ***agua limitada*** *f.* bound water || ***agua magmática*** *f.* magmatic water || ***agua marginal*** *f.* edge water || ***agua metamórfica*** *f.* metamorphic water || ***agua meteórica*** *f.* meteoric water || ***agua mineral*** *f.* mineral water || ***agua natural*** *f.* spring water || ***agua oxígenada*** *f.* oxygenated water || ***agua pluvial*** *f.* rainwater || ***agua potable*** *f.* potable water, drinkable water || ***agua profunda*** *f.* deep water || ***agua regia*** *f.* aqua regia || ***agua residual*** *f.* waste water || ***agua salada*** *f.* brine, saltwater || ***agua salobre*** *f.* brackish water || ***agua singenética*** *f.* connate water || ***agua somera*** *f.* shallow water; [Arg.] surface water || ***agua subálvea*** *f.* streambed water || ***agua subterránea*** *f.* groundwater || ***agua endicada*** *f.* perched groundwater || ***agua subyacente*** *f.* bottom or edge water || ***agua sucia*** *f.* unpotable water || ***agua superficial*** *f.* surface water || ***agua superior*** *f.* top water || ***agua surgente*** *f.* artesian water || ***agua telúrica*** *f.* [Arg.] groundwater || ***agua termal*** *f.* thermal water || ***agua vadosa, vadosiana*** *f.* vadose water || ***agua viva*** *f.* spring water || ***bajo el agua*** *a.* subaqueous

aguacero *m.* rain shower, downpour

aguachento *a.* saturated

aguachinar *vb.* to saturate

aguada *f.* flood, saturation, inundation of a mine; [Ch.] spring

aguaducho *m.* conduit, flood

agualotal *m.* [Arg., Cu.] bog, swamp

agualita *f.* shallow water

agualluvia *f.* rainwater

aguamarina *f.* aquamarine

aguanieve *f.* sleet

aguantar *vb.* to sustain, support, endure

aguar *vb.* to dilute

aguarrás *f.* turpentine

aguarse *vb.* to be flooded

aguas *f.* mineral waters || ***aguas abajo*** *a.* downstream || ***aguas aprovechables*** *f.* water resources || ***aguas arriba*** *a.* upstream || ***aguas broncas*** *f.* [Méx.] flood waters || ***aguas cloacales de infiltración*** *f.* ground water || ***aguas de afloramiento*** *f.* upwelling || ***aguas de cabecera*** *f.* [PR] headwaters || ***aguas estiales*** *f.* low water || ***aguas inmundas*** *f.* waste water || ***aguas pandas*** *f.* shallow waters || ***aguas residuales*** *f.* waste water || ***aguas termales*** *f.* hot spring(s) || ***aguas vertientes*** *f.* runoff || ***aguas vivas*** *f.* flowing water

aguatal *m.* [Ec.] pool, swamp

aguazal *m.* swamp, bog, marsh

agudo *a.* acute, sharp, pointed

aguijón *m.* packer, stinger

agüilla *f.* water, fluid

aguja *f.* peak, pinnacle, spire, pillar, needle, horn; compass needle, marker pin or arrow, (min.) [Sp.] small vein, [Col.] small vein or stringer || ***aguja acimutal*** *f.* azimuthal compass || ***aguja de calibración*** *f.* metering pin || ***aguja de marear*** *f.* compass || ***aguja de medición*** *f.* metering pin || ***aguja glaciaria*** *f.* glacial horn || ***aguja magnética*** *f.* magnetic needle

agujerar *vb.* to perforate, burrow

agujereado *a.* burrowed; leaky

agujero *m.* drill hole, fissure, well mouth; burrow || ***agujero descubierto*** *m.* open hole || ***agujero desviado*** *m.* deviated hole

agujón *m.* surveying instrument with compass

aguosidad *f.* water, fluid

aguoso *a.* watery

aguzado *a.* tapering, sharp

ahebrado *a.* fibrous

ahechar *vb.* to sift, sieve

aherrumbrarse *vb.* to become rust-colored, to be iron-stained

ahitar *vb.* to set monuments

ahogado *a.* flooded, drowned, submerged; [Méx.] embedded

ahogar *vb.* to submerge, drown, choke, damp || ***ahogar con inyección*** *vb.* to mud off, kill a well with mud || ***ahogar un pozo*** *vb.* to kill a well

ahogarse *vb.* to be submerged || ***ahogarse el oro*** *vb.* [Col.] to lose gold carried off by water

ahondadura *f.* excavation, digging

ahondamiento, ahonde *m.* excavation, digging

ahondar *vb.* to sink a shaft, deepen a well

ahonde *m.* a shaft that establishes mining title

ahorcado *a.* choked, narrow, tight, hung

ahorquillado *a.* forked

ahorros *m.* savings || ***ahorros anuales brutos*** *m.* annualized pretax savings

ahuesado *a.* bone-colored, bone hard

ahusado *a.* tapered

ahusamiento *a.* tapering

ahusar *vb.* to taper

aire *m.* air || ***compresor de aire*** *m.* air compressor || ***aire comprimido*** *m.* compressed air || ***depurador de aire*** *m.* air filter || ***enfriado por aire*** *a.* air-cooled || ***inyección de aire*** *m.* air injection || ***aire libre*** *m.* free air || ***aire puro*** *m.* fresh air || ***aire viciado*** *m.* bad air, exhaust air

aireado *a.* aereated

aislación *f.* insulation; isolation

aislado *a.* isolated

aislador *m.* insulator

aislamiento *m.* isolation, insulator, insulation

aislar *vb.* to isolate, to insulate

aiza *f.* [Bol.] caving

ajaquefa *f.* cave

ajarafe *m.* tableland

ajustable *a.* adjustable

ajustado *a.* tight, fitted, exact, adjusted

ajustador *m.* adjuster, fitter

ajustar *vb.* adjust, fit, regulate, balance

ajuste *m.* adjustment; (com.) agreement, contract, covenant || ***ajuste de instrumentos*** *m.* instrument calibration

al aire *a.* subaerial

ala *f.* limb, side, flank; wing, hinge leaf

alabandina *f.* alabandite

alabastro, alabastrino *m.* alabaster

alabeado *a.* warped

alabear *vb.* to warp

alabeo *m.* warp || ***alabeo de carga*** *m.* pressure solution surface || ***alabeo hacia abajo*** *m.* downwarp || ***alabeo hacia arriba*** *m.* upwarp

ala de mosca *f.* [Pe.] granite, hard rock

alado *a.* flanged

alagadizo *a.* flood-prone

alagado *a.* flooded land

alagar *vb.* to flood

alagunar *vb.* [Ch.] to flood

alajite *m.* [Méx.] altered rhodonite

alambique *m.* still

alambre *m.* metal wire, [Arg.] aerial tram

alanita *f.* allanite

alanoca *f.* [Pe.] advance on wages

alargado *a.* oblong, elongate

alargamiento *m.* elongation

alasquita *f.* alaskite

albañal *m.* drain, sewer; muck, mess

albar, albaro *a.* white, white-hot

albarazado, albarizo *a.* whitish

albardón *m.* levee || ***albardón semilunar*** *m.* point bar deposit

albariza *f.* lagoon, dry white earth

albayaldado *a.* cerussite-covered

albayalde *m.* cerussite

albedo *m.* albedo

Albense, Albiense, Albiano *m. & a.* Albian

alberca *f.* pond, pool, [Méx.] basin

albero *m.* white earth, [Sp.] yellow sand used for construction

albertita *f.* asphaltic coal, ash coal, albertite

albica *f.* white earth

albina *f.* salt pool, marsh; albite, apophyllite || ***albina de marisma*** *f.* white dry lake or marsh deposits

albita *f.* albite

albitización *f.* albitization

albufera *f.* lagoon

albuhera *f.* pond

álcali *f. & a.* alkali

alcalinidad *f.* alkalinity

alcalino *a.* alkaline || ***basalto alcalino*** *m.* alkali basalt || ***metal alcalino*** *m.* alkali metal || ***riolita alcalino*** *f.* alkali rhyolite || ***traquita alcalino*** *f.* alkali trachyte

alcalinotérreo *m.* alkaline earth

alcance *m.* scope, range, drift, coverage, reach, fetch; balance due || ***alcance de ola*** *m.* fetch of a wave

alcancía *f.* loading chute

alcano *m.* alkane

alcantarilla *f.* culvert, drain, storm sewer

alcaparosa *f.* melanterite, copperas || ***alcaparosa azule*** *f.* chalcanthite

alcarría *f.* plateau

alcohilación, alcoilación *f.* alkylation

alcohílico *a.* alkylic

alcohilo *m.* alkyl

alcohol *m.* alcohol; (min.) galena, [Bol.] antimony || ***alcohol de alferos*** *m.* cubic galena || ***alcohol de cereales*** *m.* grain alcohol || ***alcohol de granos*** *m.* grain alcohol || ***alcohol de madera*** *m.* wood alcohol || ***alcohol denaturalizado*** *m.* denatured alcohol || ***alcohol de oja*** *m.* [Sp.] galena with little silver || ***alcohol etílico*** *m.* ethyl alcohol || ***alcohol metílico*** *m.* methyl alcohol || ***alcohol propílico*** *m.* propyl alcohol

alcor *m.* hill

alcotana *f.* pick, stonecutter's hammer

alcubilla *f.* basin, pond

al detalle *a.* retail
álea *m.* risk, randomness || *álea numérico m.* random variable
aleación *f.* alloy
aleaje *m.* alloy
alear *vb.* to alloy
aleatorio *a.* random
alejado *a.* distal
alejamiento *m.* latitude
alejandrina, alejandrita *f.* alexandrite
alero *m.* overhang, horizontal bed
aleta *f.* fin (of rock), blade
alevante *m.* [Méx.] overhand stope
alfagra *f.* [Sp.] channel, gorge
alfaguara *f.* abundant spring
alfaque *m.* shoal, bar
alfiler de oro *m.* [Col.] acicular gold
alga, algas *f.* algae || *alga marina f.* seaweed
algáceo, algoso *a.* algal
algaida *f.* sand dune
alginita *f.* alginite
algino, de algas *a.* algal
Algomiano, Algomiense *m. & a.* Algomian
Algonkiano, Algonquiano, Algonquico, Agnostozóico *m. & a.* Algonkian
algoritmo *m.* algorithm
alianza *f.* alliance
alicuota *f.* aliquot
alidada *f.* alidade
aligación *f.* (chem.) alloy, bond
alifático *a.* aliphatic
alimentación *f.* nourishment, feed || *retro alimentación f.* feedback
alimentador *a.* feeder || *dique alimentador m.* feeder dike || *tronco alimentador m.* feeder stock
alimentar *vb.* to feed
alimonados *m.* silver sulfides, silver salts
alindar *vb.* to mark boundaries
alinderar *vb.* [Arg., Ch.] to mark boundaries
alineación *f.* lineation, alignment
alineamiento *m.* alignment
alinéar *vb.* to align || *alinear a ojo vb.* to eyeball
alipús *m.* gad
alisadora *f.* grader
alisadura *f.* grading, leveling
alisar *vb.* to grade, level
alisios *m.* trade winds
aliviadero *m.* overflow spillway
alivio *m.* relief
aljez, aljor *m.* gypsum
aljezar *m.* gypsum quarry, gypsum deposit
aljibe *m.* well, pool, [Col.] spring
allanita *f.* orthite, allanite
alma *f.* core
almacén *m.* warehouse, reservoir
almacenadora *f.* reservoir unit (véase depósito, yacimiento)
almacenaje, almacenamiento *m.* storage
almadaneta, almádena, almaína *f.* rock hammer
almadén *m.* mine, mineral deposit
almadía *f.* raft
almagra, almagre *f.* red ochre
almanaque *m.* almanac
almandina *f.* almandite
almarjal *m.* marsh
almártaga *f.* litharge
almeja *f.* clam
almendrilla *f.* fine gravel, fine coal, [Méx.] fragments of silicified limestone, quartz matrix of copper veins
almiar *m.* hum, haystack hill
almirez *m.* mortar
almoceda *f.* [Sp.] water rights
almohada *f.* pillow || *almohada de sal f.* salt pillow || *lava en almohada f.* pillow lava
almohadilla *a.* pillow lavas
alnoita *f.* alnoite
alóctono *m. & a.* allochthon, allochthonous
alófana *f.* allophane
alojamientos *m.* living quarters
a lo largo de *a.* along
a lo largo del rumbo *a.* along strike
alomado *a.* undulating
alometamorfismo *m.* contact metamorphism, dynamic metamorphism

alotipo *m.* allotype
alotriomórfico *a.* anhedral
alotriomorfo *a.* allotriomorphic
alotropía *f.* allotropy
alotrópico *a.* allotropic
al peso *a.* by weight
alpestre *a.* alpine, mountainous
Alpino, Alpinense *m. & a.* Alpine
al por mayor *a.* wholesale
al por menor *a.* retail
alqueno *m.* alkene
alquifol *m.* cubic galena, graphite
alquilación *f.* alkylation
alquílico *a.* alkylic
alquilo *a.* alkyl
alquitrán *m.* tar, coal tar
alrededores *m.* environs
alsa *f.* [Pe.] gallery roof
altearse *vb.* to uplift, elevate
alteración *f.* alteration
alterado *a.* altered
alternado *a.* staggered
altillano *m.* [Col.] plateau
altillanura *f.* plateau, high plain
altillo *m.* small hill
altimetría *f.* topographic survey
altimétrico, altimetro *a.* altimetric
altímetro *m.* altimeter
altiplanicie *f.* plateau, high plain, highland
altiplano *m.* plateau, high plain
altitud *f.* altitude, elevation
alto *a.* high; *m.* height, hill, elevation; hanging wall || ***alto calidad*** *f.* high grade || ***alto estructural*** *m.* structural high || ***alto frequencia*** *a.* high frequency || ***alto grado*** *a.* high rank || ***alto horno*** *m.* blast furnace || ***alto resolución*** *a.* high resolution || ***alto topográfico*** *m.* topographic high
altocúmulo *m.* altocumulus
altos *m.* (geog., geol., seis.) highs
altozano *m.* small hill
altura *f.* height, summit, altitude, elevation || ***altura de aspiración*** *f.* pump lift || ***altura de carga*** *f.* hydraulic head || ***altura de impulsión*** *f.* hydraulic head || ***altura hidráulica*** *f.* hydraulic head || ***altura hidrostática*** *f.* hydraulic head || ***altura piezométrica*** *f.* hydraulic head
alud *f.* avalanche; hanging glacier || ***alud de cenizas*** *f.* ash flow || ***alud de escombros*** *f.* debris avalanche || ***alud de fondo*** *f.* landslide || ***alud de hielo*** *f.* ice avalanche || ***alud de rocas*** *f.* rockslide
alumbrar *vb.* to discover, strike, find; (hydr.) to emerge
alumbre *m.* alum || ***alumbre de piedra*** *m.* alunite
alumbrera *f.* alum quarry
alúmbrico *a.* alum-bearing
alumbroso *a.* aluminous
alúmina *f.* alumina
aluminífero *a.* aluminum-bearing
aluminio *m.* aluminum
aluminita *f.* [Ch.] potassium-aluminum sulfate ore
aluminoso *a.* aluminous
alunita *f.* alunite
alutación *f.* gold nuggets found on weathered surfaces
aluvial, aluvional *a.* alluvial || ***abanico aluvial*** *m.* alluvial fan
aluvión, aluvio *m.* alluvium, wash, outwash || ***aluvión fluvio-glacial*** *m.* glacial outwash || ***aluvión metalifero*** *m.* placer || ***aluvión soplado*** *m.* [Col.] fluvial sediments deposited by strong currents
aluvionamiento *m.* alluviation, silting up, accretion
aluvionario *m.* [Arg.] alluvium
álveo *m.* riverbed, channel, river source
alveolado *a.* pitted
alveolar *a.* alveolar
alvéolo *m.* alveole, cell
alza *f.* [Bol.] washing of gold
alzaguera *f.* [Sp.] spring
alzar *vb.* to raise, hoist, lift, elevate
amagamiento *m.* [Col.] spring, source
amalgama *f.* amalgam
amalgamación *f.* amalgamation

amalgamar *vb.* to amalgamate

amarillo *m.* [Méx.] gossan

amarar tonga *vb.* [Col.] to lessen the head of water in a sluice

amarrar *vb.* to tie, link, correlate

amatista *f.* amethyst

amazonita *f.* amazonite

ámbar *m.* amber

ambiental *a.* ambient

ambiente *m.* environment ǁ ***ambiente de deposición*** *m.* depositional environment ǁ ***ambiente tectóncio*** *m.* tectonic setting

ámbito *m.* perimeter, area ǁ ***ámbito geológico*** *m.* geologic setting

ambligonita *f.* amblygonite

América Central *f.* Central America

América del Norte *f.* North America

América del Sur *f.* South America

amianto *m.* asbestos

amígdala, amígdula *f.* amygdule

amigdaloide *a.* amigdaloidal; *m.* amygdaloid

ameba *f.* amoeba

amina *f.* amine

aminoácido *m.* amino acid

amitosis *f.* amitosis

amojonar *vb.* (survey) to set monuments

amoladera *f.* [Bol.] sandstone

amolador *m.* grinder

amoladora *f.* grinder

amoladura *f.* grinding ǁ ***amoladura glacial*** *f.* galcial striae

amolar *vb.* to grind

amolinar *vb.* to collect ore for processing

amoníaco *m.* ammonia

amonio *m.* ammonium

amonita *f.* ammonite

amontonar *vb.* to pile up, make heaps

amorfo *a.* amorphous

amortiguador *m.* buffer, cushion, damper

amortiguamiento *m.* damping

amortiguar *vb.* to damp, to buffer, temper

amortización *f.* payout, amortization

amparar *vb.* to obtain title to property ǁ ***amparar en la posesión*** *vb.* to keep property by doing assessment

amparo *m.* maintain right of ownership by continued use

ampelita *f.* cannel coal, ampelite

ampliación *f.* enlargement

ampliar *vb.* to enlarge

amplificación *f.* magnification

amplificar *vb.* to enlarge, extend

amplio *a.* wide, gentle

amplitud *f.* amplitude

ampolla *f.* ore pocket, lava blister

ampolloso *a.* porous, vuggy

anaeróbio, anaeróbico *a.* anaerobic

anáglifo *m.* anaglyph

anal *a.* anal

analcima *f.* analcime

anabergita *f.* annabergite

anacarado *a.* pearly

análisis *m.* analysis ǁ ***análisis cualitativo*** *m.* qualitative analysis ǁ ***análisis cuantitativo*** *m.* quantitative analysis ǁ ***análisis de costo*** *m.* cost analysis ǁ ***análisis elemental*** *m.* elemental analysis ǁ ***análisis granulométrico*** *m.* grain size analysis ǁ ***análisis isotópico*** *m.* isotopic analysis ǁ ***análisis químico*** *m.* chemical analysis

analista *m. & f.* analyst

analítico *a.* analytical

analizador *m.* analyzer

analizar *vb.* to analyze

analógico *m.* analog

análogo *m. & a.* analogy

anamorfismo *m.* anamorphism

anarquia de ríos *f.* stream derangement

anastomosado *a.* anastomosing, braided

anastomosarse *vb.* anastomose

anatasa *f.* anatase

anatexia, anatexis *f.* anatexis

anaxil *a.* anaxial

anco *m.* [Bol.] native silver; [Ch., Pe.] silver ore

ancón *m.* bay, cove

anconada *f.* large bay

ancho *a.* wide, broad; *m.* width, breadth

anchura *f.* width, widening of a vein, open terrain

anchurón *m.* chamber, room in a mine

anchuroso *a.* wide, large

andalusita, andalucita *f.* andalucite

andana *f.* lineation of hills, etc.

andarivel *m.* aerial tram

andén *m.* [Bol., Pe.] terrace

andesina *f.* andesine

andesita *f.* andesite

Andino *a.* Andean

andradita *f.* andradite

andurrial *m.* claypit

anegable *a.* flood-prone

anegado *a.* inundated

anegar *vb.* to flood, inundate, submerge

anegarse *vb.* to be flooded, to become inundated

anelasticidad *f.* anelasticity

anelástico *a.* anelastic

anélido *m.* annelid

anemómetro *m.* anemometer

anémona *f.* anemone

aneroide *a.* aneroid

anexo *m.* attachment

anfibio *m.* amphibian

anfiteatro *m.* amphitheater

anfitrión, roca anfitriona *m. & f.* (min.) host rock

anfíbol *m.* amphibol

anfibolesquisto *m.* hornblende schist

anfibolita *f.* amphibolite

anfibolización *f.* amphibolization

anfingena *f.* leucite

angiosperma *f.* angiosperm

angla *f.* cape

anglesita *f.* anglesite

angosto *a.* narrow

angostarse *vb.* to thin, narrow

angostura *f.* narrowness, pass, gorge, river narrows

angra *f.* cove, small bay

anguena *f.* [Sp.] silver ore

angular *a.* angular

angström *m.* ángstrom (10^{-10} meter)

ángulo *m.* angle, corner || ***ángulo agudo*** *m.* acute angle || ***ángulo crítico*** *m.* critical angle || ***ángulo de buzamiento*** *m.* dip || ***ángulo de cabeceo*** *m.* pitch || ***ángulo de cizalla*** *m.* angular shear || ***ángulo de cizalladura*** *m.* shear angle || ***ángulo de charnela*** *m.* axial hinge angle || ***ángulo de depresión*** *m.* depression angle || ***ángulo de desviación*** *m.* deflection angle, deviation angle || ***ángulo de echado*** *m.* dip || ***ángulo de escalón inicial*** *m.* initial step-up angle || ***ángulo de extinción*** *m.* extinction angle || ***ángulo de fricción*** *m.* angle of friction || ***ángulo de incidencia*** *m.* incidence angle || ***ángulo de inclinación*** *m.* dip || ***ángulo de pendiente*** *m.* slope angle || ***ángulo de polarización*** *m.* polarization angle || ***ángulo de reflexión*** *m.* reflection angle || ***ángulo de refracción*** *m.* refraction angle || ***ángulo de reposo*** *m.* angle of repose || ***ángulo de rumbo*** *m.* strike || ***ángulo de rotación*** *m.* rotation angle || ***ángulo final de truncamiento*** *m.* final cutoff angle || ***ángulo inicial de truncamiento*** *m.* initial cutoff angle || ***ángulo recto*** *m.* right angle || ***ángulo obtuso*** *m.* obtuse angle || ***ángulo óptico*** *m.* optic angle || ***ángulo recto*** *m.* right angle

angulosidad *f.* angularity

anguloso *a.* angular

anhedral *a.* anhedral

anhídrico *a.* anhydrous

anhídrido, anhidro *m.* anhydride || ***anhídrido, anhidro sulfuroso*** *m.* liquid sulfur dioxide

anhidrita *f.* anhydrite

anhidro *a.* anhydrous

anidado *a.* nested

aniego, aniegue *m.* [Méx.] flooding

anihédrico *a.* allotriomorphic

anilina *f.* aniline, phenylamine

anillo *m.* ring || ***anillo bencénico*** *m.* benzene ring || ***anillo de empuje*** *m.* thrust collar || ***nafténico anillo*** *m.* naphthene ring

animal *m.* animal

Animiquiano, Animiquiense *m. & a.* Animikian

anión *m.* anion

Anisiense, Anisiano *m. & a.* Anisian

anisométrico *a.* anisometric

anisopericlasa *f.* fracture parallel to a fold

anisotropía *f.* anisotropy

anisotrópico, anisótropo *a.* anisotropic

a nivel *a.* at grade

ankerita, anquerita *f.* ankerite

anabergita *f.* annabergite

ano *m.* anus

ánodo *m.* anode

anomalía *f.* anomaly || ***anomalía de aire libre*** *f.* free air gravity anomaly || ***anomalía gravimétrica*** *f.* gravity anomaly || ***anomalía magnética*** *f.* magnetic anomaly

anómalo *a.* anomalous

anormal *a.* abnormal

anormalidad *f.* abnormality

anorganógeno *a.* inorganic

anorogénico, anorogenético *a.* anorogenic, nonorogenic

anórtico *a.* asymmetric

anortita *f.* anorthite

anortoclasa *f.* anorthoclase

anortosita *f.* anorthosite

anotación *f.* note, annotation, record

anotar *vb.* to record, write

anóxico *a.* anoxic

anquería *f.* [Pe.] silver ore

anquerita *f.* ankerite

anta *f.* [Pe.] copper || ***anta chacra*** *f.* copper mine

Antártica, Antártida *f.* Antarctica

antártico, antártido *a.* antarctic

antearrecife *m.* forereef

antecambriano *a. & m.* Precambrian

antecedente *a.* antecedant, existing data || ***antecedente de un caso*** *m.* case history

antecubeta *f.* foredeep

antefosa *f.* foreland basin, foredeep

antena *f.* antenna, aerial

anteojo *m.* telescope || ***anteojo buscador*** *m.* viewfinder, guide scope

antepaís *m.* foreland

antepecho *m.* mine bench

anteplaya *f.* shoreface

antepozo *m.* foreshaft, derrick cellar, cellar

anteproyecto *m.* preplanning

anterior *a.* anterior, frontal

antiácido *a.* acidproof

anticiclón *m.* anticyclone

anticlinal *m.* anticline || ***anticlinal abierto*** *m.* breached anticline || ***anticlinal acostado*** *m.* recumbent anticline || ***anticlinal aportillado*** *m.* breached anticline || ***anticlinal asimétrico*** *m.* asymmetric anticline || ***anticlinal colgado (o colgante)*** *m.* hanging wall anticline || ***anticlinal cornudo*** *m.* doubly overturned anticline || ***anticlinal desmantelado*** *m.* breached anticline || ***anticlinal entallado*** *m.* breached anticline || ***anticlinal fracturado*** *m.* breached anticline || ***anticlinal invertido*** *m.* overturned anticline || ***anticlinal recostado*** *m.* overturned anticline || ***anticlinal recumbente*** *m.* recumbent anticline || ***anticlinal simétrico*** *m.* symmetric anticline || ***anticlinal tumbado*** *m.* overturned anticline || ***anticlinal volcado*** *m.* recumbent or inverted anticline || ***cierre de anticlinal*** *m.* anticlinal closure || ***eje anticlinal*** *m.* anticlinal axis || ***nariz de anticlinal*** *m.* anticlinal nose

anticlinorio *m.* anticlinorium

anticongelante *m. & a.* antifreeze

anticorrosivo *a.* rust-resisting

anticuado *a.* obsolete, old-fashioned

antiduna *f.* antidune

antiforma *f.* antiform

antigorita *f.* antigorite

antigüedad *f.* relic

antiguo *a.* ancient, old

antiíncrustante *m.* scale remover

antimoniado *a.* antimony-bearing

antimonio *m.* antimony

antímonita *f.* stibnite

antióndula *f.* antiripple

antioxidante *m.* antioxidant

antiparras *f.* goggles

antipertita *f.* antiperthite

antirizadura *f.* antiripple

antitético *a.* antithetic || ***falla antitético*** *f.* antithetic fault

antlerita *f.* antlerite

antofilita *f.* anthophyllite

antracita *f.* anthracite

antracito *a.* black

antracitoso *a.* anthracite-bearing

antraconita *f.* anthraconite

antro *m.* cave, cavern

anual *a.* annual

anular *a.* annular

ánulo *m.* annulus

anuncio *m.* advertisement, announcement, notice

añadir *vb.* to add

a ojo descubierto *a.* to the naked eye

apacheta *f.* [Pe.] summit, pass

apagado *a.* extinct, extinguished

apagar *vb.* extinguish, quench

Apalache *a.* Appalachian

apalancar *vb.* to lever, get ore from a hopper

apantanar *vb.* to flood, inundate

aparato *m.* apparatus, system; (comp.) device || ***aparato perforador*** *m.* perforator || ***aparato volcánico*** *m.* volcanic system

aparentar *vb.* to manifest, to appear

aparente *a.* apparent

aparejar *to* rig up

aparejo *m.* apparatus, drill rig

aparente *a.* apparent

aparición *f.* appearance

apariencia *f.* appearance

aparinar *vb.* to indicate ore

apartado *m.* sorting ore; smelting, smelter; refining of gold || ***apartado a mano*** *m.* hand sorting || ***casa del apartado*** *f.* assay house

apartador *m.* smelter, ore picker

apartamiento *m.* mineral extraction

apartar *vb.* to sort ore, extract minerals; refine during assay

apatita, apatito *m.* & *f.* apatite

apeador *m.* land surveyor

apeadura *f.* surveying

apear *vb.* to survey; to shore, brace

apéndice *m.* appendix, outgrowth

apeo *m.* survey; timbering, bracing

aperos *m.* mining supplies

apertura *f.* opening, rift, aperture; opening a mine || ***apertura diafragmatica*** *f.* iris diaphragm

apical *a.* apical

ápice *m.* apex

apilamiento *m.* (seis.) stack, stacking; arrangement of atoms in a crystal lattice

apilar *vb.* to pile

apique *m.* trench, pit, excavation; *m.* [Col.] shaft, winze

apireo *m.* [Ch.] remove ore in baskets

apiri *m.* ore carrier

apizarrado *a.* slaty, having slaty cleavage, slate colored

apizarramiento *m.* slaty cleavage

aplanado *a.* leveled

aplanamiento *m.* levelling, flattening

aplanar *vb.* to grade, level

aplano *a.* planar

aplastado *a.* caked, crushed

aplastamiento *m.* collapsing, crushing, flattening, slumping

aplastar *vb.* to flatten, crush

aplastarse *vb.* collapse

a pleno flujo *m.* open-hole flow

aplicar *vb.* to apply

aplicación *f.* application

aplita *f.* aplite

aplomado *a.* lead-colored

aplomado *m.* [Méx.] lead-colored copper ores

a plomo *a.* plumb

apofilita *f.* apophyllite

apófisis *m.* apophysis

apogeo *m.* climax

apolvillados *m.* superior grade ores

aporcelanado *a.* porcellaneous

aportación *f.* inflow, runoff

aportar *vb.* to supply

aporte *m.* runoff, inflow

aportillar *vb.* to breach

apostar *vb.* to stake

apoyar *vb.* to support

apreciación *f.* valuation, estimate, appraisal

apreciar *vb.* to appraise, value; to see

aprecio *m.* appraisal, estimate

apresador *m.* catcher || ***apresador de testigos*** *m.* core catcher

apretado *a.* tight (folding), compressed

aprisionado *a.* stuck

aprisionarse *vb.* to stick, freeze-up

aprovechamiento *m.* use, exploitation, development, reclamation

aprovechar *vb.* to use, develop, improve, exploit, take advantage of, capitalize on

aproximar(se) *vb.* to approach, approximate

a prueba de agua *a.* waterproof

a prueba de intemperie *a.* weatherproof

Aptense, Aptiense, Aptiano *m. & a.* Aptian

apuntamiento *m.* small outcrop

apurar *vb.* to purify metal, concentrate ore

aquilatación *f.* assay, alloy

aquilatamiento *m.* assay

aquilatar *vb.* to assay, to weigh, to alloy

Aquitaniense, Aquitaniano *m. & a.* Aquitanian

arácnido *m.* arachnid

aragonita, aragonito *m. & a.* aragonite

arancel *m.* tariff

arastrar *vb.* to drag

árbol *m.* tree; mast, shaft || ***árbol de conexiones*** *m.* Christmas tree || ***árbol de navidad*** *m.* Christmas tree

arboleda *f.* grove

arbusto *m.* shrub, bush

Arcadiense, Arcadiano *m. & a.* Arcadian

Arcáico *m.* Archean

arcilita, arcillita *f.* claystone, mudstone

arcilla *f.* clay || ***arcilla de batán*** *f.* Fuller's earth || ***arcilla de frente deltáico*** *f.* prodelta clay || ***arcilla decolorante*** *f.* Fuller's earth || ***arcilla endurecida*** *f.* claystone || ***arcilla figulina*** *f.* potter's clay || ***arcilla gredosa*** *f.* gouge || ***arcilla laminar*** *f.* [Vz.] shale || ***arcilla para inyección*** *f.* drilling clay || ***arcilla pedregosa*** *f.* boulder clay, till || ***arcilla pegajoso*** *f.* gumbo || ***arcilla plástica*** *f.* gumbo || ***arcilla refractoria*** *f.* refractory clay, fire clay || ***arcilla sapropel*** *f.* sapropellic clay

arcilloso *a.* argillaceous, argillic, clayey

arco *m.* arc, arch || ***arco Beaman*** *m.* Beaman stadia arc || ***arco de islas*** *m.* island arc || ***arco insular*** *m.* island arc || ***arco natural*** *m.* natural arch || ***arco volcánico*** *m.* volcanic arc

arcosa *f.* arkose

arcósico *a.* arkosic

archipiélago *m.* archipelago

archivar *vb.* to file

archivo *m.* file

área *f.* area || ***área alimentadora*** *f.* watershed || ***área comprobada*** *f.* proven territory || ***área de drenaje*** *f.* drainage area || ***área de escurrimiento*** *f.* [Col.] watershed || ***área de exploración*** *f.* exploration area || ***área de fracturación*** *f.* fracture zone || ***área disponible*** *f.* open acreage || ***área fuente*** *f.* fetch, source area || ***área no probada*** *f.* unproven area || ***área recreativa*** *f.* recreation area || ***área útil*** *f.* effective area || ***área vertiente*** *f.* watershed

areg, arej *m.* erg

arena *f.* sand || ***arena asfáltica*** *f.* tar sand || ***arena aurífera*** *f.* gold-bearing sand || ***arena bituminosa*** *f.* bituminous sand || ***arena bruja*** *f.* fine sand || ***arena cordoniforme*** *f.* shoestring sands || ***arena de oro*** *f.* gold dust || ***arena de playa*** *f.* beach sand || ***arena eólica*** *f.* blown sand, dune sand || ***arenas filiformes*** *f.* shoestring sands || ***arena franca*** *f.* loamy sand || ***arena gasífera*** *f.* gas sand || ***arena gorda*** *f.* coarse sand || ***arena gruesa*** *f.* grit, coarse sand || ***arena movediza*** *f.* quicksand || ***arena petrolífera*** *f.* oil sand || ***arena productiva*** *f.* pay sand || ***arena recia*** *f.* coarse sand || ***arena sucia*** *f.* ratty sand || ***banco de arena*** *f.* sandbank || ***cava de arena*** *f.* sand pit || ***fosa de arena*** *f.* sand pit || ***guija de arena*** *f.* sand pit || ***hoyo de arena*** *f.* sand pit || ***llenar de arena*** *vb.* sand up

arenáceo *a.* arenaceous, sandy

arenal *m.* sandbar, sandbank, [Méx.] dune || ***arenal bajio*** *m.* sandbar

Arenigense, Arenigiano *m. & a.* Arenigian

arenilla *f.* grit, fine sand, [Méx.] tailings, [Vz.] black magnetite sand, [Col.] titaniferous iron ore, [Ch.] copper matte mixed with slag

arenisca *f.* sandstone || ***arenisca continua*** *f.* blanket sand || ***arenisca lajosa*** *f.* flagstone || ***arenisca sucia*** *f.* ratty sand

arenisco *a.* arenaceous, sandy

arenita *f.* arenite || ***arenita de cuarzo*** *f.* quartz arenite

arenoso *a.* arenaceous, sandy, gritty

areola *f.* areole

arête *m.* arête

argarita *f.* marcasite

argayo *m.* talus

argentado, argénteo, argentino *a.* silvery

argentífero, argentoso *a.* silver-bearing

argentita *f.* argentite

argila, argilla *f.* clay

argilita *f.* argillite

argilolita *f.* claystone

argiloso *a.* argillaceous

argirita *f.* argentite, [Bol.] marcasite

argirosa *f.* argentite, pyrargirite

argón *m.* argon

Argoviense, Argoviano *m. & a.* Argovian

aridez *f.* aridity

árido *a.* arid, dry

árido *m.* aggregate

arisco *a.* (geog.) rough, uneven

arista *f.* arête, right angle crystal face intersection, rib || ***arista hidrográfica*** *f.* [Vz.] watershed divide

armador *m.* builder, erector

armadura *f.* framework, [Méx.] host or country rock

armar *vb.* to be hosted by, occur in; to assemble

armazón *f.* frame, framework, truss, trestle

armoniaco *m.* sal ammoniac

armónico *a.* harmonic

aromático *a.* aromatic

arqueada *a.* cuspate, arcuate; *f.* arc-shaped vein branch

arqueado *a.* arcuate

arqueamiento *m.* arching

arquear *vb.* to arch, bend

Arqueano, Arcáico *m. & a.* Archean

arqueociátido *m.* archaeocyathid

arqueología *f.* archaeology

arqueólogo *m.* archaeologist

Arqueozóico *m. & a.* Archaeozoic

arquería *f.* [Ch.] silver amalgam

arquitectura *f.* (geol.) architecture || ***arquitectura de la secuencia*** *f.* sequence architecture || ***arquitectura paisajista*** *f.* landscaping

arrambarse *vb.* to be covered by sand

arrancar *vb.* to start, to begin, to crank up, break ground, start digging

arranque *m.* breaking ground, digging, mining, extraction, processing, drill pattern for blasting

arrasamiento *m.* stripping

arrasar *vb.* to strip

arrastrado *a.* dragged, transported, entrained

arrastrar *vb.* to drag, haul, scour

arrastre *m.* the act of dragging, hauling; slope of a shaft; silver mill; drag mill used to crush ore; washout; haulage, footwall || ***pliegue de arrastre*** *m.* drag fold

arrastres *m.* silt, suspended matter

arrecifal *a.* reefal

arrecife *m.* reef, ridge of rock || ***arrecife coralígeno*** *m.* atoll reef || ***arrecife costero*** *m.* fringing reef || ***arrecife de barra, de barrera*** *m.* barrier reef || ***arrecife franjeante*** *m.* fringing reef || ***arrecife litoral*** *m.* fringing reef || ***arrecife marginal*** *m.* fringing reef || ***arrecife parche*** *m.* patch reef

arreglar *vb.* to arrange

arreglo *m.* arrangement, pattern, framework || ***arreglo de cinco pozos*** *m.* five-spot well pattern

arremolinamiento *m.* agglomeration

arremolinarse *vb.* to agglomerate
arrendador *m.* lessor, landlord
arrendamiento *m.* lease
arrendar *vb.* to lease
arrendatario *m.* lessee, leaseholder
arriada *f.* flood
arriarse *vb.* to flood
arriba *a.* up || ***hacia arriba*** *adv.* upwards
arribera *f.* river valley
arriendo *m.* lease
arriesgar *vb.* to risk
arriñonado *a.* botryoidal, nodular
arriscado *a.* craggy
arrojado *a.* expelled
arrojar *vb.* to eject, extrude
arroyado *a. & m.* gullied; watercourse
arroyar *vb.* to form gullies
arroyo *m.* gully, stream, creek, brook || ***arroyo reséquente*** *m.* resequent stream || ***arroyo superpuesto*** *m.* superposed stream
arroyuelo *m.* small brook, creek
arruga *f.* small fold, crease, wrinkle
arrugación *f.* folding of strata
arrugado *a.* folded, wrinkled
arrugamiento *m.* minute folding
arrugar *vb.* to fold
arrugia *f.* gold mine, gold gravel works
arrumbamiento *m.* strike, bearing
arrumbar *vb.* to outcrop
arseniato *m.* arsenate
arsénico *m.* arsenic
arseniuro *m.* arsenide
arsenoblenda amarilla *f.* orpiment
arsenopirita *f.* arsenopyrite
artesa *f.* goldpan, trough; [Col.] river basin || ***artesa de criba*** *f.* screen pan || ***artesa oscilante*** *f.* cradle to wash placer ore
artesiano *a.* artesian || ***pozo artesiano*** *m.* artesian well
artesón *m.* wash trough for placers
ártico *m. & a.* arctic
articulación *f.* articulation, hinge
articulado *a.* articulated, jointed
articular *vb.* to articulate
artificio *m.* artifice; artifact
artifacto *m.* artifact
Artinskiano, Artinskiense *m. & a.* Artinskian
artrópodo *m.* arthropod
asbestino *a.* made of asbestos
asbesto *m.* asbestos
asbolana *f.* cobalt ocher
ascender *vb.* to arise, to rise
ascenso *m.* ascent, raise
ascensor *m.* cage, lift, elevator
asdic *m.* asdic, sonar
asemejar *vb.* to resemble
asentado *a.* caved
asentamiento *m.* settling, slumping, caving
asentar *vb.* to rest on || ***asentar cañería vb.*** to set pipe
asentarse *vb.* to settle
asentista *m.* contractor
aserrado *a.* serrated
asesor *m.* consultant
asexual *a.* asexual
asfalteno *a. & m.* asphaltene
asfáltico *a.* asphaltic
asfaltífero *a.* asphalt-bearing
asfaltita *f.* asphaltite
asfalto *m.* asphalt, pitch
Ashgiliense, Ashgiliano *m. & a.* Ashgillian
Asia *f.* Asia
asiento *m.* sediment; concentrate from panning || ***asiento de minas*** *m.* mineralized area || ***asiento mineral*** *m.* mineralized region
asimetría *f.* asymmetry
asimétrico *a.* asymmetric || ***anticlinal asimétrico*** *m.* asymmetric anticline || ***pliegue asimétrico*** *m.* asymmetric fold
asimilar *vb.* assimilate
asimilación *f.* assimilation
a simple vista *a.* at first sight
asíntota *f.* asymptote
asintótico *a.* asymptotic
asísmico *a.* aseismic

asociación *f.* assemblage || ***asociación de fósiles*** *f.* fossil assemblage

asomar *vb.* to crop out

asombrarse *vb.* to become flatter, more bed-like

asomo *m.* outcrop, inlier, precipice edge

aspa *f.* x-shaped; [Pe.] vein intersection

aspar *vb.* to cross in the shape of an x

aspecto *m.* appearance

aspereza *f.* roughness, uneveness

áspero *a.* rough, rugged, harsh, coarse

asperón *m.* sandstone

aspillera *f.* dissolution surface on carbonate rock

aspiración *f.* suction, indraft

aspirar *vb.* to inhale

asta de venado *f.* [Pe.] anglesite

astenósfera *f.* asthenosphere

asteria *f.* cat's eye chalcedony; asterated sapphire

asterismo *m.* asterism

asteroide *m.* (astron.) asteroid

asteroideo *m.* asteroid, (paleo.) sea star

astiense, astiano *m. & a.* Astian

astilla de piedra, astilladura *f.* spall, splinter, chip

astillamiento *m.* spalling off

astillar *vb.* to splinter, spall

astilloso *a.* hackly

astracanita *f.* astrakanite

astrago *m.* soil horizon

astrolabio *m.* astrolabe

astronomía *f.* astronomy

astronómico *a.* astronomical

asunto *m.* subject, business, affair, matter, theme

atacamita *f.* atacamite

atacar con ácido *vb.* to etch with acid

atajo *m.* trail

ataque químico *m.* etching

ataques *m.* rubbish, garbage, waste

atascadero *m.* swamp

atascado *a.* stuck, clogged

atasco *m.* obstruction

atectónico *a.* atectonic

atenuación *f.* attenuation

atenuarse *vb.* to pinch out

aterramiento *m.* silting up, alluvium

aterrar *vb.* to silt up

atierre *m.* silting up

atierres *m.* waste rock and dirt in a mine

atíncar *m.* refined borax

atípico *a.* atypical

atlas *m.* atlas

atmósfera *f.* atmosphere

atmosférico *a.* atmorpheric

atol, atolón *m.* atoll

atolladero *m.* swamp

atómico *a.* atomic || ***número atómico*** *m.* atomic number || ***peso atómico*** *m.* atomic weight

átomo *m.* atom

atorado *a.* stuck

atracción *f.* attraction || ***atracción gravitacional*** *f.* gravitational attraction || ***atracción interfacial*** *f.* interfacial tension || ***atracción magnetica*** *f.* magnetic attraction

atrancado *a.* stuck

atrapar *vb.* capture, trap

atrapado *a.* trapped

atrás *a.* behind

atravesado *a.* crosscut

atravesar *vb.* to crosscut; to penetrate, to pass through, span, pierce

atravieso *m.* [Ch.] gap, pass

atrición *f.* attrition

atrito *m.* attritus

Aturiense, Aturiano *m. & a.* Aturian

atutía *f.* metal oxide crust on furnaces

augengneis *m.* augen gneiss

augita *f.* augite

aulacógeno *m.* aulacogen

aumentar *vb.* to enlarge, to magnify, to raise, augment, add to

aumento *m.* magnification; accretion; increase; raise, enlargement, addition

auquis *m.* [Pe.] mine drillers

auramalgama *f.* gold amalgam

áureo *a.* golden

aureola *f.* aureol || ***aureola de contacto*** *f.* contact aureole, chilled margin

auricalcita *f.* aurichalcite

áurico *a.* golden, gold

aurífero *a.* auriferous, gold-bearing

Australia *f.* Australia

autigénesis *f.* authigenesis

autigénico, autígeno *a.* authigenic

autoclástico, auticlastico *f.* autoclastic

autoclasto *m.* autoclast

autocorrelación *f.* autocorrelation

autóctono *a.* *& m.* autochthonous; autochthon

autógeno *a.* autogenetic

autoinducción *f.* self-induction

autolito *m.* autolith

autometamorfismo *m.* autometamorphism

automorfo *m.* idiomorphic

autoporo *m.* autopore

autopotencial *a.* self-potencial

Autuniense, Autuniano *m. & a.* Autunian

autunita *f.* autunite

auxiliar *m.* (comp.) peripheral

auxilio *m.* credit

avalancha *f.* avalanch; [Arg.] flood

avaluar *vb.* to appraise, value, estimate

avalúo *m.* valuation, appraisal

avance *m.* advance, rise, encroachment || ***avance del agua marginal*** *m.* water encroachment || ***avance del mar*** *m.* transgression || ***avance de tiempo*** *m.* time lead

avanzado *a.* late

avenamiento *m.* drainage, draining (véase drenaje) || ***avenamiento anárquico*** *m.* deranged drainage || ***avenamiento anastomosado*** *m.* braided drainage || ***avenamiento antecedente*** *m.* antecedent drainage || ***avenamiento anular*** *m.* annular drainage || ***avenamiento centrípetro*** *m.* internal drainage || ***avenamiento dendrítico*** *m.* dendritic drainage || ***avenamiento desordenado*** *m.* aimless or deranged drainage || ***avenamiento en lengüeta*** *m.* barbed drainage || ***avenamiento paralelo*** *m.* parallel drainage || ***avenamiento reticulado*** *m.* trellis drainage || ***avenamiento radial*** *m.* radial reservoir drainage

avenar *vb.* to drain

avenc *m.* [Sp.] sinkhole

avenida *f.* flood

aventadero *m.* [Pe.] gold placer; [Arg.] shallow Quaternary placer; [Col.] placer above the riverbank

aventurina *a. & f.* aventurine

averno *m.* gas pocket in a coal mine

avetado *a.* veined, seamed

aviador *m.* financier

aviadoro *a.* subject to assessment

aviar *vb.* to finance a mine, [U.S.] grubstake

avío *m.* provision, implement, funds or goods, grubstake, hereditary lease

aviculario *m.* avicularium

avisador *m.* monitor

aviso *m.* advertisement, announcement, warning, notice || ***aviso sísmico*** *m.* foreshock

avistar *vb.* to view

avolcanado *a.* volçanic

axial, axil *a.* axial

axilar *a.* axillary

axinita *f.* axinite

axiolita *f.* axiolite

azabache *m.* jet

azada *f.* stonecutting tool, hoe

azadón *m.* hoe

azanca *f.* spring

azar *m. & a.* chance, at random

azarcón *m.* crocoite

azarnefe *m.* orpiment

azimut *m.* azimuth

azimutal *a.* azimuthal

azoato *a. & m.* nitrate

ázoe *m.* nitrogen

azogado, azogamiento *m.* mercury poisening

azogue *m.* mercury, ore amenable to amalgamation; [Méx.] second class ore, silver ores

azoguería *f.* amalgamation; mercury storehouse

azoguero *m.* [Bol.] operator, renter, or operator of mine or mill

azóico *m. & a.* nitric, azoic

azótico *a.* nitric

azolvarse *vb.* to silt up

azolve *m.* silt

azúcar *f.* [Col.] granular calcite matrix to gold ore

azufrado *a.* sulfurous; *m.* [Col.] yellow ocher; [Pe.] yellow lead ore, sulfide ore; [Ch.] nitrate of soda

azufral *m.* solfatera

azufre *m.* sulfur

azufrera *f.* sulfur mine

azufrón *m.* sulfide ore

azufroso *a.* sulfurous

azulaque *m.* bitumen; [Méx.] argentite, disseminated silver and copper ores

azulillo *m.* silver-bearing carbonate ore

azurita *f.* azurite

azurronado *m.* [Méx.] silver-bearing pyrite

B

babbitt *m.* bearing metal

bacilar *a.* fibrous

bacteria *f.* bacteria

bacterial *a.* bacterial

bacular *a.* rod-shaped

báculo *m.* [Méx.] transit rod, Jacob's staff

bacha *f.* barge

bache *m.* rut, pothole

badén *m.* gully

bagazo *m.* drilling mud

bahía *f.* bay

baja *f.* drop, fall || ***baja frequencia*** *f.* low frequency || ***baja ley*** *a.* low grade || ***baja presión*** *f.* low pressure

bajada *f.* descent, slope, drop, ebb, bahada || ***bajada pluvial*** *f.* downspout, leader

bajamar *m.* low tide, ebb tide, low water mark

bajamarea *f.* lowstand

bajante *m.* low water, low tide, leader, downspout

bajar(se) *vb.* to descend, sink, lower, drop, subside; drawdown

bajial *m.* low ground

bajío *m.* bank, shoal, sandbar, lowland, [U.K.] swale

bajo *p.* low || ***bajo cero*** *a.* subzero || ***de bajo ley*** *a.* low grade || ***bajo el agua*** *a.* subaqueous || ***bajo el nivel del mar*** *a.* subsea || ***bajo estructural*** *m.* structural low || ***bajo grado*** *a.* low rank

bajo *m.* footwall; shoal

Bajociano, Bajocense, Bajociense *m. & a.* Bajocian

bajoescurrimiento *m.* underthrust

bakelita *f.* bakelite

balaje *m.* balas ruby, red spinel

balance *m.* balance; fluctuation, vibraton || ***balance de áreas*** *m.* area balancing || ***balance de energía*** *m.* energy balance || ***balance de la longitud de líneas estratales*** *m.* bed-length balancing || ***balance de materia*** *m.* material (mass) balance

balanceado *a.* balanced

balancear *vb.* to balance

balanceo por área *m.* area balance || ***balanceo por línea (ó longitud)*** *m.* line length balance

balancín *m.* working beam, balance beam, rocker arm, walking beam || ***balancín de la brújula*** *m.* gimbal || ***balancín de válvulas*** *m.* rocker arm

balanza *f.* balance, scale || ***balanza de torción*** *f.* torsion balance

balasto, balastro, balaste *m.* ballast

balate *m.* survey marker, boundary marker, terrace

balde *m.* bucket, pail || ***balde de arrastre*** *m.* dragline bucket || ***balde de extracción*** *m.* mine bucket

baldequin *m.* canopy || ***baldequin de sal*** *m.* salt canopy

balicero *m.* [Méx.] rodman

baliza *f.* buoy; beacon, marker, pole

balneario *m.* & *a.* spa; thermal, medicinal

balón *m.* flask, baloon flask

balsa *f.* pool

bálsamo *m.* balsam

baluarte *m.* bulwark

bambolear *vb.* wobble

bamboleo *m.* swinging, swaying, rolling, wobbling

banca *f.* bench

bancada *f.* stope, bench

bancal *m.* berm, bench, terrace

banco *m.* riverbank, mine bench, thick bed, stratum; shelf || ***banco barrera*** *m.* barrier beach || ***banco competente*** *m.* competent bed || ***banco de arena*** *m.* sandbank, sandbar || ***banco de avíos*** *m.* bank lending funds to mines || ***banco de barrera*** *m.* barrier bar || ***banco de coral*** *m.* reef || ***banco de cota fija*** *m.* bench mark || ***banco de datos*** *m.* database || ***banco de derrumbe*** *m.* landslide bench || ***banco de grava*** *m.* gravel bank || ***banco de hielo*** *m.* ice floe || ***banco de referencia*** *m.* marker bed, marker horizon || ***banco guía*** *m.* key bed, marker bed || ***banco permeable*** *m.* pervious bed || ***banco petrolífera*** *m.* pay sand || ***banco platero*** *m.* [Méx.] argentiferous pyrite || ***banco productivo*** *m.* pay sand

bancos delgados *m.* thin beds || ***en bancos delgados*** *a.* thinly bedded

bancos gruesos *m.* thick beds || ***en bancos gruesos*** *a.* thickly bedded

bancos potentes *m.* thick beds; *a.* thickly bedded

bancos rojos *m.* redbeds

banda *f.* belt, band; (petrol.) streak || ***banda transportadora*** *f.* conveyor belt

bandado *a.* layered

bandeado *a.* banded, striped

bandeamiento *m.* banding

bandejón *m.* floe, ice floe

bandera *f.* (survey, levan.) flag

banquear *vb.* to level, [Col., Vz.] to excavate

bañado *m.* swamp, marsh

baño *m.* coating

bara *f.* bar (unidad de presión; unit of measure)

barbecho *a.* fallow

barcana *f.* barchan dune

barcaza *f.* barge || ***barcaza tiendetubos*** *f.* pipelaying barge

barco *m.* ship || ***barco perforador*** *m.* drill ship || ***barco petrolero*** *m.* tanker || ***barco tanquero*** *m.* tanker

barequear *vb.* [Col.] to extract placer ore without disturbing overburden

baratina *f.* barite

barchan *m.* barchan dune

bario *m.* barium

barisfera *f.* barysphere

barita *f.* barite

barján *m.* barchan dune

barlovento *a.* windward

barniz *m.* varnish || ***barniz del desierto*** *m.* desert varnish

baro *m.* marl

barógrafo *m.* barograph

barométrico *a.* barometric

barómetro *m.* barometer

barra *f.* bar, rod, ingot; sandbank || ***barra arenosa*** *f.* sandbar || ***barra aviada*** *f.* limited liability share || ***barra aviadora*** *f.* share paying assessments || ***barra costera*** *f.* beach bar || ***barra de bahía*** *f.* bay bar || ***barra de boca*** *f.* bay mouth bar || ***barra de bombeo*** *f.* sucker rod || ***barra de canal, de cauce*** *f.* channel bar || ***barra de delta*** *f.* delta bar || ***barra de desembocadura de bahía*** *f.* bay-mouth bar || ***barra de mina*** *f.* hand drill || ***barra de plata*** *f.* [Méx.] 75-lb silver ingot || ***barra de río*** *f.* shoal, river bar || ***barra de sondeo*** *f.* drill pipe || ***barra de suspensión*** *f.* beam hanger, rod hanger || ***barra en desembocadura*** *f.* bay-mouth bar || ***barra litoral*** *f.* sandbar, beach bar || ***barra maestra*** *f.* drill stem, sinker bar, auger stem || ***barra prelitoral*** *f.* longshore bar || ***barra submarina*** *f.* submarine bar

barranca, barranco *f.* & *m.* ravine, gorge, gully, cliff, bluff, gap, [U.S.] gulch, wash, arroyo, [Arg.] riverbank || ***barranca de erosión marina*** *f.* wave-cut cliff || ***barranca de falla*** *f.* fault scarp || ***barranca de falla con erosión*** *f.* fault line scarp

barranco *m.* gorge, fissure, cliff, landslide scar, [Vz.] small hole to work shallow ore

barrancón *m.* gorge, canyon

barrancoso *a.* broken, rough terrain

barranquero *m.* placer miner

barranquilla *f.* [Méx.] gully

barreal *m.* clay pit, mud flat, bog, [U.S.] playa, [U.K.] fen

Barremiano, Barremense, Barremiense *m. & a.* Barremian

barrena *f.* hand drill, drill, drill bit, crowbar, auger, (perf.) bore (véase taladro, trepano) || ***barrena de cable*** *f.* cable drill || ***barrena de cateo*** *f.* earth borer || ***barrena de chorro*** *f.* jet bit || ***barrena de discos*** *f.* disk bit || ***barrena de mano*** *f.* hand drill || ***barrena espiral*** *f.* auger bit || ***barrena giratoria de disco*** *f.* rotary disk bit || ***barrena para hoyos de poste*** *f.* post-hole auger || ***barrena percutente*** *f.* cable drilling bit || ***barrena principiadora*** *f.* spudder || ***barrena rotatoria*** *f.* rotary bit, rotary drill || ***barrena sacamuestras*** *f.* core drill || ***barrena sacanúcleos*** *f.* core drill || ***barrena sacatestigos*** *f.* core drill || ***barrena tubular*** *f.* core drill

barrenar *vb.* to drill; to fire a round of holes, (perf.) to bore

barreno *m.* shot, drill, drill hole, (perf.) bore || ***barreno de fondo de pozo*** *m.* rathole

barrera *f.* clay pit; barrier, dam || ***barrera arrecifal*** *f.* reef buildup, barrier reef || ***barrera litoral*** *f.* barrier bar, bay mouth bar || ***barrera prelitoral*** *f.* longshore bar

barrero *m.* clay deposit, clay pit

barretero *m.* miner at the working face

barretón *m.* heavy crowbar, [Col.] miner's pick

barrial *m.* clay pit

barrido *m.* scanner, sweeper, scanning, sweep || ***barrido de gases de combustión*** *m.* gas scavenging || ***barrido lateral*** *m.* sideswipe

barril *m.* barrel (158,984 liters) || ***barriles de petróleo equivalente*** *m.* barrels of oil equivalent (BOE) || ***barriles de petróleo por día*** *m.* barrels of oil per day (BOPD)

barrilla *f.* [Bol.] disseminated native copper grains; copper or tin ore concentrate

barrizal, barrial *m.* bog, [U.K.] mire, fen

barro *m.* clay, loam, mud, earth, silt, marl; [Col.] placer, overburden || ***barro de olleros*** *m.* potter's clay || ***barro de perforación*** *m.* drilling mud || ***barro glaciariaco*** *m.* till || ***barro refractorio*** *m.* [Cu.] fire clay

barroso *a.* muddy, clayey, silty

bartolina *f.* mine entrance

Bartoniano, Bartoniense *m. & a.* Bartonian

basal *m. & a.* basement; basal

basáltico *a.* basaltic || ***meseta basáltico*** *f.* basalt plateau

basalto *m.* basalt || ***basalto alcalino*** *m.* alkali basalt || ***basalto prismático*** *m.* columnar basalt

basamento *m.* basement

basanita *f.* basanite

báscula *f.* scale, balance

basculamiento *m.* tilting

base *m.* base, bottom; basis || ***base de erosión*** *m.* base level || ***base de licitación*** *m.* bidding information || ***base de triangulación*** *m.* baseline || ***base del tren de olas*** *m.* wave base || ***base geográfica relacional*** *m.* geographic relational database || ***base topográfico*** *m.* baseline || ***base vidrioso del pórfido*** *m.* groundmass, matrix

bascularse *vb.* to incline, dip, plunge

básico *a.* basic

basitas *f.* [Arg.] basic rocks

bastidor *m.* frame || ***bastidor de corredera*** *m.* skid frame

basto *a.* coarse

basura *f.* dirt, trash, waste, garbage, (min.) muck

batea *f.* wooden gold pan, trough; tray || ***batea de aceite*** *f.* oil pan; oil sump

bateado *m.* panning

bataque *m.* [Méx.] sinter

batería *f.* battery || ***batería de tanques*** *f.* tank battery

batial *a.* bathyal

batidor *m.* beater, shaker

batiente de llanura *m.* outwash apron

batimetría *f.* bathymetry

batimétrico *a.* bathymetric

batipelágico *a.* bathypelagic

batir *vb.* to churn, to mix; [Col.] to work placers

batisfera *f.* bathysphere

batolito *m.* batholith

Batoniano, Batonense, Batoniense *m. & a.* Bathonian

bauleo *m.* [Col.] pyrite cubes

bauxita *f.* bauxite

Bayociano, Bayociense *m. & a.* Bajocian

bazofía *f.* [Pe.] waste rock

benceno *m.* benzene

bencina *f.* benzene

beneficiable *a.* millable, workable

beneficiado *a.* processed

beneficiador *m.* miller, mill operator

beneficiar *vb.* to smelt, refine, work a mine, extract metals from ore or concentrate, to profit from working a mine

beneficio *m.* working of a mine, profits derived from working a mine, smelting, productive ore || ***beneficio de cazo*** *m.* [Méx.] hot amalgamation process of reducing silver ore || ***beneficio de patio*** *m.* cold amalgamation process of silver ore || ***beneficio por cianuración*** *m.* cyanide processing || ***beneficio por cloruración*** *m.* chlorine process || ***beneficio por fuego*** *m.* direct smelting of ore

beneficios *m.* benefits

bencina *f.* gasoline, benzene

benitoita *f.* benitoite

bentónico *a.* benthic, benthonic

bentoníta *f.* bentonite

bentonítico *a.* bentonitic

bentos *m.* benthos

benzol *m.* benzene

bergschrund *m.* bergschrund

berilio *m.* beryllium

berilo, berilio *m.* beryl || ***berilo verdemar*** *m.* aquamarine

berma *f.* berm

bermellón *m.* cinnabar

Berriasiano, Berriasiense *m. & a.* Berriasian

berroqueña *f.* granite

bertierita *f.* berthierite

betumen *m.* bitumen

betún *m.* bitumen, asphalt, pitch

biaxial, biáxico *a.* biaxial

bibliografía *f.* bibliography

bicarbonato *m.* bicarbonate

bicloruro *m.* bichloride

bidón *m.* jerry can; beaker

bienes *m.* assets, goods, benefits, profits || ***bienes de capital*** *m.* capital assets || ***bienes inmuebles*** *m.* real estate || ***bienes raíces*** *m.* real estate

bifurcación *f.* fork, branch of a river or vein, splay

bifurcado *a.* forked, bifurcated, split, branched

bifurcarse *vb.* to branch, fork

billete *m.* banknote

billón (o mil millónes) de pies cúbicos *m.* BCF

bimodal *a.* bimodal

binario *m.* binary || ***binario dígito*** *m.* (comp.) bit

binche *m.* [Arg.] pyrite crystals in gold conglomerates

binoculares *a.* binoculars

binóculo *m.* binoculars, field glasses

bioclástico *a.* bioclastic

bioclasto *m.* bioclast

bioestratigrafía *f.* biostratigraphy

bioestratigráfico *a.* biostratigraphic

biofacies *f.* biofacies

biogenesis *f.* biogenesis

biogenético *a.* biogenic

bioherma *f.* bioherm

biohermal *a.* biohermal

biolita *f.* biolith

biolitita *f.* boundstone

biolito *m.* organic sediment

biomarcador *m.* biomarker

bioperturbado *a.* bioturbated, burrowed

bioperturbar *vb.* to bioturbate

bioquímico *a.* biochemical

bioremediación *f.* bioremediation

biósfera *f.* biosphere

biostroma *f.* biostrome

biostromal *a.* biostromal

biota *f.* biota

biótico *a.* biotic

biotita *f.* biotite

bioturbado *a.* bioturbated, burrowed

bioturbar *vb.* to bioturbate

bióxido *m.* dioxide || ***bióxido carbónico*** *m.* carbon dioxide || ***bióxido de carbono*** *m.* carbon dioxide

bipolo *m.* dipole

birrefringencia *f.* birefríngence

birrefringente *a.* birefringent

bisagra *f.* hinge, hinge line

bisecar *vb.* to bisect

bisector *m.* bisector

bisectriz *f.* bisector, bisectrix

biselación *f.* bevelling

biselado *a.* bevelled

biselamiento *m.* bevelling, truncation

biselar *vb.* to bevel

bismalito *m.* bysmalith

bismutina, bismutinita *f.* bismuthenite

bismutita *f.* bismuthite

bismuto *m.* bismuth || ***bismuto blenda*** *f.* eulytite || ***bismuto ocre*** *f.* bismuth oxide or carbonate

bisulfato *m.* bisulfate

bisulfuro *m.* disulfide

bitounita *f.* bytownite

bitumen *m.* bitumen

bitumenífero, bituminoso *a.* bituminous

bitumenización *f.* bituminization

bivalente *a.* bivalent

bivalvo *m.* pelecypod

bixbyita *f.* bixbyite

blanco y negro *a.* salt and pepper

blando *a.* soft

blandura *a.* softness; *f.* soft earth

blanquear *vb.* to bleach

blanqueo *m.* bleaching

blanquizal *m.* pipe clay

blastogranítico *a.* blastogranitic

blastoideo *a.* blastoid

blastoporfirítico *m.* blastoporphyritic

blenda *f.* blende, sphalerite || ***blenda arsenical*** *f.* orpiment || ***blenda cadmífera*** *f.* greenockite || ***blenda córnea [Méx.]*** *f.* hornblende || ***blenda de plata*** *f.* pyrargyrite || ***blenda de zinc*** *f.* sphalerite

blocoso *a.* blocky

bloque *m.* block || ***bloque alto*** *m.* hanging wall || ***bloque bajo*** *m.* footwall || ***bloque cabalgante*** *m.* thrust sheet || ***bloque de desenganche*** *m.* knock-off block || ***bloque de falla*** *m.* fault block || ***bloque diagrama*** *m.* block diagram || ***bloque errático*** *m.* glacial erratic, erratic boulder || ***bloque hundido*** *m.* footwall || ***bloque inclinado*** *m.* tilted block || ***bloque levantado*** *m.* hanging wall || ***bloque volcado*** *m.* tilted block || ***bloque yacente*** *m.* footwall

bloquear *vb.* to block, trap; (min.) to block out ore

bloqueo *m.* blockade, blockage

bloqui *m.* ingot

bobina *f.* (elect.) coil

bobo *a.* [Ch.] sterile, barren of ore; dump

boca *f.* inlet of a river; mouth or collar of a shaft, well, mine, or adit; gap, pass, or saddle || ***boca de barrena*** *f.* drill bit || ***boca de glaciar*** *f.* glacial outlet, ice cave || ***boca de pique*** *f.* collar || ***boca de pozo*** *f.* wellhead || ***boca del río*** *f.* mouth of the river, estuary || ***boca mina*** *f.* adit, portal

bocal *m.* mouth, entrance; coal mine pit head

bocamejora *f.* auxilliary shaft

bocamina *f.* mine shaft collar

bocana *f.* [Col.] river mouth || ***bocana de flujo de mareas*** *f.* tidal inlet

bocarrena *f.* geode

bocarte *m.* stamp mill, crusher, drag mill

bocartear *vb.* to crush ore

bodega *f.* chamber deposit; warehouse; storeroom

bodoque *m.* argentite, [Méx.] silver chloride

bogue *m.* [Sp.] boghead coal

boleo *m.* vein ore nodule, waste rock dump

boliche *m.* bowl concentrator for patio process of silver amalgamation

bólido *m.* meteorite

bolita *f.* pellet ‖ ***bolita fecal*** *f.* fecal pellet, castings

bolitas de cojinete *f.* ball bearings

bolón *m.* cobble, boulder

bolsa *f.* pocket of ore; stock exchange ‖ ***bolsa de valores*** *f.* stock market

bolsada, bolsilla *f.* richest part of a mine, rich ore pocket, vug lined with druse

bolsón *m.* internal basin, [U.S.] bolson; ore pocket

Bollandiano, Bollandiense *m. & a.* Bollandian

bollo *m.* [Pe.] ore pocket

bomba *f.* pressure bomb, pump ‖ ***bomba alternativa*** *f.* reciprocating pump ‖ ***bomba anclada en la entubación*** *f.* casing pump ‖ ***bomba aspirante*** *f.* lifter pump, suction pump ‖ ***bomba centrífuga*** *f.* centrifugal pump ‖ ***bomba de arena*** *f.* sand pump ‖ ***bomba de balancín*** *f.* beam pump ‖ ***bomba de circulación*** *f.* circulating pump ‖ ***bomba de émbolo*** *f.* reciprocating pump ‖ ***bomba de extracción*** *f.* lifting pump ‖ ***bomba de inyección*** *f.* mud pump, slush pump ‖ ***bomba del fondo de pozo*** *f.* bottomhole pump ‖ ***bomba de vacío*** *f.* vacuum pump ‖ ***bomba eléctrica*** *f.* electric pump ‖ ***bomba eléctrica sumergible*** *f.* submersible electric pump ‖ ***bomba para lodos*** *f.* mud pump ‖ ***bomba recolectora*** *f.* sump pump ‖ ***bomba reforzadora*** *f.* booster pump ‖ ***bomba turbina*** *f.* turbine pump ‖ ***bomba volcánica*** *f.* volcanic bomb

bombeador *m.* pumper

bombear *vb.* to pump

bombeo *m.* pumping

bonanza *f.* ore chute, rich deposit

bonificación *f.* cash bonus

bono *m.* (com.) bonus; bond ‖ ***bono de caja*** *m.* cash bonus

Bononiano, Bononiense *m. & a.* Bononian

boquete *m.* pass, gap, opening, hole

boracita *f.* boracite

boratera *f.* [Arg., Bol., Ch.] borax deposit

bórax *m.* borax

borato *m.* borate; [Ch.] ulexite

borde *m.* rib, edge, rim, fringe, border; riverbank, levee, dike ‖ ***borde continental*** *m.* shelf break, continental borderland ‖ ***borde de ataque*** *m.* leading edge ‖ ***borde posterior*** *m.* trailing edge

bórico *a.* boric

bornita *f.* bornite, erubescite

boro *m.* boron

borra *f.* [Méx.] matrix, gangue, barren rock ‖ ***borra de veta*** *f.* soft, altered rock

borrador *m.* first draft, rough copy, preliminary sketch

borrasca *f.* unprofitable mine, barren rock ‖ ***borrasca de la capa*** *f.* tonguing of beds into adjacent beds, intertonguing

boruro *m.* boride

bosque *m.* forest, woods ‖ ***bosque lluvioso*** *m.* rain forest

bosquecillo *m.* scrub, brush

bosquejar *vb.* to sketch, outline

bosquejo *m.* outline, sketch

bostonita *f.* bostonite

bota *f.* leather ore bag

botadero *m.* dump

botaguas *m.* water table

botalón *m.* [Vz.] bench mark; derrick boom

botamiento *m.* fault, dislocation

botar *vb.* to dump; to launch

bote *m.* boat ‖ ***bote salvavidas*** *m.* lifeboat

botella *f.* flask ‖ ***botella de ácido*** *f.* acid bottle

botiquín *m.* first aid kit

botón *m.* button, plug, knob

botrioidal, botrioide *a.* botryoidal

boudín *m.* boudin

boulangerita *f.* boulangerite

bournonita *f.* bournonite

bóveda *f.* cave, cavern, dome, cupola; vaulted stope, arch, warp, swell ‖ ***bóveda de sal*** *f.* salt dome

bovedón *m.* [Pe.] chamber

boya *f.* buoy; [Pe.] rich vein or ore pocket

boyante *a.* buoyant ‖ ***lo boyante*** *m. & f.* buoyancy

boyar *vb.* to buoy

Bradfordiano, Bradfordiense *m. & a.* Bradfordian

braquiópodo *m.* brachiopod

brasa *f.* red hot coal; [Sp.] Mercury ore reduction

braunita *f.* braunite

braza, brazada *f.* fathom (1.67 m)

brazaje *m.* water depth

brazo *m.* arm, branch, distributary || ***brazo de gobierno*** *m.* control arm || ***brazo de mar*** *m.* arm of the sea || ***brazo de palanca*** *m.* lever arm || ***brazo de un río*** *m.* distributary, arm, branch || ***brazo muerto*** *m.* oxbow lake || ***brazo volado*** *m.* cantilever arm

brea *f.* pitch, tar, asphalt || ***brea de alquitrán*** *f.* coal tar || ***brea de carbón*** *f.* coal tar || ***brea de hulla*** *f.* coal tar || ***brea mineral*** *f.* heavy oil, maltha, asphalt

breccia, brecha *f.* breccia, opening, break, gap, crevasse, breach; trail || ***breccia conchífera*** *f.* shell hash || ***breccia de escurrimiento*** *f.* flow breccia || ***breccia de falla*** *f.* fault breccia || ***breccia intrusiva*** *f.* intrusive breccia || ***breccia sedimentaria*** *f.* sedimentary breccia || ***breccia tectónica*** *f.* tectonic breccia || ***breccia volcánica*** *f.* volcanic breccia

brechación *f.* brecciation

brechiforme, brechoide, brechoso *a.* brecciated

brezal *m.* [U.K.] heath

brida *f.* lifter, flange

bridado *a.* flanged

brigada *f.* party, gang, crew, squad || ***brigada de campo*** *f.* field party

brillante *a.* brilliant, glossy

brillantez *f.* brillance of gems

brillo *m.* luster, brightness || ***brillo anacarado*** *m.* pearly luster || ***brillo ceroso*** *m.* waxy luster || ***brillo diamantino*** *m.* adamantine luster || ***brillo grasiento, graso, grasoso*** *m.* greasy luster || ***brillo mate*** *m.* dull luster || ***brillo metálico*** *metallic* luster || ***brillo nacarado*** *m.* pearly luster || ***brillo resinoso*** *m.* resinous luster || ***brillo vítreo*** *m.* glassy luster

briozoario *m.* bryozoan

brisa *f.* breeze

brizna *f.* chip

broca *f.* auger, [Col.] drill bit (véase barrena)

brocado *m.* [Pe.] low-grade ore

brocal *m.* collar of a shaft, adit entrance

brocantita *f.* brochantite

bromargira, bromargirita *f.* bromargyrite

bromo *m.* bromine

bromuro *m.* bromide

bronce *m.* bronze; pyrite; chalcopyrite

broncita *f.* bronzite

bronco *m.* greenhorn, uproven employee; [Méx.] loose rock

bronquear *vb.* to hammer or pry loose rock

bronzado *a.* containing copper and iron sulfides

brookita, broquita *f.* brookite

broza *f.* [Bol.] low-grade ore

broseo *m.* [Bol.] weathering/decomposition of ore

brotadero *m.* seep, spring, source

brotar *vb.* to outcrop; to seep, emerge, flow

brote *m.* [Cu.] outcrop

broza *f.* [Bol. Pe.] barren part of a vein

brucita *f.* brucite

brújula *f.* compass || ***brújula Brunton*** *f.* Brunton compass, pocket transit

bruma *f.* haze, mist, fog

bruto *a.* coarse, rough, crude, (com.) gross

bucle de realimentación *m.* feedback loop

buche *m.* [Col.] ore pocket

buchita *f.* buchite

(de) buena calidad *a.* high grade

bufa *f.* cliff

buje *m.* bushing, sleeve

bulón *m.* bolt, dowel pin, dowel

bulto *m.* package, parcel

bulla *f.* noise

buncenita *f.* bunsenite

buque *m.* ship || ***buque de perforación*** *m.* drill ship || ***buque tanques*** *m.* tanker

burbuja *f.* bubble

Burdigaliano, Burdigaliense *m. & a.* Burdigalian

burga *f.* hot spring

buro *m.* chalk, marl

burocracia *f.* bureaucracy

busca, búsqueda *f.* search, exploration

buscado *a.* explored

buscador *m.* prospector

buscar *vb.* to search, prospect

buscón, busconero *m.* prospector

búsqueda *f.* search

bustamita *f.* bustamite

butano *m.* butane

buteno *m.* butene

butilacetileno *m.* butylacetylene

butilbenceno *m.* butylbenzene

butileno *m.* butylene, butene

butilo *m. & a.* butyl

butlerita *f.* butlerite

buzamiento *m.* plunge, dip, inclination, pitch || ***buzamiento aparente*** *m.* apparent dip || ***buzamiento arriba*** *m.* updip || ***buzamiento axil*** *m.* axial plunge || ***buzamiento del eje*** *m.* axial plunge || ***buzamiento empinado*** *m.* steep dip || ***buzamiento inverso*** *m.* reverse dip || ***buzamiento invertido*** *m.* overturned dip || ***buzamiento regional*** *m.* regional dip

buzando, buzante *a.* plunging

buzar *vb.* to plunge, dip

buzo *m.* diver

buzón *m.* ore chute, ore pass

byte *m.* byte

C

cabalgar *vb.* to thrust, overlap

cabalgador *m.* nappe

cabalgadura, cabalgamiento *f.* thrust fault, overthrust || ***cabalgadura oblicua*** *f.* oblique thrust

caballete, caballo *m.* horse; waste rock || ***caballo de bombeo*** *m.* pump jack || ***caballo portapoleas*** *m.* crown block

caballero *m.* spoil bank

caballo, caballo de piedra *m.* (geol.) horse

caballón *m.* ridge

caballos de fuerza (cdf) *m.* horsepower (hp)

cabecera *f.* headwaters of a river; hill summit; header; [Bol.] working in upper part of a mine

cabeza *f.* [Méx.] outcrop; [Col.] upper part of a placer || ***cabeza de pozo*** *f.* well head || ***cabeza golpeadora*** *f.* drive head || ***cabeza para cañería de bombeo*** *f.* tubing head || ***cabeza terminal de entubación*** *f.* casing head

cabezal *m.* header

cabezo *m.* reef, submerged rock

cabezón *m.* rapids, boulder, eddy, whirlpool

cabezuela *f.* gold and silver ore concentrate

cable *m.* cable, rope, line, hawser || ***cable de alambre*** *m.* wire line || ***cable de arrastre*** *m.* dragline || ***cable de cuchareo*** *m.* sand line, bailing line || ***cable de explosión*** *m.* shot line || ***cable de perforación*** *m.* drilling cable || ***cable tieso*** *m.* taut line

cablevía *f.* tramway, cableway

cabo *m.* cape, headland

cabrestante *m.* drawworks, winch, capstan

cabrio *m.* chevron

cabuchón, cabujón *m.* rough, unpolished ruby

cachalonga *f.* variety of opal

cachí *m.* [Bol.] barite, siderite; [Pe.] salt; [Ch.] calcite

cachimba *f.* [Arg.] shallow well, spring; [Méx.] miner's lamp

cachuela *f.* rapids

cadena *f.* chain, mountain belt || ***cadena de agrimensor*** *f.* surveyor's chain || ***cadena de montañas*** *f.* mountain belt || ***cadena volcánica*** *f.* volcanic belt || ***reacción en cadena*** *f.* chain reaction

cadenada *f.* (survey, levan.) length of one chain

cadenear *vb.* to chain (survey, levan.)

cadmio *m.* cadmium

caducación *f.* expiration of mining claim

caducar *vb.* to expire; (min.) forfeit a mining claim

caducado *a.* expired, invalid

Caenozóico, Cainozóico, Cenozóico *m. & a.* Cenozoic

caer *vb.* (comp.) to crash

cagafierro *m.* slag, scoria

caída *f.* fall, drop, downdropped side of a fault, dip, (hydr.) head || ***caída de agua*** *f.* waterfall || ***caída de detritos*** *f.* debris fall || ***caída de hielo*** *f.* ice fall || ***caída de roca*** *f.* rock fall

caídos *m.* fallen or caved ground

caíman *m.* [Méx:] ore chute

caja *f.* wall rock to a vein or deposit; [Ch.] riverbed || ***caja de la esclusa*** *f.* sluice box || ***caja de extracción*** *f.* mine cage

cajón *m.* canyon, gorge, gully; narrow or box canyon

cal *f.* lime || ***cal apagada*** *f.* slaked lime || ***cal de estaño*** *f.* cassiterite || ***cal viva*** *f.* quicklime

cala *f.* bay, cove, creek, bight; [Sp.] prospect pit || ***cala de prueba*** *test* bore, exploration well

Calabrés, Calbriano, Calabriense *m. & a.* Calabrian

calado *m.* openwork

calamaco *m.* large rock difficult to break

calamina *f.* calamine, smithsonite; [Pe.] hydrozincite

calamita *f.* magnetite, lodestone, siderite

calar *a.* calcareous, limy; *m.* limestone quarry

calaverita *f.* calaverite

calcalcalino *a.* calc-alkaline

calcantita *f.* chalcanthite

calcarenita *f.* calcarenite

calcáreo *a.* calcareous, limy; *m.* limestone

calcedonia *f.* chalcedony

cálcico *a.* calcic, calcium

calciesponja *f.* calcisponge

calcífero *a.* calcareous

calcificación *f.* calcification

calcificar *vb.* to calcify

calcilodita *f.* mudstone

calcilutita *f.* calcilutite

calcinación *f.* calcination

calcinador *m.* calciner, roaster

calcinar *vb.* to roast, calcine

calcio *m.* calcium

calcipelita *f.* calcilutite

calcita *f.* calcite

calcítico *a.* calcitic, calcic

calco *m.* cast, tracing || ***calco de carga*** *m.* load cast || ***calco de corriente*** *m.* flow cast || ***calco de flujo*** *m.* flow cast || ***calco de roce*** *m.* bounce cast || ***calco de surco*** *m.* groove cast || ***calco interno*** *m.* internal cast

calcocita *f.* chalcocite

calcopirita *f.* chalcopyrite

calcosina, calcosita *f.* chalcocite

calcotriquita *f.* chalcotrichite

calculación *f.* calculation

calcular *vb.* to calculate, compute

calculista *f.* computer

cálculo *m.* calculation, estimate

calda *f.* heating

caldas *f.* hot springs

caldear *vb.* to heat, weld

caldera *m.* caldera, volcanic crater; boiler, cauldron, kettle

calderilla *f.* winze

caldero *m.* cauldron

Caledoniano, Caledoniense *m. & a.* Caledonian

caledonita *f.* caledonite

calefacción *f.* heating, heat

calefaccionar *vb.* to heat

calentado *a.* heated

calentador *m.* [Sp.] furnace used to calcine ore

calentamiento *m.* heating

calentar *vb.* to heat

calera *f.* lime (cement) pit, quarry, or factory; lime kiln

calero *a.* calcareous, limestone

calero *m.* lime burner, pit operator

calesa, caleza *f.* [Méx.] ore bucket

caleta *f.* creek, cove, inlet

calibita *f.* chalybite, siderite

calibración *f.* calibration; sorting

calibrado *a.* calibrated; [Col.] sorted

calibrador *m.* gauge, calipers

calibrar *vb.* to calibrate

calibre *m.* caliper, gauge

calicata *f.* [Arg.] test pit, bore, drill hole

calichal *m.* [Bol.] kaolin; [Méx.] silver ore

caliche *m.* saltpeter, [U.S.] caliche, hardpan

calichoso *a.* calcareous

calidad *f.* quality

cálido *a.* hot, warm

caliente *a.* hot, warm

calificar *vb.* to grade, classify

californita *f.* californite

calina *f.* haze

caliza *f.* carbonate rock, limestone || ***caliza cristalina*** *f.* crystalline limestone || ***caliza granular*** *f.* grainstone || ***caliza lodo-granular*** *f.* packstone || ***caliza lodosa*** *f.* wackestone || ***caliza oolítica*** *f.* oolitic limestone

calizo *a.* calcareous, limy

calmas ecuatoriales *f.* doldrums

calocar *m.* white clay

calor *m.* heat || ***calor de fusión*** *m.* heat of fusion || ***calor de reacción*** *m.* heat of reaction || ***calor específico*** *m.* specific heat || ***calor latente de fusión*** *m.* latent heat of fusion || ***calor perdido*** *m.* waste heat || ***calor producido*** *m.* produced heat || ***calor propio*** *m.* geothermal temperature || ***calor que sale*** *m.* heat output

caloría *f.* calorie

calorífero, calorífico *a.* heat-producing

calorimetría *f.* calorimetry

calorímetro *m.* heat probe, calorimeter

calota *f.* cap || ***calota de hielo*** *f.* ice cap

Caloviano, Caloviense *m. & a.* Callovian

caluma *f.* pass, gap, gorge

calva *f.* glacial calving

calvo *a.* barren, bald

calzo, calza, calce *m.* wedge; shim

callana *f.* [Pe.] government refinery or mint

cama, camada *f.* layer, bed, stratum || ***cama, camada de roca*** *f.* rock shelf

cámara *f.* chamber, room, ore shoot || ***cámara magmática*** *f.* magma chamber || ***cámara fotográfica*** *f.* camera || ***cámara fotográfica aérea*** *f.* aerial camera || ***cámara y pilar*** *a.* room and pillar

cambiar *vb.* to change

cambio *m.* change, exchange || ***cambio de facies*** *m.* facies change || ***cambio de la delta*** *m.* delta switching

Cambriano, Cámbrico *m. & a.* Cambrian

caminamiento *m.* [Ec.] traversing

camino *m.* [Bol.] gallery, drift

caminos de migración *m.* migration paths

camión *m.* truck || ***camión de registro*** *m.* recording truck || ***camión de taladro*** *m.* drilling truck || ***camión tanque*** *m.* tank truck, tanker

camisa *f.* mold, jacket, case, liner, lining || ***camisa de agua*** *f.* water jacket

campamento *m.* camp || ***campamento minero*** *m.* mining camp || ***campamento petrolero*** *m.* oil camp

campanario *m.* rugged ridge

campaña *f.* level country; campaign, expedition, survey, work trip

Campaniano, Campaniense *m. & a.* Campanian

campanil *m.* [Sp.] siderite, hematite

campero *m.* foreman

campilita *f.* campylite

campista *f.* lessor

campo *m.* field; small mining lease; countryside || ***campo aurífero*** *m.* gold field || ***campo de gravedad*** *m.* gravitational field || ***campo de hielo*** *m.* ice field, pack ice || ***campo de kames*** *m.* kame field || ***campo de observación*** *m.* field of view || ***campo de un microscopio*** *m.* field of view || ***campo de pastoreo*** *m.* grazing field || ***campo del petróleo*** *m.* oil field || ***campo electromagnético*** *m.* electromagnetic field || ***campo magnético*** *m.* magnetic field || ***campo minero*** *m.* mining camp || ***campo petrolero*** *m.* oil field || ***campo petrolífero*** *m.* oil field || ***campo raso*** *m.* flat, open country

camptonita *f.* camptonite

Canadiano, Canadiense *m. & a.* Canadian

canal *m.* trough, trench, canal, channel, flume; (elect.) channel || ***canal de derivación*** *m.* diversion canal || ***canal de desagüe*** *m.* drainage ditch || ***canal de ingreso de marea*** *m.* tidal inlet || ***canal de marea*** *m.* tidal channel || ***muestras de canal*** *f.* (perf.) cuttings

canaleta *f.* trough, channel, flume, ditch

canalización *f.* channeling

canalizar *vb.* to channel

canalón *m.* ground sluice

cancrinita *f.* cancrinite

cancha *f.* ore bin; [Pe., Ch.] mine dump

canchal *m.* rocky place, moraine

cancho *m.* boulder

candencia *f.* incandescence

candente *a.* incandescent

candil de minero *m.* miner's lamp

canel *m.* [Sp.] cannel coal

caney *m.* small bay

canga *f.* red lateritic soil

cangagua *f.* [Col.] adobe clay

canahua *f.* [Ec.] hardpan

cangalla *f.* tailings

cangilón *m.* bucket; [Ec.] mud hole; [PR] small waterfall

cangrejo *m.* crab; spear || ***cangrejo de circulación*** *m.* wash down spear

canguilón *m.* [Ch.] bucket

canón *m.* (com., min.) royalty, tax || ***canón bruto*** *m.* gross royalty || ***canón de adición*** *m.* overriding royalty || ***canón del propietario*** *m.* fee royalty || ***canón de participación*** *m.* royalty interest || ***canón exento de cargos*** *m.* nonparticipating royalty || ***canón fijo*** *m.* fixed royalty

cantal *m.* stone, pebble, spall

cantalita *f.* pitchstone, cantalite

cántaro *m.* [Pe.] retort for reducing Mercury ore

canteamiento *m.* tilting

cantear *vb.* to cut stone

cantera, cantería *f.* quarry, pit; building stone

cantero *m.* stonemason

cantidad *f.* quantity, amount, slug || ***cantidad grande*** *a.* bulk || ***cantidad única*** *f.* single slug

cantil *m.* cliff, shelf

cantillo *m.* stone, pebble

cantizal *m.* stony ground

canto *m.* edge, border; stone, pebble, boulder || ***canto rodado*** *m.* boulder

cantorral *m.* stony ground

cantos rodados medianos *m.* cobbles

canturrón *m.* [Col.] manganese oxide

cana *f.* gallery || ***caño*** *m.* gallery

cañada *f.* valley, gorge, ravine, [U.K.] dell

cañadón *m.* canyon, gorge

cañería *f.* pipe, piping, conduit (véase tubería) || ***cañería agujereada*** *f.* perforated pipe || ***cañería aisladora*** *f.* casing || ***cañería ascendente*** *f.* pipe riser || ***cañería colectora*** *f.* gathering line || ***cañería de captación*** *f.* gathering line || ***cañería de derivación*** *f.* bypass line || ***cañería de descarga*** *f.* discharge pipe || ***cañería de entubación*** *f.* casing || ***cañería de flujo ascendente*** *f.* pipe riser || ***cañería de revestimiento*** *f.* casing || ***cañería de surgencia*** *f.* flow string || ***cañería filtro*** *f.* perforated liner || ***cañería guía*** *f.* surface pipe, surface casing || ***cañería pérdida*** *f.* liner || ***cañería pérdida perforada*** *f.* perforated liner || ***cañería pérdida ranurada*** *f.* slotted liner || ***cañería perforada*** *f.* perforated pipe || ***cañería vertical de alimentación*** *f.* standpipe

cañista *m. & f.* pipe fitter, pipe layer

caño *m.* pipe, conduit; gutter; stream; gallery || ***caño de perforación*** *m.* drill pipe || ***caño de ventilación*** *m.* ventilation pipe || ***caño filtro*** *m.* perforated liner || ***caño punzonado*** *m.* perforated liner

cañón *m.* canyon, gorge; perforator gun || ***cañón de aire*** *m.* air gun || ***cañón perforador*** *m.* gun perforator || ***cañón submarino*** *m.* submarine canyon

cañutillo *m.* [Col.] fine emerald

caolín *m.* kaolin, China clay

caolinita *f.* kaolinite, kaolin

caolinización *f.* kaolinization

capa *f.* rock bed, layer, strata, seam; ply; vein || ***capa acuífera*** *f.* aquifer || ***capa arable*** *f.* topsoil || ***capa basal*** *f.* bottomset bed || ***capa clave*** *f.* key bed, marker bed || ***capa de almacenamiento*** *f.* reservoir bed || ***capa de carbón*** *f.* coal bed || ***capa de hielo*** *f.* ice cap || ***capa dorsal*** *f.* topset bed || ***capa dura*** *f.* hardpan || ***capa freática*** *f.* aquifer || ***capa frontal*** *f.* foreset bed || ***capa gasífera*** *f.* gas cap || ***capa guia*** *f.* key bed, marker bed || ***capa índice*** *f.* key bed, marker bed || ***capa inferior*** *f.* bottomset bed || ***capa intrusiva*** *f.* sill || ***capa invertida*** *f.* overturned bed || ***capa lenticular*** *f.* lens || ***capa mineralífera*** *f.* ore layer || ***capa petrolífera*** *f.* oil pay, oil horizon || ***capa por capa*** *a.* bed-on-bed || ***capa protectora*** *f.* caprock || ***capa roja*** *f.* redbed || ***capa superficial*** *f.* weathered layer || ***capa superior*** *f.* topset bed || ***capa volcada*** *f.* overturned bed

capacidad *f.* capacity, ability, capability || ***capacidad asignada*** *f.* rated capacity || ***capacidad calculada*** *f.* design capacity || ***capacidad calórica*** *f.* heat capacity || ***capacidad de almacenamiento*** *f.* (comp.) memory || ***capacidad de flujo*** *f.* flow capacity || ***capacidad de infiltración*** *f.* infiltration capacity || ***capacidad de producción*** *f.* production capacity || ***capacidad diaria*** *f.* daily capacity || ***capacidad eléctrica*** *f.* capacitance || ***capacidad estimada*** *f.* rated capacity || ***capacidad nominal*** *f.* rated capacity || ***capacidad normal*** *f.* rating || ***capacidad térmica*** *f.* heat capacity

capacitación *f.* training

capacitado *a.* qualified

capacitar *vb.* to train, educate

capacho *m.* bucket

caparazón *m.* carapace, shell

caparrosa *f.* melanterite

capataz *m.* foreman

capas *f.* bedding, layering, strata || ***capas basales*** *f.* bottomset beds || ***capas deltaicas inferiores*** *f.* bottomset beds || ***capas deltaicas superiores*** *f.* topset beds || ***capas dorsales*** *f.* topset beds || ***capas entrecruzadas*** *f.* crossbedding || ***capas frontales*** *f.* foreset beds

capaz *a.* large; adequate; capable, competent

capilar *a.* capillary

capilaridad *a.* capillarity

capital *m.* (com.) capital, investment moneys || ***capital activo*** *m.* working capital; capital assets || ***capital arriesgado*** *m.* venture capital || ***capital de inversión*** *m.* equity investment, investment capital || ***capital de riesgo*** *m.* venture capital || ***capital de trabajo*** *m.* working capital || ***capital físico*** *m.* capital assets || ***capital propio*** *m.* equity || ***capital propio de los accionistas*** *m.* shareholder equity

capitalizar *vb.* to capitalize on

capitán *m.* foreman || ***capitán de patio*** *m.* surface boss

caporal *m.* [Vz.] foreman

capote *m.* [Col.] topsoil

capotero *a.* shallow

cápsula *f.* blasting cap; (paleo.) test

captación, captura *f.* capture, piracy || ***captación del río*** *f.* river piracy || ***zona de captación*** *f.* catchment area

captor *m.* sensor

capturar *vb.* to seize

capturista *m.* & *f.* (comp.) data entry person

capuchón *m.* cap rock

cara *f.* face, crystal facet, fissure wall || ***cara de deslizamiento*** *f.* slip face || ***cara del crucero*** *f.* cleavage face

cárabe *m.* amber

carácter *m.* character, feature

característico *m.* & *a.* characteristic

caracol *m.* snail

caracol *a.* spiral

característica *f.* characteristic, property, trait

característico *a.* characteristic

caracterizar *vb.* characterize

Caradociano, Caradociense *m.* & *a.* Caradocian

carámbano *m.* icicle

carapacho *m.* carapace

carbón *m.* coal, carbon, charcoal, graphite || ***carbón aglomerante*** *m.* coking coal || ***carbón aglutinante*** *m.* coking coal || ***carbón antracita*** *f.* anthracite || ***carbón antracitoso*** *m.* anthracite || ***carbón bituminoso*** *m.* bituminous coal || ***carbón brillante*** *m.* bright coal || ***carbón cannel*** *m.* cannel coal || ***carbón de cok*** *m.* coking coal || ***carbón de piedra pasada*** *m.* coking coal || ***carbón graso*** *m.* coking coal || ***carbón lavado*** *m.* dressed coal || ***carbón mate*** *m.* dull coal || ***carbón oscuro*** *m.* dull coal || ***carbón pardo*** *m.* lignite || ***carbón semigraso*** *m.* semibituminous || ***carbón subbitminoso*** *m.* sub-bituminous coal

carbonáceo *a.* carbonaceous

carbonatado *a.* carbonate

carbonatita *f.* carbonatite

carbonato *m.* carbonate || ***carbonato azul de cobre*** *m.* azurite || ***carbonato de calcio*** *m.* calcium carbonate, limestone || ***carbonato de hierro*** *m.* siderite || ***carbonato de sodio*** *m.* soda, sodium carbonate || ***carbonato de zinc*** *m.* smithsonite || ***carbonato magnésico*** *m.* magnesite || ***carbonato plomizo*** *m.* cerussite || ***carbonato verde de cobre*** *m.* malachite

carbonera *f.* coal bin, coal bunker, coal mine

carbónico *a.* carbonic

Carbonífero *m. & a.* Carboniferous; coal-bearing

carbonización *f.* carbonization

carbono *m.* carbon || ***carbono 14*** *m.* carbon-14 || ***carbono orgánico total (COT)*** *m.* total organic carbon (TOC) || ***proporción de carbono*** *f.* carbon ratio

carbonoso *a.* carbonaceous, coaly

carborundo *m.* corundum

carburo *m.* carbide || ***carburo de calcio*** *m.* calcium carbide

cárcamo *m.* (min.) drain; flume; culvert

cárcava *f.* gully, chute, ravine, ditch || ***cárcava de derrubios*** *f.* scree slope || ***cárcava de nieve*** *f.* snow gully

carcavón *m.* gorge

carcomar *vb.* to erode, undermine

cardenillo *m.* ruby silver, copper carbonate

carear *vb.* to face stone

careta antigás *f.* gas mask

carga *f.* charge, load, loading, (hydr.) head; cargo || ***calco de carga*** *m.* load cast || ***carga de alimentación*** *f.* feedstock || ***carga de trabajo*** *f.* work load || ***carga de hidrocarburo*** *f.* hydrocarbon charge || ***carga de presión*** *f.* pressure head || ***carga del lecho*** *f.* bed load || ***carga diaria*** *f.* daily throughput, daily charge || ***carga disuelta*** *f.* dissolved load || ***carga explosiva*** *f.* explosive charge || ***carga hidrostática*** *f.* head, hydrostatic head || ***carga máxima*** *f.* peak load || ***carga suspendida*** *f.* suspended load

cargado *a.* loaded, heavy, dense, charged

cargador de película *m.* film cartridge, canister

cargamento *m.* load, shipment, cargo

cargar *vb.* to load; to charge; to carry

cargo *m.* burden, load

carguero *m.* cargo boat; pack animal

carhua *f.* [Pe.] pyrite

carnalita *f.* carnallite

carne de vaca *m.* [Bol., Pe.] galena

carneola *f.* chalcedony

Carniano, Carniense *m.* Carnian, Karnian

carnotita *f.* carnotite

carrascal *f.* honeycombed quartz

carrera *f.* course, trip || ***carrera de sacada y bajada*** *f.* round trip

carretilla *f.* wheelbarrow, hand truck

carro *m.* car, wagon, truck; mine car || ***carro cisterno*** *m.* tank car || ***carro de mano*** *m.* hand cart || ***carro de plataforma*** *m.* flat car || ***carro tanque*** *m.* tank car

carroñero *m.* scavenger

carst *m.* karst

carstenita *f.* karstenite, anhyrite

cárstico *a.* karstic

carstificación *f.* karstification

carta *f.* map, chart; letter, document || ***carta de crédito*** *f.* letter of credit || ***carta de crédito permanente*** *f.* revolving letter of credit || ***carta de pago*** *f.* receipt || ***carta de porte*** *f.* bill of lading, cover letter || ***carta náutica*** *f.* nautical chart

cartera *f.* portfolio

cartografía *f.* mapping || ***cartografía aéria*** *f.* aerial mapping

cartografiar *vb.* to survey, to make maps

cartográfico *a.* cartographic

cartógrafo *m.* cartographer

cartucho *m.* cartridge || ***cartucho de voladura*** *m.* blasting cartridge

casa de comisión *f.* brokerage

casar metales *vb.* to mix ore for smelting or amalgamation

casarón *m.* large stope

cascada *f.* waterfall, cascade

cascajal, cascajar *m.* gravel deposit, gravel pit

cascajero *m.* gravel bed, worked out mine

cascajo *m.* gravel deposit, quarry waste, gravel, rubble

cascajoso *a.* gravelly

cáscara *f.* mud cake

casco *m.* hard hat; shell, casing, cask

cascote *m.* rubble, waste, muck

caseta del cabrestante *f.* doghouse, drawworks shed

casi *adv.* almost, nearly || ***casi vertical*** *a.* almost vertical

casimba *f.* well, spring; [Cu.] sinkhole

casíterita *f.* cassiterite

caso *m.* occurrence, case || ***caso fortuito*** *m.* act of God, contingency

casquete *m.* cap, helmet; shell || ***casquete de gas*** *m.* gas cap

casquijo *m.* gravel, gravel bar

casquillo *m.* blasting cap

castaña *f.* abandoned mine

castigar *vb.* [Méx.] to dress or smooth stone

castillete *m.* headframe

castina *f.* limestone flux

castor *m.* petalite

cata *f.* shallow mine, placer mine, test pit, prospect pit, mine under litigation

cataclasis *f.* cataclasis

cataclasita *f.* cataclasite

cataclástico *a.* cataclastic

cataclismo *m.* cataclysm

catadora *f.* auger

catálisis *f.* catalysis

catalítico *a.* catalytic

catalizador *m.* catalyst

catalizar *vb.* to catalyze

catamorfismo *m.* katamorphism, catamorphism

catarata *f.* waterfall, cataract

catarroca *m. & f.* rock prospector

catastral *a.* cadastral

catastro *m.* land register, cadastre, inventory

catastrófico *a.* catastrophic

catastrofismo *m.* catastrophism

catazona *f.* katazone

cateado *a.* (min.) explored

cateador *m.* (min.) prospector, sampler

cateando *m.* (min.) prospecting

catear *vb.* to look for, search, prospect, explore

cateo *m.* prospecting, (min.) exploration, sampling; claim, prospect, lease

catión *f.* cation

catódico *a.* cathodic

cátodo *m.* cathode

catogénico, catógeno *a.* katogenic

cauce *m.* riverbed, channel, irrigation ditch

caucho *m.* foam rubber

caudal *m.* fortune, wealth; rate, discharge, volume

caudaloso *a.* in full flood

causa *f.* cause

cáustico *a.* caustic

cava *f.* digging, ditch, pit, excavation, rill

cavador *m.* digger, excavator

cavadora *f.* excavator || ***cavadora de zanjas*** *f.* trenching machine

cavadura *f.* digging

cavar *vb.* to dig, excavate

cavas *f.* rills

caverna *f.* cave, cavern, hollow

cavernoso *a.* cavernous, vuggy, vugular

cavidad *f.* cavity, vug, opening || ***cavidad milimétrica llena de calcita*** *f.* fenestrae

cayo *m.* cay, shoal, small island, key

caz *f.* flume, canal

cazeador *m.* amalgamator

cazo *m.* vessel for heating ore

cebar *vb.* to smelt ore

ceca *f.* royal mint

cedazo *m.* sieve, screen

cedencia *f.* yield

ceder *vb.* to yield

cefalópodo *m.* cephalopod

ceja *f.* ledge, summit

celda *f.* cell; bin

celestina, celestita *f.* celestite

célula *f.* cell

cellisca *f.* sleet

cementación *f.* cementation, setting casing || ***cementación de fondo*** *f.* plug back

cementado *a.* cemented

cementar *vb.* to cement

cemento *m.* cement || ***cemento armado*** *m.* reinforced concrete || ***cemento de fragüe lento*** *m.* slow-setting cement || ***cemento de fragüe rápido*** *m.* quick-setting cement

cenagal, [PR] cienegal *m.* swamp, bog

cenagoso *a.* swampy

cenefa *f.* rim

cenit *m.* zenith

cenital *a.* azimuthal

ceniza *f.* ash, slag, cinder

Cenomanense, Cenomaniano, Cenomaniense *m. & a.* Cenomanian

cenote *m.* [Méx.] sinkhole

Cenozóico, Caenozóico *m. & a.* Cenozoic

cenozona *f.* assemblage zone

centellador *m.* scintillometer

centenaja *f.* [Col.] pothole

centiárea *m.* square meter

centígrado *a.* centigrade

centila *f.* percentile

centígramo *m.* centigram

centilitro *m.* centiliter

centímetro *m.* centimeter

centinela *m. & f.* watchman, guard

central *f. & a.* headquarters; central || ***central de bombeos*** *f.* fire station || ***central de energía*** *f.* power plant, powerhouse || ***central eléctrica*** *f.* power plant || ***central hidroeléctrica*** *f.* power station

centrar *vb.* center, align

centrífugo *m. & a.* centrifuge; centrifugal

centro *m.* center || ***centro de curvatura*** *m.* center of curvature || ***centro de gravedad*** *m.* center of gravity

ceolita *f.* zeolite

cera *f.* wax

cerámica *f.* ceramic

cerámico *a.* ceramic

cerargirita *f.* cerargyrite

cercado de tierra *a.* landlocked

cercana a la playa *a.* nearshore

cercanía(s) *f.* neighborhood, environs, surroundings || ***cercanía(s) de costa*** *f.* foreshore

cercano *a.* near, nearby, proximal

céreo *a.* waxy

cerio *m.* cerium

cernedor, cernidor *m.* sieve, screen

cerner, cernir *vb.* to sift, to sieve

cerniduras *f.* screenings

cero *m.* zero || ***cero absoluto*** *m.* absolute zero

ceroso *a.* waxy

cerrado *a.* shut-in, closed

cerrar *vb.* to close, to shut-in || ***cerrar un levantamiento*** *vb.* to tie in a survey

cerrarse *vb.* to thin out, pinch out

cerril *a.* rough, uneven, hilly

cerrillada *f.* range of low hills

cerrillo *m.* hillock

cerro *m.* peak, summit, hill || ***cerro almiar*** *m.* hum, haystack hill || ***cerro pepino*** *m.* hum, haystack hill || ***cerro sepultado*** *m.* buried hill || ***cerro testigo*** *m.* butte, monadnock

certificación *f.* certification

certificado *m.* certificate || ***certificado de pesos y medidas*** *m.* certificate of weights and measures

cerusa, cerusita *f.* cerussite

cervantita *f.* cervantite

cesantías *f.* severance pay

cesio *m.* cesium

césped *m.* turf, grass

cesta de pesca *f.* drilling fishing basket

cetona *f.* ketone

cetrino *m.* cetrine

cianita *f.* kyanite

cianuración *f.* cianuration

cianuro *m.* cyanide

cibairo *m.* blue or green copper mineral

cicatriz *f.* scar

cíclico *a.* cyclic, cyclical

ciclo *m.* cycle || ***ciclo de erosión*** *m.* erosional cycle || ***ciclo de la vida*** *m.* life cycle || ***ciclo geomórfico*** *m.* geomorphic cycle || ***ciclo vital*** *m.* life cycle

ciclón *m.* cyclone

ciclopentano *m.* cyclopentane

ciclopenteno *m.* cyclopentene

ciclotema *f.* cyclothem

ciclotrón *m.* cyclotron

ciego *a.* blind

cielo *m.* sky; roof, ceiling || ***a cielo abierto*** *m.* open pit || ***trabajar de cielo*** *vb.* to overhead stope

ciénaga *f.* marsh, swamp, bog

ciencia *f.* science

ciencias naturales *f.* natural sciences

cieno *m.* mud, sludge, ooze

científico *m.* scientist

científico *a.* scientific

cierre *m.* shut-in, shutdown; closure; shutoff || ***cierre estructural*** *m.* structural closure || ***cierre contra la falla*** *m.* rollover || ***cierre de seguridad*** *m.* blowout preventer || ***cierre yacimiento*** *m.* closure

cigüairo, civairo *a.* [Pe.] peacock-colored

cilíndrico *a.* cylindrical, rod-shaped

cima *f.* summit, top, crest, height, peak; (seis.) pick

cimborrio *m.* dome

cimbrozano *m.* [Vz.] earthquake

cimentar *vb.* to ground, to lay a foundation

cimento *m.* cement

cimiento *m.* foundation

cimofana *f.* chrysoberyl, cymophane

cinabarita *f.* cinnabar

cinabrio *m.* cinnabar

cinc *m.* zinc

cincel *m.* chisel; [Pe.] massive native silver

cincelar *vb.* to chisel

cincífero *a.* zinc-bearing

cincita *f.* zincite

cinemático *a.* kinematic

cinerita *f.* ash flow

cinerítico *a.* cinder, ash

cinético *m. & a.* kinetic

cinética *f.* kinetics

cinta *f.* measuring tape, tape; (min.) ore streak || ***cinta de medir*** *f.* measuring tape || ***cinta magnética*** *f.* (comp.) magnetic tape || ***cinta métrica*** *f.* measuring tape

cinteada *f.* streak

cinteado *a.* banded, ribboned

cintilómetro *m.* scintillometer

cinturón *m.* belt || ***cinturón corrido*** *m.* [Arg.] thrust belt || ***cinturón de pliegues*** *m.* fold belt || ***cinturón de montañas*** *m.* mountain belt || ***cinturón de seguridad*** *m.* safety belt || ***cinturón de volcánicos*** *m.* volcanic belt || ***cinturón plegado*** *m.* fold belt

circo *m.* cirque

circón *m.* zircon

circonio *m.* zirconium

circuito *m.* circuit, hookup || ***circuito de explosión*** *m.* firing circuit || ***circuito eléctrico*** *m.* electrical circuit, wiring

circulación *f.* circulation || ***circulación de la inyección*** *f.* mud circulation

circulado *a.* circulated

circular *a.* circular

circular *vb.* to circulate

círculo *m.* circle || ***círculo polar ártico*** *m.* arctic circle || ***círculo polar antártico*** *m.* antarctic circle || ***gran círculo*** *m.* great circle

circunferencia *f.* circumference

circunvalación *f.* perimeter

cirro *m.* cirrus

cirrocúmulo *m.* cirrocumulus

cirrostrato *m.* cirrostratus

cisco *m.* broken coal, coal dust

cisura *f.* flaw

cisterna *f.* cistern, tank, reservoir

citar *vb.* to record, to cite

citrina *f.* citrine

cizalla *f.* shear || ***cizalla de Riedel*** *f.* Riedel shear || ***cizalla flexural*** *f.* flexural slip, layer-parallel shear || ***cizalla inclinada*** *f.* inclined shear || ***cizalla pura*** *f.* pure shear || ***cizalla simple*** *f.* simple shear || ***ondas de cizalla*** *f.* shear waves

cizallar *vb.* to shear

cizallamiento *m.* shearing

ckallar *vb.* to highgrade by selective mining

ckollpa *f.* [Bol.] trona

claraboya *f.* monitor

clarificador *m.* clarifier

clarificar *vb.* to clarify

clarita *f.* clarain

claro *a.* clear, light colored

clase *m.* class, grade, rank

clasificable *a.* classifiable

clasificación *f.* classification, sorting

clasificado *a.* classified

clasificador *m.* classifier, sorter

clasificadora *f.* sorter, classifier

clasificar *vb.* to classify, class, sort, grade

clástico *a.* clastic

clasto *m.* clast || ***clastos rasgados*** *m.* ripup clasts

claustalita *f.* clausthalite

cláusula *f.* clause

clavada *f.* winze

clavado *a.* [Col.] vertical

clavaque *m.* quartz pebbles

clavar *vb.* to drive, to set || ***clavar la entubación en el fondo*** *vb.* to set pipe on bottom

clave *a.* key; (legal) code

clave *f.* keystone; legal code

clavo *m.* (min.) ore, rich ore, pay streak, ore chute, spike

cleavelandita *f.* cleavelandite

cliachita *f.* bauxite

cliente *m.* client, customer

clima *m.* climate, weather

climático *a.* climatic

climatología *f.* climatology

clímax *m.* climax

clinoclasa *f.* clinoclase

clinocloro *m.* clinochlore

clinoenstatita *f.* clinoenstatite

clinoforma *f.* clinoform

clinómetro *m.* clinometer

clinopiroxenita *f.* clinopyroxenite

clinozoisita *f.* clinozoisite

clivaje *m.* cleavage || ***clivaje de crenulación*** *m.* crenulation cleavage || ***clivaje de disyunción*** *m.* fracture cleavage || ***clivaje de flujo*** *m.* flow cleavage || ***clivaje de fractura*** *m.* fracture cleavage || ***clivaje de lápiz*** *m.* pencil cleavage || ***clivaje de plano axial*** *m.* axial plane cleavage || ***clivaje de transposición*** *m.* slip cleavage || ***clivaje pizarreño*** *m.* slaty cleavage || ***clivaje pizarroso*** *m.* slaty cleavage

clizoglifo *m.* mud crack

cloaca *f.* sewer

cloantita *f.* cloanthite

cloración *f.* chlorination

clorato *m.* chlorate

clorhidrato *m.* hydrochloride

clorhídrico *a.* hydrochloric

clórico *a.* chloric

clorinación *f.* chlorination

clorita *f.* chlorite

clorítico *a.* chloritic

cloritización *f.* chloritization

cloritoide *m.* chloritoid

cloro *m.* chlorine

clorocalcita *f.* hydrophilite

cloruro *m.* chloride || ***cloruro de amonio*** *m.* ammonium chloride || ***cloruro de calcio*** *m.* calcium chloride || ***cloruro de polivinilo*** *m.* polyvinyl chloride, PVC || ***cloruro de sodio*** *m.* sodium chloride, salt || ***cloruro de vinilo*** *m.* vinyl chloride || ***cloruro de zinc*** *m.* zinc chloride

coagulación *f.* coagulation

coagulador *m.* coagulator

coagulante *m.* coagulant

coagular *vb.* to coagulate

coalescencia *f.* coalescence, fusion

coalescer(se) *vb.* to coalesce

cobáltico *m.* cobaltic

cobaltina, cobaltita *f.* cobaltite

cobalto *m.* cobalt

cobertura *f.* covering, cover, overburden

cobijadura *f.* alluvium; overthrust

Coblenciano, Coblenciense, Coblentziense *m. & a.* Coblenzian, Coblentzian, Coblencian

cobre *m.* copper || ***cobre abigarrado*** *m.* bornite || ***cobre amarillo*** *m.* copper pyrites || ***cobre azul*** *m.* azurite || ***cobre gris*** *m.* tetrahedrite || ***cobre nativo*** *m.* native copper || ***cobre roso*** *m.* red copper oxide || ***cobre verde*** *m.* malachite || ***cobre virgen*** *m.* native copper

cobrizo *a.* cupriferous, copper-bearing

cobro contra entrega *m. & a.* cash on delivery (COD)

cobros *m.* collections, income, revenue

cocarda *f.* rosette

cocción *f.* roasting

coceo *m.* [Méx.] thrust

cocer metal *vb.* to roast ore

cociente *m.* ratio

cocina *f.* kitchen

cocolita, cocolito *f.* coccolith

cocha *f.* [Bol., Ec.] lake, pond

cochada *f.* (ref.) batch, refinery run

cochano *m.* [Vz.] nugget

coche *m.* rock crusher

cochina *f.* rock crusher

cochiso, cochizo *m.* proustite, pyrargyrite

cochura *f.* roasting

codificar *vb.* to encode; to codify

código *m.* code of laws; code

codo *m.* bend, [U.S.] dog leg

coeficiente *m. & a.* coefficient || ***coeficiente de almacenaje del pozo*** *m.* wellbore storage coefficient || ***coeficiente de confianza*** *m.* confidence level || ***coeficiente de contracción*** *m.* contraction coefficient || ***coeficiente de correlación*** *m.* correlation coefficient || ***coeficiente de elasticidad*** *m.* elastic modulus || ***coeficiente de expansión*** *m.* coefficient of expansion || ***coeficiente de fricción*** *m.* coefficient of friction || ***coeficiente de productividad*** *m.* productivity index || ***coeficiente de reflexión*** *m.* reflection coefficient || ***coeficiente de rigidez*** *m.* stiffness coefficient || ***coeficiente de seguridad*** *m.* factor of safety

coexistente *a.* concurrent, coexistent

coffinita *f.* coffinite

cogemuestras *m.* sampler

coherencia *f.* coherence

cohesión *f.* cohesion

cohete *m.* fuse; blast

cok *m.* coke

col *m.* col

colación *f.* percolation

colada *f.* flow, melt; *a.* strained, filtered

coladero *m.* winze, ore chute, ore pass

colado *a.* leaked; *m.* cast

colador *m.* strainer, percolator || ***colador de lodo*** *m.* shale shaker, mud screen || ***colador vibratorio*** *m.* shale shaker

coladura *f.* straining, filtering, percolation

colagón *m.* [Méx.] conduit, canal

colar(se) *vb.* to seep, percolate; strain; melt

colas *f.* tailings

coleccionar *vb.* to collect

colector *m.* collector || ***colector de aire*** *m.* air trap || ***colector de arena*** *m.* sand trap || ***colector de barro*** *m.* mud trap || ***colector de gas*** *m.* gas trap || ***colector de inyeccion*** *m.* mud trap

colemanita *f.* colemanite

colero *m.* boss, foreman

colgado *a.* suspended, hung

colgante *a.* hanging, suspended

colicuación *f.* smelting

colillas *f.* tailings

colimación *f.* collimation

colimador *m.* collimator

colimar *vb.* to collimate

colina *f.* hill, ridge, hillock, [U.K.] fell || ***colina aislada*** *f.* (geog.) knob || ***colina de presion*** *f.* pressure ridge

colindante *a.* adjacent

colmado *a.* piled in heaps

colmarse *vb.* to fill

colmo *m.* height, summit

colocado *a.* laid, placed

colocar *vb.* to locate, place

colofana *f.* collophane

colofonita, colófano *f. & m.* collophane

coloformo *a.* colloform

coloidal, coloide, coloideo *a.* colloidal

coloide *m. & a.* coloid; colloidal

colomorfo *a.* coloform

color *m.* trace of gold; mineral stain || ***color de ante*** *m.* buff

colorados *m.* red or oxidized ores

colorante *m.* dye

colorimétrico *a.* colorimetric

colorímetro *m.* colorimeter

colpa *f.* copperas; [Pe.] silver ore; iron sulfate

colpas *f.* [Ch.] lump ore

colqui *m.* [Bol., Pe.] silver

columbita *f.* columbite

columna *f.* stratigraphic column, ore pipe || ***columna conductora*** *f.* surface pipe || ***columna de petróleo*** *f.* oil column || ***columna de sondeo*** *f.* drill string || ***columna estratigráfica*** *f.* columnar section || ***columna geológica*** *f.* geologic column, geologic section

columnario *a.* columnar

coluvial *a.* colluvial

coluvión *f.* colluvium

collado *m.* col, saddle, pass, gap, hill, [U.K.] fell

collpa *f.* [Bol., Pe.] nitrate, trona

Comancheano, Comanchense *m. & a.* Comanchean

comando *m.* (comp.) command

comarca *f.* territory, district, region, border, boundary, limit, study area || ***comarca minera*** *f.* mining district

comba *f.* warp, bulge; [Pe.] sledge hammer

combado *a.* warped

combar(se) *vb.* to warp, bend

combustible *a. & m.* combustible; fuel

combustión *f.* combustion

combustoleo *m.* fuel oil

comerciable *a.* marketable

comercial *a.* commercial

comercio *m.* commerce, trade, business

comercialización *f.* marketing

comezcla *f.* admixture

comillo *m.* reverberatory furnace

comisa *f.* overhang

comisión *f.* [Col.] survey party; brokerage

comisura estratigráfica *f.* bedding plane

compacción *f.* compaction || ***compacción diferencial*** *f.* differential compaction

compacidad *f.* [PR, Vz.] compaction, compactness

compactación *f. & a.* compaction, compacting, consolidated

compactado *a.* compact, dense

compacto *a.* dense, compact

compañía (Cía.) *f.* company (Co.) || ***compañía en comandita*** *f.* limited partnership || ***compañía matriz*** *holding* company || ***compañía operadora*** *f.* operator

comparación *f.* comparison

compás *m.* drawing compass

compensación *f.* compensation, adjustment || ***compensación isostática*** *f.* isostatic compensation

compensador *m.* buffer; *a.* compensating

compensar *vb.* to adjust

competencia *f.* competence; competition

competente *a.* competent

complejo *m.* complex ‖ ***complejo basal*** *m.* basement complex ‖ ***complejo de barrera*** *m.* barrier complex ‖ ***complejo de base*** *m.* basement complex ‖ ***complejo nuclear metamórfico*** *m.* metamorphic core complex

componente *m.* component, constituent ‖ ***componente lógico*** *m.* software

componer con madera *vb.* to timber a mine

comportamiento *m.* behavior

composición *f.* composition ‖ ***composición de color*** *f.* color composite

compra *f.* purchase

comprador *m.* purchaser

comprensivo *a.* comprehensive

compresibilidad *f.* compressibility

compresible *a.* compressible

compresión *f.* compression

compresivo *a.* compressive

compreso *a.* compressed

comprimible *a.* compressible

comprimido *a.* compressed

comprimir *vb.* to compress

comprobar *vb.* to check, verify, prove

compromiso *m.* commitment

compuerta del vertedero *f.* sluice

compuesto *m.* compound; *a.* composite ‖ ***compuesto iónico*** *m.* ionic compound ‖ ***compuesto orgánico*** *m.* organic compound ‖ ***compuesto químico*** *m.* chemical compound

computador *m.* computer

computadora *f.* computer

computar *vb.* to compute, calculate

computista *m. & f.* accountant, computer

cómputo *m.* computation, calculation

común *a.* common, standard ‖ ***de común acuerdo*** *m.* joint agreement

cóncavo *a.* concave

concentración *f.* concentration

concentrado *m. & a.* concentrate; concentrated

concentrador *m.* concentrator

concentrar *vb.* to concentrate

concéntrico *a.* concentric

concepto *m.* concept

concesión *f.* concession, grant, lease

concesionar *vb.* to grant a concession

concesionario *m.* concession holder, grantee

concoidal, concoide, concoideo *a.* conchoidal

concordancia *f.* conformity

concordante *a.* concordant, conformable

concordantemente *adv.* concordantly

concreción *f.* concretion

concrecionado *a.* precipitated

concrecionar *vb.* to sinter

concreto *m.* concrete ‖ ***concreto reforzado*** *m.* reinforced concrete

concurrente *a.* concurrent

concurso de propuestas *m.* call for bids

concha *f.* shell

conchifero *a.* shelly

cónchula *f.* foram test

condensación *f.* condensation ‖ ***condensación retrógrada*** *f.* retrograde condensation

condensado *m.* condensate, distillate

condensador *m.* (petrol.) microscope condenser; (ref.) condenser

condensar *vb.* to condense

condicione *f.* condition

condiciones *f.* terms, conditions ‖ ***condiciones climáticas*** *f.* climatic conditions ‖ ***condiciones del mercado*** *f.* market conditions ‖ ***condiciones de operación*** *f.* operating conditions

condrita *f.* chondrite

condrodita *f.* chondrodite

cóndrulo *m.* chondrule

conducción *f.* conduction; transportation

conductancia *f.* conductance

conductabilidad, conductividad *f.* conductivity ‖ ***conductabilidad eléctrica*** *f.* electrical conductivity ‖ ***conductabilidad térmica*** *f.* thermal conductivity

conductible, conductivo *a.* conductive

conducta *f.* conduct, behavior

conducto *m.* conduit, duct, flume, chute, volcanic vent

conductor *m.* conveyor, leader, conduit

conductos *m.* burrows

conectar *vb.* to connect, to hook up ‖ ***conectar a tierra*** *vb.* to ground

confeccionar *vb.* to prepare, make out, compile, write up

confiabilidad *f.* reliability, confidence

confiable *a.* reliable

configuración *f.* configuration, pattern, array

conflación *f.* smelting

confluencia *f.* confluence, fork of a river, junction

confluente *a.* confluent, tributary

confluir *vb.* to converge, meet, come together

conformar *vb.* to form, shape, mold; conform, adjust, adapt

conforme *a.* alike, similar

congelación *f.* freezing, solidifying

congelado *a.* frozen, solid

congelar *vb.* to freeze, solidify

conglomerádico, conglomerado *a.* conglomeratic

conglomerado *m.* conglomerate ‖ ***conglomerado basal*** *m.* basal conglomerate ‖ ***conglomerado torrencial*** *m.* fanglomerate

conglomerante *m.* binding material

conglomerático *a.* conglomeratic

conglutinar(se) *vb.* to coalesce, coagulate, cake

congosto *m.* pass, canyon, gorge

congruente, congruo *a.* congruent

Coniaciano, Coniaciense *m. & a.* Coniacian

cónico *a.* conical

conífero *a.* conifer

conificación del agua *f.* water coning

conjugado *m. & a.* conjugate

conjuntado *a.* united, joined, combined

conjuntar *vb.* to unite, join, combine

conjunto *m.* suite, assemblage, set ‖ ***conjunto de diques*** *m.* dike swarm ‖ ***conjunto de estratos*** *m.* sedimentary complex, bedset ‖ ***conjunto de estratos escalonados*** *m.* shingled bedsets ‖ ***conjunto de pozo inyector y seis productores circundantes*** *m.* seven-spot well pattern ‖ ***conjunto de pozo inyector y cuatro productores circundantes*** *m.* five-spot well pattern ‖ ***conjunto de pozo inyector y tres productores circundantes*** *m.* four-spot well pattern

cono *m.* cone, volcanic cone ‖ ***cono aluvial*** *m.* apron, alluvial cone ‖ ***cono compuesto*** *m.* composite cone, stratovolcano ‖ ***cono de agua*** *m.* water coning ‖ ***cono de cenizas*** *m.* cinder cone ‖ ***cono de deyección*** *m.* alluvial fan, talus fan ‖ ***cono en cono*** *a.* cone-in-cone ‖ ***cono entre cono*** *a.* cone-in-cone ‖ ***cono volcánico*** *m.* volcanic cone

conocimiento *m.* knowledge ‖ ***conocimiento de embarque*** *m.* bill of lading

conodonto *m.* conodont

consanguinidad *a.* consanguinity, relationship

consciente de la seguridad *a.* safety minded

consecuente *a.* consequent

consejo *m.* council, board, advisor ‖ ***consejo asesor*** *m.* advisory board ‖ ***consejo consultivo*** *m.* advisory board ‖ ***consejo de síndicos*** *m.* board of trustees ‖ ***consejo directivo*** *m.* board of directors

conservación *f.* upkeep, maintenance; conservation

consignación *f.* consignment

consignador *m.* consignor

consignatorio *m.* consignee

consistencia *f.* consistency

consola *f.* console

consolidación *f.* consolidation, compaction

consolidado *a.* consolidated

consolidar *vb.* to consolidate, to compact

consorcio *m.* consortium

constante *m., f. & a.* constant ‖ ***constante de conversión*** *f.* conversion constant ‖ ***constante dieléctrica*** *f.* dielectric constant ‖ ***constante elástica*** *f.* elastic constant

constituido *a.* built, composed of

constituir *vb.* to constitute, to build

constituyente *m.* constituent

constricción *f.* constriction

construcción *f.* construction || ***construcción hacia abajo*** *f.* downbuilding || ***construcción hacia arriba*** *f.* upbuilding

constructivo *a.* constructional

constructor *m.* builder, maker

construido *a.* built, composed of

construir *vb.* to build, construct, erect, make

consumidor *m.* consumer

consumir *vb.* to consume, to wear

consumirse *vb.* to wear away, to wear out

consumo *m.* consumption

contabilidad *f.* accounting

contacto *m.* bedding contact, layer boundary || ***contacto gas–agua*** *m.* gas–water contact || ***contacto gas–petróleo*** *m.* gas–oil contact || ***contacto petróleo–agua*** *m.* oil–water contact

contador *m.* counter, meter; accountant || ***contador de agua*** *m.* water meter || ***contador de centelleo*** *m.* scintillometer || ***contador de flujo*** *m.* flowmeter || ***contador de gas*** *m.* gas meter || ***contador de petróleo*** *m.* oil meter || ***contador eléctrico*** *m.* electric meter || ***contador geiger*** *m.* Geiger counter

contaminación *f.* pollution, contamination

contaminar *vb.* to pollute

contemporáneo *a.* contemporaneous

contener *vb.* to contain, hold, include; to restrain

contenido *m.* concentration, grade; content

conteo de puntos *m.* point count

contextura *f.* structure, texture

continental *a.* continental

continente *m.* continent

contingente *m.* share, quota, allowable

continuidad *f.* continuity

contínuo *a.* continuous

contorsión *f.* contortion; fold

contorsionado *a.* contorted

contorno *m.* contour, outline

contracción *f.* shortening, contraction, shrinkage

contracorriente *f.* countercurrent, eddy

contráctil *a.* contractile

contraer(se) *vb.* to shrink, contract

contraflujo *m.* countercurrent, eddy

contrafuerte *m.* buttress, abutment, spur

contrapendiente *f.* acclivity

contrapesar *vb.* to balance, to counterbalance

contrapeso *m.* balance weight, counterweight

contrapresión *f.* back pressure

contrario a las agujas (o manecillas) del reloj *a.* counterclockwise || ***contrario al orden cronométrico*** *a.* counterclockwise

contraseña *f.* password

contraste *m.* assay; contrast

contratar *vb.* to hire, to contract

contratista *m. & f.* contractor

contrato *m.* contract || ***contrato de producción compartida*** *m.* production sharing contract

contratope *m.* backstop of a critical wedge

contraviento de alambre *m.* guy wire

contribución *f.* tax; contribution; assessment || ***contribución predial*** *f.* property tax

control *m.* control || ***control a distancia*** *m.* remote control || ***control automático*** *m.* automatic controller || ***control geodético*** *m.* geodetic control

controlable *a.* controllable

controlado *a.* controlled

controlador *m.* controller

controlar *vb.* to control

convección *f.* convection

convenio *m.* agreement || ***convenio comercial*** *m.* trade agreement

convergencia *f.* convergence

convergente *a.* converging

conversión *f.* conversion || ***constante de conversión*** *f.* conversion constant || ***tabla de conversión*** *f.* conversion table || ***unidad de conversión*** *f.* conversion unit

convexo *a.* convex

convolución *f.* convolution

convoluto *a.* convolute

coordenada *f.* map coordinate

coordinación *f.* coordination

coordinado *a.* coordinated

coordinadora *m. & f.* coordinator

coordinar *vb.* to coordinate

copa *f.* cup, beaker

copal *m.* copalite, mineral resin

copalillo *m.* sphalerite

copaquira *f.* melanterite, copper or nickel sulfide

copalita *f.* copalite

copelilla *f.* lead carbonate

copete *m.* top, summit, crest

copey *m.* [Ec.] bitumen

copia *f.* copy || ***copia azul*** *f.* blueprint || ***copia heliografía*** *f.* blueprint

copiapita *f.* copiapite

copolimerización *f.* copolymerization

coprolito *m.* coprolite

coque *m.* coke

coqueo *m.* coking

coquería *f.* coking plant

coquificación, coquización *f.* coking

coquificar, coquizar *vb.* to coke

coquina *f.* coquina

coral *m.* coral

Coraliano, Coraliense *m. & a.* Corallian

corálico *a.* [Vz.] coral

coralífero *a.* coraliferous

coraliforme *a.* coraliform

coralígeno *a.* coral

coralino *a.* coralline

corazón *m.* core; heart

corazonar *vb.* to core

cordierita *f.* cordierite

cordillera *f.* mountain range, [U.S.] cordillera

cordón *m.* [Arg. Ch.] mountain range || ***cordónes estratigráficos*** *m.* shoestring sands || ***cordón interoceánico*** *m.* mid-ocean ridge || ***cordón litoral*** *m.* barrier beach, beach bar, offshore bar, barrier bar

cori *m.* [Bol. Pe.] gold

coribronce *m.* chalcopyrite

corindón *m.* corundum

corneana *f.* hornfels

cornisa *f.* cornice

cornubianita *f.* hornblende

corona *f.* rim, halo || ***corona de diamante*** *f.* diamond drill || ***corona de mastil*** *f.* mast head

corondón *m.* carborundum

corrasión *f.* corrasion

corrección *f.* correction || ***corrección al nivel de referencia*** *f.* datum correction, normalizing || ***corrección de aire libre*** *f.* free air correction || ***corrección de Bouger*** *f.* Bouger correction || ***corrección de deriva*** *f.* drift correction || ***corrección de fase*** *f.* phase correction || ***corrección de temperatura*** *f.* temperature correction || ***corrección dinamica*** *f.* normal moveout (NMO)

correcto *a.* accurate, correct

corredor *m.* jobber, broker

corredura *f.* flow, overflow

corregir *vb.* to revise, correct

correlación *f.* correlation || ***correlación estratigráfica*** *f.* stratigraphic correlation

correlacionar *vb.* to correlate

corrental *m.* flood, current

correntera *f.* rapids

correntilíneo *a.* streamlined

correr *vb.* to run, flow

corresponder *vb.* to correspond

corrido *m.* strike of a vein

corriente *a.* running, flowing

corriente *f.* current, stream || ***corriente abajo*** *a.* downstream || ***corriente alternada*** *f.* alternating current || ***corriente arriba*** *a.* upstream || ***corriente ajustada*** *f.* adjusted stream || ***corriente de barro*** *f.* mudflow || ***corriente de convección*** *f.* convection current || ***corriente de densidad*** *f.* density current, turbidity flow, turbidite || ***corriente de Humboldt*** *f.* Humboldt Current || ***corriente de lava*** *f.* lava flow || ***corriente de línea*** *a.* streamlined || ***corriente de marea*** *f.* tidal current || ***corriente de resaca*** *f.* rip current || ***corriente de***

turbeidad o turbidez *f.* turbidity current || ***corriente del Golfo*** *f.* Gulf Stream || ***corriente directa*** *f.* direct current || ***corriente ecuatoriales*** *f.* equatorial current || ***corriente eléctrica*** *f.* electric current || ***corriente estacional*** *f.* seasonal current || ***corriente laminar*** *f.* laminar flow || ***corriente litoral*** *f.* longshore current || ***corriente marginal*** *f.* longshore current || ***corrientes parásitas*** *f.* eddy currents || ***corriente turbia, turbidítica*** *f.* turbidity current || ***corriente turbulenta*** *f.* turbulent flow

corrimiento *m.* landslide; overthrust, thrust

corroer *vb.* to corrode

corrosión *f.* corrosion

corrosivo *a.* corrosive

corroyente *a.* corrosive, abrasive

corrugación *f.* crinkle, corrugation

corrugado *a.* corrugated, crinkled

cortado *a.* cut, slabbed, incised || ***cortado longitudinalmente*** *a.* slabbed

cortador de aceite *m.* oil jobber

cortafierro, cortafrío, cortahierro *m.* cold chisel

cortanúcleos *m.* core bit, core drill

cortar *vb.* to cut; intersect || ***cortar longitudinalmente*** *vb.* to slab || ***cortar pilar*** *vb.* to form mine pillars

corta testigos *m.* core bit

cortatubos *m.* casing cutter, pipe cutter

cortaveta *f.* crosscut

corte *m.* cut, working face; cutoff; cross section; slice; (min.) shaft or drift || ***corte de caja*** *m.* balance sheet, account balance || ***corte de camino*** *m.* roadcut || ***corte delgado*** *m.* thin section || ***corte estratigráfico*** *m.* stratigraphic cross section || ***corte estructural*** *m.* structural cross section || ***corte final*** *m.* (ref.) end fraction || ***corte liviano*** *m.* (ref.) light ends || ***corte longitudinal*** *m.* slab || ***corte transversal*** *m.* cross section

corteza *f.* crust || ***corteza continental*** *f.* continental crust || ***corteza oceánica*** *f.* oceanic crust

cortina *f.* dam

corto *a.* short || ***corto circuito*** *m.* short circuit || ***corto plazo*** *a.* short-term

corundo *m.* corundum

corvadura *f.* curvature

cósmico *a.* cosmic

costa *f.* coast || ***costa afuera*** *f.* offshore || ***costa fuera*** *a.* offshore

costado *m.* side, flank

costalera *f.* ore sack

costanero *a.* coastal

costanera *f.* slope, flank; jetty

coste *m.* cost (véase costo) || ***coste de extracción*** *m.* lifting cost, extraction cost || ***coste de operación*** *m.* operating cost || ***coste de producción*** *m.* production cost || ***coste diario*** *m.* per diem || ***coste intangible*** *m.* intangible cost

costeable *a.* economic, profitable

costeño, costero *a.* coastal, nearshore

costero *m.* side of a mineral deposit

costilla *f.* rib

costo *m.* cost (véase coste) || ***costo eficiente*** *a.* cost-efficient || ***costos de desarrollo*** *m.* development costs

costra *f.* crust, scale, slag; mud cake || ***costra endurecida*** *f.* hard ground

costura *f.* seam

costurón *m.* scar, welt

cota *f.* altitude, elevation || ***cota de comparación*** *f.* datum || ***cota de referencia*** *f.* bench mark, datum

cotación *f.* elevation

cotización *f.* quotation

coto *m.* boundary marker

coulée *m.* coulée

covacha *f.* small cave, crevice

covadera *f.* guano deposit

covalente *a.* covalent

covarianza *f.* covariance

covelina, covelita *f.* covellite

coyote *m.* agent who buys and sells mining shares

cracidolita *f.* cracidolite

Cracoiano, Cracoiense *a.* Cracoian

cráckíng *m.* (ref.) cracking || ***cráckíng catalítico*** *m.* catalytic cracking || ***cráckíng térmico*** *m.* thermal cracking

craqueado *a.* cracked

craquear *vb.* (ref.) to crack petroleum

craqueo *m.* [Arg.] (ref.) cracking ‖ ***craqueo a presión elevado*** *m.* high pressure cracking ‖ ***craqueo catalítico*** *m.* catalytic cracking ‖ ***craqueo térmico*** *m.* thermal cracking

cráter *m.* crater

cratógeno *m.* craton

crecer *vb.* to accrue; to grow, increase

crecida *f.* flood

crecido *a.* flooded

creciente *m. & a.* flood, rise, high tide; growing

crecimiento *m. & a.* flooding, rising, growth ‖ ***crecimiento excesivo*** *m.* overgrowth ‖ ***crecimiento secundario*** *m.* secondary growth ‖ ***falla de crecimiento*** *f.* growth fault

crédito *m.* credit, loan

cremallera *f.* (mech.) rack

crenulado *a.* crenulated, crenulate

cresta *f.* ridge, crest ‖ ***cresta acrecionada*** *f.* accretionary ridge ‖ ***cresta de espuma*** *f.* whitecap ‖ ***cresta de gas*** *f.* gas cap ‖ ***cresta de presión*** *f.* pressure ridge ‖ ***cresta dentada*** *f.* hogback ‖ ***cresta monoclinal*** *f.* hogback

crestería *f.* hogback, ridge, watershed divide

crestón *m.* outcrop; crest

creta *f.* chalk

Cretáceo, Cretácico *m. & a.* Cretaceous

cretoso *a.* chalky

criadero *m.* ore deposit, vein, seam, chute; oil pool ‖ ***criadero de contacto*** *m.* skarn ‖ ***criadero de petróleo*** *m.* oil pool ‖ ***criadero detrítico*** *m.* alluvial deposit

criba *f.* screen, grizzly, riddle, sieve

cribadura *m.* sieving, screening, jigging of ore

cribar *vb.* to screen, sieve, jig, sift

crinoide, crinoideo *m.* crinoid

crioconito *m.* cryoconite

criolita *f.* cryolite

criptocristalino *a.* cryptocrystalline

criptógeno *a.* cryptic

criptomagmático *a.* cryptomagmatic

criptómera *f.* cryptomere, aphanitic rock

criptón *m.* krypton

criptovolcánico *a.* cryptovolcanic

crisoberilo *m.* chrysoberyl

crisocola *f.* chrysocolla, borax

crisol *m.* assay or smelting crucible, hearth

crisolita, crisólito *f. & m.* chrysolite

crisoprasa *f.* chrysoprase

crisotilo *m.* chrysotile

cristal *m.* crystal

cristalinidad *f.* crystallinity

cristalino *a.* crystalline

cristalización *f.* crystallization

cristalizador, cristalizante *a.* crystallizing

cristalizar *vb.* to crystallize

cristaloblástico *a.* crystalloblastic

cristalografía *f.* crystallography

cristalográfico *a.* crystallographic

cristobalita *f.* cristobalite

criterio *m.* criteria

crocoísa, crocoisita, crocoita *f.* crocoite

cromato *m.* chromate

cromatografía *f.* cromatography ‖ ***cromatografía de gas–espectrometría en masa*** *f.* gas chromatography–mass spectrometry (GCMS)

cromita *f.* chromite ore

cromito *m.* chemical chromite

cromo *m.* chromium, chrome

cromoferrita *f.* chromite

cronoestratigrafía *f.* chronostratigraphy

cronoestratigráfico *a.* chronostratigraphic

cronógrafo *m.* chronograph

cronolito *m.* chronolite

cronología *f.* chronology

cronológico *a.* chronological

cronómetro *m.* timer, clock, watch

croquis *m.* sketch, diagram

cruce *m.* crossing, point of intersection

crucero *m.* jointing, cleavage; cross-cut ‖ ***crucero astilloso*** *m.* pencil cleavage ‖ ***crucero de plano axial*** *m.* axial plane cleavage ‖ ***crucero ondulado*** *m.* crenulation cleavage ‖ ***crucero pizarroso*** *m.* slaty cleavage ‖ ***crucero verdadero*** *m.* slaty cleavage

crudo *m. & a.* crude oil; crude, raw || ***crudo agrio*** *m.* sour crude || ***crudo dulce*** *m.* sweet crude || ***crudo dulce liviano*** *m.* light sweet crude || ***crudo liviano*** *m.* light crude, high gravity crude || ***crudo parafínico*** *m.* waxy crude || ***crudo pesado*** *m.* heavy crude, heavy oil || ***crudo sulfuroso*** *m.* sour crude || ***crudo virgen*** *m.* raw crude

cruzada *f.* crosscut

cruzado *a.* crossed, crosscut

crudeza *f.* water hardness

cruzamiento *m.* crosscutting vein, structure, etc.

cruzante *m.* crosscutting vein or structure

cruzar *vb.* to cross, cut; traverse; intersect

cuaderno *m.* field book, notebook

cuadrado *m. & a.* square

cuadrángulo *m.* quadrangle map

cuadrante *m.* quadrant; instrument dial

cuadrícula *f.* grid, graticule

cuadriculado *a.* gridded

cuadricular *vb.* to grid

cuadrilla *f.* grid; gang, party, crew

cuadro *m.* table of figures; shaft; square || ***cuadro de maniobras*** *m.* hoist, drawworks || ***cuadro periódico*** *m.* periodic table

cuajar *vb.* to evaporate brine

cuajarse *vb.* to coagulate, cake

cualidad *f.* quality

cualitativo *a.* qualitative

cuantificación *f.* quantification

cuantificar *vb.* to quantify

cuantitativo *a.* quantitative

cuanto *m.* quantum

cuarcífero *a.* quartziferous

cuarcita *f.* quartzite

cuarcítico *a.* quartzitic

cuarteadura *f.* crack, split, fissure

cuartel general *m.* headquarters

cuartear *vb.* to break, to quarter, crack, split

cuarteo *m.* crack, fissure

cuarto *m.* a shift, turn at work; quart || ***cuarto oscuro*** *m.* darkroom

cuartón *m.* boulder; [Méx.] flagstone

cuarzo *m.* quartz || ***cuarzo ahumado*** *m.* smoky quartz || ***cuarzo aurífero*** *m.* gold-bearing quartz || ***cuarzo lechoso*** *m.* milky quartz, [U.S.] bull quartz || ***cuarzo rosado*** *m.* rose quartz

cuarzodiorita *f.* quartz diorite

cuarzomonzonita *f.* quartz monzonite

cuarzoso *a.* quartzose

Cuaternario *m. & a.* Quaternary

cuba *f.* cask, tub, vat

cubaje *m.* volume

cubertura *f.* cover, overburden

cubeta *f.* basin, bucket, pail || ***cubeta sedimentaria*** *f.* trough, basin

cubetada *f.* bucketfill

cubetadraga *f.* dragline bucket

cubicaje *m.* volume

cubicar *vb.* to block out ore, calculate volume

cúbico *a.* cubic

cubierta *f.* cover, overburden, roof rock

cubierto *m. & a.* cover, covered; capped || ***cubierto de nieve*** *a.* snowcapped

cubilete *m.* beaker

cubilote *m.* cupola; smelting furnace

cubo *m.* cube

cubreobjeto *m.* cover slip

cubrimiento *m.* (seis.) [Arg.] fold

cubrir *vb.* to cover, drape over

cucúrbita *f.* retort

cucurucho *m.* peak, summit, top

cuchara *f.* bailer

cucharear *vb.* to bail

cuchareo *m.* bailing of a well

cuchilla *f.* ridge, mountain crest, hogback, razorback

cuelga *f.* [Ch., Col.] slope, grade of a river

cuello *m.* collar || ***cuello de botella*** *m.* bottleneck || ***cuello de perforación*** *m.* drill collar || ***cuello volcánico*** *m.* volcanic neck

cuenca *f.* basin, drainage area, watershed || *cuenca acuesta f.* piggyback basin || *cuenca cerrada f.* closed basin || *cuenca colectora f.* catchment basin || *cuenca de acumulación f.* constructional basin || *cuenca de destrucción f.* destructional basin || *cuenca de drenaje f.* drainage basin || *cuenca de fallas f.* graben || *cuenca empobrecida f.* starved basin || *cuenca estructural f.* structural basin || *cuenca fluvial f.* drainage basin || *cuenca hidrográfica f.* drainage basin || *cuenca obstruida f.* barred basin || *cuenca post-arco f.* backarc basin || *cuenca sedimentaria f.* sedimentary basin || *cuenca transportada f.* piggyback basin || *cuenca tributaria f.* watershed

cuenco *m.* pool, basin || *cuenco de amortiguación m.* spillway basin

cuenta *f.* account, bill, invoice || *cuentas por pagar f.* accounts payable

cuerda *f.* land measure about one acre; mine property limits; cord

cuerpo *m.* rock body, rock mass, orebody || *cuerpo de mena m.* orebody || *cuerpo efusivo m.* extrusive body || *cuerpo eruptivo m.* eruptive body || *cuerpo ígneo m.* igneous body || *cuerpo intrusivo m.* intrusive body || *cuerpo magmático m.* magmatic body || *cuerpo mineralizado m.* ore body

cuesco *m.* core; coarse ore; cemented rock fragments

cuesta *f.* hogback, hill, rise, slope, grade || *cuesta arriba f.* uphill

cueto *m.* rocky peak, high crag

cueva *f.* cave

Cuisiano, Cuisense *m. & a.* Cuisian

Culm *m. & a.* Culm

culminación *f.* culmination

culombio *m.* coulomb

cultura *f.* culture

cumbre *f.* summit, top, culmination, ridge, peak, crest

cumbrera *f.* summit, top, ridge, crest

cumulativo *a.* cumulative

cumulitos *m.* cumulites

cúmulo *m.* accumulation, pile, heap

cumulonimbo *m.* cumulonimbus

cuneta *f.* small trench, gutter, ditch || *cuneta de desagüe f.* ditch, trench

cuña *f.* wedge || *cuña critica f.* critical wedge, taper || *cuña de cuarzo f.* quartz wedge || *cuña de falla f.* fault wedge || *cuña desviadora f.* (perf.) whipstock || *cuña para cañería de entubación f.* casing dog || *cuña tectónica f.* tectonic wedge

cuñas a fricción *f.* friction slips

cuota *f.* allowable, quota, share

cúprico *a.* cupric

cuprífero *a.* copper-bearing, cupriferous

cuprita *f.* cuprite

cuproso *a.* cuprous

cúpula *f.* cupola, dome, roof || *cúpula gasífera f.* gas cap || *cúpula salina f.* salt dome

curador *m.* trustee, property guardian

curso *m.* course || *curso desplazado m.* offset stream || *curso ramificado m.* braided stream || *curso sobrepuesto m.* superimposed stream || *curso subsecuente m.* subsequent stream

cursor *m.* cursor

curtosis *m.* kurtosis

curva *f.* curve, bend || *curva de declinación f.* decline curve || *curva de dilatación f.* expansion bend || *curva de expansión f.* expansion bend || *curva de nivel f.* contour || *curva de nivel de la estructura f.* structure contour || *curva de nivel de referencia f.* contour datum || *curva de producción f.* production curve

curvacaños *m.* pipe bender

curvado *a.* bent, curved

curvar *vb.* to bend

curvas *f.* curves || *curvas de desviación f.* drift lines

curvatubos *m.* pipe bender

curvatura *f.* curvature, flexure

cuspidado *a.* cuspate

cúspide *f.* peak, summit, apex, cusp, crest

cutinita *f.* cutinite

CH

chabarco *m.* [Sp.] pond, lake
chabasia, chabasita *f.* chabazite
chaco *m.* [Bol., Par.] plain
chacuaco *m.* furnace
chacurruscar *vb.* [Pe.] to mix several kinds of ore
Chadroniano, Chadroniense *m. & a.* Chadronian
chafalonía *f.* wrought silver; scrap gold or silver
chafariz *m.* pool, pond
chalán *m.* barge
chalcopirita *f.* chalcopyrite
chalcosina *f.* copper sulfide
chalchihuitl *m.* [Méx.] jadeite, turquoise, any green precious stone
chalmersita *f.* cubanite
chalybita *f.* siderite
challantita *f.* ferricopiapite
chamarcal *m.* seasonal swamp
chamba *f.* [Méx.] job
chameolito *m.* humic coal
chamoisita *f.* chamosite
chamorro *a.* vegetation-free hilltop
champa *f.* dirt clod, turf
chamuscar *vb.* [Pe.] superficial roasting of ore
Champlainiano, Champlainiense *m. & a.* Champlainian
chanca *f.* crushing, sorting of ore
chancado *m.* crushed stone
chancadora *f.* crusher
chancal *m.* glacial moraine
chancar *vb.* to crush, grind
chango *m.* [Méx., Col.] roughneck, derrickman
chanta *f.* flagstone, slate
chañarcillita *f.* huntilite
chapapote, chapopote *m.* [Méx.] asphalt, tar, bitumen
chaparrón *m.* cloudburst, downpour
chapatal *m.* swamp, muddy ground
chapopote *m.* [Méx.] tar
chapopotera *f.* [Méx.] oil seep
chapuro *m.* [Arg., Cu.] asphalt, tar
charca *f.* small lake, pond
charcal *f.* [Arg., Cu.] pond
charco *m.* pool, pond, water hole
Charmouthiano, Charmoutense, Charmoutiense *m. & a.* Charmouthian
charnela *f.* fold hinge
Charniano, Charniense *m. & a.* Charnian
charnockíto *m.* charnockite
charquear *vb.* to dewater a mine
charquero *m.* drainage ditch
chassignito *m.* chassignite
chatarra *f.* scrap iron, scrap, slag, junk
Chatiense, Chattiano *m. & a.* Chattian
chato *a.* [Arg.] flattened
Chautauquiense, Chautauquiano *m.* Chautauquan
Chaziense, Chaziano *m. & a.* Chazyan
chekao *m.* barite
Chemungiano, Chamungiense *m. & a.* Chemungian
chengbolita *f.* moncheite
chernozem *m.* chernozem
chert *m.* chert
chertificación *f.* chertification
chessylita *f.* azurite
Chesteriano, Chesteriense *m. & a.* Chesterian
chiastolita *f.* chiastolite
chicana *f.* baffle
chicuite *m.* small ore bucket
chicharrones *m.* tailings
chichicle, chichitle *m.* [Méx.] amethyst
Chideruano *m. & a.* Chideruan
chifladero *m.* ore hopper
chifle *a.* stagnant, still
chiflón *m.* chamber in a mine, drift; flume; rockslide, incline, chute
chile *m.* [Pe.] greatest depth in a mine
chile-loeweita *f.* humberstonite
chiltonita *f.* prehnite

chimborazita *f.* aragonite

chimenea *f.* vent, chimney, pipe, shaft

china *f.* stone, cobble, pebble

chinarro *m.* boulder; cobble; pebble

chinazo *m.* cobble, stone

chingana *f.* cave, underground passage

chingarse *vb.* (min.) to be cut off or peter out

chinidina *f.* thin section glue

chinkolobwita *f.* sklodowskite

chino *m.* pyrite; chalcopyrite; cobble

chinoita *f.* libethenite

chiscarra *f.* deep-weathered limestone

chispa *f.* small diamond; fine gold ore; spark

chispiador *m.* [Pe.] gold washer in a placer operation

chispero *m.* flame arrester, spark arrester

chladnita *f.* chladnite

chloantita *f.* chloanthite

chocar *vb.* to crash, collide, shock, encounter, knock

choclo de oro *m.* massive gold in matrix

chocolón *m.* hole remaining after a blast

cholla *f.* opening or hollow space

chompa *f.* [Vz.] churn drill

chondrítico, chóndrulo *a.* chondritic

chondrito *m.* chondrite

chondrodita *f.* chondrodite

choque *m.* crash, collision, encounter, shock, jolt || ***choque sísmico*** *m.* seismic shock, tremor

chorrillo *m.* brook

chorlo *m.* schorl, black tourmaline

chorreadero *m.* ore chute

chorrera *f.* ore shoot; loose rock

chorro *m.* jet, spring, source || ***chorro de agua a presión*** *m.* jet of water

chorroborro *m.* alluvium

chortal *m.* small spring-fed lake or pool

chozola *f.* river bend

christensenita *f.* tridymite

christophita *f.* marmatite

chubascos *m.* rain showers

chulano *m.* hole drilled upwards

chumbe *m.* [Bol., Pe.] sphalerite

chumpi *m.* magnetite; [Bol.] platinum

chungita *f.* shungite

chupador *m.* suction pipe; (perf.) swab

chupón *m.* glory hole

churumbella *f.* [Col.] altered felsite wallrock

chuzo *m.* [Ch.] crowbar

D

dacita *f.* dacite

dacítico *a.* dacitic

dador *m.* grantor

danburita *f.* danburite

Danense, Danés, Daniano, Daniense *m. & a.* Danian

daño *m.* damage || ***daño de la formación*** *m.* formation damage

dar cuele *vb.* to drive a mine level

darcy *m.* (meas.) darcy

data *f.* date

datación *f.* dating, determining age

datar *vb.* to date

dato *m.* fact, datum, record, information

datos *m.* data || ***banco de datos*** *m.* (comp.) database || ***conjunto de datos*** *m.* data set || ***datos de entrar*** *m.* (comp.) input || ***datos de salir*** *m.* (comp.) output || ***gestión de datos*** *f.* data management || ***proceso de datos*** *m.* (comp.) data processing || ***recopilación de datos*** *f.* data gathering

datolita *f.* datolite

debajo de *p.* under, below

débil *a.* weak

debilidad *f.* weakness

debutanización *f.* debutanization

debutanizador *m.* debutanizer

debutanizar *vb.* to debutanize

decano *m.* decane

decantación *f.* decantation

decantador *m.* decanter, settler

decantando *a.* settling, decanting

decantar *vb.* to settle

decapitación *f.* beheading, capture, piracy

decapitar *vb.* to behead, capture

decibel, decibelio *m.* decibel

deciduo *a.* deciduous

declararse en huelga *vb.* to declare a strike, to go on strike

declinación *f.* declination

declive *m.* inclination, slope, incline dip, pitch

declividad *f.* slope

decoloración *f.* bleaching

decolorar *vb.* to bleach

deducción *f.* deduction

deducir *vb.* to deduce

defecto *m.* defect

defectuoso *a.* defective

deficiente *a.* deficient, poor

definición *f.* definition

definido *a.* defined

definir *vb.* to define

deflación *f.* deflation

deflector *m.* baffle, deflector

deflegmador *m.* separator

deflexión *f.* deflection

deformación *f.* deformation, strain, distortion || ***deformación elástica*** *f.* elastic deformation || ***deformación en plano*** *f.* plain strain || ***deformación global*** *f.* bulk strain || ***deformación infinitesimal*** *f.* infinitesimal strain || ***deformación interna*** *f.* strain || ***deformación interna por cizalla*** *f.* shear strain || ***deformación mecánica*** *f.* mechanical strain || ***deformación penetrante*** *f.* penetrative or pervasive deformation || ***deformación plana*** *f.* plane strain || ***deformación plástica*** *f.* plastic strain, plastic deformation

deformado *a.* strained, deformed

deformar *vb.* to deform, strain || ***deformar permanentamente*** *vb.* to strain

degradación *f.* degredation, erosion

degradar *vb.* to degrade

delessita *f.* delessite

delgado *a.* thin

delineación *f.* delineation; debanding, destriping

delineado *a.* blocked-out ore

delineamiento *m.* contouring

delineante *m.* draftsman

delinear *vb.* to block out ore, delineate

delta *m.* delta || ***delta constructivo*** *m.* constructive delta || ***delta de cabeza de bahia*** *m.* bay head delta || ***delta de marea*** *m.* tidal delta || ***delta destructivo*** *m.* destructive delta || ***delta digitado*** *m.* bird-foot delta || ***delta pie de pajaro*** *m.* bird-foot delta

deltáico *a.* deltaic

deluvio *m.* deluvium

demarcación *f.* demarcation

demarcar *vb.* to demarcate, delimit, mark out

demulsificación *f.* demulsification

dendriforme, dendrítico, dendroide, dendroideo *a.* dendritic

dendrita *f.* dendrite

denotar *vb.* to denote, indicate, show, express

densidad *f.* density || ***densidad óptica*** *f.* optical density || ***densidad relativa*** *f.* specific gravity

densitómetro *m.* densitometer

denso *a.* dense, heavy, thick

dentado, dentellado *a.* jagged, serrated, dentate

denticulado *a.* toothed

dentro *a.* within, inside || ***dentro del pozo*** *a.* downhole

denudación *f.* erosion, denudation

denudado *a.* denuded

denudar *vb.* to erode, denude

denuncia *f.* petition for a mining claim

denunciar *vb.* (min., legal) to obtain legal possession of a mining claim

departamento *m.* department; state || ***departamento de compras*** *m.* purchasing department || ***departamento de ventas*** *m.* sales department

dependiente *m. & a.* clerk, salesman; dependent

depleción *f.* depletion

depocentro *m.* depocenter

deposición, depositación *f.* deposition

deposicional, depositacional *a.* depositional

depositado *a.* deposited; depositional

depositar(se) *vb.* to deposit, to be deposited

depositario *m.* trustee, receiver, consignee

depósito *m.* deposit, sediment, deposition (véase almacenadora, yacimiento) || ***depósito aluvial*** *m.* alluvial deposit || ***depósito de asentamiento*** *m.* settling pond || ***depósito de borde*** *m.* rimstone || ***depósito de lodos*** *m.* mud pit || ***depósito de placer*** *m.* placer deposit || ***depósito de relleno de cauce*** *m.* channel fill deposit || ***depósito mineral*** *m.* mineral deposit, ore body || ***depósito petrolífero*** *m.* oil pool || ***depósito por deslizamiento*** *m.* slump || ***depósito por gravidad*** *m.* gravity deposit || ***depósito residual*** *m.* lag deposit

depreciación *f.* depreciation

depresión *f.* depression, hollow, gap, pass

de prove *a.* proven

depurador *m.* scrubber, cleaner, purifier; *a.* purifying

depurar *vb.* to purify, scrub, filter

derecho *m.* grant, concession; true; law || ***derechos*** *m.* (legal) rights || ***derechos de aduana*** *m.* tariff, customs duty || ***derechos de importación*** *m.* import tarrifs || ***derechos industriales*** *m.* royalties || ***derechos mineros*** *m.* mining rights

derecho *a.* straight

derechos *m.* ore tax, royalty

deriva *f.* drift, deviation || ***deriva continental*** *f.* continental drift || ***deriva litoral*** *f.* longshore drift

derivación *f.* bypass, diversion, shunt, overflow; offset

derivado *m.* derivative; *a.* derived

derivar *vb.* to calculate; to drift, bypass

derramar(se) *vb.* to pour out, spill, leak, overflow, to runoff

derrame *m.* spill, leak, leakage, overflow, runoff; slope, incline, declivity; sill || ***a punto de derrame*** *a.* filled to spill point

derramiento *m.* runoff; spilling, pouring out

derretido *a.* molten

derretimiento *m.* melting, thawing

derretir *vb.* to melt, thaw

derribar *vb.* to break ground

derribo *m.* breaking ground

derrocado *a.* caved

derrubiar *vb.* to erode, undermine, wash out

derrubio(s) *m.* scree, talus, alluvium, debris || ***derrubio de faldeo*** *m.* slope wash

derrumbadero *m.* cliff, precipice

derrumbado *a.* caved

derrumbamiento *m.* caving, landslide, collapse, slump

derrumbarse *vb.* to cave, fall in, collapse, slide

derrumbe *m.* caving, collapse, landslide, washout, failure, slip

derrumbo *m.* precipice

desacoplamiento cortical *m.* crustal decoupling

desagregación *f.* desaggregation, disintegration

desagregado *a.* disintegrated

desagregar *vb.* disintegrate

desaguadero, desaguador *m.* drain, outlet, water pipe

desaguar *vb.* to drain, empty, runoff, flow; desiccate

desagüe *m.* dewatering, draining; outlet, drainage, gutter, channel, sewer, drain

desahogo *m.* relief

desalación, desalinización *f.* desalination

desalar, desalinizar *vb.* to desalinate

desamparar *vb.* to abandon

desamparo *m.* abandonment of a mine or working

desaparecer *vb.* to peter out, disappear

desarenador *m.* desander

desarmar *vb.* to take apart

desarrollado *a.* developed

desarrollar *vb.* to develop, unroll, unfold

desarrollo *m.* development

desasfaltación *f.* deasphalting

desazogadera *f.* retort for recovering mercury
desazogar *vb.* to volatalize and recover mercury
desazufrar el petróleo *vb.* to sweeten oil, to remove sulfur
desbarrancadero *m.* precipice
desbastar *vb.* to scour
desbocado *a.* overflowing
desbordamiento *m.* overflow, overflowing
desbordar(se) *vb.* to overflow; to stope
desborde *m.* overflowing; underhand stope
descabezamiento *m.* decapitating
descalcificación *f.* decalcification
descalcificar *vb.* to decalcify
descamando *m.* scaling off
descansar *vb.* to rest on
descantillar *vb.* to chip
descapote *m.* [Col.] cleared material
descarbonatar *vb.* to decarbonate
descarbonización *f.* decarbonization
descarburación *f.* decarbonation
descarburar *vb.* to decarbonize
descarga *f.* discharge
descargar *vb.* to flow into, discharge, exhaust, unload || ***descargar flúido a presión*** *vb.* to blow off
descartar *vb.* to unload, dump, discharge
descascaramiento *m.* flaking
descascararse *vb.* spalling off
descendente *a.* downslope, descending
descendiendo *a.* sloping, descending
descenso *m.* subsidence, descent, fall, decline; drawdown, drop || ***descenso de nivel*** *m.* drawdown
descoagulante *a.* liquifying
descoagular *vb.* to liquify, dissolve
descolorar, descolorir *vb.* to bleach
descolorido *a.* bleached
descolorimiento *m.* bleaching
descomponer *vb.* to decompose
descomposición *f.* decomposition
descompresión *f.* decompression
descomprimir *vb.* to decompress
descompuesto *a.* decomposed
desconcharse *vb.* to exfoliate, spall
desconectador de barrena *m.* bit breaker
desconectar *vb.* disconnect
descongelar *vb.* to thaw, defrost, melt
desconvolución *f.* deconvolution
descopetada *a.* breached
descostrarse *vb.* spall, scale
describir *vb.* to describe
descripción *f.* description
descuajar *vb.* to liquify, dissolve
descubierto *a.* discovered, uncovered
descubridor *a.* discovered || ***pozo descubridor*** *m.* discovery well
descubrimiento *m.* discovery
descubrir *vb.* to discover, uncover
descuento *m.* discount, rebate || ***descuento de flujo neto futuro de caja*** *m.* discounted future net cash flow
desdoblado *a.* unfolded
desecación *f.* desiccation, drying
desecado *a.* dried, desiccated
desecador *m.* desiccator
desecamiento *m.* drying, draining a marsh
desecante *a.* drying, desiccating
desecar *vb.* to dry, desiccate, drain, dewater
desechar *vb.* to junk, scrap; to dispose, to reject, refuse
desecho *m.* waste rock, low grade ore, residue; (survey, levan.) offset
desechos *m.* tailings, wastes, debris
desembarcadero *m.* dock, landing
desembocadura *f.* river mouth, outlet
desembocar *vb.* to flow, run
desembotellar *vb.* to debottleneck
desempeñar *vb.* to redeem, recover
desempeño *m.* performance; redemption, payment
desenganchar un pozo *vb.* to bump-off a well
desenganche *m.* knock-off
desenglazamiento *m.* deglaciation
desentubar cañería *vb.* to pull casing
desentubar tubería de bombeo *vb.* to pull tubing

desértico *a.* arid, desert-like

desescombrar *vb.* to muck

desescombro *m.* mucking

desfiladero *m.* notch, gorge, defile, pass, canyon, narrows

desgarramiento *m.* tear, tear fault

desgarre *m.* rift, tear fault

desgasificación *f.* degasification, degassing

desgasificar *vb.* degassify

desgasificador *m.* degasser

desgastado *a.* eroded, abraded

desgastando *m.* mass wasting, slumping

desgastar *vb.* to waste, slump, weather, abrade, erode, wear away

desgaste *m.* weathering, slumping, corrosion, erosion, wear and tear, attrition

deshelar(se) *vb.* to melt, thaw

deshidratación *f.* dehydration

deshidratador *m.* dehydrator

deshidratar *vb.* dehydrate

deshidrogenación *f.* dehydrogenation

deshiele, deshielo *m.* thaw

deshumedecer *vb.* dehumidify

desierto *m.* desert

desigual *a.* uneven

desilificación *f.* desilicification

desilificar *vb.* desilicify

desimanación, desimantación *f.* demagnetization

desimanar, desimantar *vb.* to demagnetize

desincrustante *m. & a.* scale remover

desincrustar *vb.* to descale

desintegración *f.* disintegration

desintegrador de vapor *m.* steam cracker

desintegrar *vb.* to disintegrate; (ref.) to crack

desinversión *f.* divestment

desinvertir *vb.* to divest

desistimiento *m.* abandonment of a mining claim

deslava *m.* erosion, alluvium

deslavable *a.* erodible

deslavar *vb.* to erode

deslave *m.* tailings, landslide, rockfall

desleído *a.* dissolved; diluted

desleimiento *m.* dilution

desleir *vb.* to dissolve; to dilute

desleznable *a.* friable, crumbly

desleznamientos *m.* slakings

desleznar *vb.* to crumble

deslindador *m.* land surveyor

deslindamiento *m.* demarcation, delimitation

deslindar *vb.* to set boundaries

deslinde *m.* demarcation

desliz *f.* landslide, slip

deslizamiento *m.* slip, landslide ‖ ***deslizamiento brilloso*** *m.* slickenside ‖ ***deslizamiento capa-a-capa*** *m.* bedding plane slip ‖ ***deslizamiento de escombros*** *m.* debris slide ‖ ***deslizamiento de faldeo*** *m.* landslide ‖ ***deslizamiento de lodo*** *m.* mudflow ‖ ***deslizamiento de tierra*** *m.* landslide ‖ ***deslizamiento flexural*** *m.* flexural slip ‖ ***deslizamiento por translación*** *m.* translation gliding ‖ ***deslizamiento y subida de dislocaciones*** *m.* dislocation glide and climb

deslizar(se) *vb.* to glide, slip, slide

deslustrado *a.* frosted

deslustrar(se) *vb.* to tarnish, frost

desmagnetización *f.* demagnetization

desmagnetizar *vb.* to demagnetize

desmantelar *vb.* to dismantle

desmenuzable *a.* friable

desmenuzamiento *a.* chipping, crumbling

desmenuzar(se) *vb.* to chip, to crumble

desmontar *vb.* to strip of overburden, to clear land of brush or forest

desmonte *m.* overburden, waste rock, rubble; ground clearing ‖ ***desmonte y relleno*** *m. & a.* cut-and-fill

desmoronadizo *a.* crumbly

desmoronamiento *m.* riverbank erosion, mass wasting, slumping, cave in, landslide

desmoronarse *vb.* to crumble, cave in

desmuestre *m.* coring, sampling

desmulsificación *f.* demulsification

desmulsificador *m.* demulsifier

desmulsificar *vb.* to demulsify

desnaftado *m.* redistillation

desnaftadora *a.* naptha scrubber

desnatación *f.* skimming

desnatado *m.* skimming

desnatador *m.* skimmer

desnatar *vb.* to skim

desnivel *m.* difference in height, uneveness of ground, relief

desnivelado *a.* sloping

desorientado *a.* unoriented, disoriented

despachante *m.* broker, dispatcher || ***despachante de aduana*** *m.* customs agent

despacho *m.* office, station; dispatch, shipment

despajar *vb.* [Méx.] to concentrate ore; remove waste rock

despaje *m.* concentration tailings

desparafinación *f.* dewaxing

desparafinado *a.* dewaxed

desparafinar *vb.* to dewax

despedir *vb.* to lay off

despegar *vb.* to detach, separate

despegue *m.* detachment surface, decollement

despejar *vb.* to clear

despeñadero *m.* precipice, cliff; *a.* steep, precipitous

despensa *f.* store room

desperdicio *m.* waste, spoil || ***desperdicio perjudicial para la salud*** *m.* hazardous waste

despilar *vb.* to rob pillars

despilamiento *m.* robbing of pillars

desplazado *a.* displaced

desplazamiento *m.* displacement, slip, offset || ***ángulo de desplazamiento*** *m.* displacement angle, rake || ***desplazamiento aparente*** *m.* apparent slip || ***desplazamiento a presión*** *m.* pressure drive || ***desplazamiento de falla*** *m.* fault displacement || ***desplazamiento de inclinación*** *m.* dip slip || ***desplazamiento de rumbo*** *m.* strike slip || ***desplazamiento diferencial*** *m.* differential displacement || ***desplazamiento horizontal*** *m.* heave || ***desplazamiento polar*** *m.* polar wandering || ***desplazamiento por agua*** *m.* water drive || ***desplazamiento por fase miscible*** *m.* miscible drive || ***desplazamiento por gas*** *m.* gas drive || ***desplazamiento según el buzamiento*** *m.* throw || ***desplazamiento radial*** *m.* radial displacement || ***desplazamiento vertical*** *m.* throw

desplazamientos *m.* displacements; (perf.) stepouts

desplazar *vb.* to displace

desplegable *m.* folder, brochure

desplegar *vb.* display

despliegue *m.* deployment; unfolding; (seis.) spread || ***despliegue de cruz*** *m. & a.* cross spread || ***despliegue de detectores*** *m.* detector spread

desplomado *a.* caved, collapsed

desplomar(se) *vb.* to cave in, collapse

desplome *m.* collapse || ***desplome de montaña*** *m.* rockfall, landslide

despoblar *vb.* to suspend mining

despojado *a.* stripped

despojadora *f.* stripper tower

despojar *vb.* to strip, to trim

despojo *m.* spoil, waste, rubbish; [Bol.] debris; [Col.] (min.) extraction

desprender *vb.* to separate from

desprendimiento *m. & a.* landslide, slip; loosening

desprovisto *a.* devoid of, without, lacking

después de apilamiento *a.* post-stack

despuntador *m.* pick, pickaxe

desquebrajado *a.* cracked, split

desquebrajar *vb.* to crack, split, fracture

destaje, destajo *m.* contract; [Bol.] contract work; [Pe.] open cut

destellador *m.* flare

destemplar *vb.* anneal

destilación *f.* distillation || ***destilación al vacío*** *f.* vacuum distillation || ***destilación a presión*** *f.* pressure distillation || ***destilación a vapor*** *f.* steam distillation || ***destilación destructiva*** *f.* dry distillation || ***destilación diferencial*** *f.* differential distillation || ***destilación estabilizadora*** *f.* stripping || ***destilación fraccionada*** *f.* fractional distillation || ***destilación ini-***

cial f. topping, skimming || ***destilación por cargas*** *f.* batch distillation || ***destilación por conchadas*** *f.* batch distillation || ***destilación por expansión instantánea*** *f.* flash distillation || ***destilación primaria*** *f.* topping || ***destilación relámpago*** *f.* flash distillation || ***destilación selectiva*** *f.* selective distillation

destilado *m.* distillate; *a.* distilled || ***destilado fraccional*** *m.* fractional distillate || ***destilado integro*** *m.* strainght-run distillate

destiladora *f.* still

destilados de la parte superior de la torre *m.* tops

destilar *vb.* to distill

destilería *f.* refinery, distillery

destreza *f.* skill

destrozo *m.* breakage, destruction

destrucción *f.* destruction

destruir con explosivo *vb.* to blast

destruir por corrosión *vb.* to corrode

destufar *vb.* to sweeten gasoline

desulfuración *f.* desulfurization

desulfurar *vb.* to sweeten, to remove sulfur

desvenar *vb.* to extract ore

desviación *f.* deviation, deflection, sidetrack, bypass || ***desviación estándar*** *f.* standard deviation

desviar *vb.* to deflect, divert, switch

desviarse *vb.* to deviate, branch off

desvío *m.* wandering, deviation; bypass, detour; baffle

desviación *f.* deviation, deflection; (perf.) kick || ***punto de desviación*** *m.* (perf.) kick point

desviado *a.* deviated, crooked, twisted, bypassed

desviador *m.* deflector, baffle; (perf.) whipstock

desviar *vb.* to deviate, (perf.) to sidetrack bypass, whipstock

desvío *m.* bypass; detour; baffle; deviation, (perf.) sidetrack

desviómetro *m.* drift meter

desvitrificación *f.* devitrification

desvitrificar *vb.* to devitrify

deszafrar *vb.* to carry waste from a mine

detallado *a.* detailed

detallar *vb.* to detail, to itemize

detalle *m.* detail

detallista *m., f. & a.* retail dealer; retail

detección *f.* detection

detector *m.* detector || ***detector de gas*** *m.* gas detector || ***detector de hidrógeno sulfurado*** *m.* hydrogen sulfide detector || ***detector sísmico*** *m.* geophone pickup, siesmometer || ***detector sismográfico*** *m.* seismometer || ***detector térmico*** *m.* hot-wire detector

detergente *m. & a.* detergent

determinación *f.* determination, test

determinar *vb.* to determine

detonación *f.* explosion

detonador *m.* detonator, blaster, squib

detonante *m. & a.* detonator, explosive

detrítico *a.* detrital

detrito *m.* detritus

detritos *m.* cuttings, debris

deuda *f.* debt

deudor *m.* debtor; *a.* debit balance, balance due

deutérico *a.* deuteric

deuterio *m.* deuterium

Devoniano, Devónico *m. & a.* Devonian

dextral *a.* right-lateral

dextroso *a.* dextral

deyección *f.* ejecta, debris

diabasa *f.* diabase

diaclasa *f.* joint (véase fractura, grieta) || ***diaclasa columnar*** *f.* columnar jointing || ***diaclasa de cizalla*** *f.* shear joint || ***diaclasa de compresión*** *f.* compression joint || ***diaclasa de corte*** *f.* shear joint || ***diaclasa de relajación*** *f.* release joint || ***diaclasa de tensión*** *f.* tension joint || ***diaclasa principal*** *f.* master joint || ***juego de diaclasas*** *m.* joint set

diaclasado *a.* jointed

diaclasado, diaclasamiento *m.* jointing

diaclasas conjugadas *f.* conjugate joints

diacrónico *a.* diachronous

diafanidad *f.* transparency

diaftoresis *m.* diaftoresis

diagénesis *f.* diagenesis, induration

diagenétco *a.* diagenetic

diagrafía *f.* logging

diafragma *m.* diaphragm

diagnosis *f.* diagnosis

diagnóstica *f.* diagnostics

diagnosticar *vb.* to diagnose

diagnóstico *m.* & *a.* diagnosis; diagnostic

diagrama *m.* diagram || ***diagrama de bloques*** *m.* block diagram || ***diagrama de circulación*** *m.* flow diagram || ***diagrama de elaboración*** *m.* flow diagram || ***diagrama de fases*** *m.* phase diagram || ***diagrama de flujo*** *m.* flow diagram || ***diagrama de sectores*** *m.* pie diagram || ***diagrama estereográfico*** *m.* block diagram || ***diagrama por bloques*** *m.* block diagram || ***diagrama roseta*** *m.* rose diagram

diálaga *f.* diagolite

diálisis *f.* dialysis

diamagnético *a.* diamagnetic

diamantado *a.* diamond-like

diamante *m.* diamond || ***diamante bruto*** *m.* raw diamond, rough diamond || ***diamante negro*** *m.* bort || ***diamante tallado*** *m.* cut diamond

diamantífero *a.* diamond-bearing

diamantino *a.* diamond-like

diamantista *m.* diamond cutter, diamond merchant

diámetro *m.* diameter

diapirismo *m.* diapirism

diapiro *m.* diapir, salt dome || ***diapiro arcilloso*** *m.* mudlump, shale diapir || ***diapiro de lodo*** *m.* mudlump, mud diapir

diapositiva *f.* slide

diario *m.* logbook, journal, record; *a.* daily

diáspora *f.* diaspore

diásporo *m.* diaspore

diasquístico *a.* differentiated

diastema *f.* diastem

diastrófico *a.* diastrophic

diastrofismo *m.* diastrophism

diastroma *f.* diastrophism

diatomáceo *a.* diatamaceous

diatomea *f.* diatom

diatomita *f.* diatomite

diatrema *f.* diatreme, maar

diaxial *a.* biaxial

dibujante *m.* & *f.* draftsman, designer

dibujar *vb.* to sketch, draw

dibujo *m.* sketch, drawing

dickita *f.* dickite

dicroísmo *m.* dichroism

dicroita *f.* cordierite

diedro *m.* & *a.* dihedron; dihedral

dieléctrico *m.* & *a.* dielectric

diente *m.* tooth; tusk; fang

dientes de perro *m.* scalenohedral calcite, dogtooth spar

diesel *m.* & *a.* diesel

diestro *a.* dextral, right; capable

dietilo *m.* & *a.* diethyl

difenilo *m.* & *a.* diphenyl

diferencia *f.* difference || ***diferencia de altura*** *f.* elevation difference, relief || ***diferencia de cota*** *f.* elevation difference, relief || ***diferencia de esfuerzo*** *f.* stress difference

diferenciación *f.* differentiation || ***diferenciación magmática*** *f.* magmatic differentiation

diferencial *a.* differential, distinctive

diferenciado *a.* differentiated

diferenciar *vb.* to differentiate, distinguish

difracción *f.* diffraction || ***difracción fantasma*** *f.* phantom diffraction || ***difracción radiográfica*** *f.* x-ray diffraction

difractar *vb.* to diffract

difractivo, difrangente *a.* diffractive

difundido *a.* diffuse

difundir *vb.* to diffuse, scatter

difusión *f.* diffusion, scattering

difusividad *f.* diffusivity

difuso *a.* diffused, pervasive

difusor *m.* diffuser, spreader

digestión *f.* assimilation

digitación *f.* interfingering
digitado *a.* digitate
digital *a.* digital
digitizado *a.* digitized
digitizar *vb.* to digitize
dígito *m.* digit || ***dígito binario*** *m.* binary digit, bit
dilatación *f.* dilation || ***dilatación por congelación*** *f.* frost heaving || ***dilatación termal*** *f.* thermal expansion
dilatado *a.* dilated
dilatar *vb.* to dilate, expand
dilución *f.* dilution
diluído *a.* dilute
diluir *vb.* to dilute
diluvial *a.* diluvial
diluvio *m.* flood, deluge
diluvión *m.* diluvium
diluyente *m.* flux, thinner
dimanar *vb.* to flow, run
dimensión *f.* dimension
dimetilbenceno *m.* dimethylbenzene
dimetilbutano *m.* dimethylbutane
dimetilpropano *m.* dimethylpropane
dina *f.* dyne
dinámica *f.* dynamics; dynamic range
dinámico *a.* dynamic
dinamita *f.* dynamite
dinamitación *f.* dynamiting, (seis.) shooting || ***dinamitación de un pozo*** *f.* shooting a well
dinamitar *vb.* to dynamite
dínamo, dinamo *f.* dynamo
dinamometamórfico *a.* dynamometamorphic
dinamometamorfismo *m.* dynamometamorphism
dinamómetro *m.* dynamometer
Dinantiano, Dinantiense *m. & a.* Dinantian
dinero *m.* money, cash, currency
dinosaurio *m.* dinosaur
diodo *m.* diode
dionisia *f.* specularite, bloodstone, jasper
diópsida, diópsido *f. & m.* diopside
dioptasa *f.* dioptase
dioptra *f.* diopter
diorita *f.* diorite
dióxido *m.* dioxide || ***dióxido de carbono*** *m.* carbonic acid, carbon dioxide || ***dióxido de silicio (o sílice)*** *m.* silicon dioxide, quartz || ***dióxido de sulfuro*** *m.* sulfur dioxide
dipirámide *m. & a.* bipyramid; bipyramidal
dipolo *m.* dipole
dique *m.* dike, levee, mound; dock, weir; dam, embankment || ***dique alimentador*** *m.* feeder dike || ***dique anular*** *m.* ring dike || ***dique clástico*** *m.* clastic dike || ***dique concordante*** *m.* sill || ***dique de arco*** *m.* arch dam || ***dique de machones*** *m.* buttress dam || ***dique de terraplén*** *m.* embankment dam, earthfill dam || ***dique ígneo*** *m.* igneous dike || ***dique lateral*** *m.* levee || ***dique pegmático*** *m.* pegmatite dike || ***dique radial*** *m.* radial dike || ***dique terraplén*** *m.* causeway || ***enjambre de diques*** *m.* dike swarm
dirección *f.* strike, trend, direction, bearing
directorio *m.* board of directors
dirigido de ___ a ___ *a.* trending from ___ to ___
dirigirse *vb.* (geol.) to strike, trend
disarmónico *a.* disharmonic
discernimiento *m.* discrimination
discernir *vb.* to discern, distinguish, discriminate
disco *m.* disk || ***disco duro*** *m.* hard drive || ***disco flexible*** *m.* floppy disk || ***disco magnético*** *m.* magnetic disk || ***unidad de disco*** *f.* disk drive
discoidal *a.* disk-shaped
disconformidad *f.* unconformity
discontinuidad *f.* discontinuity || ***discontinuidad de Mohoro*** *f.* Mohorovicic discontinuity (Moho)
discontínuo *a.* discontinuous
discordancia *f.* discordance, unconformity || ***discordancia angular*** *f.* angular unconformity || ***discordancia regional*** *f.* regional unconformity
discordante *a.* discordant, unconformable

diseminación *f.* dissemination
diseminado *a.* disseminated
diseminar *vb.* to disseminate
diseño *m.* design, sketch
disfrute *m.* exploitation
disgregación *f.* disintegration
disgregamiento *m.* slacking
disgregar *vb.* to disintegrate, separate
disimetría *f.* disymmetry
disipar *vb.* dissipate
diskette *m.* (comp.) diskette, floppy disk
dislocación *f.* dislocation, throw of a fault, fault
dislocado *a.* displaced, faulted, disrupted
dislocar *vb.* to displace, fault
disminuación *f.* decrease
disminuir *vb.* to decrease
disociación *f.* dissociation, separation || ***disociación térmica*** *f.* thermal cracking
disolubilidad *f.* solubility
disoluble *a.* soluble
disolución *f.* dissolution, dissolving, solution || ***disolución por presión*** *f.* pressure solution
disolvente *m.* solvent
disolver *vb.* to dissolve
disparar *vb.* to shoot (shot holes, drill holes)
disparo *m.* shot, discharge, explosion; trip
dispersado *a.* dispersed
dispersar *vb.* to disperse, scatter
dispersión *f.* scattering, dispersion || ***dispersión coloidal*** *f.* colloidal dispersion
disperso *a.* scattered, sparse
disponerse en *vb.* to be found, to occur
disposición *f.* disposition, arrangement
dispositivo *m.* layout, arrangement; (comp.) device || ***dispositivo de tiro*** *m.* shot pattern || ***dispositivos sedimentarios superiores*** *m.* topsets || ***dispositivo simétrico*** *m.* split spread
distal *a.* distal, distant
distancia *f.* distance || ***distancia entre curvas de nivel*** *f.* contour interval || ***distancia focal*** *f.* focal distance || ***distancia horizontal de afloramientos*** *f.* normal horizontal separation || ***distancia vertical*** *f.* vertical separation
distanciamiento *m.* spacing
distante *a.* distant, distal, remote
distena *f.* disthene
disteno *f.* disthene
distensión *f.* extension, distension
distensivo *a.* extensional
distorsión *f.* distortion, strain
distorsionar *vb.* to distort, strain
distribución *f.* distribution; (seis.) spread
distribuido *a.* distributed || ***distribuido ampliamente*** *a.* widespread || ***distribuido esporádicamente*** *a.* patchy, sporadic
distribuidor *m.* distributer || ***distribuidor comercial*** *m.* dealer
distributario *m.* distributary
distributivo *m.* & *a.* distributary
distrito *m.* district || ***distrito minero*** *m.* mining district
distrófico *a.* dystrophic
disturbio *m.* disturbance
disuelto *a.* dissolved
disyunción *f.* fracture, jointing; separation (véase diaclasa, fractura, grieta) || ***disyunción prismática*** *f.* columnar jointing || ***disyunción tabular*** *f.* tabular jointing
diurno *a.* diurnal
divagación *f.* meandering
divalente *a.* divalent
divergencia *f.* divergence
divergente *a.* divergent
divergir *vb.* to diverge
diversificación *f.* diversification
diversificar *vb.* to diversify
dividendo *m.* dividend
divisas *f.* foreign exchange, currency || ***divisas fuertes*** *f.* hard currency
división *f.* joint; division, branch, divide || ***división de aguas*** *f.* drainage divide
divisoria *f.* divide || ***divisoria de aguas*** *f.* watershed || ***divisoria de vertientes*** *f.* watershed
divorcio de aguas *m.* drainage divide, ridge
dobla-caños *m.* pipe bender
doblado *a.* bent, folded
dobladura *f.* bending

doblar *vb.* to fold, bend

doble *m. & a.* duplex; double

doblez *m.* fold, crease, kink, bend, dogleg ‖ ***doblez pata de pero*** *m.* dogleg

documento *m.* document; bank note

dodecaedro *m.* dodecahedron

dodecano *m.* dodecane

Dogger *m. & a.* Dogger

dolerita *f.* dolerite, trap rock

dolina *f.* doline, karst, sinkhole

dolomía, dolomita *f.* dolomite

dolomítico *a.* dolomitic

dolomitizar *vb.* to dolomitize

Domeriano, Domeriense *m. & a.* Domerian

doméstico *a.* domestic

dominio *m.* domain ‖ ***dominio de inclinación*** *m.* dip domain ‖ ***dominio de inclinación frontal*** *m.* forward dip domain ‖ ***dominio de inclinación dorsal*** *m.* rear dip domain

dominó *m.* domino

domo *m.* dome, cupola ‖ ***domo cúmulo*** *m.* volcanic dome ‖ ***domo de exfoliación*** *m.* exfoliation dome ‖ ***domo de lava*** *m.* lava dome ‖ ***domo de penetración*** *m.* piercement dome ‖ ***domo de sal*** *m.* salt dome ‖ ***domo gasífero*** *m.* gas cap ‖ ***domo salino*** *m.* salt dome ‖ ***domo volcánico*** *m.* volcanic dome

dorsal *f.* ridge; *a.* back, dorsal ‖ ***dorsal oceánica*** *f.* mid-ocean ridge

dosificación *f.* titration, dosage, proportioning

dosificar *vb.* to proportion

dosis *f.* dose, (chem.) proportion

dotación *f.* crew ‖ ***dotación lógica*** *f.* software

Downtoniano, Downtoniense *m. & a.* Downtonian

draga *f.* dredge, dragline

dragado *m.* dredging

dragador *m. & a.* dredger, dredging

dragaje *m.* dredging

dragalina, dragalínea *f.* dragline

dragar *vb.* to dredge

drenaje *m.* drainage (véase avenamiento) ‖ ***drenaje acostillado*** *m.* trellis drainage ‖ ***drenaje anular*** *m.* annular drainage ‖ ***drenaje de mina*** *m.* mine drainage ‖ ***drenaje dendrítico*** *m.* dendritic drainage ‖ ***drenaje efectivo*** *m.* net drainage ‖ ***drenaje por expansión*** *m.* depletion drive drainage ‖ ***drenaje por expansión de gas*** *m.* gas drive ‖ ***drenaje por gravedad*** *m.* gravity drainage ‖ ***drenaje subterráneo*** *m.* underground drainage

drenar *vb.* to drain, remove, siphon off

dromocrona *f.* (seis.) travel time curve

drumlin *m.* drumlin

drusa *f.* druse ‖ ***en drusa*** *a.* drusy

dúctil *a.* ductile

ductilidad *f.* ductility

dudoso *a.* uncertain

dueño *m.* owner

dulce *a.* sweet, unsulfured

dumortierita *f.* dumortierite

duna *f.* dune ‖ ***duna de arena*** *f.* sand dune ‖ ***duna falciforme*** *f.* barchan ‖ ***duna longitudinal*** *f.* longitudinal dune ‖ ***duna parabólica*** *f.* parabolic dune ‖ ***duna seif*** *f.* seif dune ‖ ***duna transversal*** *f.* transverse dune ‖ ***duna trepadora*** *f.* climbing dune

dunita *f.* dunite

duplex *m.* duplex

durabilidad *f.* durability

duracion *f.* duration, length of time

dureza *f.* hardness ‖ ***dureza del agua*** *f.* water hardness

duricostra *f.* duricrust

durita *f.* durain

duro *a.* hard, tough

E

ebonita *f.* ebonite
ebullición *f.* boiling
eclinómetro *m.* clinometer
eclogita *f.* eclogite
eco *m.* echo
ecología *f.* ecology
economía *f.* economy
económica *f.* economics
económico *a.* economic
ecosistema *m.* ecosystem
ecosonda *f.* echo sounder
ecuación *f.* equation || ***ecuación de onda*** *f.* wave equation
ecuador *m.* equator
ecuatorial *a.* equatorial
echado *m. & a.* [Méx.] dip; inclined || ***echado abajo*** *m. & a.* downdip || ***echado arriba*** *m. & a.* updip
echar a andar *vb.* to start
edad *f.* age || ***edad absoluta*** *f.* absolute age
edafología *f.* pedology
edenita *f.* edenite
educción *f.* discharge
efectivo *m.* cash; *a.* effective
efecto *m.* effect || ***efectos de barreras sin flujo*** *m.* no-flow boundary effects || ***efectos del límite de presión constante*** *m.* constant pressure boundary effects || ***efecto superficial en el pozo*** *m.* wellbore skin effect
eficacia *f.* effectiveness, efficiency || ***eficacia de barrido*** *f.* sweep efficiency
eficaz *a.* effective, efficient
eficiencia *f.* efficiency
eficiente *a.* efficient
efimero *a.* ephemeral
eflorescencia *f.* efflorescence; [Pe.] outcrop
eflorescente *a.* efflorescent
efluente *m.* effluent, outflow
efusión *f.* eruption, extrusion || ***efusión de lava*** *f.* lava flow || ***efusión de lodo*** *f.* mud flow || ***efusión de petróleo bajo de la influencia de un solvente*** *f.* streaming cut || ***efusión por grietas*** *f.* fissure flow || ***efusión térmica*** *f.* hot spot
efusivo *a.* extrusive
egirina *f.* aegirine
egirita *f.* aegirite
Eifeliano, Eifeliense *m. & a.* Eifelian
einstenio *m.* einsteinium
eje *m.* axis; axle, shaft; core || ***eje anticlinal*** *m.* anticlinal axis || ***eje cristalográfico*** *m.* crystallographic axis || ***eje de coordenadas*** *m.* coordinate axis || ***eje de plegamiento*** *m.* fold axis || ***eje de revolución*** *m.* axis of rotation || ***eje de simetría*** *m.* axis of symmetry || ***eje geométrico*** *m.* center line || ***eje óptico*** *m.* optic axis || ***eje principal de deformación*** *m.* principal strain axis || ***ejes principales máximo y mínimo*** *m.* maximum and minimum principal axes
ejecución *f.* performance, workmanship
ejemplar *f.* sample, specimen
ejemplo *m.* example
elaboración *f.* processing || ***elaboración petróleo*** *f.* oil refining
elaborar *vb.* elaborate, process
elasticidad *f.* elasticity
elástico *a.* elastic
elastómero *m.* elastomer
elaterita *f.* elaterite
electricidad *f.* electricity || ***electricidad estática*** *f.* static electricity
eléctrico *a.* electric, electrical
electro *m.* amber; electrum
electrodo *m.* electrode
electroestático *a.* electrostatic
electrofiltración *f.* electrofiltration
electrogalvánico *a.* electrogalvanic
electrogenerador *m.* electric generator
electroimán *m.* electromagnet
electrólisis *f.* electrolysis
electrolito *m.* electrolyte
electrolización *f.* electrolysis
electromagnético *a.* electromagnetic
electrometalurgía *f.* electrometallurgy
electromotriz *a.* electromotive
electrón *m.* electron

electronegativo *a.* electronegative
electrónico *a.* electronic
electroósmosis *f.* electroosmosis
electropositivo *a.* electropositive
electroquímico *a.* electrochemical
elemento *m.* element || ***elemento matricial*** *m.* matrix element, pixel || ***elemento trazo*** *m.* trace element
eleolita *f.* nepheline
elevación *f.* elevation, height || ***elevación suave*** *f.* swell
elevador *m.* elevator, hoist || ***elevador de petróleo por presión del gas*** *m.* gas lift || ***elevador para barras de sondeo*** *m.* drill pipe elevator || ***elevador para tubería*** *m.* tubing elevator
elevar *vb.* to lift, to hoist
elevarse *vb.* to rise, to aggrade
eliminación *f.* disposal || ***eliminación de esfuerzo*** *f.* stress relief
eliminado *a.* removed
eliminar *vb.* to dispose
elipse *f.* ellipse
elipsoidal *a.* ellipsoidal
elipsoide de deformación *m.* strain ellipse || ***elipsoide de deformación infinitesimal*** *m.* infinitesimal strain ellipse
elíptico *a.* elliptical
elongación *f.* elongation
elutriación *f.* elutriation
eluviación *f.* eluviation
eluvial *a.* eluvial
eluvión *f.* eluvium, talus
emanación *f.* seep, emanation, exhalation || ***emanación de petróleo manantial*** *f.* seep
embalado *a.* packed
embalaje *m.* packing
embalar *vb.* to pack
embalsar *vb.* to dam, impound
embalse *m.* reservoir; embayment
embarcadero *m.* loading platform, wharf
embarque *m.* shipment
embargar *vb.* to seize
embargo *m.* seizure
embate *m.* swash
embocadura *f.* entrance, mouth of a river
embolita *f.* embolite
émbolo *m.* piston, plunger || ***émbolo buzo*** *m.* plunger
emboquillado *m.* mine entrance
emborrascarse *vb.* to pinch out
embozar *vb.* to wrap
embudo *m.* sink hole; funnel, hopper
emergencia *f.* emergence
emergente *a.* emergent
emerger *vb.* to project, jut out, emerge
eminencia *f.* rise
emisión *f.* emission; issuance, output
emisor *m.* transmitter; transceiver
empalmar *vb.* to splice, join
empalme *m.* jointing
empapar *vb.* to soak
empaque *m.* packing
empaqueatado *a.* packed
empaquetador *m.* packer || ***empaquetador de cementación*** *m.* cementing packer || ***empaquetador de fondo*** *m.* bottom hole packer || ***empaquetador para cañería perdida*** *m.* liner packer
empaquetadura *f.* packing, gasket
empaquetar *vb.* to pack
emparejar *vb.* grade, level
empastamiento de trépano *m.* (perf.) balling of the bit
empastar *vb.* to paste, gum up, ball up
empinado *a.* steep
emplazado *a.* seated, emplaced
emplazamiento *m.* emplacement
emplazar *vb.* to emplace
empleado *m.* employee
empleador *m.* employer
emplear *vb.* to employ
empleo *m.* employment
empobrecer *vb.* to peter out, impoverish, deplete
empobrecido *a.* impoverished, depleted, starved
empobrecimiento *m.* depletion of ore, impoverishment of a deposit

empotrado *a.* embedded; fixed

empotrar *vb.* to embed, fix

empresa *f.* business, company, enterprise || ***empresa conjunta*** *f.* joint venture

empresario *m.* jobber, contractor, merchant

empréstito *m.* loan

empujadora niveladora *f.* bulldozer

empujar *vb.* to push

empuje *m.* pressure, drive, push || ***empuje de agua*** *m.* water drive || ***empuje hidráulico*** *m.* water drive || ***empuje hidrostático*** *m.* water drive || ***empuje por gas*** *m.* gas drive

Emscheriano, Emscheriense *m. & a.* Emscherian

Emsiano, Emsiense *m. & a.* Emsian

emulsificación *f.* emulsification

emulsificador *m.* emulsifier

emulsionado *a.* emulsified

emulsificar *vb.* to emulsify

emulsión *f.* emulsion

emulsionador *m.* emulsifier

emulsionar *vb.* to emulsify

en *p.* in, on

enajenada *f.* change of ownership

enarenar(se) *vb.* to sand up

enargita *f.* enargite

encañado *a.* (geog.) dissected

encapado *a.* cloaked, coated, encased; capped

encargado *m. & a.* superintendent; in charge

encargo *m.* job, assignment; consignment

encastrado *a.* embedded, built in

encastrar *vb.* to embed, encase

encauzamiento *m.* channeling

encauzar *vb.* to channel

encender(se) *vb.* to ignite

encentrar *vb.* to open a vein along strike

encerrar *vb.* to enclose

encerrado *a.* enclosed

encima *p.* above

enclavar cañería *vb.* to set pipe

enclave *m.* inclusion

encogimiento *m.* shrinkage

en construcción *a.* under construction

encontrar petróleo (o minerales) *vb.* to strike oil (or ore)

encontrarse *vb.* to be found

en corriente *a.* on-line, on-stream

encorvado *a.* curved, crooked

encorvadura *f.* bending, curvature

encorvamiento hacia abajo *m.* downwarp || ***encorvamiento hacia arriba*** *m.* upwarp

encorvar *vb.* to curve, bend

encostramiento *m.* (perf.) mud caking

encrespadura *f.* ripple mark

encuellador *m.* [Vz.] derrickman

encuesta *f.* survey

encharcamiento *m.* waterlogging

enchufar *vb.* telescope, nest; mesh; plug-in

enchufe *m.* joint, connection, socket, sleeve || ***enchufe de lodo*** *m.* mud crib, mud box

endémico *a.* endemic

endentado *a.* serrate

enderezado *a.* appropriate, favorable, opportune

enderezadora *f.* pipe straightener

enderezar *vb.* to align, straighten

endicamiento *m.* impounding

endicar *vb.* to impound

endliquita *f.* endlichite

endogénesis *f.* endogenesis

endogenético, endógeno *a.* endogenetic, endogenous

endomórfico *a.* endomorphic

endomorfismo *m.* endomorphism

endoplasma *f.* endoplasm

endosar *vb.* to endorse

endoso *m.* endorsement

endotérmico *a.* endothermic, endothermal

endulzar petróleo *vb.* to sweeten oil, to desulfur

endurecer(se) *vb.* to harden

endurecido *a.* indurated

endurecimiento *m.* induration

en el lugar *a.* in place, in situ

en el sentido de las agujas del reloj *a.* clockwise

en el sentido opuesto de las agujas del reloj *a.* counterclockwise

energía *f.* energy, power, force || ***energía cinética*** *f.* kinetic energy || ***energía de activación*** *f.* activation energy || ***energía deficiente*** *f.* low energy, energy deficient || ***energía específica*** *f.* specific energy || ***energía libre*** *f.* free energy || ***energía mecánica*** *f.* mechanical energy || ***energía potencial*** *f.* potential energy || ***energía química*** *f.* chemical energy || ***energía radiante*** *f.* radiant energy || ***energía térmica*** *f.* thermal energy

en escalón *a.* en echelon

en existencia *a.* in stock

enfermedades laborales *f.* occupational diseases

enfocar *vb.* to focus

enfoque *m.* focus, grasp, approach

en forma de *a.* -shaped || ***en forma de U*** *a.* U-shaped

en fragüe de *a.* waiting on

enfriador *m.* cooler

enfriamiento *m.* cooling || ***enfriamiento por evaporación*** *m.* evaporative cooling

enfriar *vb.* to cool

en funcionamiento *a.* on-stream, on-line, in operation

enganchador *m.* derrickman

engastar *vb.* to embed, to set

englazado *a.* glaciated

englobado *a.* enclosed

englobar *vb.* to enclose

engolfamiento *m.* embayment

engranando *a.* intertonguing

engrasar *vb.* to grease

engrosamiento *a.* thickening, swelling

engrosarse *vb.* to swell, to thicken

enigmatita *f.* aenigmite

enjambre de diques *m.* dike swarm

enlace *m.* bond, tie || ***enlace atómico*** *m.* atomic bond

en la estructura *a.* on-structure

enlame *m.* silting

enlazar *vb.* to connect, join, bind, tie, link, loop

enlazimiento *m.* connection, link

enmallamiento *m.* grid || ***enmallamiento agrimensura*** *m.* survey grid || ***enmallamiento sísmico*** *m.* seismic grid

enmascarar *vb.* to obscure

en montaje *a.* under construction

en operación *a.* on-line, on-stream, operating

enrejado *m. & a.* lattice; latticed || ***enrejado cristalino*** *m.* crystal lattice

enriquecimiento *m.* enrichment || ***enriquecimiento secundario*** *m.* secondary enrichment

enrollado *a.* convoluted

ensamblar *vb.* to abut, join, connect, couple

ensanchado *a.* widening, flaring

ensanchador *m.* expander, reamer || ***ensanchador de caños*** *m.* pipe expander || ***ensanchador para pozo piloto*** *m.* rat hole reamer

ensanchamiento *m.* swelling, enlargement, widening

ensanchar *vb.* to ream, expand

ensancharse *vb.* to become wider, widen

ensanchatubos *m.* pipe expander

ensanche *m.* widening

ensayar *vb.* to assay, test

ensayador *m.* assayer

ensaye, ensayo *m.* assay, test; assay office || ***ensayo con ácido*** *m.* acid test || ***ensayo de acidez*** *m.* acid test || ***ensayo de capas*** *m.* drill stem test || ***ensayo de pozo*** *m.* well test || ***ensayo de pozo estratigrafico*** *m.* stratigraphic test || ***ensayo de surgencia libre*** *m.* open flow test || ***ensayo de surgencia restringida*** *m.* restricted flow test || ***ensayo de vibración*** *m.* vibration test || ***ensayo sobre el terreno*** *m.* field test

ensenada *f.* bay, cove, embayment, inlet, creek, [U.S.] bayou

en servicio *a.* in service, on line, on stream

en sitio *a.* in situ, in place

enstatita *f.* enstatite

entallar *vb.* to sculpture
entalpía *f.* enthalpy
entendimiento *m.* understanding
enterrado *a.* buried
enterrar *vb.* to bury || ***enterrar cañería*** *vb.* to lower pipe
entibado *m.* cribbing
en tierra *a.* onshore
entierro *m.* burial
entrada *f.* entrance, recess, adit, intake; cash receipts || ***entrada de la marea*** *f.* tidal inlet
entrampar *vb.* to trap
en tránsito *a.* in transit
entrecortar *vb.* intersect
entrecrecer *vb.* to intergrow
entrecrecido *a.* intergrown
entrecrecimiento *m.* intergrowth
entrecruzado *a.* cross-bedded
entreduna *a.* interdune
entrega *f.* delivery
entregar *vb.* to deliver
entrelaminado *a.* interbedded, interlaminated
entrelazado *a.* interlocking
entremontaña *a.* intermontain
entrenamiento *m.* training
entrenar *vb.* to train, teach
entresuelo *m.* gallery between two mine levels
entreverado *m. & a.* stockwork
entropía *f.* entropy
entubación *f.* setting pipe, casing a well
entubamiento *m.* casing
entubar *vb.* to run casing || ***entubar a presión*** *vb.* to snub-in
enumerar *vb.* to list
envasar *vb.* to fill
envejecimiento *m.* aging
enviado por cañería *a.* piped
enviar *vb.* to send
envoltura *f.* casing, wrapping
envolvente *a.* involute
envolver *vb.* to surround, encase
envuelto *a.* encased, surrounded
Eocambriano, Eocámbrico *m. & a.* Eocambrian
Eoceno, Eocénico *m. & a.* Eocene
eólico, eolio *a.* eolian, aeolian
eón *m.* aeon
Eozóico *a.* Paleozoic
epéirico, epírico *a.* epeiric
epicentro *m.* epicenter
epiclástico *a.* epiclastic
epicontinental *a.* epicontinental, neritic
epidota, epidoto *f. & m.* epidote
epigénesis *f.* epigenesis
epigenético, epigénico *a.* epigenetic
epipelágico *a.* epipelagic
epírico *a.* epeiric
epirogénesis *f.* epeirogenesis, epeirogeny
epirogenético, epirogénico *a.* epeirogenetic
epirogenia *f.* epeirogenesis, epeirogeny
epirogénico *a.* epeirogenic
episodio *m.* episode, event
epitermal *a.* epithermal
epizona *f.* epizone
época *f.* epoch || ***época fría*** *f.* ice age || ***época glacial*** *f.* ice age
epsomita *f.* epsomite
equigranular *a.* even-grained
equilibrio *m.* equilibrium; material balance || ***equilibrio de masa*** *m.* material balance, mass balance || ***equilibrio dinámico*** *m.* dynamic equilibrium || ***equilibrio entre sólido y líquido*** *m.* solid–liquid equilibrium || ***equilibrio estable*** *m.* stable equilibrium || ***equilibrio inestable*** *m.* unstable equilibrium || ***equilibrio isostático*** *m.* isostatic equilibrium
equinoccio *m.* equinox || ***equinoccio de otoño*** *m.* autumnal equinox || ***equinoccio de primavera*** *m.* spring equinox
equinoideo *m.* echinoid
equipo *m.* equipment, gear, rig; shift; team || ***equipo de ensayo*** *m.* testing equipment || ***equipo de primeros auxilios*** *m.* first aid kit || ***equipo de producción*** *m.* production equipment || ***equipo físico*** *m.* hardware || ***equipo para elaboración*** *m.* processing

equipment || ***equipo perforador*** *m.* drilling rig

equipotencial *a.* equipotential

equivalente *a.* equivalent

era *f.* era; salty evaporite pond

eregir *vb.* erect

ergio *m.* (phys., física) erg

erguimiento *m.* uplift

eritrina *f.* erythrite

erodado *a.* eroded

erosión *f.* erosion, weathering || ***erosión de marea*** *f.* wave erosion || ***erosión diferencial*** *f.* differential erosion || ***erosión eólica*** *f.* wind erosion || ***erosión fluvial*** *f.* river erosion || ***erosión glacial*** *f.* glacial erosion || ***erosión mecánico*** *f.* mechanical erosion || ***erosión químico*** *f.* chemical erosion || ***erosión retrocedente*** *f.* headward erosion || ***erosión y relleno de cauce*** *f.* scour and fill

erosionado *a.* eroded

erosionar *vb.* to erode

erosivo *a.* erosive

errático *m. & a.* erratic

error *m.* mistake, error, aberration || ***error cuadrático medio*** *m.* mean square error || ***error de redondeo*** *m.* rounding error || ***error de truncamiento*** *m.* truncation error

erubescita *f.* bornite, erubescite

erupción *f.* eruption

eruptivo *a.* eruptive, volcanic

escabroso *a.* rugged

escala *f.* scale; ladder || ***escala de tiempo*** *f.* time scale || ***escala gráfica*** *f.* bar scale

escalenoédrico *a.* scalenohedral

escalenoedro *m.* scalenohedron

escalera *f.* ladder, stairway

escalón *m.* step; stope || ***escalón de falla*** *m.* fault scarp || ***escalón de fractura*** *m.* fault scarp

escalonado *a.* staggered, en echelon, stepped, shingled

escalonamiento *m.* steps, stepping

escalonar *vb.* to stagger

escama *f.* slice, flake; scale; fault, sliver

escamiforme, escamoso *a.* scaly, platy

escamociones *f.* incrustations, scale

escamoso *a.* scaly, flaky

escanear *vb.* to scan

escape *m.* leak, outlet, exhaust

escapolita *f.* scapolite

escarapela *f.* rosette

escarcha *f.* frost

escariador *m.* reamer

escariar *vb.* to ream

escarpa *f.* slope, talus, scarp, escarpment, bluff || ***escarpa de falla*** *f.* fault scarp || ***escarpa de fractura*** *f.* fault scarp || ***escarpa de línea de falla*** *f.* fault line scarp || ***escarpa tectónica*** *f.* tectonic scarp

escarpado *a.* steep

escarpe *m.* slope, escarpment || ***escarpe de playa*** *m.* beach slope

escasez *f.* scarcity, shortage, lack

escaso *a.* rare, infrequent, sparse

escintilómetro *m.* scintillometer

Escitiano, Escitiense *m. & a.* Scythian

esclusa *f.* sluice

escoger *vb.* to sort, choose || ***escoger a mano*** *vb.* to hand sort || ***escoger lo mejor*** *vb* to highgrade

escogido *a.* selected

escolecodonto *m.* scolecodont

escollera *f.* cliff; pier, breakwater, jetty

escollo *m.* shoal, reef

escombrera *f.* slag heap, waste heap, dump

escombro *m.* waste rock, detritus, overburden, debris || ***escombro de granito*** *m.* granite wash || ***escombro glacial*** *m.* glacial outwash

escopetar *vb.* to work a mine

escoria *f.* scoria, slag, dross, clinker

escorial *m.* slag heap, volcanic ash, lava bed

escota *f.* miner's pick

escotadura *f.* indentation

escotilla *f.* shaft, hatchway

escrepa de empuje *f.* bulldozer

escribir *vb.* to write

escrobiculado *a.* pitted

escudo *m.* shield || ***escudo precambrico*** *m.* Precambrian shield

esculpir *vb.* to carve, sculpture

escurrente *m. & a.* exhalant, effluent

escurrimiento *m.* drainage, runoff, flow; slip, glide, creep || ***escurrimiento del cristal*** *m.* crystal gliding || ***plano de escurrimiento*** *m.* glide plane

escurrirse *vb.* to runoff, leak; to glide, slip

esexita *f.* essexite

esfalerita *f.* sphalerite

esfena *f.* sphene

esfeno *m.* sphene

esfera *f.* sphere

esférico *a.* spherical

esferoide *m.* spheroid

esferolítico *a.* spherulitic

esferolito *m.* spherulite

esferulita *f.* spherulite

esfuerzo *m.* stress, force; effort || ***esfuerzo cizallante*** *m.* shear stress || ***esfuerzo cortante*** *m.* shear stress || ***esfuerzo de compresión*** *m.* compressive stress || ***esfuerzo de corte*** *m.* shear stress || ***esfuerzo de escurrimiento*** *m.* creep stress || ***esfuerzo de flexión*** *m.* bending stress || ***esfuerzo de rotura*** *m.* ultimate stress || ***esfuerzo de tensión*** *m.* tensile stress || ***esfuerzo de torsión*** *m.* torque || ***esfuerzo de tracción*** *m.* tensile stress || ***esfuerzo directo*** *m.* normal stress || ***esfuerzo flexor*** *m.* bending stress || ***esfuerzo límite*** *m.* ultimate stress || ***esfuerzo normal efectivo*** *m.* effective normal stress || ***esfuerzo oblicuo*** *m.* stress deviator || ***esfuerzo por compresión*** *m.* compressive stress || ***esfuerzo restante*** *m.* residual stress

esker *m.* esker

eslamosa *f.* nagyagite

esmaltina, esmaltita *f.* smaltite

esmanil *m.* sphalerite

esmeralda *f.* emerald

esmeril *m.* emery

esmitsonita *f.* smithsonite

espaciado *m.* spacing

espaciador *m.* spacer, separator, packing

espaciamiento *m.* spacing || ***espaciamiento de pozos*** *m.* well spacing || ***espaciamiento intersticial*** *m.* void space || ***espaciamiento poral*** *m.* pore space || ***espaciamientos sísmicos*** *m.* shotpoint groupings, shotpoint spacing

espaciar *vb.* to space

espacio *m.* space, spacing

espalto *m.* spalt, fluorspar flux

esparagmita *f.* sparagmite

esparcir *vb.* to spread, scatter

Esparnaciano, Esparnaciense *m. & a.* Sparnacian

espático *a.* sparry, spathic

espato *m.* spar || ***espato adamantino*** *m.* corundum || ***espato amargo*** *m.* magnesite || ***espato azul*** *m.* lazulite || ***espato bruno*** *m.* dolomite, magnesite || ***espato calcáreo, cálcico*** *m.* calcite || ***espato calizo, de cal*** *m.* calcite || ***espato de cinc*** *m.* smithzonite || ***espato de flúor*** *m.* fluorite || ***espato de hierro*** *m.* siderite || ***espato de Islandia*** *m.* Iceland spar || ***espato de itrio*** *m.* xenotime || ***espato de magnesio*** *m.* rhodochrosite, rhodonite, or dialogite || ***espato en tablas*** *m.* wollastonite || ***espato ferrífero*** *m.* siderite || ***espato frambuesa*** *m.* rhodochrosite || ***espato lustroso*** *m.* satin spar || ***espato pesado*** *m.* barite || ***espato romboidal*** *m.* dolomite || ***espato satinado*** *m.* satin spar

especie tipo *m.* type species

especies *m.* species

especificación *f.* specification

espécimen *m.* specimen

espectro *m.* spectrum

espectrómetro *m.* spectrometer || ***espectrómetro colectivo*** *m.* mass spectrometer || ***espectrómetro de rayos infrarrojos*** *m.* infrared spectrometer || ***espectrómetro global*** *m.* mass spectrometer

especular *a.* specular

especularita *f.* specular hematite

espejado *m.* [Pe.] galena

espejo *m.* mirror; [Col.] slickenside || ***espejo concavo*** *m.* concave mirror || ***espejos de fricción*** *m.* slickensides || ***espejo oscilante*** *m.* oscillating mirror

espejuelo *m.* mica; transparent talc

espeleología *f.* spelunking, caving, speleology

espesamiento *m.* thickening

espesar(se) *vb.* to thicken

espesartina, espesartita *f.* spessartite

espeso *a.* thick, dense, heavy

espesor *m.* thickness

espesura *f.* thickness

espetón *m.* spit

espícula *f.* spicule

espiga *f.* bay bar

espigón *m.* jetty

espilita *f.* spilite

espilítico *a.* spilitic

espina de pez *a.* herringbone

espinazo *m.* hogback

espinela *f.* spinel || ***espinela cincífera*** *f.* gahnite, franklinite

espinelo *m.* spinel

espiral *a.* spiral

espirífero *m.* spirifer

espodito *m.* white volcanic ash

espodumen *m.* spodumene

espodumena *f.* spodumene

espolón *m.* spur, appendage, (geol.) nose || ***espolón del anticlinal*** *m.* anticlinal nose || ***espolón truncado*** *m.* truncated spur, faceted spur

esponja *f.* stockwork of veins

esponjoso *a.* spongework, spongy

espora *f.* spore

esporádico *a.* sporadic, occasional

esporinita *f.* sporinite

espuma *f.* meerschaum; gossan; magnesia; foam || ***espuma de caucho*** *f.* foam rubber || ***espuma de hule*** *f.* foam rubber || ***espuma de mar*** *f.* sepiolite, meerschaum

espumado *a.* skimmed

espumar *vb.* to skim

esqueletal *a.* skeletal

esqueleto *m.* skeleton, frame

esquema *m.* sketch, classification; diagram, plan || ***esquema gráfico*** *m.* graph

esquirla *f.* splinter

esquisto *m.* schist, slate || ***esquisto anfibolítico*** *m.* amphibolite schist || ***esquisto bituminoso*** *m.* oil shale || ***esquisto clorítico*** *m.* chlorite schist, greenschist || ***esquisto micáceo*** *m.* mica schist || ***esquisto talcoso*** *m.* talc schist || ***esquisto verde*** *m.* greenschist

esquistóideo *m.* schistoid

esquistocidad *f.* schistocity, cleavage

esquistoso *a.* schistose, slaty

essexita *f.* essexite

essonita *f.* hessonite

estabilidad *f.* stability

estabilización *f.* stabilization

estabilizador *m.* stabilizer

estabilizadora *f.* stripper tower

estabilizante *m. & a.* stabilizer; stabilizing

estabilizar *vb.* to stabilize

estable *a.* stable

establecer *vb.* to establish

estaca *f.* (survey, levan.) stake, pole

estacar *vb.* to stake

estación *f.* season; station, plant || ***estación acotada*** *f.* bench mark || ***estación auxiliar*** *f.* booster station || ***estación de base*** *f.* base station || ***estación de bombas*** *f.* pump station || ***estación de control*** *f.* control station || ***estación de gasolina*** *f.* gas station || ***estación de servicio*** *f.* service station, gas station || ***estación de trabajo*** *f.* (comp.) workstation || ***estación generadora*** *f.* powerhouse, generating station || ***estación recolectora*** *f.* gathering station

estacional *a.* seasonal

estacionario *a.* stationary

estadal *m.* stadia, level rod; measure of length of 3.3 m; area measure of 11.2 m^2

estadalero *m.* rodman

estadia *f.* stadia, leveling rod

estadio *m.* furlong

estadística *f.* statistics

estado *m.* stage, state || ***estado estacionario*** *m.* steady-state || ***estado juvenil*** *m.* youth stage || ***estado natural*** *m.* natural state || ***estado original*** *m.* original state || ***estado senil*** *m.* old age stage || ***estado virgen*** *m.* natural state

Estaffordiano, Estaffordiense *m. & a.* Staffordian

estalactita *f.* stalactite

estalagmita *f.* stalagmite

estallar *vb.* to blast

estallido por presión *m.* rock burst

Estampiano, Estampiense *m. & a.* Stampian

estancado *a.* stagnant

estanco *a.* airtight, watertight, weatherproof

estándar, estándard *m. & a.* standard

estandarización, estandardización *f.* standardization

estandarizar, estandardizar *vb.* to standardize

estanekita *f.* stannekite

estanina, estannita *f.* stannite

estanoso *a.* stannous

estanque *m.* pond, basin, water tank || ***estanque para colas*** *m.* tailings pond

estañífero *a.* stannous, tin-bearing

estañina *f.* stannite

estaño *m.* tin, tin ore, cassiterite

estañoso *a.* stannous

estatalización *f.* nationalization

estática *f.* (seis.) statics || ***estáticas de refracción*** *f.* refraction statics

estático *a.* static

estaurolita *f.* staurolite

este *m.* east.

estearina *f.* stearine

estearato *m.* stearate

esteatita *f.* steatite, soapstone

Estefaniano, Estefaniense *m. & a.* Stephanian

estefanita *f.* stephanite

estelar *a.* stellar

estelión *f.* olivine dolerite

estepa *f.* steppe

estequiometría *f.* stoichiometry

éster *m.* ester

estereo *m. & a.* stereo

estereográfico *a.* stereographic

estereograma *m.* stereogram, block diagram

estereómetro *m.* stereometer

estereoscopía *f.* stereoscopy

estereoscópico *a.* stereoscopic

estereoscopio *m.* stereoscope

estepa *f.* steppe

esteparío *a.* steppe-like

estéril *m. & a.* gangue; barren, sterile

estero *m.* estuary, inlet, bog, bayou, marsh; [Arg., Méx.] swamp; [Bol., Ch.] creek; [Vz.] pond

estiaje *m.* low water mark

estibina, estibnita *f.* stibnite

estilbita *f.* stilbite

estilo *m.* style || ***estilo tectonico*** *m.* tectonic style

estilolita, estilolito *f. & m.* stylolite

estimable *a.* valuable

estimación *f.* estimation

estimador *m.* estimator

estimar *vb.* to estimate

estimarse *vb.* to suggest to be, be taken to be

estimulación *f.* stimulation

estimular *vb.* to stimulate

estipulaciónes *f.* stipulations, terms

estirar *vb.* stretch, pull, strain

estiramiento *m.* stretching || ***estiramiento por presión*** *m.* extrusion

estireno *m.* styrene

estoperol *m.* tubing head

estorbo *m.* impingement

estrangulación *f.* narrowing of a vein, bottleneck, stoppage

estrangulado *a.* choked

estrangulador *m.* choke

estrangulamiento *m.* narrowing of a vein, stoppage, bottleneck

estrangular *vb.* to cut off, choke

estrategia *f.* strategy

estratégico *a.* strategic

estratificación *f.* strata, bedding, stratification, layering || ***estratificación clasificada*** *f.* graded bedding || ***estratificación cruzada*** *f.* crossbedding || ***estratificación delgada*** *f.* & *a.* thin bedding, thin bedded || ***estratificación diagonal*** *f.* crossbedding || ***estratificación entrecruzada*** *f.* crossbedding || ***estratificación espesa*** *f.* & *a.* thick-bedded || ***estratificación gradada, graduada, gradual*** *f.* graded bedding || ***estratificación intraplegada*** *f.* convolute bedding || ***estratificación laminar*** *f.* laminar bedding || ***estratificación lenticular*** *f.* lenticular bedding || ***estratificación masiva*** *f.* massive bedding || ***estratificación ondulada o ondulítica*** *f.* wavy crossbedding || ***estratificación potente*** *f.* & *a.* thick bedded

estratificado *a.* stratified

estratificar *vb.* stratify

estratiforme *a.* stratiform

estratigrafía *f.* stratigraphy || ***estratigrafía de secuencias*** *f.* sequence stratigraphy || ***estratigrafía secuencial*** *f.* sequence stratigraphy

estratígrafo *m.* stratigrapher

estratigráfico *a.* stratigraphic

estrato *m.* strata, bed, layer; stratus clouds (véase capa, horizonte, lecho) || ***estrato anual*** *m.* varve || ***estrato carbonífero*** *m.* coal measure || ***estrato clave*** *m.* key bed || ***estrato competente*** *m.* competent bed || ***estrato cruzado*** *m.* crossbed || ***estrato cruzante*** *m.* crossbed || ***estrato de almacenamiento*** *m.* reservoir bed || ***estrato de cobertura*** *m.* cap rock || ***estrato freático*** *m.* aquifer || ***estrato generador*** *m.* source rock || ***estrato guía*** *m.* key bed || ***estrato impermeable*** *m.* impervious bed, aquiclude, aquitard || ***estrato incompetente*** *m.* incompetent bed || ***estrato índice*** *m.* key bed || ***estrato madre*** *m.* source rock || ***estrato mineralizado*** *m.* ore bed || ***estrato productivo*** *m.* reservoir, pay interval || ***estratos frontales*** *m.* foreset beds

estratocúmulo *m.* stratocumulus

estratos entrecruzados *m.* crossbedding

estratósfera *f.* stratosphere

estratovolcán *m.* strato-volcano

estrechamiento *m.* thinning or pinching

estrecharse *vb.* to thin

estrechez *f.* narrowness; gap, pass

estrecho *m.* & *a.* strait, narrows; narrow

estrechón *m.* narrowness

estrechura *f.* narrows, gap, pass; bottleneck

estrellarse la veta *vb.* [Pe.] to peter out, become lean

estría(s) *f.* striation(s), stria, striae, groove(s), furrow, gouge, flute

estriación *f.* striation

estriado *a.* striated, fluted, slickensided

estriadura *f.* striation

estriar *vb.* to sort ore; to groove, flute, gouge

estribación *f.* foothills, spur

estribo *m.* abutment, pillar

estribor *a.* starboard

estromatolita *f.* stromatolite

estromeyerita *f.* stromeyerite

estroncianita *f.* strontianite

estroncio *m.* strontium

estropeado *a.* worn out

estructura *f.* structure || ***estructura abajo*** *a.* down structure || ***estructura alveolar*** *f.* honeycomb structure || ***estructura amigdaloide*** *f.* amygdaloid structure || ***estructura arriba*** *a.* upstructure || ***estructura atómica*** *f.* atomic structure || ***estructura cerrada*** *f.* closed structure || ***estructura columnar*** *f.* columnar structure || ***estructura criptovolcánica*** *f.* cryptovolcanic structure || ***estructura flaser*** *f.* flaser structure' || ***estructura de carga*** *f.* load structure || ***estructura de flujo*** *f.* flow structure || ***estructura de llama*** *f.* flame structure || ***estructura de presión*** *f.* load structure || ***estructura de tortuga de sal*** *f.* salt turtle structure || ***estructura en rosario*** *f.* boudinage || ***estructura espigada*** *f.* herringbone structure || ***estructura imbricada*** *f.* imbricate structure || ***estructura laminar*** *f.* planar structure || ***estructura molecular*** *f.* molecular structure || ***estructura moniliforme*** *f.* boudinage || ***estructura primaria*** *f.* primary structure || ***estructura sedimentaria*** *f.* sedimentary structure

estructuración *f.* structure, structuring

estructural *a.* structural
Estruiano, Estruiense *m. & a.* Strunian
estuarino *a.* estuarine
estuario *m.* estuary
estudio *m.* study, survey || ***estudio catastral*** *m.* cadastral survey
etanato de etilo *m.* ethyl acetate
etano *m.* ethane
etapa *f.* stage, phase
eteno *m.* ethylene
éter *m.* ether
etileno *m.* ethylene
etino *m.* acetylene
etilo *m.* ethyl
etiqueta *f.* label, tag
eufótico *a.* euphotic
eugeosinclinal *m.* eugeosyncline
euhedral *a.* euhedral
Eurasia *f.* Eurasia
Europa *f.* Europe
europio *m.* europium
eustasia *f.* eustasy
eustático *a.* eustatic
eutéctica *f.* eutectic
eutéctico *a.* eutectic
eutrófico *a.* eutrophic
euxenita *f.* euxenite
euxínico, euxénico *a.* euxinic
evacuación *f.* evacuation
evacuado *a.* evacuated
evacuar *vb.* to evacuate
evaluación *f.* evaluation, assay, assessment
evaluar *vb.* to evaluate, assess
evaporación *f.* evaporation
evaporador *m.* evaporator
evaporar, evaporarse *vb.* to evaporate
evaporita *f.* evaporite
evaporítico *a.* evaporitic
evapotranspiración *f.* evapotranspiration
evento *m.* event, episode
evitar *vb.* to avoid, elude, preclude
evolución *f.* evolution
evolutivo *a.* evolutionary
exactitud *a.* accuracy, precision
exacto *a.* exact, accurate
exageración *f.* exaggeration
exagonal *a.* hexagonal
examen *m.* examination, exam; survey
excavación *f.* excavation || ***excavación a cielo abierto*** *f.* open pit
excavado *a.* excavated, burrowed
excavador *a.* excavating, burrowing
excavadora *f.* excavating machine, excavator || ***excavadora de rueda de cangilones*** *f.* bucket wheel excavator
excavar *vb.* to dig out, excavate, scour
excedente *m. & a.* excess, surplus || ***excedente de ganancias*** *f.* excess profit
excentricidad *f.* eccentricity
excéntrico *a.* offset; eccentric
exceso *a.* excess, surplus || ***exceso de utilidades*** *f.* excess profit
excluir *vb.* to seal off
excrecencia *f.* excrescence
excursión *f.* excursion, trip
exegir *vb.* to order
exención *f.* exemption
exfoliación *f.* exfoliation
exfoliar, exfoliarse *vb.* to exfoliate
exhalación *f.* exhalation
exhalante *a.* exhalant
exhalar *vb.* to exhale, give off
exhibición *f.* exhibition, review; assessment
exhumación *f.* exhumation
exhumado *a.* exhumed, uncovered, dug up
exina *f.* exine
exinita *f.* exinite
existencia *f.* existence, occurrence; in stock
exitela *f.* valentinite
exogénesis *f.* exogenesis
exogenético *a.* exogenetic
exógeno *a.* exogenous
exomórfico *a.* exomorphic
exomorfismo *m.* exomorphism
exósfera *f.* exosphere
exotérmico *a.* exothermic
exótico *a.* exotic, transported
expandible *a.* expandable

expandidor de tubos *m.* tube expander, pipe expander

expandir *vb.* to expand

expansión *f.* spreading, widening, expansion || ***expansión del fondo oceanico*** *f.* seafloor spreading || ***expansión del histograma*** *f.* histogram stretching || ***expansión termal*** *f.* thermal expansion

expansor *m.* expander

expectativa *f.* objective, expectation

expedición *f.* expedition

expeler *vb.* to expel, eject

experiencia *f.* experience

experimental *a.* experimental

experimento *m.* experiment

expertaje *f.* expertise

experto *m.* expert

expirar *vb.* to expire, to die

explanación *f.* grading, leveling

explanar *vb.* to grade, level

explicación *f.* legend, explanation

exploración *f.* exploration, [U.S.] wildcatting || ***exploración y producción*** *a.* exploration and production; (com.) upstream

explorado *a.* explored

explorador *m.* explorer, explorationist || ***explorador de barrido*** *m.* scanner

explorar *vb.* to explore

exploratorio *a.* exploratory

explosión *f.* explosion, blowout, blast

explosivo *m.* explosive

explotación *f.* exploitation, mining

explotado *a.* developed, exploited

explotar *vb.* to exploit, work, mine, develop; to explode

exponente *m.* exponent

exportación *f.* export

exportador *m.* exporter

exportar *vb.* to export

exposición *f.* exposure, exposition, display

expresión *f.* expression || ***expresión topográfica*** *f.* topogrphic expression

expropiación *f.* expropriation

expropiar *vb.* to expropriate, to commandeer

expuesto *a.* exposed, expressed

expulsanúcleos *m.* core pusher

expulsar *vb.* to expel

expulsatestigos *m.* core pusher

expulsión *f.* expulsion

exsolución *f.* exsolution

extender *vb.* to extend, stretch, renew

extendido ampliamente *a.* widespread

extensión *f.* extension; range, areal extent

exterior *m.* exterior, outside; aboveground

externo *a.* external, exterior

extinción *f.* extinction || ***extinción paralela*** *f.* (petrol.) parallel extinction

extinguido *a.* extinct

extinguidor *m.* extinguisher

extinguirse *vb.* to become extinct

extinto *a.* extinct

extracción *f.* extraction, production, mining, plucking, lifting || ***extracción artificial por gas*** *f.* gas lift || ***extracción con pistón*** *f.* swabbing || ***extracción por aire*** *f.* airlift || ***extracción por solvente*** *f.* solvent extraction || ***extracción vigorizada*** *f.* enhanced recovery

extracto *m.* extract

extractor *m.* puller, extractor || ***extractor de herramientas*** *m.* tool extractor || ***extractor de muestras*** *m.* sampler || ***extractor de testigo*** *m.* core puller

extraer *vb.* to extract

extraible *a.* recoverable, minable, producible

extraño *a.* strange, extraneous

extrapolación *f.* extrapolation

extrapolar *vb.* to extrapolate

extremo *m.* end; final cut

extrusión *f.* extrusion

extrusivo *m. & a.* extrusive

extumescencia *f.* squeeze-up

exudación *f.* seep

exudar *vb.* to exude

eyección *f.* ejecta

eyecto *m.* ejectum, ejecta

eyector *m.* ejector || ***eyector a vapor*** *m.* steam ejector || ***eyector de gas*** *m.* gas exhauster

F

fábrica *f.* fabric; factory || ***fábrica de gas*** *f.* gas plant || ***fábrica S–C*** *f.* S–C fabric

fabricación *f.* manufacture, fabrication

fabricado por encargo *a.* custom built

fabricante *m.* maker, manufacturer

fabricar *vb.* to make, manufacture

faceta *f.* facet

facie *f.* crystal face

facies *f.* sedimentary facies

facilidad *f.* facility, plant; *a.* ease

facoidal *a.* lenticular

facilitación *f.* expediting

facolito *m.* phacolith

factible *a.* feasible, likely

factor *m.* factor || ***factor constante de equilibrio*** *m.* equilibrium constant || ***factor de carga*** *m.* load factor || ***factor de conversión*** *f.* conversion factor || ***factor de corrección*** *f.* correction factor || ***factor de recuperación*** *f.* recovery factor || ***factor de seguridad*** *f.* factor of safety

factura *f.* invoice, bill

facturar *vb.* to invoice, to bill

faena *f.* work, labor, task

faja *f.* band, girdle, belt, zone, strip, trend || ***faja capilar*** *f.* capillary fringe || ***faja de meandros*** *f.* meander belt || ***faja de meteorización*** *f.* belt of weathering ***faja de oro*** *f.* [Mex.] Golden Lane || ***faja metalífera*** *f.* mineral belt || ***faja montañosa*** *f.* mountain belt || ***faja plegada*** *f.* fold belt

fajado *m.* prop of mine timber

fajeado *a.* banded

falda *f.* slope, hillside

faldeo *m.* flank, slope, hillside

falsificar el mineral *vb.* to salt a mine

falso *a.* treacherous ground, false, pseudo-

falla *f.* fault || ***falla acostada*** *f.* thrust fault || ***fallas al tres bolillo*** *f.* echelon faults || ***falla antitética*** *f.* antithetic fault || ***falla cabalgadura*** *f.* thrust fault || ***falla corriente*** *f.* strike-slip fault || ***falla de alto ángulo*** *f.* high angle fault || ***falla de altoempuje*** *f.* thrust fault || ***falla de arrastre*** *f.* drag fault || ***falla de ascenso*** *f.* normal fault || ***falla de bajo ángulo*** *f.* low angle fault || ***falla de buzamiento*** *f.* dip-slip fault || ***falla de cabalgamiento*** *f.* thrust fault || ***falla de corrimiento*** *f.* thrust fault || ***falla de corte*** *f.* shear fault, wrench fault || ***falla de crecimiento*** *f.* growth fault || ***falla de descenso*** *f.* reverse fault || ***falla de desgarre o desgarramiento*** *f.* tear fault, strike-slip fault || ***falla de despegue*** *f.* detachment fault || ***falla de desplazamiento en el rumbo*** *f.* strike-slip fault || ***falla de escurrimiento*** *f.* thrust fault || ***falla de gravedad*** *f.* gravity fault || ***falla de rasgadura*** *f.* tear fault || ***falla de relajación*** *f.* release fault || ***falla de rumbo*** *f.* strike-slip fault || ***falla de sobrecorrimiento*** *f.* thrust fault || ***falla de transcurrencia*** *f.* strike-slip fault, wrench fault || ***falla directa*** *f.* normal fault || ***falla en clave de arco*** *f.* keystone fault || ***falla en pivote*** *f.* rotational fault || ***falla entre capas*** *f.* bedding plane fault || ***falla gravitacional*** *f.* normal fault || ***falla giratoria*** *f.* rotational fault || ***falla gran angular*** *f.* high-angle fault || ***falla inferida*** *f.* inferred fault || ***falla inversa*** *f.* reverse or thrust fault || ***falla invertida*** *f.* reverse fault || ***falla lateral*** *f.* strike-slip fault, wrench fault || ***falla lístrica*** *f.* listric fault || ***falla longitudinal de rumbo*** *f.* strike-slip fault || ***falla marginal*** *f.* boundary fault || ***falla normal*** *f.* normal fault || ***falla pivotal*** *f.* scissors fault, rotational fault || ***falla por desgarramiento*** *f.* tear fault || ***falla por empuje*** *f.* thrust fault || ***falla por rotación*** *f.* rotational fault || ***falla probable*** *f.* inferred fault || ***falla ramificada*** *f.* branching fault || ***falla rotativa*** *f.* rotational fault, scissors fault || ***falla rumbodeslizante*** *f.* strike-slip fault || ***falla según buzamiento*** *f.* dip-slip fault || ***falla sellante*** *f.* sealing fault || ***falla transcurrente*** *f.* strike-slip fault, wrench fault || ***falla transformante*** *f.* transform fault || ***falla transversal de buzamiento*** *f.* dip-slip fault || ***falla transversal de rumbo*** *f.* strike-slip fault

fallado *a.* faulted

fallamiento *m.* faulting

fallar *vb.* to fault

fallas en escalón *f.* en echelon faults

fallas escalonadas *f.* en echelon faults

Fameniano, Fameniense *m. & a.* Famennian

familia *f.* (seis.) gathers

familias CDP *f.* (seis.) CDP gathers

fanega *v.* dry measure, usually 90.815 liters; area measure usually 3.5663 hectares

fanerítico, fanerómero *a.* phaneritic

fangal *m.* marsh, bog

fanglomerado *m.* fanglomerate

fango *m.* mud, slime, ooze, sludge

fangolita *f.* mudstone

fangoso *a.* muddy, silty

farallón *m.* stack, bluff; [Bol.] wallrock; [Pe.] vein outcrop

farero *m.* derrickman

farmacolita *f.* pharmacolite

faro *m.* light, lighthouse; derrick

fase *f.* phase, aspect

fatiga *f.* fatigue

fauna *f.* fauna || ***asociación de fauna*** *f.* faunal assemblage

fáunico, faunístico *a.* faunal

fayalita *f.* fayalite

faz *f.* front, face; aspect || ***faz de laboreo*** *f.* working face

fecha *f.* date

fechamiento *m.* age dating

fechar *vb.* to date, to assign an age

feldespático *a.* feldspathic

feldespato *m.* feldspar || ***feldespato calizo*** *m.* anorthite || ***feldespato nacarado*** *f.* adularia || ***feldespato potásico*** *m.* potash feldspar || ***feldespato sódico*** *m.* albite || ***feldespato vítreo*** *m.* sanidine

feldespatoide *m.* feldspathoid

félsico, felsítico *a.* felsic

felsita *f.* felsite

fémico *a.* mafic

fenacita, fenaquita *f.* phenakite

fenestra *f.* inlier, window

fenilamina *f.* phenylamine, aniline

fenoclástico *a.* phenoclastic

fenoclasto *m.* phenoclast

fenocristal *m.* phenocryst

fenol *m.* phenol, carbolic acid

fenolftaleína *f.* phenolphthalein

fenómeno *m.* phenomenon

fergusita *f.* fergusite

fermentación *f.* fermentation

ferrato *m.* ferrate

férreo *a.* iron-bearing, ferrous

ferrería *f.* iron works

ferrete *m.* copper sulphate

férrico *a.* ferric

ferrífero *a.* iron-bearing

ferrocarril *m.* railroad, railway, train

ferrogusita *f.* fergusite

ferromagnesiano, ferromagnésico, ferromagnesio *m. & a.* ferromagnesian, mafic

ferromagnético *a.* ferromagnetic

ferrometeorito *m.* iron meteorite

ferroníquel *m.* ferronickel

ferroso *a.* ferrous

ferrugíneo, ferruginoso *a.* ferruginous, iron-bearing

fétido *a.* fetid

fiador *m.* bondsman; (com.) surety, retainer

fianza *f.* (com.) bond, bail, guarantee, security, surety

fibra *f.* fiber, vein, filament || ***fibras de carbono*** *f.* carbon fibers

fibrolita *f.* fibrolite

fibroso *a.* fibrous

ficha *f.* index card, filing card, record

fichero *m.* card file, records

fidelidad *f.* accuracy

fiebre del petróleo *f.* oil boom

fiebre del oro *f.* gold rush

fierro *m.* iron

figura *f.* figure, form; illustration || ***figura de interferencia*** *f.* interference figure

fijado *a.* fixed

fijar *vb.* to fix, to attach

fijo *a.* attached, stationary, fixed

fila *f.* row

filamento *m.* filament, stringer

filamentoso *a.* filamentous

filiforme *a.* thread-like

filita *f.* phyllite

filo *m.* ridge, divide

filogenia *f.* phylogeny

filón *m.* vein, lode; mineralized seam or bed, dike, streak (véase vena, veta) || ***filón capa*** *m.* sill, sheet || ***filón metalífero*** *m.* ore vein || ***filón mineralizado*** *m.* lode

filoniano *a.* vein-like

filonita *f.* phyllite-mylonite

filonitización *f.* veining

filtración *f.* seep, seepage, percolation, filtration || ***filtración en frío*** *f.* cold pressing

filtrado *a. & m.* filtrate, filtering; filtered

filtrador *m.* filter

filtrante *a.* filtering, porous

filtrar *vb.* to filter, percolate, strain

filtro *m.* filter; perforated pipe || ***filtro de agua*** *m.* water filter || ***filtro de grava*** *m.* gravel pack || ***filtro de pedregullo*** *m.* sand screen, gravel pack || ***filtro direccional*** *m.* directional filter || ***filtro para pozos*** *m.* well screen pipe

filum *m.* phyllum

final *a.* end, final, ultimate

finalización de un pozo *f.* well completion

financiación *f.* financing

financiar *vb.* to finance

financiero *a.* financial

financista *m., f. & a.* financier, financial expert; financial

finanzas *f.* finances

fineza *f.* fineness, purity, quality

finiquito *m.* quitclaim

fino *a.* fine, polished, pure || ***de grano fino*** *a.* fine-grained

finos *m.* fines, ore dust

finura *f.* fineness; grain

fiordo *m.* fjord, [U.K.] firth

firmante *m. & f.* undersigned, signatory

firme *m. & a.* firm, steady, secure, stable; surface; fissile

firmeza *f.* firmness, hardness, strength

firn *m.* firn

fisco nacional *m.* national treasury

física *f.* physics

físico *m. & a.* physicist; physical

fisicoquímico *a.* physiochemical

físil *a.* fissile

fisilidad *f.* fissility

fisión *f.* fission

fisionable *a.* fissionable

fisiografía *f.* physiography

fisiográfico *a.* physiographic

fisura *f.* crevice, fissure, seam

fisurable *a.* fissile

fisuración *f.* spalling

fisurarse *vb.* to crack

fitógeno *a.* phytogenic

flabeliforme *a.* fan-shaped

flanco *m.* flank, limb, hillside || ***flanco anterior*** *m.* forelimb || ***flanco delantero*** *m.* forelimb || ***flanco del pliegue*** *m.* fold limb || ***flanco invertido*** *m.* overturned limb || ***flanco posterior*** *m.* backlimb || ***flanco trasero*** *m.* backlimb || ***flanco volcado*** *m.* overturned limb

flanqueado *a.* flanked, on the side of

flanquear *vb.* to flank

flaser *m. & a.* flaser

fleco *m.* fringe

flecha *f.* spit of land; deflection; arrow || ***flecha litoral*** *f.* spit

flete *m.* charter, freight, cargo

flexibilidad *f.* flexibility

flexible *a.* flexible

flexión *f.* bending, flexure; fold

flexura *f.* flexure; flexure folding

floculación *f.* flocculation

floculador *m.* flocculator

floculento *m.* flocculent

flogopita *f.* phlogopite

flojo *a.* loose

flor *f.* (min.) bloom; (botan.) bloom

flora *f.* flora

(que) flota libremente *a.* free-floating

flotabilidad, flotación *f.* buoyancy, flotation

flotar *vb.* to float

flotador *m.* float

flotante *a.* floating

fluctuación *f.* fluctuation, jitter

fluctuar *vb.* to fluctuate, oscillate

fluencia *f.* flowage, flow || ***fluencia elástica*** *f.* elastic flow

fluente *a.* fluid, flowing

fluidez *f.* fluidity

fluído *m. & a.* fluid || ***fluído húmedo*** *m.* wetting fluid

fluir *vb.* to flow

flujo *m.* flow, stream, flux; creep, ebb; bleeding || ***flujo calórico*** *m.* heat flow || ***flujo de caja*** *m.* cash flow || ***flujo de detritos*** *m.* debris flow || ***flujo de efectivo*** *m.* cash flow || ***flujo de escombros*** *m.* debris flow || ***flujo de fango*** *m.* mud flow || ***flujo de fractura*** *m.* frac fluids || ***flujo de gravedad*** *m.* gravity flow || ***flujo de lava*** *m.* lava flow || ***flujo de lodo*** *m.* mudflow || ***flujo elástico*** *m.* elastic flow || ***flujo en manto*** *m.* sheet flow || ***flujo futuro de caja*** *m.* future cash flow || ***flujo laminar*** *m.* laminar flow, sheet flow || ***flujo magnético*** *m.* magnetic flux || ***flujo natural*** *m.* primary drive || ***flujo plástico*** *m.* plastic flow || ***flujo termico*** *m.* heat flow || ***flujo turbulento*** *m.* turbulent flow || ***flujo viscoso*** *m.* viscous flow

flúor *m.* fluorine

fluorescencia *f.* fluorescence || ***fluorescencia del petróleo*** *f.* oil bloom

fluorescente *a.* fluorescent

fluorhidrato *m.* hydrofluoric acid

fluorhidrico *a.* hydrofluoric

fluorina, fluorita *f.* fluorite

fluoruro *m.* fluoride

fluvial *a.* fluvial

fluvioglaciario *a.* fluvioglacial

fluviomarino *a.* fluviomarine

fluxión *f.* fluxion

flysch *m.* flysch

focal *a.* focal

focalizar *vb.* to focus on

foco *m.* focus || ***foco cubierto*** *m.* buried focus || ***foco de velocidad*** *m.* velocity focus || ***foco sísmico*** *m.* hypocenter

fogata *f.* blasting fumes

foliación *f.* foliation, schistocity

foliado *a.* foliated

foliatura *f.* foliation

folleto *m.* brochure

fomentar *vb.* to promote

fomento *m.* development

fondo *m.* background, bottom || ***fondo del mar*** *m.* sea bottom || ***fondo del pozo*** *m.* total depth (TD) || ***fondo del rio*** *m.* riverbed

fonolita *f.* phonolite

fonolito *m.* phonolite

foraminífero *m.* foraminifer, foram

forja *f.* forge

forjado *a.* forged

forjar *vb.* to forge

forma *f.* form, mode, shape || ***forma aerodinámica*** *a.* streamlined

formación *f.* formation, unit || ***formación acuífera*** *f.* aquifer || ***formación conductor*** *f.* carrier bed, migration horizon || ***formación petrolífera*** *f.* oil-bearing formation || ***formación que le circundan*** *f.* host rock

formar *vb.* to form, shape, fashion, make

formarse *vb.* to be formed, take form

formas de destrucción *f.* destructional landforms

formas fisiográficas *f.* landforms

formas de acumulación *f.* constructional landforms

formas de erosión *f.* destructional landforms

formatear *vb.* to format

formateo *m.* formatting

formato *m.* format

fórmico *a.* formic

fórmula *f.* formula

forrado *a.* lined

forrar *vb.* to line

forro *m.* well casing, cylinder liner, lining

forsterita *f.* forsterite

fortaleza *f.* strength, toughness

forzado *a.* forceful, forced

fosa *f.* trough, graben, trench, deep, pit, drain (véase foso) || ***fosa abisal*** *f.* ocean trench || ***fosa de hundimiento*** *f.* sinkhole || ***fosa oceánica*** *ocean* trench || ***fosa posterior*** *f.* back deep || ***fosa tectónica*** *f.* graben, fault trough, downthrown block

fosfático *a.* phosphatic

fosfato, fosfeno *m.* phosphate

fosfeno *m.* phosphite

fosforado *a.* phosphoric

fosforecer *vb.* phosphoresce

fosforescencia *f.* phosphorescence

fosforescente *a.* phosphorescent

fosfórico *m.* phosphoric

fosforita *f.* phosphorite

fósforo *m.* phosphorous

forgenita *f.* phosgenite

fósil *m.* fossil || ***fósil guía*** *m.* index fossil || ***fósil indicador*** *m.* index fossil || ***fósil índice*** *m.* index fossil

fosilífero *a.* fossiliferous

fosilización *f.* fossilization

fosilizar(se) *vb.* to fossilize, become fossilized

foso *m.* moat, ditch, trench, hole, pit (véase fosa) || ***foso de inyección*** *m.* mud pit, slush pit || ***foso de lodo*** *m.* mud pit, slush pit, sump hole

fótico *a.* photic

fotocalcar *vb.* to blueprint

fotocalco azul *m.* blueprint

fotocopiar *vb.* to photocopy

fotogeología *f.* photogeology

fotografía, foto *f.* photography, photograph || ***foto aérea*** *f.* aerial photograph, airphoto || ***foto aérea oblicua con horizonte sensible*** *f.* high oblique airphoto || ***foto aérea oblicua sin horizonte sensible*** *f.* low oblique airphoto || ***foto aérea vertical*** *f.* vertical airphoto || ***foto del micróscopio*** *f.* photomicrograph

fotográfico *a.* photographic

fotogrametría *f.* photogrammetry

fotoíndice *m.* photoindex

fotómetro *m.* photometer

fotomicrografía *f.* photomicrography

fotomicrográfica *m.* photomicrograph

fotomosaico aéreo *m.* aerial photomosaic, airphoto mosaic

fotón *m.* photon

fowlerita *f.* fowlerite

foyaita *f.* foyaite

fracaso *m.* failure

fracción *f.* fraction, (ref.) cut

fraccionación *f.* fractionation, cracking

fraccionado *a.* fractionated

fraccionamiento *m.* fractionation, cracking

fraccionar *vb.* to crack, fractionate

fraccionario *a.* fractional

fracciones livianas *f.* light fraction, light ends

fractil *f.* fractile

fractura *f.* fracture, rupture, breaking (véase diaclasa, falla, grieta) || ***escarpa de fractura*** *f.* fault scarp || ***fractura astillosa*** *f.* hackly fracture || ***fractura concoidal*** *f.* conchoidal fracture || ***fractura de corte*** *f.* shear fracture || ***fractura de extensión*** *f.* extension fracture || ***fractura de tensión*** *f.* tension fracture || ***fractura escalonada*** *f.* en echelon fractures || ***fractura según el clivaje*** *f.* fracture cleavage || ***fractura sigmoidea*** *f.* sigmoidal fracture

fracturación *f.* fracturing; (perf.) frac job

fracturado *a.* fractured

fracturamiento *m.* fracturing (véase agrietamiento)

fracturar *vb.* to fracture, break

frágil *a.* fragil, brittle

fragilidad *f.* brittleness

fragmentación *f.* frac job, fragmentation

fragmentar *vb.* to fragment, frac

fragmento *m.* fragment

fragmentoso *a.* crumbly

fragua *f.* forge

fraguar *vb.* to forge

fraguado *m.* & *a.* forging; forged

fragüe *m.* setting

francio *m.* francium

franco *a.* free, clear, open || ***franco a bordo*** *a.* free on board (FOB) || ***franco de comisión*** *a.* free of brokerage || ***pozo franco*** *m. & a.* open hole

franja *f.* fringe, belt, strip of land, band, zone || ***franja capilar*** *f.* capillary fringe || ***franja de meandros*** *f.* meander belt

franklinita *f.* franklinite

franquear *vb.* to free, liberate, exempt, clear

franquicia *f.* franchise; exemption

frasco *m.* flask

Frasniano, Frasniense *m. & a.* Frasnian

freático *a.* phreatic, subsurface

frecuencia *f.* frequency

freibergita *f.* freibergite

frenita *f.* phrenite

frente *m.* front, face, heading || ***frente cálido*** *m.* warm front || ***frente de corte*** *m.* mine face || ***frente de deyección*** *m.* alluvial apron || ***frente deltaico*** *m. & a.* delta front || ***frente de playa*** *m.* foreshore || ***frente de playa alta*** *m.* upper shoreface || ***frente de playa baja*** *m.* lower shoreface || ***frente de playa media*** *m.* middle shoreface || ***frente de onda*** *m.* wave front || ***frente de ribera*** *m.* shore face || ***frente estacionario*** *m.* stationary front || ***frente frío*** *m.* cold front || ***frente tectónico*** *m.* tectonic front

fresa, fresadora *f.* bit, milling tool (véase barrena)

fresado *a.* milled

fresar *vb.* to mill, to face

fresco *a.* cool

friabilidad *f. & a.* friable; friability, embrittlement

friable *a.* friable

fricción *f.* friction

frigoría *f.* kilocalorie

frio *m. & a.* cold

frontal *a.* frontal

frontera *f.* boundary, border, frontier, borderland

frontón *m.* cliff, pediment; working face

frústulo *m.* diatom test

fruto *m.* ore, product, yield

ftanita *f.* chert

fucsita, fuchsita *f.* fuchsite

fucoide *m.* fucoid

fuego *m.* fire

fuelle *m.* bellows

fuente *f.* fountain, spring, source, well || ***fuente artesiana*** *f.* artesian spring || ***fuente caliente*** *f.* hot spring || ***fuente de calibración*** *f.* calibration source || ***fuente de emisión*** *f.* seismic reflection source || ***fuente de gas*** *f.* gas seep || ***fuente de origen*** *f.* source, provenance || ***fuente mineral*** *f.* mineral spring || ***fuente sísmica*** *f.* seismic source || ***fuente termal*** *f.* thermal spring, hot spring

fuera de *a.* out of, off of || ***fuera de control*** *a.* out of control, wild || ***fuera de la estructura*** *a.* off structure || ***fuera de operación*** *a.* out of operation || ***fuera de secuencia*** *a.* out of sequence

fuerte *a.* stong, heavy grade, sharp angle, steep incline

fuerza *f.* force, stress, power, strength || ***fuerza centrífuga*** *f.* centrifugal force || ***fuerza centrípeta*** *f.* centripetal force || ***fuerza compresiva*** *f.* compressive strength || ***fuerza compresora*** *f.* compressive strength || ***fuerza de adherencia*** *f.* soil stress || ***fuerza de flotación*** *f.* buoyant force || ***fuerza de rotura*** *f.* yield stress || ***fuerza electromotriz*** *f.* electromotive force || ***fuerza motriz*** *f.* power

fuga *f.* leak, escape || ***punto de fuga*** *m.* spill point || ***a punto de fuga*** *a.* fill-to-spill

fugacidad *f.* fugacity

fulcro *m.* fulcrum

fulgurita *f.* fulgurite

fulminante *m.* blasting cap

Fuloniano, Fuloniense *m. & a.* Fullonian

fumarola *f.* fumarole

función *f.* function || ***en función de*** as a function of

funcionamiento *m.* operation, performance

funcionar *vb.* to function, work, go

fundación *f.* foundation

fundamento *m.* foundation

fundar *vb.* to erect, build, found, establish

fundente *a.* melting

fundente *a.* melting; *m.* flux, melt

fundible *a.* fusible, easily melted

fundición *f.* smelting, casting, melting; foundry

fundido *a.* molten

fundidor *m.* smelter

fundir *vb.* to melt, smelt

fundirse *vb.* to coalesc

furo *m.* channel

fusalina *f.* smelter

fuselado *a.* streamlined

fusibilidad *f.* fusibility

fusible *m. & a.* fuse; fusible

fusinita *f.* fusinite

fusión *f.* fusion, melting || ***fusión residual*** *f.* residual melt || ***fusión corporativa*** *f.* merger

fusita *f.* fusain

fusulina *f.* fusulinid

G

gabarra *f.* barge

gabarro *m.* xenolith; large ore chunks

gábrico *a.* gabbroic

gabro *m.* gabbro

gabroíco *a.* gabbroic

gadolinita *f.* gadolinite

gadolino *m.* gadolinium

gafas protectoras *f.* goggles

gahnita *f.* gahnite

gal *m.* gal, unit of gravitational acceleration

galáctica *f.* fuller's earth

galápago *m.* pig of copper, lead, or tin

galemador *m.* furnace for roasting ore

galemar *vb.* to roast silver ore

galeme *m.* reverberatory furnace

galena, galenita *f.* galena

galera *f.* irregular-shaped ore deposit

galería *f.* trench, gallery || ***galería de excavación*** *f.* adit || ***galería intermedia*** *f.* sub-level, subdrift || ***galería lateral*** *f.* side drift || ***galería principal*** *f.* mine main level

galga *f.* boulder

galio *m.* gallium

galmeíta *f.* carbonate or silicate of zinc

galón *m.* gallon || ***galón americano*** *m.* 3.7853 liters || ***galón inglés*** *m.* 4.543 liters

galvánico *a.* galvanic

galvanizado *a.* galvanized

galvanómetro *m.* galvanometer

gallos *m.* rich specimens

gama *f.* variety, gamut, range || ***gama de análisis*** *f.* variety of analyses || ***gama de tonos de gris*** *f.* gray scale || ***gama dinámica*** *f.* dynamic range

gambusino *m.* prospector

gamella *f.* gold pan

gamma *f.* gamma, unit of magnetic attraction

gammagrafía *f.* gamma ray log

ganancias *f.* gain, earnings, profit, proceeds, returns

ganar *vb.* to gain, to earn, to profit

gancha *f.* hook, drag

gancho *m.* hook; gab

ganchudo *a.* hackly

gándara *f.* low wasteland

gandinga *f.* fine ore, concentrates

gang *m.* dike; vein

ganga *f.* gangue, matrix

ganita *f.* gahnite

garantía *f.* guarantee

garantizar *vb.* guarantee, vouch for

garbillar *vb.* to screen minerals

garbillo *m.* screen

garganta *f.* gorge, narrows, groove

garlancha *f.* spade

garnierita *f.* garnierite

garúa *f.* drizzle

gas *m.* gas || ***gas acetileno*** *m.* acetylene gas || ***gas ácido*** *m.* sour gas || ***gas amargo*** *m.* sour gas || ***gas asfixiante*** *m.* asphyxiating gas || ***gas asociado*** *m.* associated gas || ***gas combustible*** *m.* fuel gas || ***gas comprimido*** *m.* compressed gas || ***gas con azufre*** *m.* sour gas, sulfur-bearing gas || ***gas condensado*** *m.* gas condensate || ***gas corrosivo*** *m.* corrosive gas || ***gas de boca de pozo*** *m.* casinghead gas || ***gas de carbón*** *m.* coal gas || ***gas de chimenea*** *m.* flue gas || ***gas de desecho*** *m.* waste gas || ***gas de desperdicio*** *m.* waste gas || ***gas de entubación*** *m.* casing head gas || ***gas de pantano*** *m.* swamp gas, methane || ***gas de escape*** *m.* exhaust gas || ***gas disuelto*** *m.* solution gas || ***gas dulce*** *m.* sweet gas || ***gas en solucion*** *m.* solution gas || ***gas grisú*** *m.* methane, coalbed methane || ***gas húmedo*** *m.* wet gas, casinghead gas || ***gas libre*** *m.* unassociated gas || ***gas licuado*** *m.* condensate || ***gas natural*** *m.* natural gas, methane || ***gas natural comprimido (GNC)*** *m.* compressed natural gas (CNG) || ***gas natural licuado (GNL)*** *m.* liquified natural gas (LNG) || ***gas no asociado*** *m.* unassociated gas || ***gas ocluído*** *m.* occluded gas || ***gas seco*** *m.* dry gas || ***gas sulfuroso*** *m.* sour gas

gaseoso *a.* gaseous

gasífero *a.* gas-bearing

gasificación *f.* gasification

gasificar *vb.* to gasify

gasoducto *m.* gas pipeline

gasóleo *m.* diesel oil, gas oil

gasolina *f.* gasoline || ***gasolina de aviación*** *f.* aviation gas || ***gasolina natural*** *f.* casinghead gasoline, natural gasoline

gasometría *f.* soil gas analysis

gastable *a.* erodible

gastado *a.* worn out

gastar *vb.* to wear away, erode; to spend

gasterópodo *m.* gastropod

gasto *m.* discharge, flow volume; expense, cost || ***gasto sólido*** *m.* bed load

gastos *m.* costs, charges, expenses, expenditures || ***gastos de capital*** *m.* capital expenditures || ***gastos de iniciación de operación*** *m.* startup costs || ***gastos fijos*** *m.* overhead, fixed expenses

gastrolito *m.* gastrolith

gato *m.* jack || ***gato de bombeo*** *m.* pump jack

gausio, gauss *m.* gauss

gaylussita *f.* gaylussite

geanticlinal *m.* geanticline

Gedinniano, Gedinniense *m. & a.* Gedinnian

géiser *m.* geyser, gusher

geiserita *f.* geyserite

gel *m.* gel

gelatina *f.* gel

gelatinoso *a.* colloid, gelatinous

gelificarse *vb.* to gel

gelisol *m.* frozen ground

geloide *m.* firn ice

gema *f.* gem

gemelo *m. & a.* twin, double

gemelos *m.* binoculars || ***gemelos de campaña*** *m.* field glasses

geminación *f.* twinning

gemmología *f.* gemmology

generación *f.* generation

generador *m.* generator

generar *vb.* to generate

genérico *a.* generic

género *m.* genus

génesis *f.* genesis

genético *a.* genetic

genotipo *m.* genotype

geociencia *f.* geoscience

geocronita *f.* geochronite

geocronología *f.* geochronology

geocronológico *a.* geochronological, geochronologic

géoda *f.* geode

geodepresión *f.* geodepression

geodesía *f.* geodesy

geodésica *f.* geodetics

geodético *a.* geodetic

geodinámico *a.* geodynamic

geofísica *f.* geophysics

geofísico *m. & a.* geophysicist; geophysical

geófono *m.* geophone

geognosia *f.* geology

geográfico *a.* geographic

geografía *f.* geography

geógrafo *m.* geographer

geohidrología *f.* geohydrology

geoide *m.* geoid

geología *f.* geology || ***geología de campo*** *f.* field geology || ***geología de pozo*** *f.* well site geology || ***geología económica*** *f.* economic geology || ***geología regional*** *f.* regional geology

geológico *a.* geologic

geólogo *m.* geologist || ***geólogo asesor*** *m.* consulting geologist || ***geólogo consultor*** *m.* consulting geologist || ***geólogo de campo*** *m.* field geologist || ***geólogo de minas*** *m.* mining geologist || ***geólogo de petróleo*** *m.* petroleum geologist || ***geólogo de pozo*** *m.* well site geologist || ***geólogo económico*** *m.* economic geologist

geomagnético *a.* geomagnetic, geomagnetics

geometría *f.* geometry

geométricamente *adv.* geometrically

geométrico *a.* geometric

geomórfico *a.* geomorphic

geomorfología *f.* geomorphology

geoquímica *f.* geochemistry

geoquímico *m. & a.* geochemist; geochemical

Georgiano, Georgiense *m. & a.* Georgian

geosinclinal *m. & a.* geosyncline; geosynclinal

geosutura *f.* geosuture

geotectónica *f.* tectonics

geotectónico *a.* geotectonic, tectonic

geotérmica *f.* geothermics

geotérmico *a.* geothermal

geotita *f.* goethite

geo-referencia *f.* geo-reference

geo-referenciado *a.* geo-referenced

gerencia *f.* management

gerente *m.* manager, director

gersdorfita *f.* gersdorfite

gestor *m.* promoter, agent || ***gestor de servidumbres (o permisos)*** *m.* permitter

giba *f.* hump

gibsita *f.* gibbsite

gigante *m.* (hydr., min.) giant (used in hydraulic placering)

gigapiés cubicos *m.* trillion cubic feet (TCF)

gilsonita *f.* gilsonite

giobertita *f.* magnesite

gipsita *f.* gypsite

gira *f.* drill trip, twist, rotation, spin, turn

girar *vb.* to spin, revolve, turn, rotate

girasol *m.* fire opal

giro *m.* turn, revolution, rotation

giroscopio, giróscopo *m.* gyroscope

gis *m.* chalk

Givetiano, Givetiense *m. & a.* Givetian

glaciación *f.* glaciation || ***afectado por glaciación*** *a.* glaciated

glaciado *a.* glaciated

glacial *a.* glacial

glaciar *m.* glacier || ***glaciar alpino*** *m.* alpine or mountain glacier || ***glaciar colgado, colgante*** *m.* hanging glacier || ***glaciar continental*** *m.* continental ice sheet || ***glaciar de montaña*** *m.* mountain glacier || ***glaciar de roca*** *m.* rock glacier || ***glaciar de valle*** *m.* valley glacier || ***glaciar rocoso*** *m.* rock glacier

glaciario *a.* glacial

glaciarismo *m.* glaciology

glacifluvial *a.* glaciofluvial

glacilacustre *a.* glaciolacustrine

glaciología *f.* glaciology

glauberita *f.* glauberite

glaucodot *m.* glaucophane

glaucofana *f.* glaucophane

glauconia, glauconita *f.* glauconite

glauconítico *a.* glauconitic

glauconitización *f.* glauconitization

gleba *f.* horst || ***gleba tectónica*** *f.* horst, uplifted block

glicerina *f.* glycerine

glicol *m.* glycol

gliptolita *f.* ventifact

global *a.* global; complete, total, bulk

globigerina *f.* globigerina

globo *m.* globe, sphere || ***globo terrestre*** *m.* earth

globolito *m.* microlite

globoso, globuloso, globular *a.* globular

glóbulo *m.* globule

glomeroporfirítico *a.* glomeroporphyritic

glosópteris *m.* glosopteris

gmelita, gmelinita *f.* gmelinite

gneis *m.* gneiss || ***gneis ojiforme*** *m.* augen gneiss

gnéisico *a.* gneissic

gobierno *m.* government

goethita *f.* goethite

golfo *m.* gulf || ***golfo de México*** *m.* Gulf of Mexico

golpe *m.* slug || ***golpe de agua*** *m.* water slug

golpetear *vb.* to blast

golpeteo *m.* blasting || ***golpeteo de arena*** *m.* sandblasting, sandblast

gollete de botella *m.* bottleneck

goma *f.* modeling or foam rubber, glue

gongorado *a.* vuggy

goniógrafo *m.* [Sp.] plane table

goniómetro *m.* goniometer

gosan *m.* gossan

goslarita *f.* goslarite

gota *f.* drop

gotear *vb.* to drip, leak

gotera *f.* precipitate

gotero *m.* pipette

Gotlandiense, Gotlándico *m. & a.* Silurian

gozne *m.* hinge

grabado *a.* etched, recorded

grabar *vb.* to record; to engrave, carve, emboss, etch

grabenfosa *f.* graben, trough, downdropped block

grada *f.* stope

gradación *f.* grading, gradation

gradado *a.* graded

gradiente *m.* gradient, grade || ***gradiente de presión*** *m.* pressure gradient || ***gradiente geotérmico*** *m.* geothermal gradient || ***gradiente regional*** *m.* regional gradient || ***gradiente térmico*** *m.* thermal (or temperature) gradient

grado *m.* degree, step, gradient, grade, rank || ***alto grado*** *a.* high grade || ***bajo grado*** *a.* low grade || ***grado de libertad*** *m.* degrees of freedom || ***grado de selección*** *m.* degree of sorting || ***grado geotérmico*** *m.* geothermal gradient || ***grado metamórfico*** *m.* metamorphic grade

graduación *f.* graduation, calibration; grading, grade

graduado *a.* graduated

gradual *a.* gradational, gradual

graduar *vb.* to grade, classify, gauge

gráfica *f.* diagram, graph, graphics, chart

graficado *a.* plotted, graphed

graficador *m.* plotter

graficar *vb.* to plot, graph

gráfico *m. & a.* graph, chart; graphic || ***gráfico de ___ contra ___*** *m.* crossplot

grafítico *a.* graphitic

grafito *m.* graphite

grahamita *f.* grahamite, asphalt

gramo *m.* gram

granada *f.* garnet

granalla *f.* small metal grains, usually gold

granate *f.* garnet

granatífero *a.* garnet-bearing

grancilla *f.* screened coal

granipórfido *m.* granophyre

granítico *a.* granitic

granitización *f.* granitization

granito *m.* granite; small grain

granitoide *m.* granitoid

granitoideo *a.* granitoid

granizo *m.* hail

grano *m.* grain, (meas.) grain = 0.06 gram || ***grano fino*** *a.* fine-grained || ***grano grueso*** *a.* coarse-grained || ***grano mediano*** *a.* medium-grained

granoblástico *a.* granoblastic

granocreciente hacia arriba *a.* coarsening upward

granocreciente hacia abajo *a.* coarsening downward

granodecreciente hacia arriba *a.* fining upward

granodiorita *f.* granodiorite

granofírico *a.* granophyric

granofiro, granófido *m.* granophyre

granosidad *f.* granularity

granoso, granudo *a.* granular

granulación *f.* granulation

granulado *m. & a.* granulation; granulated

granular *a.* granular

granulita *f.* granulite

gránulo *m.* granule

granulometría *f.* grain size analysis

granulosidad *f.* granularity

granuloso *a.* granular, granulose

granza *[Arg.]* crushed stone; [Cu.] gravel; [Méx.] ore fragments

granzón *m.* gravel

grapa, grampa *f.* clamp

graptolito *m.* graptolite

grasa *f.* grease; slag

grasero *m.* slag pile

graso, grasoso *a.* greasy

gratis *a.* free, at no cost

grauvaca, grauwaca *f.* graywacke, greywacke

grava *f.* gravel, grit || ***grava residual*** *f.* lag gravel

gravador *m.* recorder

gravamen *m.* burden, obligation, lien, tax, encumbrance

gravedad *f.* gravity || ***gravedad API*** *f.* API gravity || ***gravedad específica*** *f.* specific gravity

gravera *f.* gravel pit

gravilla *f.* fine gravel, lag gravel || ***gravilla de desierto*** *f.* lag gravel

gravimetría *f.* gravimitry

gravimétrico *a.* gravimetric

gravímetro *m.* gravimeter

gravitación *f.* gravitation

gravitacional, gravitativo *a.* gravitational

greca *f.* irregular workings

greda *f.* marl, chalk, fuller's earth

gredal *f.* clay pit

gredoso *a.* chalky, marly, clayey

greenockita *f.* greenockite

grena *f.* undressed ore

grenoquita, greenockita *f.* greenockite

gremio *m.* society, guild, corporation, trade union, cartel, labor union

gres *m.* sandstone; potter's clay; greisen

greta *f.* fuller's earth, litharge

grieta *f.* joint, crevasse, fissure, crack, seam (véase diaclasa, fractura) || ***grieta capilar*** *f.* hairline crack || ***grieta circense*** *f.* bergschrund || ***grieta de desecación*** *f.* mud crack || ***grieta de deslizamiento*** *f.* slump scar || ***grieta de enfriamiento*** *f.* cooling joint || ***grieta de erupción*** *f.* fissure vent || ***grieta de exfoliación*** *f.* exfoliation crack || ***grieta de tracción*** *f.* tension gash || ***grieta glaciar*** *f.* crevasse

grietarse *vb.* to crack

grieteado *a.* cracked

grieteo *m.* cracking, fissuring

grietoso *a.* cracked

grisú *m.* firedamp || ***gas grisú*** *m.* coalbed methane

grilla *f.* grid

grosero *a.* rank

grosor *m.* thickness

grosularia, grosularita *f.* grossularite

grúa *f.* crane; derrick; hoist, boom

grueso *m. & a.* thickness; thick, fat, coarse

grupo *m.* group, cluster

grus *m.* grus

gruta *f.* cave, cavern, grotto

guad *m.* wadi, dry wash, gully, [U.S.] arroyo

guadaloso *a.* swampy

guanero, guano *m.* guano

guantes *m.* gloves

guarda *m. & f.* guard || ***guarda raya*** *f.* landmark, monument

guardar *vb.* (comp.) to save

guardián *m. & f.* guard

guaya *f.* [Vz.] cable

guía *f.* richest part of an orebody; guide, indicator

guíabarrena *f.* whipstock

guía de ondas de alimentación *f.* feeder waveguide

guiar *vb.* to guide, pilot

guíasondas *f.* whipstock

guija *f.* granule, pebble, cobble, gravel; matrix, gangue || ***guija facetada*** *f.* faceted pebble

guijarra *f.* stony place

guijarro *m.* pebble, cobble, stone || ***guijarro aplanado*** *m.* flat pebble || ***guijarro facetado*** *m.* faceted pebble

guijarroso *a.* stony, pebbly

guijo *m.* gravel

guijón *m.* cobble

guijoso *a.* pebbly, gravelly, quartzy, stony, gritty

guinche *m.* winch, hoist

guíngaro *m.* pickaxe

guiñada *f.* (aeron., naut.) yaw

guiñar *vb.* (aeron., naut.) to yaw

güiris *m.* mining expert

gumbo *m.* gumbo, black mud

gunita *f.* gunite

gutolita *f.* dripstone

guyot *m.* guyot, seamount

H

haba *f.* nodule

habilitación *f.* outfit, equipment; outfitting

habilitar *vb.* to equip, outfit; provide funds for

habitaciónes *f.* living quarters, rooms

hábitat *m.* habitat || ***hábitat de la fauna*** *m.* wildlife habitat

hábito *m.* habit

hacer girar *vb.* to spin, twist, rotate

hacia abajo *a. & adv.* downward

hacia adentro *a. & adv.* inward

hacia afuera *a. & adv.* outward

hacia arriba *a. & adv.* upward

hacia el mar *a. & adv.* seaward || ***hacia la costa*** *a.* shoreward

hacienda *f.* smelter

hacha *f.* axe

hachuras *f.* hachures, hatching

hacienda nacional *f.* national treasury

hacinar *vb.* to stack

hade *m.* hade

halita *f.* halite, rock salt

halmirógeno *a.* halmyrogenic

halmirólisis *m.* halmyrolysis

halo *m.* halo || ***halo de hidrocarburos*** *m.* hydrocarbon halo

halocínesis *f.* halokinesis, salt tectonics

halógeno *m. & a.* halogen; halogenous

halóideo *m.* haloid

haloisita *f.* halloysite

halotriquita *f.* halotrichite

haluro *m.* halide

hallar(se) *vb.* to be found

hallazgo *m.* discovery, occurrance

halloisita *f.* halloysite

hamada *f.* butte, stoney desert

harina *f.* powdered ore or metal; rock flour || ***harina de falla*** *f.* fault gouge || ***harina fósil*** *f.* diatamaceous earth

hardware *m.* hardware

harzburgita *f.* harzburgite

hastial *m.* lateral wall of an excavation

hastingsita *f.* hastingsite

hauerita *f.* hauerite

hausmanita *f.* hausmannite

Hauteriviano, Hauteriviense *m. & a.* Hauterivian

haüyna *f.* hauynite, hauyne

haz *f.* bundle; group of deposits || ***haz de diques*** *f.* dike swarm || ***haz de geófonos*** *f.* geophone spread || ***haz explorador*** *f.* scanner

hebilla *f.* clasp, buckle

hebra *f.* vein; layer, bed, stratum

heces *f.* sludge, dregs, sediment, feces

hectara, hectárea *f.* hectare (2.471 acres; 0.00386 square miles)

hechura *f.* workmanship, making
hedenberguita *f.* hedenbergite
hedionda *a.* fetid
helada *f.* frost
helado *a.* frosted, frozen
helar *vb.* to freeze
helero *m.* glacier, ice sheet
hélico *a.* helical
helicoidal *a.* helicoidal
helio *m.* helium
heliodoro *m.* heliodor, gem beryl
heliografía *f.* blueprint
heliotropo *m.* heliotrope, bloodstone
helipuerto *m.* helideck, heliport
Helveciano, Helveciense, Helvetiano, Helvetiense *m. & a.* Helvetian
helvina *f.* helvite
hematita *f.* hematite
hemera *f.* hemera
hematítico *a.* hematitic
Hembergiense, Hembergiano *m. & a.* Hembergian
hembra *f.* female
hemimórfico *a.* hemimorphic
hemimorfita *f.* hemimorphite
hemipelágico *a.* hemipelagic
hemisférico *a.* hemispheric
hemisferio *m.* hemisphere
hendedura *f.* fissure, crevice, crack, split, rift
hender *vb.* to cleave, split, crack
hendible *a.* fissile, cleavable
hendido *a.* caved in, cleft
hendidura *f.* fissure, crevice, crack, split, rift, fracture
hendimiento *m.* splitting, cleaving, cracking
heptano *m.* heptane
Herciniano *m. & a.* Hercynian
herecinita *f.* hercynite
hercio *m.* Hertz
hermético *a.* airtight, hermetic
herramienta *f.* tool || ***herramienta de cable*** *f.* cable-tool || ***herramienta de pesca*** *f.* fishing tool || ***herramienta desviadora*** *f.* sidetracking tool
herrumbre *f.* rust
herrumbroso, herrumbrado *a.* rusty
hervidero *m.* boiling spring; mud volcano
hesita *f.* hessite
heteroblástico *a.* heteroblastic
heterocrónico *a.* heterochronous
heterodonto *m.* heterodont
heterogeneidad *f.* heterogeneity
heterogénio *a.* heterogeneous
heteromorfismo *m.* heteromorphism
heterópico *a.* heteropic
heterópodo *m.* heteropod
heterostrofía *f.* heterostrophy
heterotópico *a.* heterotopic
heterozooide *m.* heterozooid
Hettangiense, Hettangiano *m. & a.* Hettangian
heulandita *f.* heulandite
hexágonal *a.* hexagonal
hexágono *m.* hexagon
hexano *m.* hexane
hexeno *m.* hexene
hez *f.* sludge, dregs, sediment, feces
hialina *f.* hyaline
hialino *a.* hyaline
hialita *f.* hyalite
hialodacita *f.* hyalodacite
hialofano *m.* hyalophane
hialofítico *a.* hyalophitic
hialopilítico *a.* hyalopilitic
hiato *m.* hiatus
híbrido *m. & a.* hybrid
hidatógeno *a.* hydatogenic
hidrargillita *f.* hydrargillite
hidratación *f.* hydration
hidratar *vb.* to hydrate
hidrato *m.* hydrate || ***hidrato de gas*** *m.* gas hydrate, clathrate
hidráulica *f.* hydraulics
hidráulico *a.* hydraulic
hidrico *a.* hydrous
hidrobiología *f.* hydrobiology
hidroboracita *f.* hydroboracite

hidrocarburo *m.* hydrocarbon || ***hidrocarburo alifático*** *m.* aliphatic hydrocarbon || ***hidrocarburo aromático*** *m.* aromatic hydrocarbon || ***hidrocarburo clorinado*** *m.* chlorinated hydrocarbon || ***hidrocarburo de cadena abierta*** *m.* straight chain hydrocarbon || ***hidrocarburo de cadena cerrada*** *m.* closed chain hydrocarbon || ***hidrocarburo no saturado*** *m.* unsaturated hydrocarbon || ***hidrocarburo saturado*** *m.* saturated hydrocarbon

hidrocincita *f.* hydrozincite

hidroclástico *a.* hydroclastic

hidroclasto *m.* hydroclast

hidrocloruro *m.* hydrogen chloride

hidrocraqueo *m.* hydrocracking

hidrodesintegración *f.* hydrocracking

hidrodinámica *f.* hydrodynamics

hidrodinámico *a.* hydrodynamic, streamlined

hidroeléctrico *a.* hydroelectric

hidrófilo *a.* hydrophilic

hidrófugo *a.* water-repellent, waterproof

hidrogenación *f.* hydrogenation

hidrogenado *a.* hydrogenated

hidrogenar *vb.* to hydrogenate

hidrógeno *m.* hydrogen || ***hidrógeno pesado*** *m.* deuterium || ***hidrógeno sulfurado*** *m.* hydrogen sulfide

hidrogeología *f.* hydrogeology

hidrografía *f.* hydrography

hidrográfico *a.* hydrographic

hidrógrafo *m.* hydrograph

hidrolacolito *m.* pingo

hidrólisis *f.* hydrolysis

hidrolizador *vb.* hydrolyzer

hidrolita *f.* hydrolith

hidrología *f.* hydrology

hidrológico *a.* hydrologic

hidrólogo *m.* hydrologist

hidrómetro *m.* stream gauge, hydrometer, water meter

hidromuscovita *f.* illite

hidroneumático *a.* hydraulic-pneumatic

hidrósfera *f.* hydrosphere

hidrosilicato *m.* hydrosilicate

hidrosol *m.* hydrosol

hidrostático *a.* hydrostatic

hidrotermal, hidrotérmico *a.* hydrothermal

hidróxido *m.* hydroxide || ***hidróxido de amonio*** *m.* ammonium hydroxide || ***hidróxido de calcio*** *m.* calcium hydroxide

hidroxilo *m.* hydroxyl

hidruro *m.* hydride

hielo *m.* ice, frost || ***banco de hielo*** *m.* ice field || ***hielo a la deriva*** *m.* ice pack, drift ice || ***hielo atascado*** *m.* ice jam || ***hielo fijo*** *m.* fast ice, coastal ice || ***hielo seco*** *m.* dry ice || ***hielos perpetuos*** *m.* ice cap || ***tempano de hielo*** *m.* iceberg

hierro *m.* iron, iron ore || ***hierro acicular*** *m.* goethite || ***hierro al molibdeno*** *m.* ferromolybdenum || ***hierro arsenical*** *m.* arsenopyrite || ***hierro azul*** *m.* vivianite || ***hierro blanco*** *m.* arsenopyrite || ***hierro brillante*** *m.* specularite || ***hierro colado*** *m.* cast iron || ***hierro cromado*** *m.* chromite || ***hierro de pantano*** *m.* bog iron || ***hierro dulce*** *m.* wrought iron || ***hierro en lingotes*** *m.* pig iron || ***hierro espático*** *m.* siderite || ***hierro forjado*** *m.* wrought iron || ***hierro fundido*** *m.* cast iron || ***hierro galvanizado*** *m.* galvanized iron || ***hierro magnético*** *m.* magnetite || ***hierro oligisto*** *m.* hematite || ***hierro pardo*** *m.* limonite || ***hierro redondo*** *m.* reinforcing bars || ***hierro rezago*** *m.* scrap iron || ***hierro viejo*** *m.* scrap iron || ***mineral de hierro*** *m.* iron ore

hierroníquel *m.* nickel iron

higrometría *f.* hygrometry

higrómetro *m.* hygrometer

hijuela *f.* ore seam; small drain

hilo *m.* small vein or thread of ore

hilos de estadía *m.* stadia cross-hairs

hilos taquimétricos *m.* stadia hairs

hinca pilotes *m.* pile driver

hincar *vb.* to drive into, push in, sink in, thrust in

hinchado *a.* swollen, bulging

hinchar(se) *vb.* to swell, inflate

hipautomórfico *a.* hypautomorphic, hypidiomorphic, subhedral

hipérbola *f.* hyperbola

hiperbólico *a.* hyperbolic

hiperstena, hipersteno *f. & m.* hypersthene

hipidiomorfo *a.* hypidiomorphic

hipoabisal *a.* hypabyssal

hipocentro *m.* hypocenter

hipoclorito *m.* hypochlorite

hipocristalino *a.* hypocrystalline

hipogénico *a.* hypogene

hipolimnio *m.* hypolimnion

hipotermal *a.* hypothermal

hipótesis *f.* hypothesis || ***hipótesis de trabajo múltiple*** *f.* multiple working hypotheses

hipsógrafo *m.* hypsograph

hipsómetro *m.* hypsometer

histéresis *f.* histeresis, creep || ***histéresis viscosa*** *f.* viscous creep

histograma *m.* histogram

historia *f.* history || ***historia de producción*** *f.* production history || ***historia natural*** *f.* natural history

historial *m.* production history || ***historial clinico*** *m.* case history

hito *m.* survey monument

hocino *m.* narrows, canyon, gorge

hofmanita *f.* hofmannite

hogar *m.* hearth, furnace

hoja *m.* slab, sheet; nappe; ply

hojoso *a.* laminated

holmio *m.* holmium

holmquistita *f.* holmquistite

Holocénico, Holoceno *m. & a.* Holocene

holocristalino *a.* holocrystalline

holohialino *a.* holohyaline

holotipo *m.* holotype

holla de acarreo *f.* (geog.) drift hollow

hollín *m.* soot; fumes, furnace smoke

homeoblástico *a.* homeoblastic

homoclinal *m.* homocline

homogeneidad *f.* homogeneity

homogen(e)izar *vb.* to homogenize, smooth

homogéneo *a.* homogeneous

homogenización *f.* homogenization

homólogo *a.* homologous

homotáxico *a.* homotaxial

hondamente *adv.* deeply, to a great depth

hondo *m. & a.* bottom, depth; deep, low

hondonada *f.* trough, depression; ravine, dip, gap, basin, gully, deep valley, lowland, [U.K.] swale, [U.S.] arroyo, gulch

honorarios *m.* fee

hora *f.* hour || ***hora de Greenwich*** *f.* Greenwich mean time || ***horas de trabajo*** *f.* man-hours, work hours || ***horas extras*** *f.* overtime

horadar *vb.* to bore

horario *m.* schedule

horizontal *a.* horizontal

horizonte *m.* horizon, layer, stratum, bed (véase capa, estrato) || ***horizonte A, B, o C*** *m.* A, B, or C soil horizon] || ***horizonte acuífero*** *m.* aquifer || ***horizonte de correlación*** *m.* correlation horizon || ***horizonte de referencia*** *m.* datum, marker horizon, key horizon || ***horizonte fantasma*** *m.* phantom horizon || ***horizonte guía*** *m.* key layer, marker bed, datum || ***horizonte índice*** *m.* key horizon || ***horizonte imaginario*** *m.* phantom horizon || ***horizonte madre*** *m.* source bed || ***horizonte petrolífero*** *m.* reservoir bed, oil horizon || ***horizonte productivo*** *m.* productive horizon || ***horizonte reflector*** *m.* reflector horizon

hormigón *m.* concrete

hormiguero monticular *m.* ant hill

hornablenda, hornblenda *f.* hornblende

hornabléndico *a.* hornblende

hornblendita *f.* hornblendite

hornacho *m.* excavation

hornada *f.* melt

hornaguear *vb.* to mine hard coal

hornaguera *f.* hard coal

hornaza *f.* small furnace

hornfelsa *f.* hornfels

hornfelsítico *a.* hornfelsic

hornillo *m.* small furnace

horno *m.* furnace, kiln || ***alto horno*** *m.* blast furnace || ***horno de cal*** *m.* lime kiln || ***horno de calcinación*** *m.* calcining furnace || ***horno de coqueo*** *m.* coking oven || ***horno de fundición*** *m.* blast furnace || ***horno de reverbero*** *m.* reverberatory furnace

horqueta de un río *f.* fork of a river

horrura *f.* scoria, dross

horst *m.* horst, uplifted block

horsteno *m.* chert

hoya *f.* deep excavation; pit, ditch, hole, trough; basin || ***hoya de falla*** *f.* fault trough, graben || ***hoya glacial*** *f.* kettle hole || ***hoya hidrográfica*** *f.* watershed

hoyada *f.* depression, hollow

hoyar *vb.* to dig holes

hoyo *m.* hole, pit; well || ***hoyo abajo*** *a.* downhole || ***hoyo circular*** *m.* pothole || ***hoyo de disolución*** *m.* solution pit || ***hoyo de disparo*** *m.* shot hole || ***hoyo de la explosión*** *m.* shot hole || ***hoyo de marmita*** *m.* pothole || ***hoyo de remolino*** *m.* pothole

hoyoso *a.* pitted

hoyuelo *m.* pit

hoz *f.* gorge, ravine

huaico *m.* alluvium

huaiqueria *f.* badlands

huebnerita *f.* huebnerite

huecarrera *f.* cavern with stalagtite ore

hueco *m. & a.* vug; hollow, void, cavity || ***hueco de descanso*** *m.* rathole

huelga *f.* strike, labor stoppage

huella *f.* track, trail; trace || ***huellas de deslizamiento*** *f.* landslide scars || ***huella de fisión*** *f.* fission tracks

hueso *m.* bone

huincha *f.* [Bol., Ch.] measuring tape

huinche *m.* winch

hule *m.* rubber

hulla *f.* bituminous coal || ***hulla menuda*** *f.* [U.K.] culm

hullera *f.* coal bed; coal mine, [U.K.] colliery

hullero *m. & a.* coal miner; coal

hullífero *a.* coal-bearing

hullita *f.* anthracite

humectabilidad *f.* wettability

humectante *a.* wetting

humedad *f.* humidity

humedecer *vb.* to moisten, humidify

humedecimiento *m.* damping

húmedo *a.* humid, wet, damp, moist

humero *m.* flue

húmico *a.* humic

humita *f.* humite

humo *m.* smoke

humulita *f.* humulith

humus *m.* humus

hundido *m. & a.* dipping; caved; buried, submerged

hundiendose *vb.* sinking, pitching, plunging

hundimiento *m.* subsidence, caving, dip, pitch || ***hundimiento en caldero*** *m.* cauldron subsidence

hundir *vb.* to subside, sink

hundirse *vb.* subside; plunge

huracán *m.* hurricane

Huroniano, Huroniense *m. & a.* Huronian

huso horario *m.* time zone

I

ibón *m.* tarn

iceberg *m.* iceberg

Iceniano, Icenense *m. & a.* Icenian

ida y vuelta *f.* (perf.) round trip

identificación *f.* identification

idioblástico *a.* idioblastic

idiocromático *a.* idiochromatic

idiógeno *a.* syngenetic

idiomorfo *a.* idiomorphic

idocrasa *f.* idocrase

igarapés *m.* branches of a river

ígneo *a.* igneous

ignífugo *a.* refractory, fireproof

ignimbrita *f.* ignimbrite, welded tuff

Igualadense, Igualadiano *m. & a.* Igualadian

ijolita *f.* ijolite

ilegal *a.* (legal) illegal, illicit

ilícito *a.* (legal) illegal, illicit

ilita *f.* illite

ilmenita *f.* ilmenite

iluminación *f.* illumination

iluminar *vb.* to illuminate; to enlighten

ilustración *f.* illustration, figure, picture, drawing

ilustrar *vb.* to illustrate, draw; explain, make clear

iluviación *f.* illuviation

iluvial *a.* illuvial

Illinoisiano, Illinoisiense *m. & a.* Illinoisan

ilvaita *f.* ilvaite

imagen *f.* image, picture

imán *m.* magnet, lodestone

imanación, imantación *f.* magnetization

imanado *a.* magnetized

imanar, imantar *vb.* to magnetize

imbibición *f.* absorption

imbricación *f.* imbrication, overlap

imbricado *a.* imbricated, overlapping, shingled, staggered, stepped

imbricar *vb.* to imbricate, overlap

impacción *f.* impact

impactar *vb.* to impact, affect

impacto *m.* impact

impar *m. & a.* odd number; odd, uneven

impedancia *f.* impedance || ***impedancia acústica*** *f.* acoustic impedance

imperfección *f.* imperfection

imperfecto *a.* imperfect

impermeabilidad *f.* impermeability

impermeable *a.* impermeable, impervious, waterproof

imperturbado *a.* undisturbed

ímpetu *m.* impetus, impulse; momentum

impiderreventones *m.* blowout preventers

implementación *f.* implementation, introduction

implementar *vb.* to implement, introduce

implemento *m.* tool, implement

importación *f.* imports; importation

importar *vb.* to import

impregnación *f.* impregnation

impregnado *a.* impregnated

impregnar *vb.* to impregnate

impresición *f.* mistake, error

impresión *f.* impression, imprint, print || ***impresión azul*** *f.* blueprint || ***impresión de ola*** *f.* ripple mark

impresor *m.* printer

impresora *f.* printer

imprimir *vb.* to print

improductivo *a.* non-producing, barren || ***pozo improductivo*** *m.* dry hole

impronta *f.* impression, imprint, cast, mark

impsonita *f.* impsonite

impuesto *m.* tax || ***impuesto al consumo del petróleo crudo*** *m.* crude oil excise tax (COET) || ***impuesto a los réditos*** *m.* income tax || ***impuesto de producción*** *m.* severence tax || ***impuesto indirecto*** *m.* excise tax || ***impuesto interior*** *m.* excise tax

impulsar *vb.* to drive

impulsión *f.* drive || ***impulsión por agua*** *f.* water drive || ***impulsión por gas disuelto*** *f.* solution gas drive || ***impulsión por gravedad*** *f.* gravity drive

impulso *m.* impulse || ***impulso eléctrico*** *m.* electric drive || ***impulso momentáneo*** *m.* transient || ***impulso transitorio*** *m.* transient

impureza *f.* impurity

impurificación *f.* contamination

impurificado *a.* impure, contaminated

impurificar *vb.* to contaminate

impuro *a.* impure

inactivo *a.* inactive, inert

inagotable *a.* inexhaustible

inalterado *a.* unaltered, unweathered

inarmónico *a.* disharmonic

incandescencia *f.* glow, incandescence

incandescente *a.* incandescent

incapaz *a.* incompetent

incarbonización *f.* coalification

incendio *m.* fire

incertidumbre *f.* uncertainty, doubt

incierto *a.* uncertain

incidencia *f.* incidence || ***incidencia normal*** *f.* normal incidence || ***incidencia oblicua*** *f.* oblique incidence || ***angulo de incidencia*** *m.* incidence angle

incidente *m.* incident; *a.* incidental

incineración *f.* incineration

incipiente *a.* incipient

incisión *f.* incision

inciso *a.* incised

inclinación *f.* inclination, dip, slope, gradient, tilt, pitch || ***inclinación abajo*** *f.* downdip || ***inclinación al veedor*** *f.* layover distortion || ***inclinación aparente*** *f.* apparent dip || ***inclinación arriba*** *f.* updip || ***inclinación de las rampas en el primer escalón*** *f.* fundamental step-up angle || ***inclinación dorsal*** *f.* rear dip || ***inclinación frontal*** *f.* forward dip || ***inclinación original*** *f.* original dip || ***inclinación regional*** *f.* regional dip

inclinado *a.* inclined

inclinar(se) *vb.* to incline, slope, dip

inclinómetro *m.* dipmeter, clinometer

incluído *a.* included, imbedded

incluir *vb.* to enclose

inclusión *f.* inclusion || ***inclusión de arcilla*** *f.* clay gall || ***inclusión de fluidos*** *f.* fluid inclusion

incoherente *a.* unconsolidated

incoloro *a.* colorless

incolumita *f.* flexible sandstone

incombustible *a.* fireproof

incompetente *a.* incompetent

incondensable *a.* incondensable

incongruente *a.* incongruous

inconsecuente *a.* inconsequent

inconsistente *a.* unstable, loose, uncompacted

incorporación *f.* incorporation, embodiment

incorporar *vb.* to incorporate

incremento *m.* increment, recharge

incrustación *f.* incrustation, inlay; scale, sinter || ***incrustación calcárea*** *f.* flowstone

incrustado *a.* incrusted; embedded

incrustante *a.* incrusting, scale-forming

incrustar *vb.* to incrust, inlay, embed

indefinido *a.* indefinite

indeformado *a.* undeformed

indemnización *f.* severance pay

indescrito *a.* nondescript, undescribed

indeterminable *a.* indeterminable

indicación *f.* indication || ***indicación de gas (o petróleo)*** *f.* kick, indication

indicador *m.* tracer, indicator; gage || ***indicador de caudal*** *m.* flow gage || ***indicador de deformación*** *m.* strain gage || ***indicador de desviación*** *m.* drift indicator || ***indicador de flujo*** *m.* flow indicator || ***indicador de presión*** *m.* pressure gage

indicatriz *f.* indicatrix

índice *m.* index; rate || ***índice de alteración térmica*** *m.* thermal alteration index || ***índice de hidrógeno*** *m.* hydrogen index || ***índice de oxígeno*** *m.* oxygen index || ***índice de producción*** *m.* production index || ***índice de refracción*** *m.* index of refraction || ***índice fotográfico*** *m.* photo index

indicios *m.* traces, indications, shows

indigación *f.* search, examination

indígena *a.* indigenous, native

indígeno *m.* native, aboriginal

indigolita *f.* indicolite

indio *m.* indium

indisolubilidad *f.* indisolubility

indistinto *a.* indistinct

inducción *f.* induction; inference

inducido *a.* induced

inducidor *m. & a.* inducer; inducing

inducir *vb.* to induce

inductancia *f.* inductance

inductivo *a.* inductive

inductor *m. & a.* inductor, inducing

induración *f.* induration

indurar *vb.* to indurate

indurado *a.* indurated

industria *f.* industry

industrial *a.* industrial

ineficaz *a.* inefficient, ineffective

inercia *f.* inertia

inerte *a.* inert

inertinita *f.* inertinite

inestabilidad *f.* instability
inestable *a.* unstable
inexactitud *f.* inaccuracy
inexacto *a.* inaccurate
inexplotable *a.* unexploitable
inexplotado *a.* unexploited
inexplorado *a.* unexplored
infecundo *a.* barren, sterile
inferencia *f.* implication
inferido *a.* inferred
inferior *a.* lower; inferior || ***inferior extremo*** *a.* lowermost
infiltración *f.* infiltration
infiltrar *vb.* to infiltrate
infiltrarse *vb.* to infiltrate into
infinitesimal *a.* infinitesimal
infinito *a.* infinite
inflación *f.* inflation; tumescence
inflado *a.* inflated
inflamable *a.* inflammable
inflamación *f.* ignition, combustion
inflamar *vb.* to ignite, inflame
inflexión *f.* inflection
influencia *f.* influence
influjo *m.* inflow; flood tide || ***influjo de agua*** *m.* watering out of the wellbore
información *f.* information, data
informador *m.* scout
informante *m.* scout, informant
informar *vb.* to report, inform
informe *m.* report, brief || ***informe de estado de cuentas*** *m.* financial statement || ***informe de perforación*** *f.* driller's report
infracámbrico *m. & a.* Infracambrian
infraescrito *m.* undersigned
infraestructura *f.* infrastructure, installations, groundwork
infrapuesto *a.* underlying
infrarrojo *m. & a.* infrared
infrayacente *a.* underlying
infrayacer *vb.* to lie below
infundible *a.* infusible
infusión *f.* infusion
infusoriolita *f.* diatomaceous earth
Ing. *m.* engineer
ingeniería *f.* engineering
ingeniero *m.* (Ing.) engineer || ***ingeniero de elaboración*** *m.* processing engineer || ***ingeniero de minas*** *m.* mining engineer || ***ingeniero de perforación*** *m.* drilling engineer || ***ingeniero de petróleo*** *m.* petroleum engineer || ***ingeniero de producción*** *m.* production engineer || ***ingeniero de yacimientos*** *m.* reservoir engineer || ***ingeniero geológico*** *m.* geological engineer || ***ingeniero mecánico*** *m.* mechanical engineer || ***ingeniero químico*** *m.* chemical engineer
ingenio *m.* mill, plant, refinery; device; ingenuity
ingrediente *m.* ingredient
ingresión *f.* ingression
ingreso *m.* entrance; earnings, revenues, income
inhibidor *m.* inhibitor
inhibir *vb.* to inhibit
inhomogeneidad *f.* inhomogeneity
inicial *a.* initial
iniciación de surgencia por balanceo de presión *f.* rocking a well
iniciar la perforación *vb.* to spud a well
inmadurez *f.* immaturity
inmaduro *a.* immature
inmersión *f.* immersion
inmerso *a.* immersed
inmiscibilidad *f.* immiscibility
inmiscible *a.* immiscible
inorgánico *a.* inorganic
inosilicato *m.* inosilicate
inquilinato *m.* leasehold
insecuente *a.* insequent
insertado *a.* inserted
insertador de tubería *m.* snubber
insertar *vb.* to insert
ínsito, in situ *a.* in situ, on the spot
insolación *f.* insolation
insolubilidad *f.* insolubility
insoluble *a.* insoluble
insondeable *a.* bottomless, abyssal
inspección *f.* inspection

inspeccionar *vb.* to inspect

inspector *m.* inspector

instalación *f.* installation, assembly, plant || ***instalacións complementarias*** *f.* offsite facilities || ***instalacións propiamente dichas*** *f.* onsite facilities || ***instalación por telemando*** *f.* remote installation

instalado *a.* installed

instalar *vb.* to install, erect, place

instantáneo *a.* instantaneous

instituto *m.* institute

instrucción *f.* instruction

instructor *m.* instructor, teacher, trainer

instrumento *m.* instrument

insula *f.* island

insularidad *f.* insularity

intacto *a.* unbroken, intact

integración *f.* integration, merging

integrado *a.* integrated

integral *a.* integral

integrante *m.* member

integrar *vb.* to integrate, merge, join

intemperismo *m.* weathering

intendencia *f.* official district

intensidad *f.* intensity

intensivo *a.* intensive

intenso *a.* intense

interarrecife *m.* interreef

interarrecifal *a.* interreef

intercalación *f.* intercalation

intercalado *a.* intercalated

intercalar *vb.* to intercalate

intercambiable *a.* interchangeable

intercambiador *m.* exchanger

intercambiar *vb.* to exchange, trade

intercambio *m.* trade, exchange

intercapa *f.* interbed

interceptar *vb.* to intercept

interceptor *m.* trap || ***interceptor al vacío*** *m.* vacuum trap

interclusión *f.* intercalation

intercorriente *a.* interstream

intercrecimiento *m.* intergrowth

intercuenca *f.* interbasin

interdigitación *f.* interfingering, intertonguing

interdigitado *a.* intertongued

interduna *f.* & *a.* interdune

interés *m.* interest || ***interés en la inversión*** *m.* working interest || ***interés minoritario*** *m.* minority interests

interestratificación *f.* interbedding

interestratificado *a.* interbedded

interestratificar *vb.* to interbed

interfaz *f.* interface

interferencia *f.* interference

interferómetro *m.* interferometer

interfluvio *m.* interfluve

interformacional *a.* interformational

interglacial *a.* interglacial

intergranular *a.* intergranular

interior *a.* internal, inner, interior

internacional *a.* international

interlaminado *a.* interlaminated

interlobulado *a.* interlobate

intermedio *a.* intermediate

intermitente *a.* intermittent

intermontaño *a.* intermontane

interno *a.* inner, interior, internal

interpartícula *f.* interparticle

interpolación *f.* interpolation

interpolar *vb.* to interpolate

interpretar *vb.* to interpret, explain

interpretación *f.* interpretation || ***interpretación fotográfica*** *f.* photointerpretation || ***interpretación sísmica*** *f.* seismic interpretation

intérprete *m.* & *f.* interpreter

interpretador fotográfico *m.* photo-interpreter

interrumpir *vb.* to interrupt

interrupción *f.* interruption, stoppage || ***interrupción en la plataforma continental*** *f.* shelf break

interruptor de encendido y apagado *m.* off–on switch || ***interruptor de encendido y apagado de reinicialización*** *m.* reset switch

intersecarse *vb.* to intersect

intersección *f.* intercept, intersection

intersticial *a.* interstitial

intersticio *m.* interstice

intervalo *m.* interval, tract || ***intervalo de grupo*** *m.* group spacing interval || ***intervalo del sistema*** *m.* system tract

interyacente *a.* interbedded || ***interyacente de lutita*** *m.* shale break

intracubetal, intracuencal *a.* intrabasinal

intradelta *f.* delta plain

intraestratal *a.* intrastratal

intrafiloniano *a.* between veins

intraformacional *a.* intraformational

intraplegado *a.* convoluted

intraplegamiento *m.* convolute folding

introducir datos *vb.* to input data

intromisión *f.* influx

intruido *a.* intruded

intruir *vb.* to intrude

intrusión *f.* intrusion, (salt diapir) upbuilding || ***intrusión acuífera*** *f.* water encroachment

intrusivo *m.* intrusive

intumescencia *f.* swelling, intumescence

inundable *a.* easily flooded

inundación *f.* flood, flooding, transgression || ***inundación artificial*** *f.* water flood || ***inundación con aire*** *f.* air flood || ***inundación con gas*** *f.* gas flood || ***inundación por agua*** *f.* water flood || ***inundación repentina*** *f.* flash flood

inundado *a.* flooded

inundar *vb.* to flood

invadido *a.* invaded by water

invasión *f.* water invasion; trespass

inventario *m.* inventory

inversión *f.* investment; inversion

inverso *a.* inverted, inverse

invertido *a.* inverted, overturned

invertir *vb.* to invest, to overturn

investigación *f.* research, investigation

investigar *vb.* to research, investigate

invierno *m.* winter

involución *f.* involution

involuta *f. & a.* involute

inyección *f.* injection, drilling mud || ***inyección de agua*** *f.* water flood || ***inyección de agua de línea media*** *f.* center line water injection || ***inyección de aire*** *f.* air injection || ***inyección de espuma*** *f.* foam injection || ***inyección de gas*** *f.* gas injection || ***inyección de vapor*** *f.* steam flood, steam injection || ***inyección magmática*** *f.* magmatic injection

inyectado *a.* injected

inyectar *vb.* to inject

inyector *m.* injector || ***inyector de lodo*** *m.* mud gun

inyoita *f.* inyoite

iodargirita *f.* iodyrite

iodo *m.* iodine

iolita *f.* iolite, cordierite

ion *m.* ion

iónico *a.* ionic

ionización *f.* ionization

ionizar *vb.* to ionize

ionógeno *a.* inorganic

ipabismal *a.* hypabyssal

iridio *m.* iridium

iridiscencia *f.* iridescence

iridiscente *a.* iridescent

iridosmina *f.* iridosmium

iris *m.* iridescent hyaline quartz

irisación *f.* iridescence

irisado *a.* iridescent

irisar *vb.* to iridesce, be iridescent

irradiación *f.* irradiation

irradiar *vb.* to irradiate

irrecuperable *a.* irrecoverable

irreducible *a.* irreducible

irregular *a.* uneven

irrevocable *a.* irrevocable

irse desgastando *vb.* to wear away, to wear out

isla *f.* island, isle || ***arco de islas*** *m.* island arc

isleo *m.* klippe; island || ***isleo cobijante*** *m.* erosional remnant

isleta, islilla, islota *f.* islet

islete, islote *m.* islet

islote litoral *m.* barrier island
isoantracita *f.* isoanthracite
isóbara *f.* isobar
isobárico *a.* isobaric
isobarométrico *a.* isobarometric
isóbata *f.* isobath, depth contour
isobutano *m.* isobutane
isocarbono *m.* isocarbon
isoclinal *m. & a.* isocline; isoclinal
isoclínico *a.* isoclinic
isoclino *a.* isoclinal
isocora *f.* isochore
isocromo *m.* isochrome
isócrona *f.* isochron
isócrono *a.* isochronous
isodina *f.* isomagnetic line
isodinámico *a.* isodynamic
isofana *f.* franklinite
isogal, isogálica *f.* isogal
isogama *f.* isogam
isogeoterma *f.* isogeotherm
isogeotermo *m.* isogeotherm
isogiro *m.* isogyre
isógono *m. & a.* isogon; isogonic
isohalina *f.* isohaline
isohieto *m. & a.* isohyet; isohyetal
isoíeta *f.* isohyet
isoípsa, isohipsa *f.* contour line
isolínea *f.* isopleth
isolita *f.* isolith
isomagnético *a.* isomagnetic
isomérico *a.* isomeric
isomerismo *m.* isomerism
isomerización *f.* isomerization
isómero *m. & a.* isomer; isomeric
isometamorfismo *m.* isometamorphism
isométrico *a.* isometric
isomórfico *a.* isomorphic
isomorfismo *m.* isomorphism
isomorfo *m.* isomorph
isopaca *f.* isopach
isopaco *m.* isopach
isoparafina *f.* isoparrafin
isopático *a.* isopachous
isopentano *m.* isopentane
isopleta *f.* isopleth
isopluvial *a.* isohyetal, isopluvial
isopora *f.* isopore
isopreno *m.* isoprene
isoprenoide *m.* isoprenoid
isoquímico *a.* isochemical
isorrumbado *a.* follows the same strike
isostasia, isostasis *f.* isostacy
isostático *a.* isostatic
isoterma *f.* isotherm
isotermal, isotérmico, isotermo *a.* isothermic, isothermal
isotípico *a.* isotypic
isotipo *m.* isotype
isotópico *a.* isotopic
isótopo *m.* isotope
isotropía *f.* isotropy
isotrópico, isótropo *a.* isotropic
isotropismo *m.* isotropism
isoyeta *f.* isohyet
ístmico *a.* isthmian
istmo *m.* isthmus
itacolumita *f.* flexible sandstone
italita *f.* leucite
ítem *m.* item
iteración *f.* iteration
iterar *vb.* to iterate
iterativo *a.* iterative
itérbico *a.* ytterbic
iterbio *m.* ytterbium
itinerario *m.* itinerary, schedule
itria *f.* yttria
itrio *m.* yttrium
itroilmenita *f.* samarskite
itztle, itztli *m.* [Méx.] obsidian
iu *m.* nephrite
izador *m.* hoist
izaje *m.* hoisting
izar *vb.* to hoist
izquierdo *a.* left || ***izquierdo lateral*** *a.* left-lateral

J

jabalcón *m.* trench brace

jabalconar *vb.* to shore, brace, prop

Jabaliense *m. & a.* Acadian

jable *f.* fine white sand

jabón *m.* [Méx.] talc ore; [Col.] slickenside || ***jabón de sastre*** *m.* steatite, soapstone || ***jabón de soldado*** *m.* smectite || ***jabón de vidrieros*** *m.* pyrolusite || ***jabón mineral*** *m.* smectite

jaboncillo *m.* clay, gouge

jacintina *f.* idocrase

jacinto *m.* red zircon

Jacksoniano, Jacksoniense *m. & a.* Jacksonian

jacksonita *f.* prehnite

jacupirangita *f.* jacupirangite

jacutinga *f.* itabirite

jade *m.* jade

jadeita *f.* jadeite

jaguadero *m.* river outlet

jagüel *m.* [Arg., Ch.] water-filled pit or trench

jais *f.* jet

jaki, jali *m.* khal

jalea de petróleo *f.* petroleum jelly

jales *m.* [Méx.] rich tailings

jalón *m.* rod, pole, survey stake, milestone, milepost || ***jalón de mira*** *m.* level rod

jalonar *vb.* to run a survey line

jalonero *m.* rodman

jalsontles *m.* [Méx.] rich tailings

jamasquita *f.* jamaskite

jamesonita *f.* jamesonite

Janitorense, Janitoriano *m. & a.* Janitorian

janosita *f.* copiapite

japanita *f.* penninite

jaquelado *m.* checkerboard pattern

jardín *m.* blemish on an emerald

jargón *m.* jargon, yellow gem zircon

Jarkov *m. & a.* Kharkov

jarosita *f.* jarosite

jaspágata *f.* agate jasper

jaspe *m.* jasper

jaspeado *a.* spotted, mottled, variegated

jaspear *vb.* to mottle, vein, streak

jasperita *f.* jasper

jasperización *f.* jasperization

jasperoide *a.* jasperoid

jaspilita *f.* jaspilite

jaspoide *m. & a.* tachylyte; jasperoid

jaspón *m.* red and yellow marble

jaula *f.* (min.) cage in a shaft || ***jaula de extracción*** *f.* lift, mine cage

jayet *m.* jet

jazido *m.* layer, bed, seam

jefe *m.* boss, chief, foreman || ***jefe de cuadrilla*** *m.* crew chief, party chief || ***jefe ejecutivo oficial*** *m.* chief executive officer

jeffersonita *f.* jeffersonite

jejo *m.* stone, rock

jentschita *f.* lengenbachite

jenzschita *f.* chalcedony

jerarquizar *vb.* to prioritize, rank, put in order of importance

jerga *f.* jargon

jeroglífico *m. & a.* hieroglyph; hieroglyphic

jeroglifo *m.* hieroglyph

Jerseyense, Jerseyiano *m. & a.* Jerseyian

jet *m.* jet

jewreinowita *f.* vesuvianite

jezekita *f.* morinite

Johannense *m. & a.* Olenidian

johnstonita *f.* vanadinite; galena

johnstrupita *f.* mosandrite

jolita *f.* iolite, cordierite

jorfe *m.* precipice

Jorgense, Jorgiano *m. & a.* Danian

jorgyanita *f.* scorodite

jornada *f.* work day; journey, trip

jornal *m.* day's work, wage

jornalar *vb.* to employ by the day

jornalero *m.* workman

Josarense, Josariano *m. & a.* lower Pontian

josefinita *f.* ferronickel

josena *f.* hartite

Jovarense, Jovariano *m. & a.* mid–middle Oligocene

joven *a.* young, (geog.) uneroded

Joviano, Joviense *m. & a.* part of the Quaternary

jovin *a.* uneroded

joya *f.* jewel

juddita *f.* Mn-hornblende

judicial *a.* judicial

juego *m.* set || ***juego de diaclasas*** *m.* joint set || ***juego de fallas*** *m.* fault set || ***juego de fracturas*** *m.* fracture set

jugar *vb.* to rise and fall (tidal)

jugo de la tierra *m.* hydrocarbon

jugo lapídeo *m.* parent magma; residual melt

julianita *f.* tennantite

julio *m.* joule

jungla *f.* jungle

junkerita *f.* siderite

junta *f.* connection, seam, joint; meeting; river junction

juntar(se) *vb.* to join; coalesce

juntura *f.* joint

junturamiento *m.* jointing

juquero *m.* [Pe.] ore thief

Jura *f.* Jurassic || ***Jura blanco*** *f.* Malm || ***Jura negro*** *f.* Lias || ***Jura pardo*** *f.* Dogger

Jurásico *m. & a.* Jurassic

jurinita *f.* brookite

jusente *m. & a.* ebb tide; that rises and falls with the ebb tide

justipreciación *f.* appraisal

justipreciador *m.* appraiser

justipreciar *vb.* to appraise, rate

justiprecio *m.* appraised value

Juvarense, Juvariano *m. & a.* Juvarian

juvenil *a.* juvenile

juventud *f.* youth

K

Kaersutita *f.* kaersutite

Kainolítico *m. & a.* Cenozoic

kakirita *f.* kakixen

Kalaviense *m. & a.* lower Algonquin

kalicamptonita *f.* cascadite

kaligranito *m.* kaligranite, granite

kalikeratófido *m.* kalikeratophyre, keratophyre

kaliofilita *f.* kaliphite

kaliqueratofiro *m.* kalikeratophyre, keratophyre

kalitraquita *f.* kalitrachyte, trachyte

kalkdiabasa *f.* amygdaloidal diabase

kalkgranito *m.* calcigranite, granite

kalkuranita *f.* uranite

kaluszita *f.* singenite

kam, kame *m.* kame

kamarezita *f.* brochantite

kammgranito *m.* kammgranite, granite

Kansan, Kansaniense *m. & a.* Kansan

kansita *f.* mackinawite

kaolín *m.* kaolin

kaolinita *f.* kaolinite

kaolinítico *a.* kaolinitic

kaolinización *f.* kaolinization

karbonatita *f.* carbonatite

karbonatitalnoíta *f.* alnoite

karlsteinita *f.* carlstainite, alkali granite

Karniense, Karniano *m. & a.* Carnian

karst *m. & a.* karst

karstenita *f.* anhydrite

kárstico *a.* karstic

karstificación *f.* karstification

kaskasia *m. & a.* Kaskasia

kataforita *f.* kataforite

katatéctico *a.* katatectic

katazona *f.* katazone

katogénico *a.* katogenic

katógeno *m.* katogene

katungita *f.* katungite

kaukasita *f.* caucasite, sanidine granite

Kazán *m. & a.* Kazanian

Keewatino *m. & a.* Keewatin
keiviíta *f.* keivyite
kelifítico, kelyfítico *a.* kelyphitic
kenngotita *f.* miargyrite
kenyita *m.* kenyite, olivine phonolitic trachyte
kerabitumen *m.* kerogen
kerargirita *f.* kerargyrite
keratita *f.* hornfels
keratofido, keratófiro *m.* keratophyre
kernita *f.* kernite
kerógeno *m.* kerogen
kerogenita *f.* oil shale
kerosén, keroseno *m.* kerosene
kerosina *f.* kerosene
kersantita *f.* kersantite
ketona *f.* ketone
Keuperiano, Kéuprico *m. & a.* Keuper
Keweenawiano, Keweenawense *m. & a.* Keweenawan
k-feldespato *m.* K-feldspar
khareadora *f.* ore sorter
khibinita *f.* mosandrite
khinganita *f.* kesterite
khuniita *f.* iranite
kieselgur *m.* kieselguhr, kiselgur
kiirunavaarita *f.* magnetite
kilita *f.* kylite, olivine theralite
kilo *m.* kilogram
kiloamperio *m.* kiloampere
kilobara *f.* kilobar
kilocaloría *f.* kilocalorie
kilociclo *m.* kilocycle
kilogramo *m.* kilogram
kilolitro *m.* kiloliter
kilometraje *m.* distance in kilometers
kilómetro *m.* kilometer
kilovatio *m.* kilowatt || ***kilovatio horas*** *f.* kilowatt hours
kilovoltio *m.* kilovolt
kimberlita *f.* kimberlite
Kimmeriano, Kimmérico *m. & a.* Kimmerian
Kimmeridgiano, Kimmeridgiense *m. & a.* Kimmeridgian
Kinderhook *m. & a.* Kinderhookian
kirunavaarita *f.* magnetite
kivita *f.* kivite
klausenita *f.* clausenite, diorite norite, lamprophyre, quartz gabbro
klippe *m.* klippe
klockmannita *f.* klockmannite
kmaita *f.* celadonite
knoxvillita *f.* copiapite
Koblenzense, Koblenziano *m. & a.* Coblenzian
kodurita *f.* kodurite, granulite
koelbingita *f.* kenigmatite
koivinita *f.* florencite
kokkonita *f.* cocconite, coccolithic limestone
kolbeckina *f.* herzenbergite
komatiíta *f.* komatiite
kongadiabasa *f.* Konga diabase, quartz dolerite
Kongurense, Konguriano *m. & a.* Kungurian
Konialnikense, Konialnikiano *m. & a.* Konialnikian, Pontian
koppita *f.* pyrochlore
kornerupina *f.* kornerupine
korund-anortosita *f.* kyschtymite, corundum anorthosite
korund-gabro *m.* corundum hornblende gabbro
kosmocloro *m.* kosmochlor, ureyite
kramerita *f.* probertite
kratógeno *m.* shield, craton
krennerita *f.* krennerite
kraurita *f.* dufrenite
kreuzbergita *f.* pleystenite
kriptón *m.* krypton
kristianita *f.* cristianite, biotite granite
kryokonita *f.* cryoconite
kulaíta *f.* kulaite
Kungur, Kunguriense, Kongurense *m. & a.* Kungurian
kunzita *f.* kunzite
kürunavaarita *f.* magnetite, kiirunvaarite
kuselita *f.* cuselite, biotite-augite lamprophyre
kusuita *f.* wakefieldite

L

labíl *a.* unstable, friable

labio *m.* edge, lip || ***labio alto*** *m.* upthrown block, high side || ***labio bajo*** *m.* downthrown block, downthrown side || ***labio de falla*** *m.* fault limb

labor *f.* miner; working face, mine workings || ***labor a cielo abierto*** *f.* open pit working || ***labor a través*** *f.* cross cut || ***labor de bancos*** *f.* underhand stoping || ***labor de beneficio*** *f.* mill, milling || ***labor de reconocimiento*** *f.* exploratory workings || ***labor de testeros*** *f.* overhaul stope, overhand stoping || ***labor escalonada*** *f.* stope

laboratorio *m.* laboratory

laboreo *m.* mining, exploitation, workings

labra *f.* working, cutting, carving, or dressing of stone

labrado *a.* worked, wrought, cut, carved

labradorita *f.* labradorite

labrar *vb.* to work a mine or metals, dress stone, carve

lacolítico *a.* laccolithic

lacolito *m.* laccolith

lacunario, lacustre *a.* lacustrine, lake

lacustre *a.* lacustrine

ladeado *a.* tilted

ladear *vb.* to tilt

ladera *f.* declivity, slope, hillside, flank, foothills || ***ladera abajo*** *f.* downhill || ***ladera arriba*** *f.* uphill || ***ladera de derrubios*** *f.* talus slope

ladero *a.* side, lateral

Ladiniano, Ladiniense *m. & a.* Ladinian

lado *m.* side

ladrillo *m.* brick || ***ladrillo refractario*** *m.* insulating brick

lago *m.* lake, [U.K.] loch || ***lago de circo*** *m.* tarn || ***lago en herradura*** *m.* oxbow lake || ***lagos en rosario*** *m.* paternoster lakes || ***lago salado*** *m.* salt lake || ***lago semilunar*** *m.* oxbow lake

láguena *f.* weathered argillaceous schist

laguillo *m.* pond

laguna *f.* lagoon

lagunal *a.* lagoonal

lahar *m.* lahar, mudflow

laja *f.* flagstone, thin flat stone, slab, bed, layer; ledge

lajoso *a.* flaggy

lama *f.* slime, mud, silt; tailings

lamelar *a.* laminated

lamelibranquio *m.* pelecypod

lameliforme *a.* lamelliform

lámina *f.* lamination, lamella; slab; scale || ***lámina acuífera*** *f.* water table || ***lámina continental*** *f.* continental plate || ***lámina delgada*** *f.* thin section

laminación *f.* lamination

laminado *a.* laminated; tabular, platy

laminar *vb.* to laminate

laminilla *f.* lamella; thin sheet; flake; rust, scale

lamoso *a.* slimy

lamprófido, lampróﬁro *m.* lamprophyre

lamprofilita *f.* lamprophyllite

Lancastriano, Lancastriense *m. & a.* Lancastrian

lancha *f.* thin, flat stone; stone slab; boat

lanchar *m.* quarry

lanchón *m.* barge

Landeniano, Landeniense *m. & a.* Landenian

lantano *m.* lanthanum

lapiez *m.* karren

lapili *f.* lapilli

lapislázuli *f.* lapis lazuli

lápiz *m.* graphite

lapizar *m.* graphite mine

laque *m.* druse

Laramide *m.* Laramide, *a.* Laramídico

lardalita *f.* laurdalite

largo *a.* long || ***largo plazo*** *m.* long-term

larguero *m.* stringer

larnita *f.* larnite

larsenita *f.* larsenite

lasca *f.* fragment, chip, spall

lascar *vb.* to spall

lastra *f.* flat stone, slab, flagstone

lastre *m.* weathered stone; ballast

lastrón *m.* large flagstone

latente *a.* dormant, latent

lateral *a.* lateral, to the side || ***lateral derecho*** *a.* right-lateral || ***lateral izquierdo*** *a.* left-lateral

laterita *f.* laterite

laterítico *a.* lateritic

laterización *f.* laterization

latilla *f.* lath

latita *f.* latite

latitud *m.* latitude; breadth, width || ***latitudes tropicales*** *m.* horse latitudes

latón *m.* brass

Lattorfiano, Lattorfiense *m. & a.* Lattorfian

laumontita *f.* laumontite

laurencio *m.* lawrencium

Laurentiano, Laurentiense *m. & a.* Laurentian

lava *f.* lava; washing of ore || ***lava aa*** *f.* aa lava || ***lava acordonada*** *f.* ropy lava || ***lava cordada*** *f.* pahoehoe lava || ***lava cordiforme*** *f.* ropy lava || ***lava de gas*** *f.* gas scrubber || ***lava en almohada*** *f.* pillow lava || ***lava escoriácea*** *f.* aa lava || ***lava pahoehoe*** *f.* pahoehoe lava

lavada *f.* washing of placers or ore

lavado *m.* washing || ***lavado a plato*** *m.* panning || ***lavado de granito*** *m.* granite wash

lavador *m.* gold pan; washer; scrubber || ***lavador de gas*** *m.* gas scrubber

lavadora *f.* gold pan

lavadura *f.* washing of placers or ore; outwash || ***lavadura del granito*** *f.* granite wash

lavaduras de falda *f.* slope wash

lavaje *m.* washing of placers or ore; denudation || ***lavaje de pendiente*** *m.* slope wash || ***lavaje en mantos*** *m.* sheet wash

lavar *vb.* to wash; (min.) to pan

lawsonita *f.* lawsonite

lazo de meandro *m.* meander loop

lazulita *f.* lazulite

lazurita *f.* lapis lazuli, lazurite

lechada *f.* slurry, paste

leche de luna *m.* calcareous clay

leche de tierra *f.* magnesia

lecho *m.* layer, stratum, bed rock; riverbed || ***lecho de arroyo*** *m.* riverbed || ***lecho de cantera*** *m.* bed of stone || ***lecho de creciente*** *m.* flood plain || ***lecho de roca*** *m.* bedrock || ***lecho del río*** *m.* riverbed || ***lecho mayor*** *m.* flood plain || ***lecho oceánico*** *m.* deep-sea floor || ***lecho rojo*** *m.* redbed

lector *m.* reader; (comp.) tape drive

lectora *f.* reader || ***lectora de cartas*** *f.* (comp.) card reader || ***lectora optica*** *f.* optical scanner

lectotipo *m.* lectotype

lectura *f.* reading || ***lectura de medición*** *f.* gage reading

lectura *f.* reading; (comp.) readout

Lediano, Lediense *m. & a.* Ledian

légamo *m.* slime, mud, ooze; loam

legamoso *a.* slimey, oozey, loamy

legislación *f.* legislation

legua *f.* league (about 5572 meters)

lejano, lejos *a.* distant, distal

lenar *m.* karren

lengua *f.* (geol.) tongue || ***lengua de arena*** *f.* spit of sand || ***lengua de piedra*** *f.* tongue of rock || ***lengua de sal*** *f.* salt tongue

lengüeta *f.* small tongue; barb

lente *m. & f.* lens; magnifying glass; lenticular rock body || ***lente bicóncava*** *f.* biconcave lens || ***lente biconvexa*** *f.* biconvex lens || ***lente cóncava*** *f.* concave lens || ***lente convergente*** *f.* converging lens || ***lente convexa*** *f.* convex lens || ***lente de amplificación*** *m.* magnifying glass || ***lente del condensador*** *m.* condenser lens || ***lente divergente*** *f.* diverging lens || ***lente menisco convergente*** *f.* positive meniscus || ***lente menisco divergente*** *f.* negative meniscus || ***lente planocóncavo*** *f.* planoconcave lens

lenteja *f.* lenticular rock, ore, or mineral mass

lentejón *m.* lense of rock

lenticular *a.* lenticular

leñoso *a.* woody

lepidoblástico *a.* lepidoblastic

lepidolita *f.* lepidolite

lepidolito *m.* lepidolite

letal *a.* lethal
letra de cambio *f.* bill of exchange
letra de crédito permanente *f.* revolving letter of credit
leucita *f.* leucite
leucito *m.* leucite
leucocrático *a.* leucocratic
leucoxeno *m.* leucoxene
levantada *f.* orogeny
levantado *a.* uplifted, up-thrown
levantamiento *m.* uplift; elevation; survey || ***levantamiento aerofotogramétrico*** *m.* aerial survey || ***levantamiento a plancheta*** *m.* plane table survey || ***levantamiento de reconocimiento*** *m.* reconnaissance survey || ***levantamiento geológico*** *m.* geologic program || ***levantamiento gravimétrico*** *m.* gravity survey || ***levantamiento magnético*** *magnetic* survey || ***levantamiento sísmico*** *m.* seismic program || ***levantamiento topográfico*** *m.* topographic mapping, land survey
levantar *vb.* to raise, uplift, elevate, erect; to survey
levée *f.* levee
levógiro *a.* sinistral
ley *f.* law; ore grade || ***de alta ley*** *f.* high grade || ***de baja ley*** *f.* low grade || ***ley de potencia (o poder) de reptación*** *f.* power law creep || ***ley en aceite*** *f.* oil content in percent || ***ley en hidrocarburo*** *f.* hydrocarbon content in percent
leyenda *f.* map legend
leyes de los gases *f.* gas laws
lezna *f.* awl
LGN (Líquidos del Gas Natural) *m.* natural gas liquids (NGL)
lherzolita *f.* lherzolite
Liás, Liásico *m. & a.* Lias, Liassic
libra *f.* pound || ***libra pié*** *m.* foot-pound
librar *vb.* to free, release || ***librar cañería aprisionada*** *vb.* to free up casing
libre *a.* free || ***libre cambio*** *m.* free trade || ***libre de cargos*** *a.* free of charge || ***libre de comisión*** *a.* free of brokerage
libreta *f.* notebook, field book
libro *m.* book
licencia *f.* license, leave, leave of absence
licenciado *m.* bachelors degree
licenciar *vb.* to grant a permit, to license
licenciatura *f.* degree; graduation
licitación *f.* bid, bid round
licitar *vb.* to bid
licuable *a.* liquefiable
licuación, licuefacción *f.* liquification
licuado *a.* liquefied
licuifacer *vb.* to liquify
lidita *f.* lydite, chert
lienza *f.* measuring tape
lievrita *f.* ilvaite
liga *f.* band, fringe (of fields, reefs); alloy
ligación *f.* bond
ligador *m.* binder
ligadura *f.* bond, tie
ligar *vb.* to alloy; to bind, tie together
ligazón *m.* bond; tie
ligero *a.* light; fast
ligeros *m.* volatiles
lignítico *a.* lignitic
lignito *m.* lignite
lima *f.* file, rasp
limadura *f.* scouring
limaduras *f.* filings, scrapings
limar *vb.* to scour
limbo *m.* limb
limburgita *f.* limburgite
limitación *f.* limitation
limitado *a.* limited
limitar *vb.* to limit
límite *m.* boundary, limit || ***límite de cedencia*** *m.* yield stress || ***límite de confianza*** *m.* confidence limit || ***límite de elasticidad*** *m.* elastic limit || ***límite de fatiga*** *m.* fatigue limit || ***límite del bosque*** *m.* timberline, tree line || ***límite de la zona montañosa*** *m.* borderland || ***límite de los árboles*** *m.* tree line, timberline || ***límite de placa*** *m.* plate boundary || ***límite de playa*** *m.* strand line || ***límite de presión de condiciones constantes*** *m.* constant pressure boundary || ***límite de resistencia*** *m.* endurance limit || ***límite de seguridad*** *m.* safety limit || ***límite elástico*** *m.* elastic limit

limites *m.* range || ***limites que no muestran flujo*** *m.* no-flow boundaries

limítrofe *a.* bordering

límnico *a.* lacustrine

limnología *f.* limnology

limo *m.* silt, mud; ooze

limolita *f.* siltstone

limolito *m.* siltstone

limonita *f.* limonite

limonítico, limonitizado *a.* limonitic

limoso *a.* muddy, silty

limpia *f.* low-grade ore, waste

limpiador *m.* ore sorter; cleaner

limpiadora *f.* sluice

limpiar *vb.* to clean; scrub || ***limpiar con diablo*** *vb.* pigging a pipeline || ***limpiar con taco*** *vb.* pigging a pipeline || ***limpiar un pozo*** *vb.* swabbing a well

limpiatubos *m.* swab, pig

limpieza *f.* cleaning || ***limpieza de tubos*** *f.* tube cleaning

limpio *m.* cobbed or sorted ore; *a.* clean

linde *m.* boundary, limit

lindero *m.* (survey, levan.) claim boundary, limit, boundary, abutment

línea *f.* line, track || ***línea alfiler*** *f.* pin line || ***línea axil*** *f.* hinge line || ***línea costera*** *f.* coastline || ***línea de alimentación*** *f.* feeder line || ***línea de articulación*** *f.* hinge line || ***línea de base*** *f.* baseline || ***línea de Becke*** *f.* Becke line || ***línea de bosques*** *f.* timberline || ***línea de captación*** *f.* gathering line || ***línea de colimación*** *f.* collimation line || ***línea de conducción*** *f.* pipeline || ***línea de costa*** *f.* coastline || ***línea de cresta*** *f.* crestline || ***línea de charnela*** *f.* hingeline || ***línea de energía eléctrica*** *f.* power line || ***línea de falla*** *f.* fault line, fault trace, scarp || ***línea de flujo*** *f.* flowline || ***línea de mira*** *f.* line-of-sight || ***línea de nivel*** *f.* contour line || ***línea de oleaje*** *f.* swash line || ***línea de pendiente*** *f.* fall line || ***línea de plomada*** *f.* plumb line || ***línea de referencia*** *f.* loose line; reference line; baseline || ***línea de registro*** *f.* (seis.) trace || ***línea de ribera*** *f.* shoreline || ***línea de vuelo*** *f.* flight line || ***línea en tiempo*** *f.* (seis.) time section || ***línea equipotencial*** *f.* equipotential line || ***línea fijada (o fija)*** *f.* pin line || ***línea internaciónal de cambio de fecha*** *f.* international date line || ***línea litoral*** *f.* shoreline || ***línea lobal*** *f.* suture || ***línea recta*** *f.* straight line || ***línea sísmica*** *f.* seismic line || ***línea suelta*** *f.* loose line || ***línea vertical*** *f.* plumb line

lineación *f.* lineation

lineal *a.* linear

lineamiento *m.* lineament

lingote *m.* ingot

liparita *f.* rhyolite

lipis *m.* copper sulfate

liptinita *f.* liptinite

liquidación *f.* liquidation

liquidar *vb.* to liquidate, liquify; to settle; to lay off

liquidez *f.* liquidity

líquido *m.* liquid

liso *a.* even, flush, smooth, flat; rock face; slickenside

lista *f.* stripe, band; list || ***lista de embalaje*** *f.* packing list || ***lista de precios*** *f.* price list

listado *a.* striped, laminated

listones *m.* riffles

lístrico *a.* listric

litarge, litargirio *m.* litharge

literatura *f.* bibliography, literature

lítico *a.* lithic

litificación *f.* lithification

litificado *a.* lithified

litio *m.* lithium

litoclasa *f.* lithoclase

litoclasto *m.* lithoclast

litoestratigráfico *a.* lithostratigraphic

litofacies *f.* lithofacies

litófilo *a.* lithophile

litofisuras *f.* lithophysae

litogénesis *f.* lithogenesis

litográfico *a.* lithographic

litología *f.* lithology

litológico *a.* lithologic

litomargo *m.* kaolin

litoral *m. & a.* shore, coast; littoral, coastal

litósfera *f.* lithosphere

litostático *a.* lithostatic

litro *m.* liter = 0.264 U.S. gallons

liviano *a.* light

lixivación *f.* leaching

lixiviar *vb.* to leach

lobulado, lobuloso *a.* lobate, lobed

lóbulo *m.* lobe

local *m. & a.* site; local

localidad *f.* location || ***localidad tipo*** *f.* type section

localización *f.* location, site || ***localización acústica*** *f.* acoustic ranging

localizador de cañería *m.* pipe finder

localizar *vb.* to locate, to fix in place

lodazal *m.* marsh, bog, [U.K.] fen

lodo *m.* mud, ooze, sludge, slush || ***lodo calcáreo*** *m.* calcareous ooze || ***lodo de circulación*** *m.* drilling mud || ***lodo de silíceo*** *m.* siliceous ooze || ***lodo de diatomeas*** *m.* diatomaceous ooze || ***lodo de perforación*** *m.* drilling mud || ***lodo de radiolarios*** *m.* radiolarian ooze

lodolita *f.* mudstone

lodoso *a.* muddy

loes, loess *m.* loess

logarítmico *a.* logarithmic

logarítmo *m.* logarithm

logística *f.* logistics

loma *f.* hill, ridge, rise

lomo *m.* ridge, hogback; shoulder || ***lomo de ballena*** *m.* whaleback || ***lomo de perro*** *m.* hogback

loncha *f.* flagstone

longitud *f.* length, longitude; easting or westing || ***longitud de onda*** *f.* wavelength || ***longitud focal*** *f.* focal length

longitudinal *a.* longitudinal

lopolito *m.* lopolith

losa *f.* flat stone, slab, flagstone

losero *m.* quarry, flagstone quarry

Lotaringiano, Lotaringiense *m. & a.* Lotharingian

lote *m.* portion, share, lot, batch; land measure about 100 hectares in México, about 400 hectares in Argentina

lubricación *f.* lubrication

lubricador *m.* lubricator

lubricante *m. & a.* lubricant; lubricating

lubricar *vb.* to lubricate, to oil, to grease

lucha por las divisorias *f.* stream piracy

Ludiano, Ludiense *m. & a.* Ludian

Ludloviano, Ludloviense *m. & a.* Ludlovian

ludwigita *f.* ludwigite

lugar *m.* place, location, spot, site || ***en su lugar*** *a.* in place

lumbre *m.* flint

lumbrera *f.* shaft, ventilation shaft

luminescente *a.* luminescent

luminiscencia *f.* luminescence

luminoso *a.* luminous

lumnita *f.* lumnite

luna *f.* moon || ***luna creciente*** *f.* crescent moon || ***luna llena*** *f.* full moon || ***luna nueva*** *f.* new moon

lupa *f.* hand lens

lurte *m.* landslide; glacier berg

Lusitaniano, Lusitaniense *m. & a.* Lusitanian

lustre, lustro *m.* luster, polish, glaze, gloss || ***lustro diamantino*** *m.* adamantine luster || ***lustro grasoso*** *m.* greasy luster || ***lustro metálico*** *m.* metallic luster || ***lustro resinoso*** *m.* resinous luster || ***lustro vítrico*** *m.* vitreous luster

lustroso *a.* shiny, glossy, lustrous

Luteciano, Luteciense *m. & a.* Lutetian

lutecio *m.* lutetium

lutita *f.* shale, lutite || ***lutita bituminosa*** *f.* oil shale, bituminous shale || ***lutita carbonífera*** *f.* carbonaceous shale || ***lutita intrusa*** *f.* shale break

lutítico *a.* shaly

lutolita *f.* mudstone

luz *f.* light; clearance || ***luz ambiente*** *f.* available light || ***luz incidente*** *f.* incident light || ***luz natural*** *f.* natural light || ***luz polarizada*** *f.* polarized light

LL

llábana *f.* smooth or polished rock slab

llama *f.* swamp, marsh; flame; pack animal

llamada a licitación *f.* call for bids

llamada de fondo *f.* upwelling

llamada de incendio *f.* fire alarm

llamargo, llamazar *m.* swamp, marsh

llambria *f.* cliff

llampera, llampo *f.* & *m.* [Bol.] talus; detrital ore; [Ch.] easily worked ore

llana, llanada *f.* plain

Llandeiliano, Llandeilense, Llandeilo, Llandilo *m.* & *a.* Llandeilian

Llandoveriano, Llandoveriense *m.* & *a.* Llandoverian

llaneza *f.* plain

llanita *f.* llanite, rhyolite

llano *m.* & *a.* plain; even, flat, smooth || ***llano alcalino*** *m.* alkali flat || ***llano de marea*** *m.* tidal flat || ***llano fangoso*** *m.* mud flat || ***llano fangoso de marea*** *m.* tidal mud flat

llanura *f.* plain, tropical plain, flatland || ***llanura abisal*** *f.* abyssal plain || ***llanura aluvial*** *f.* flood plain, alluvial plain || ***llanura costanera*** *f.* coastal plain || ***llanura costera, costanera*** *f.* coastal plain || ***llanura de agradación*** *f.* aggradation plain || ***llanura de erosión*** *f.* destructional plain || ***llanura de lodo*** *f.* mudflat || ***llanura deltaica*** *f.* delta plain || ***llanura de pedimentos*** *f.* pediplain || ***llanura troncal*** *f.* peneplain

Llanvirniense, Llanvirniano *m.* & *a.* Llanvirnian

llapa *f.* mercury used to amalgamate silver

llapar *vb.* to amalgamate silver using mercury

llave *f.* arc-shaped mine roof support; key, keystone, wedge; wrench || ***llave de cierre*** *f.* stopcock || ***llave de paso*** *f.* stopcock || ***llave de purga*** *f.* petcock || ***llave de tuerca*** *f.* wrench

llegada *f.* seismic arrival; water breakthrough

lleira *f.* pebble, gravel deposit

llena *f.* sedimentary fill; alluvium

llenado *m.* filling

llenante *f.* flood tide

llenar *vb.* to fill; to flood, overflow

llenar(se) de arena *vb.* to sand up

lleno *a.* flooded; high tides; filled, full || ***lleno de arena*** *a.* sanded up

llera *f.* quarry; rocky ground

llerón *m.* boulder-covered ground

llevar *vb.* to carry, transport, (geol.) raft || ***llevar al plano*** *vb.* to plot

llimpi *m.* [Pe.] cinnabar; red ore

llovizna *f.* drizzle

lloviznar *vb.* drizzle

lluvia *f.* rain

lluvial *a.* pluvial

lluvioso *a.* rainy

M

maar *m.* maar

macedonita *f.* macedonite

maceral *m.* maceral

maciño *m.* feldspar-rich sandstone

macizo *m.* massif, shield, mass; *a.* massive, solid || ***macizo intrusivo*** *m.* boss

macla *f.* twinning, twin, twinned crystal || ***maclas de contacto*** *f.* contact twins

maclación polisintética *f.* polysynthetic twinning

maclado *a.* twinned

macroclástico *a.* macroclastic

macrocristalino *a.* macrocrystalline

macrofósil *m.* macrofossil

macroscópico *a.* macroscopic

macrosísmico *a.* macroseismic

macrosismo *m.* macroseism

machacado *m.* [Pe.] native silver ore

machacamiento *m.* crushing

machacar *vb.* to crush
machador *f. & a.* crushing machine; grinding
machaqueo *m.* crushing
machar *vb.* to spall; to hammer
machays *m.* [Pe., Ec.] sulfur-lined cavities in volcanic rock
macho *m.* dike, pier; buttress; sledge hammer; male
machón *m.* buttress
machote *m.* underground survey stake, benchmark
machucador *m.* crusher
machucar *vb.* to crush, break, buck, or stamp ore
madera *f.* wood || ***madera fósil*** *f.* petrified wood
maderamen *m.* timber
madre *f.* river bed; mineralizing fluid
madriguera *f.* burrow
madurar *vb.* to mature
madurez *f.* maturity
maduro *a.* mature
Maestrichtiano, Maestrichtiense *m. & a.* Maestrichtian
maestro *m.* teacher, professor
máfico *a.* mafic
Magdaleniano, Magdaleniense *m. & a.* Magdalenian
magistral *m.* smelting flux
magma *f.* magma
magmático *a.* magmatic
magmatismo *m.* magmatism
magnesia *f.* magnesia, bitter earth
magnesiano *a.* magnesian
magnesio *m.* magnesium
magnesita *f.* magnesite
magnetar, magnetizar *vb.* to magnetize
magnético *a.* magnetic
magnetismo *m.* magnetism || ***magnetismo inducido*** *m.* induced magnetism || ***magnetismo fósil*** *m.* paleomagnetism || ***magnetismo remanente*** *m.* remnant magnetism
magnetita *f.* magnetite
magnetización *f.* magnetization
magnetizado *a.* magnetized
magnetizar *vb.* magnetize
magneto *m. & a.* magnet; magneto
magnetometría *f.* magnetometry
magnetómetro *m.* magnetometer
magnitud *f.* magnitude
magro *a.* lean
mainel *m.* mullion
maja *f.* pestle
majadero *m.* pestle
majamiento *m.* crushing, grinding, pounding
majar *vb.* to crush, pound, grind, break
malacate *m.* winch, hoist
malaquita *f.* malachite
mal de choco *m.* silicosis
mal funcionamiento *a.* malfunctioning
maleabilidad *f.* maleability
maleable *a.* maleable
malecón *m.* mound, dike, seawall
maleza *f.* thicket, brush, underbrush
málico *a.* malic
malla *f.* network, grid, mesh, screen || ***malla sísmica*** *f.* seismic grid
mallar *vb.* to grid
Malm *m. & a.* Malm
malpaís *f.* badlands
malquita *f.* malchite, lamprophyre
malta *f.* asphalt
mamífero *m.* mammal
mana *f.* source, spring
manadero de petróleo *m.* seep
manantial *m. & a.* spring, source || ***manantial artesiano*** *m.* artesian spring || ***manantial caliente*** *m.* hot spring || ***manantial surtidor*** *m.* artesian spring
manar *vb.* to ooze, to flow
mancomunidad *f.* pooling of properties to obtain an exploration permit
mancha *f.* (perf.) oil stain; (min.) spot of ore; (petrol.) flaw in a crystal, stain; smear
manchado *a.* spotted, stained, smeared
manchar *vb.* to smear, stain
manchita *f.* speck

mandíbula *f.* jaw

mando de tierra *m.* ground control

mandón *m.* mine foreman, boss

mandos del cable de perforación *m.* wireline logging unit

mandril de expansion *m.* tube expander

manejese con cuidado *vb.* handle with care

manejo *m.* handling

manera de presentarse *f.* mode of occurrence

manga *f.* waterspout

manganato *m.* manganate

manganesa, manganesia *f.* pyrolusite, manganese peroxide

manganeso *m.* manganese

mangánico *a.* manganic

manganita *f.* manganite

manganoso *a.* manganous

manglar, mangle *m.* mangrove

manguera para agua *f.* water hose

manifestación *f.* oil or gas seep, show, occurrence

manifestar *vb.* to manifest, demonstrate, reveal

manifiesto de aduana *m.* bill of entry

manigua *f.* bush, jungle

maniobra de pesca *f.* (perf.) fishing job

manipulación *f.* manipulation

manómetro *m.* stream gauge, pressure gauge, oil gauge

manta *f.* bedded vein or deposit; blanket deposit

manteado *a.* blanket-like

mantear *vb.* to hoist

mantener *vb.* to hold up, support, maintain, service

mantenimiento *m.* maintenance, upkeep || ***mantenimiento de presión*** *m.* pressure maintenance

manteo *m.* bedding

mantéo *m.* inclined hoist; hoisting

mantisa *f.* mantissa

manto *m.* blanket, mantle, horizontal vein, bed, layer, stratum; nappe, overthrust sheet; stratabound deposit; pocket || ***manto carbonoso*** *m.* coal seam || ***manto de arena*** *m.* sand sheet || ***manto de hielo*** *m.* ice sheet || ***manto de sobrecorrimiento*** *m.* overthrust || ***manto detrítico*** *m.* regolith || ***manto externo*** *m.* upper mantle of the earth || ***manto glacial*** *m.* ice sheet || ***manto interno*** *m.* lower mantle of the earth || ***manto sobreescurrido*** *m.* nappe, overthrust || ***manto terrestre*** *m.* earth's mantle

manual *m.* manual, guidebook, handbook; *a.* manual, by hand

manufactura *f.* manufacture

manufacturar *vb.* to make, manufacture

manufacturero *m.* & *a.* manufacturer; manufacturing

manutención *f.* maintenance

mapa *m.* map || ***mapa acotado*** *m.* contour map || ***mapa administrativo*** *m.* plat || ***mapa aéreo*** *m.* aerial map || ***mapa básico*** *m.* base map || ***mapa base*** *m.* base map || ***mapa catastral*** *m.* cadastral map || ***mapa contornado*** *m.* contour map || ***mapa de curvas de nivel*** *m.* contour map || ***mapa de reconocimiento*** *m.* reconnaissance map || ***mapa de relación arena/pelita*** *m.* sand-shale ratio map || ***mapa de subsuelo*** *m.* subsurface map || ***mapa de superficie*** *m.* surface map || ***mapa estructural*** *m.* structure map || ***mapa geológico*** *m.* geologic map || ***mapa gravimétrico*** *m.* gravity map || ***mapa índice*** *m.* index map || ***mapa isopáquico*** *m.* isopach map || ***mapa magnetométrico*** *m.* magnetic map || ***mapa temático*** *m.* thematic map || ***mapa topografico*** *m.* topographic map

mapamiento *a.* mapping

mapoteca *f.* map archive

mapear *vb.* to map

mapeo *m.* mapping || ***mapeo de superficie*** *m.* surface mapping || ***mapeo de subsuelo*** *m.* subsurface mapping

maquilar *vb.* to work ore

maquilero *m.* [Pe.] ore buyer

máquina *f.* machine, engine

maquinaria *f.* machinery

maquinista *m.* engineer, mechanic

mar *m. & f.* sea, ocean || ***mar abierto*** *a.* open marine || ***mar adentro*** *a.* offshore || ***mar Caribe*** *m.* Caribbean Sea || ***mar Caspio*** *m.* Caspian Sea || ***mar de Bering*** *m.* Bering Sea || ***mar de fondo*** *m.* surf || ***mar de Groenlandia*** *m.* Greenland Sea || ***mar de la China*** *m.* China Sea || ***mar de leva*** *m.* surf || ***mar del Norte*** *m.* North Sea || ***mar epicontinental*** *m.* epeiric sea || ***mar gruesa*** *m.* high seas || ***mar Mediterráneo*** *m.* Mediterranean Sea || ***mar Negro*** *m.* Black Sea || ***mar playo*** *m.* shallow sea || ***mar profundo*** *a.* highstand || ***mar regresivo*** *m.* regressive sea || ***mar remanente*** *m.* relic sea || ***mar Rojo*** *m.* Red Sea || ***mar transgresivo*** *m.* transgressive sea

maray *m.* crude ore mill

marca *f.* brand name; framework; district || ***marca de arrastre*** *f.* scour mark || ***marca de base*** *f.* sole mark || ***marca de desgaste*** *f.* scour mark || ***marca de fábrica*** *f.* trademark, brand || ***marca de oleaje*** *f.* ripple mark, swash mark || ***marca de percusión*** *f.* percussion mark || ***marca de rechinamiento*** *f.* chattermark || ***marca fiduciaria*** *f.* fiducial mark || ***marca lunada*** *f.* gouge mark || ***marca terrestre*** *f.* landmark

marcación *f.* bearing || ***marcación relativa*** *f.* relative bearings

marcador *m.* gauge, gage

marcadores útiles *f.* tool marks

marcasita *f.* marcasite

marco *m.* framework, frame, setting || ***marco de referencia*** *m.* datum || ***marco tectónico*** *m.* tectonic setting

marcha *f.* course

marea *f.* tide, (naut.) bore || ***canal de marea*** *f.* tidal inlet || ***marea alta*** *f.* high tide, highstand || ***marea baja*** *f.* low tide, neap tide || ***marea menguante*** *f.* ebb tide || ***marea muerta*** *f.* neap tide || ***marea roja*** *f.* red tide || ***marea viva*** *f.* spring tide

marecanita *f.* obsidian

marejada *f.* swell, surf

maremotar, maremotriz *a.* tidal

maremoto *m.* seaquake

marga *f.* marl. loam, clay-marl; marlstone, chalk

margalita, marlita *f.* marlstone

margen *m.* edge, border, margin, fringe, rim; (geog.) riverbank; (com.) margin || ***margen congelada*** *m.* chilled margen || ***margen continental*** *m.* continental margin || ***margen de seguridad*** *m.* margin of safety

márgenes *m.* financial margins

marginal *a.* marginal

margoso *a.* marly, loamy

marguero *m.* marl pit

marina *f.* shore, coast

marino *a.* marine || ***marino profundo*** *a.* deep sea

marisma *f.* swamp, salt lake

marismeño *a.* marshy

marítimo *a.* marine, maritime

marjal *m.* marsh, bog, [U.K.] fen

marmatita *f.* marmatite

marmita *f.* pothole || ***marmita glacial*** *f.* kettle

mármol *m.* marble

marna *f.* [Arg.] marl

marquesita *f.* marcasite

marro *m.* sledgehammer

marsupio *m.* marsupial

martillar *vb.* to hammer

martillo *m.* hammer

martinete *m.* pile driver

masa *f.* mass; irregular deposit, buildup || ***masa de hielo*** *f.* ice sheet || ***masa ígnea*** *f.* igneous body || ***masa terrestre*** *f.* land mass

máscara *f.* cover, mask || ***máscara antigás*** *f.* gas mask || ***máscara contra el polvo*** *f.* dust mask

masivo *a.* massive

mástil *m.* pole, post, mast

mata *f.* bush

matafuego *m.* fire extinguisher

matar un pozo *vb.* to kill a well

mate *a.* unpolished, matte finish

materia *f.* material, matter || ***materia prima*** *f.* feedstock, raw material

material *m.* material || ***material en bruto*** *m.* raw material

matloquita *f.* matlockite

matorral *m.* brush, scrub, thicket, [U.S.] chaparral

matraz *m.* flask

matriz *f.* matrix, gangue

máximo *m. & a.* high, upper part, maximum

maxvelio *m.* Maxwell

mayoria *a.* bulk, majority

mayorista *m. & f.* wholesale dealer

maza *f.* mallet

mazo *m.* sledge, maul, mallet, hammer

mazut *m.* fuel oil

meandro *m.* meander || ***barra de meandro*** *f.* meander bar || ***faja de meandro*** *f.* meander belt || ***lazo del meandro*** *m.* meander loop || ***meandro abandonado*** *m.* meander scar || ***meandro encajonado*** *m.* entrenched meander || ***meandro excavado*** *m.* incised meander || ***resección del meandro*** *f.* meander cutoff || ***rizo del meandro*** *m.* meander loop

mecánica *f.* mechanics

mecánico *m. & a.* mechanic; mechanical

mecanismo *m.* mechanism

mecha *f.* fuse, wick; torch

mechero *m.* burner; flare

medanal *m.* dune field

médano *m.* dune, sandbar

media *f.* half, mean || ***en media luna*** *a.* cuspate || ***media barreta*** *f.* [Pe.] inclined shaft || ***media vida*** *f.* half-life

mediador *m.* mediator

mediana *f.* median

mediano *a.* medium

medición *f.* measurement || ***medición a distancia*** *f.* telemetering || ***medición de producto a granel*** *f.* bulk measurement

medida *f.* measure, measurement, rule, ruler

medidor *m.* meter, measuring instrument, gauge || ***medidor de caudal*** *m.* flowmeter || ***medidor de deformación*** *m.* strain gauge

medio *a.* mid, middle, mean; half; medium || ***medio ambiente*** *m.* environment || ***medio dia*** *m.* midday || ***medio noche*** *f.* midnight

medioambiente *m.* environment

medir *vb.* to measure

Mediterráneo *a.* Mediterranean; midland

meerschaum *m.* meerschaum

megaclasto *m.* megaclast

megafósil *m.* macrofossil

mega-óndulas *f.* megaripples

megascópico *a.* megascopic

megaterio *m.* megatherium

mejoramiento *m. & a.* enhancement; enhanced

mejorar *vb.* to improve, upgrade, enhance

melaconisa *f.* tenorite

melaconita *f.* melaconite

meláfero, meláfido, meláfiro *m.* melaphyre, basalt

melange *m. & f.* melange

melanita *f.* melanite

melanocrático *a.* melanocratic

melanterita *f.* melanterite, copperas

melilita *f.* melilite

melita *f.* mellite

mellado *a.* jagged, indented

melladura *f.* indentation

membrana *f.* membrane

memoria *f.* report, record; (comp.) memory; (com.) payroll || ***memoria intermedia*** *f.* (comp.) buffer

mena *f.* vein; ore || ***mena ferruginosa*** *f.* lodestone || ***mena lucrativa*** *f.* pay ore, pay dirt

mención *f.* record

mencionar *vb.* to record, note, mention

mene *m.* seepage

menguar *vb.* to diminish, decrease

menisco *m.* meniscus

mensura *f.* survey

mensurado *a.* surveyed

mensurar *vb.* to survey

menú *m.* (comp.) menu || ***menú desplegable*** *m.* pull-down menu

mercadear *vb.* to market

mercadeo *m.* marketing

mercader *m.* merchant

mercaderías *f.* goods, commodities, merchandise

mercado *m.* market || ***mercado de exportación*** *m.* export market

mercancías *f.* merchandise

mercaptano *m.* mercaptan

merced *f.* grant, concession; benefit, reward

mercúrico *a.* mercuric

mercurio *m.* mercury

mercurioso *a.* mercurous

meridiano *m.* meridian, longitude line

meridional *a.* southern

merma *f.* waste; shrinkage || ***merma retrocedente*** *f.* backwasting

meroxeno *m.* meroxene

mesa *f.* mesa, table mountain; principal facet || ***mesa rotatoria (MR)*** *f.* rotary table (RT)

meseta *f.* plateau, mesa, tableland

mesoclasto *m.* mesoclast

mesocrático *a.* mesocratic

mesosfera *f.* mesosphere

mesotermal, mesotérmico *a.* mesothermal

mesotipo *m.* mesotype

Mesozóico *m. & a.* Mesozoic

mesozona *f.* mesozone

metacristal *m.* metacryst

metacuarcita *f.* metaquartzite

metal *m.* metal; alloy; ore || ***metal alcalino*** *m.* alkali metal || ***metal azul*** *m.* galena || ***metal base*** *m.* base metal || ***metal blanco*** *m.* cerargyrite || ***metal cenizo*** *m.* silver chloride || ***metal común*** *m.* ordinary ore || ***metal corrido*** *m.* placer ore || ***metal de ayuda*** *m.* flux || ***metal de fuego*** *m.* smelting ore || ***metal de fundición*** *m.* smelting ore || ***metal en barra*** *m.* bullion || ***metal en bruto*** *m.* ore || ***metal gráfico*** *m.* sylvanite || ***metal nativo*** *m.* native metal || ***metal negro*** *m.* sphalerite || ***metal negro argentífero*** *m.* tetrahedrite || ***metal no ferroso*** *m.* nonferrous metal || ***metal plomoso*** *m.* lead ore || ***metal polvorilla*** *m.* argentite || ***metal precioso*** *m.* precious metal || ***metal virgen*** *m.* native metal

metálico *a.* metallic

metalífero *a.* metalliferous

metalización *f.* metallization

metalogénesis *f.* metalligenesis

metalogenético *a.* metallogenic

metalurgia *f.* metallurgy

metalúrgico *m. & a.* metallurgist; metallurgical

metamíctico *a.* metamict

metamórfico *a.* metamorphic

metamorfísmo *m.* metamorphism || ***metamorfísmo cáustico*** *m.* regional metamorphism || ***metamorfísmo de contacto*** *m.* contact metamorphism || ***metamorfísmo de dislocación*** *m.* dynamic metamorphism || ***metamorfísmo de presión*** *m.* burial metamorphism || ***metamorfísmo dinámico*** *m.* dynamic metamorphism || ***metamorfísmo general*** *m.* regional metamorphism || ***metamorfísmo geotérmico*** *m.* thermal metamorphism || ***metamorfísmo hidrotérmico*** *m.* hydrothermal metamorphism || ***metamorfísmo local*** *m.* contact metamorphism || ***metamorfísmo regional*** *m.* regional metamorphism || ***metamorfísmo térmico*** *m.* thermal metamorphism

metamorfizado *a.* metamorphosed

metamorfosear *vb.* to metamorphose

metamorfosis *f.* metamorphosis, transformation

metano *m.* methane, natural gas || ***metano de carbón*** *m.* coalbed methane

metanol *m.* methanol

metasedimentario *a.* metasedimentary

metasedimento *m.* metasediment

metasomático *a.* metasomatic

metasomatismo *m.* metasomatism

metasomatosis *f.* metasomatism

metatipo *m.* metatype

metatorbernita *f.* metatorbernite

metavolcanica *f.* metavolcanic

meteorico *a.* meteoric

meteorito *m.* meteorite

meteorización *f.* weathering || ***meteorización diferencial*** *f.* differential weathering || ***meteorización mecánica*** *f.* mechanical weathering || ***meteorización química*** *f.* chemical weathering

meteorizado *a.* weathered

meteorizar *vb.* to weather

meteoro *m.* meteor

meteorología *f.* meteorology

meteorológico *a.* meteorological

meteorólogo *m.* meteorologist

metileno *m.* methylene

metilo *m. & a.* methyl

método *m.* procedure, method, process || ***métodos de ensayo*** *m.* trial and error

metodología *f.* methodology

métodos de ensayo *m.* trial and error

métrico *a.* metric

metro *m.* (meas.) meter; oil, gas, or water meter || ***metro cuadrado*** *m.* square meter

mezcla *f.* mixture, admixture, blend; melange || ***mezcla aguada*** *f.* slurry

mezclable *a.* miscible

mezclado *a.* mixed, blended, commingled

mezclador *m.* mixer, blender

mezcladora *f.* mixer, blender

mezclar *vb.* to mix, admix

mezclarse *vb.* to commingle

mezcuitlatl *m.* [Méx.] opal

miargirita *f.* miargyrite

miarolítico *a.* miarolitic

miascita *f.* miaskite, nepheline seyenite

mica *f.* mica; film, overlay || ***mica blanca*** *f.* muscovite || ***mica clara*** *f.* muscovite || ***mica de litio*** *f.* lithium mica, lepidolite || ***mica negra*** *f.* biotite || ***mica parda*** *f.* phlogopite

micáceo *a.* micaceous

micacita, micasita, mica-pizarra *f.* mica schist

micasquisto *m.* mica schist

micela *f.* micelle

micra *f.* micron

micrinita *f.* micrinite

micrita *f.* micrite

microbrecha *f.* microbreccia

microclástico *a.* microclastic

microclasto *m.* microclast

microclina *f.* microcline

microclino *m.* microcline

microcristalino *a.* microcrystalline

microelemento *m.* trace element

microestructura *f.* microstructure

micrófono *m.* microphone, geophone

microfósil *m.* microfossil

microfotografía *f.* microphotography

micrográfico *a.* micrographic

microgranítico *a.* microgranitic

microgranito *m.* microgranite

microlita *f.* microlite

micrómetro *m.* micrometer

micromilímetro *m.* micron

micrón *m.* micron

micro-onda *f.* microwave

micropaleontología *f.* micropaleontology

microperfil *f.* microlog

micropertita *f.* microperthite

micropliegue *f.* microfold

microregistro *m.* microlog

microscopía *f.* microscopy

microscópico *a.* microscopic

microscopio *m.* microscope || ***microscopio binocular*** *m.* binocular microscope || ***microscopio electrónico*** *m.* electron microscope || ***microscopio estereoscópico*** *m.* stereoscopic microscope || ***microscopio monocular*** *m.* monocular microscope || ***microscopio polarizante*** *m.* polarizing microscope

microsísmico *a.* microseismic

microsismo *m.* microseism

miembro *m.* member

migajón *m.* loam

migmatita *f.* migmatite

migmatización *f.* migmatization

migración *f.* migration || ***migración de los continentes*** *f.* continental drift || ***migración de profundidad antes de apilamiento*** *f.* depth migration before stacking || ***migración de tiempo antes de apilamiento*** *f.* time migration before stacking || ***migración en profundidad*** *f.* depth migration || ***migración en tiempo*** *f.* time migration

migrado *a.* migrated

migrar *vb.* to migrate
milerita *f.* millerita
milibar *m.* millibar
milidarcy *m.* millidarcy
miligal *m.* milligal
miligramo *m.* milligram = 0.0154 grains
mililitro *m.* milliliter
milímetro *m.* millimeter = 0.0394 inches
milímetros de mercurio *m.* millimeters of mercury
milonita *f.* mylonite
milonítico *a.* mylonitic
milonitización *f.* milonitization
mil piés cúbicos *m.* MCF, thousand square feet
milla *f.* mile || ***milla marina*** *f.* nautical mile (1,852 m) || ***milla terrestre*** *f.* statute mile (1,609 m)
millaje *m.* mileage
millón *m.* [Méx.] ore pile
millón de pies cubicos *m.* MMCF
mimatesa, mimetita, mimetesita *f.* mimetite
mimetismo *m.* mimetism
mimetita *f.* mimetite
mina *f.* mine || ***mina en fajas*** *f.* strip mine
minado *m.* underground working
minador *m.* & *a.* miner; mining
minar *vb.* to mine, undermine
minera *f.* mine
mineraje *m.* mining
mineral *m.* & *a.* mineral, ore; mineral, ore || ***mineral accesorio*** *m.* accessory mineral || ***mineral beneficiable*** *m.* pay ore || ***mineral clave*** *m.* index mineral || ***mineral de alta ley*** *m.* high grade ore || ***mineral de ganga*** *m.* gangue mineral || ***mineral de montaña*** *m.* disseminated ore || ***mineral filoniano*** *m.* gangue mineral || ***mineral guía*** *m.* index mineral
minerales pesados *m.* heavy minerals
minerales secundarios *m.* replacement minerals
mineralización *f.* mineralization
mineralizado *a.* mineralized
mineralizador *a.* mineralizing
mineralizar *vb.* to mineralize
mineralogía *f.* mineralogy
mineralógico *a.* mineralogical
mineralogista *m.* & *f.* mineralogist
minería *f.* mining, art/science of mining
minero *m.* & *a.* miner, mine owner; mineral, mining || ***minero de cuarto*** *m.* shift boss
mineta *f.* minette
mínimo *m.* cut-off, minimum
minimos cuadrados linea de correlación *f.* least-squares best fit
minimum *a.* minimum
minorista *m.* & *f.* retail dealer, retailer
minuta *f.* rough draft, first draft
minuto *m.* minute
minutas *f.* survey notes
Mioceno, Miocénico *m.* & *a.* Miocene
mióforo *m.* myophore
miogeosinclinal *m.* miogeosyncline
mira *f.* rod, level rod; rod target; sight; stream gauge || ***mira de corredera*** *f.* level rod || ***mira de nivelación*** *f.* levelling rod || ***mira taquimétrica*** *f.* stadia rod
mirada atrás *f.* back sight
miradero *m.* fenster
mirero *m.* rodman
miriápodo *m.* myriapod
mirmekita, mirmequita *f.* myrmekite
miscible *a.* miscible
Misisipiano, Misisípico, Misisipiense *m.* & *a.* Mississippian
mispíquel *m.* arsenopyrite
Misuriano, Misuriense *m.* & *a.* Missourian
mita *m.* [And.] mandatory public service or forced mine labor [Quechua]
mixto *a.* mixed
mixtura *f.* mixture, admixture, blend
mnemotécnico *a.* mnemonic
moco *m.* iron scoria, slag
mochila *f.* tool bag, backpack
moción *f.* motion
moda *f.* mode
modelaje *m.* modeling || ***modelaje del yacimiento*** *m.* reservoir modeling || ***modelaje iterativo*** *m.* forward modeling
modelar *vb.* to model

modelo *m.* model, mold, pattern || ***modelo a escala*** *m.* scale model || ***modelo estereoscópico*** *m.* stereoscopic model || ***modelo estocástico*** *m.* stochastic model || ***modelos numéricos*** *m.* numerical models

módem *m.* modem

modificador *m.* modifier

modo *m.* mode || ***modo de origen*** *m.* manner of origin

módulo *m.* modulus || ***módulo de elasticidad*** *m.* elasticity modulus || ***módulo de volumen*** *m.* bulk modulus || ***módulo de Young*** *m.* Young's modulus

mofeta *f.* mofette

mogote *m.* humock, hillock, [U.S.] mesa; pile

moho *m.* mold; rust

mohoso *a.* rusty

mojado *a.* wet

mojón *m.* claim corner, monument, landmark

mojona *f.* land survey

mojonamiento *m.* monumentation

mojonar *vb.* to set monuments

mojonera *f.* claim corner, monument, landmark

mol *m.* (chem., quím.) mole

molar *a.* (chem., quím.) molar

molaridad *f.* molarity

molasa *f.* molasse

molde *f.* mold

moldear *vb.* to mold, cast

moldeo *m.* molding

moldico *a.* moldic

moldura de rompiente *f.* cove

mole *m.* mole; large rock mass

molécula *f.* molecule

moléculagramo *m.* mol

molécular *a.* molecular

moledor *m.* grinder, mill

moledora *f.* grinder, mill

moler *vb.* to grind, comminute

molibdato *m.* molybdate

molibdenita *f.* molybdenite

molibdeno *m.* molybdenum

molibdita *f.* molybdite

molienda *f.* grinding, crushing

molino *m.* grinder; (min.) ore mill

molusco *m.* mollusc, mollusk

momentáneo *a.* transient

momento *m.* moment; momentum || ***momento clave*** *m.* moment of hydrocarbon generation || ***momento crítico*** *m.* moment of generation || ***momento de inercia*** *m.* moment of inertia || ***momento de torción*** *f.* torsion moment || ***momento magnético*** *m.* magnetic moment

monacita *f.* monacite, monazite

monadnock *m.* monadnock

monchiquita *f.* monchiquite

moneda *f.* money || ***moneda correinte*** *f.* currency || ***moneda legal*** *f.* currency

monetizar *vb.* to monetize

monitor *m.* (comp.) monitor

monitoreo *m.* monitoring

monacita *f.* monazite

monocíclico *a.* monocyclic

monoclinal *m.* monocline

monoclínico *a.* monoclinic

monocromático *a.* monochromatic

monofásico *a.* single phase

monolítico *a.* monolithic

monolito *m.* monolith, monument

monotalámico *a.* monothalamous

monovalencia *f.* monovalence

monovalente *a.* monovalent

monóxido *m.* monoxide || ***monóxido de carbono*** *m.* carbon monoxide

montacargas *f.* elevator, windlass, hoist

montaje *m.* assembly, installation

montaña *f.* mountain; country rock || ***montaña isla*** *f.* inselberg

montañoso *a.* mountainous, hilly

montar *vb.* to assemble, erect, mount, set

monte *m.* (geog.) mountain, mount, [U.K.] fell; (botan.) woodland, jungle || ***monte isla*** *m.* monadnock || ***monte marino*** *m.* seamount || ***monte residual*** *m.* monadnock || ***monte testigo*** *m.* monadnock

montecillo *m.* mound, hummock

montera *f.* outcrop

Montiano, Montiense *m. & a.* Montian

monticelita *f.* monticellite

monticulado *a.* mounded, hummocky

montículo *m.* hillock; mound, hummock || ***montículo de barro*** *m.* mud lump

montmorilonita *f.* montmorillonite

monto *m.* total amount || ***monto por cobrar*** *m.* balance due

montón *m.* heap, pile, stack

monumento *m.* survey monument

monzón *m.* monsoon

monzonita *f.* monzonite

monzonítico *a.* monzonitic

morena, morrena *f.* moraine || ***morena basal*** *f.* basal moraine, ground moraine || ***morena de ablación*** *f.* ablation moraine || ***morena de fondo*** *f.* ground moraine || ***morena de retroceso*** *f.* recessional moraine || ***morena frontal*** *f.* terminal moraine || ***morena lateral*** *f.* lateral moraine || ***morena recesional*** *f.* recessional moraine || ***morena terminal*** *f.* terminal moraine

morénico *a.* morainal

morfoestructural *a.* morphostructural

morfogénesis *f.* morphogenesis

morfografía *f.* land forms, geomorphology

morfología *f.* morphology

morfosis *f.* morphosis

morfotectónica *f.* morphotectonics

morganita *f.* morganite

morrillo *m.* cobble, boulder

morro del subsuelo *m.* buried hill

mortero *m.* mortar

mosáico *m.* mosaic || ***mosáico aerofotográfico*** *m.* airphoto mosaic || ***mosáico no regulado*** *m.* uncontrolled mosaic || ***mosáico regulado*** *m.* controlled mosaic

Moscoviano, Moscoviense *m. & a.* Moscovian

moscovita *f.* muscovite

mosor *m.* monadnock

mosqueado *a.* spotted

moteado *a.* mottled, speckled

motita *f.* speck

motón viajero *m.* traveling block

motor *m.* motor, engine

motores del cabrestante *m.* (perf.) draw works

moveout *m.* moveout || ***moveout no hiperbólico*** *m.* non-hyperbolic moveout (NMO)

mover por deslizamiento *vb.* to skid, to slide

movible, móvil *a.* movable, portable

móvil *a.* movable

movilidad *f.* mobility

movilización *f.* mobilization

movimiento *m.* movement, displacement, motion || ***movimiento alternativo*** *m.* reciprocating motion || ***movimiento aperiódico*** *m.* non-periodic motion || ***movimiento armóncio*** *m.* harmonic motion || ***movimiento bascular*** *m.* oscillation || ***movimientos de la corteza*** *m.* crustal movements || ***movimientos de masas*** *m.* mass movements || ***movimiento diferencial*** *m.* differential movement || ***movimiento eustático*** *m.* eustatic movement || ***movimiento orogénico*** *m.* orogenic movement || ***movimiento paulatino del terreno*** *m.* creep || ***movimiento periodico*** *m.* periodic motion || ***movimiento retrógrado*** *m.* regression || ***movimiento traslacional*** *m.* translational movement

moya *f.* volcanic clay

mozo *m.* roustabout

muela *f.* grindstone, whetstone; hillock

muelle *m.* wharf, pier, breakwater

muera *f.* brine

muerto *a.* dead

muesca *f.* notch, furrow || ***muesca de protección contra la grabación*** *f.* (comp.) write-protect notch

muestra *f.* sample, specimen; core, cuttings || ***muestra de canal*** *f.* cuttings || ***muestra de pared*** *f.* core sample || ***muestra de suelo*** *f.* soil sample || ***muestra de terreno*** *f.* soil sample || ***muestra lateral*** *f.* side wall sample || ***muestra manual*** *f.* hand sample

muestrador *m.* sampler

muestrear *vb.* to sample, collect

muestreo, muestro *m.* sampling || ***muestreo por canaleta*** *m.* channel sampling || ***muestreo por picado*** *m.* chip sampling

mugrón *m.* nubbin

multa *f.* (com., legal) fine

multidisciplinario *a.* multidisciplinary

multiespectral *a.* multispectral

multiplaje *m.* multiplexing

múltiple *m. & a.* multiple

multiplexación *f.* multiplexing

multiplexor *m.* multiplexor

múltiplo *m.* (seis.) multiple

multitemporal *a.* multitemporal

mundial *a.* worldwide

mundo *m.* world, earth

muñon *m.* mullion

muralla *f.* wall, cliff

muriato *a.* muriatic

muro *m.* footwall; wall || ***muro colgante*** *m.* hanging wall || ***muro de base*** *m.* footwall || ***muro de contención*** *m.* retaining wall || ***muro de defensa*** *f.* groin || ***muro de detención*** *m.* retaining wall || ***muro de sostén*** *m.* retaining wall

muscovita *f.* muscovite

museo *m.* museum

musgo *m.* moss

muy minado *a.* honeycombed

N

nácar *m.* mother of pearl, nacre

nacarado *a.* pearly

nacarita *f.* kaolinite

nacer *vb.* to emerge, be born

naciente *m.* rising; east; headwaters

nacimiento *m.* spring, headwaters

nacional *a.* national, domestic

nadir *m.* nadir

nafta *f.* crude oil, naptha

naftalina *f.* napthalene

nafteina *f.* naphthine

naftina *f.* naphthene

nagiagita *f.* nagyagite

naife *m.* uncut diamond

namaqualita *f.* cyanotrichite

Namuriano, Namurense *m. & a.* Namurian

nanofósil *f.* nanofossil

nanosegundo *m.* nanosecond

nantoquita *f.* nantokite

napa *f.* plate; aquifer || ***napa freática*** *f.* water table

napoleanita *f.* corsite

nariz (del pliegue) *f.* nose (of a fold) || ***nariz estructural*** *f.* anticlinal nose

nativo *a.* native

natroborocalcita *f.* ulexite

natrofilita *f.* natrophilite

natrolita *f.* natrolite

natural *a.* natural

naturaleza *f.* nature

naútico *a.* nautical

navegación *f.* navigation

neblina *f.* fog, mist

Nebraskiano, Nebraskiense *a.* Nebraskan

necrófago *m.* scavenger

nectón *m.* nekton

nefelina, nefelinita, nefelita *f.* nepheline

nefelino *m.* nepheline

nefrita *f.* jade, nephrite

negativo *m. & a.* negative

negociación *f.* negotiation

negociar *vb.* to negotiate, do business

negocio *m.* business, enterprise

neis *m.* gneiss || ***neis de ojos*** *m.* augen gneiss || ***neis glandular*** *m.* augen gneiss

néisico *a.* gneissic

nematoblástico *a.* nematoblastic

Nemausense *m. & a.* Nemausian, Valanginien

neoautóctono *m.* neoautochthon

Neoceno, Neógeno *m. & a.* Neogene

Neócico, Neocomiano, Neocomiense *m. & a.* Neocomian

neocrístico, neocrísto *m. & a.* neocryst

neodimio *m.* neodymium

neoformación *f.* neogenesis

neógeno *m. & a.* Neogene

neoglaciación *f.* neoglaciation

neohexano *m.* neohexane

neolítico *a.* neolithic

neosomo *m.* neosome

neotectónico *m. & a.* neotectonics; neotectonic

neotipo *m.* neotype

neovolcánco *a.* neovolcanic

Neozóico *m. & a.* Quaternary

neptunio *m.* neptunium

neptunita *f.* neptunite

nerítico *a.* neritic

Nerviense, Nerviano *m. & a.* Nervian

nesofítico *a.* diabasic

nesosilicato *m.* nesosilicate

neto *m.* (com.) net

Neudeckense, Neudeckiano *m. & a.* Neudeckian

neumatógeno, neumatolítico, neumatológeno *a.* pneumatolytic

neutral *a.* neutral

neutralizar *vb.* to neutralize

neutrón *m.* neutron

Neuvizense *m. & a.* Oxfordian

nevada *f.* snowstorm

Nevadiense, Nevadiano *m. & a.* Nevadan

nevadita *f.* nevadite

nevado *m. & a.* firnfield; snow-covered

nevar *vb.* to snow

nevasca, nevisca *f.* snowstorm, blizzard

nevero *m.* ice field, snowfield, névé

neviza *f.* firn

Niagarano, Niagarense, Niagariense *m. & a.* Niagaran

nicol *m.* nicol || ***nicoles cruzados*** *m.* crossed nicols

nicolita *f.* niccolite

nicopirita *f.* pentlandite

nicho *m.* niche, alcove || ***nicho de rompiente*** *m.* cove

nido *m.* pocket of ore

niebla *f.* fog

nieve *f.* snow

nife *m.* earth's inner core

nifesima *f.* earth's inner core

níkel *m.* nickel

niobio *m.* niobium

niobofilita *f.* niobophyllite

níquel *m.* nickel

niquelífero *a.* nickel-bearing

niquelina *f.* nickeline, niccolite

niquelocre *m.* annabergite

nitral *m.* nitrate deposit

nitratina, nitratita *f.* soda niter; [U.S.] caliche

nitrato *m.* nitrate || ***nitrato de amonio*** *m.* ammonium nitrate || ***nitrato de chile*** *m.* saltpeter || ***nitrato de plata*** *m.* silver nitrate || ***nitrato de potasio*** *m.* potassium nitrate

nitrera *f.* nitrate deposit

nítrico *a.* nitric

nitro *m.* nitrate; niter, saltpeter

nitrocelulosa *f.* nitrocellulose

nitrógeno *m.* nitrogen

nitroglicerina *f.* nitroglycerin

nitroso *a.* nitrous

nitruro *m.* nitride

nivel *m.* level; grade; elevation || ***nivel alto*** *m.* highstand || ***nivel bajo*** *m.* lowstand || ***nivel de agrimensor*** *m.* surveyor's level || ***nivel de agua freática*** *m.* water table || ***nivel de base*** *m.* base level || ***nivel de burbuja*** *m.* spirit level || ***nivel de comparación*** *m.* datum || ***nivel de confianza*** *m.* confidence level || ***nivel de cota cero*** *m.* base level, datum plane || ***nivel de creciente*** *m.* flood level || ***nivel de deslizamiento*** *m.* detachment level || ***nivel de erosión local*** *m.* base level || ***nivel de la saturación irreducible*** *m.* level of irreducible water saturation || ***nivel de referencia*** *m.* base level; datum || ***nivel del agua*** *m.* water level || ***nivel del agua freática*** *m.* water table || ***nivel del mar*** *m.* sea level || ***nivel estructural de referencia*** *m.* structural datum || ***nivel fosilífero*** *m.* fossil horizon || ***nivel freático*** *m.* water table

|| ***nivel guia*** *m.* marker bed || ***nivel hidrostático*** *m.* groundwater level || ***nivel imaginario*** *m.* phantom horizon || ***nivel medio de pleamar*** *m.* mean high water mark || ***nivel piezométrico*** *m.* hydraulic gradient, piezometric surface || ***nivel regional*** *m.* regional elevation

nivelación *f.* leveling

nivelada de atrás *f.* backsight

nivelador *m. & a.* leveller, levelman; levelling

niveladora *f.* grader

nivelar *vb.* to grade, level

niveles *m.* water-bearing strata

niveleta *f.* levelling staff

no alterado *a.* unaltered

nobelio *m.* nobelium

nobleza *f.* richness of ore

nocerita *f.* fluoborite

no comercial *a.* noncommercial

no concordancia *f.* nonconformity

no consolidado *a.* unconsolidated

nodular *a.* nodular

nódulo *m.* nodule

noduloso *a.* nodular

no endurecido *a.* unindurated

no estratificado *a.* unstratified

no explorado *a.* unexplored

no explotado *a.* unexploited, unworked

no indurado *a.* not indurated

nomenclatura *f.* nomenclature

no metálico *a.* nonmetallic

no meteorizado *a.* unweathered

nómina *f.* payroll

nomograma, nomogramo *m.* nomogram, nomograph

nontronita *f.* nontronite

no perturbado *a.* undisturbed

no polar *a.* non-polar

no poroso *a.* nonporous

norbergita *f.* norbergite

nordfieldita *f.* esmeraldite

nordmarkita *f.* nordmarkite

no renovable *a.* nonrenewable

Norfolkenese, Norfolkiano *m. & a.* Norfolk

Noriano, Noriense *m. & a.* Norian

norilsquita *f.* norilskite

norita *f.* norite

norma *f.* standard, norm; pattern

normal *a.* normal || ***normal óptica*** *f.* optic normal

normalización *f.* normalization

normalizar *vb.* normalize

normativo *a.* normative

norte *m.* north || ***norte geográfico*** *m.* true north || ***norte magnético*** *m.* magnetic north || ***norte verdadero*** *m.* true north

northfieldita *f.* esmeraldite

noseana *f.* noselite

noseanolíto *m.* noseanolith

nosena *f.* nosean

notario *m.* notary public

notas *f.* notes

no tóxico *a.* nontoxic

no única *a.* non-unique

novaculita *f.* novaculite

novaculítico *a.* novaculitic

novato *m.* inexperienced driller; novice

nube *f.* cloud || ***nube ardiente*** *f.* ash flow || ***nube de cenizas*** *f.* ash cloud || ***nube de chubascos*** *f.* squall cloud || ***nube de tormenta*** *f.* thunderhead

nublado *a.* cloudy

nubosidad *f.* cloudiness

nucleado *a.* cored

nuclear *vb.* to bring together, concentrate; to core

nuclear *a.* nuclear

núcleo *m.* core, craton, nucleus || ***núcleo continental*** *m.* craton, continental shield || ***núcleo de pared*** *m.* sidewall core || ***núcleo exterior*** *m.* earth's outer core || ***núcleo interior*** *m.* earth's inner core || ***núcleos laterales*** *m.* sidewall cores || ***núcleo orientado*** *m.* oriented core || ***núcleo terrestre*** *m.* core of the earth

nucleogénesis *f.* nucleogenesis

nuclideo *m.* nuclide

nudo *m.* kink, knot, nautical mile per hour

nudoso *a.* knotty

nuevas empresas *f.* new business ventures

nuevita *f.* samarskite

Nuevoleoniense, Nuevoleoniano *m. & a.* Nuevoleonian

númen *m.* core

numérico *a.* numerical

número *m.* number ‖ ***número al azar*** *m.* random number ‖ ***número atómico*** *m.* atomic number ‖ ***número digital*** *m.* digital number (DN)

Numulítico *a.* Paleogene

nunatak *m.* nunatak

nutriente, nutrimento *m.* nutrient

Ñ

ñakha *f.* finely crushed concentrates

ñanga *f.* swampy ground, marsh; bit, portion

ñato *a.* (geog.) flattened

ñuto *a.* cushed, ground

O

oasis *m.* oasis

obducción *f.* obduction

objectivo, objetivo *m.* objective; objective lens

objeto *m.* object, scope

oblado, oblato *a.* oblate

oblicuo *a.* oblique

obligación *f.* liability, obligation, commitment

oblongo *a.* oblong

obra *f.* work, labor; piece of work

obrador *m.* workman

obrajero *m.* foreman

obrero *m.* worker, laborer

obrizo *a.* pure, refined

obrizo *m.* fine gold

obscurecer *vb.* to obscure, hide, mask

obsequente *a.* obsequent

observación *f.* observation

observado *a.* observed

observador *m.* observer; seismograph recorder

observar *vb.* to monitor, observe

observatorio *m.* observatory

obsidiana *f.* obsidian

obstáculo *m.* obstacle, obstruction

obstrucción *f.* obstruction

obstruido *a.* obstructed

obstruir *vb.* to obstruct

obturado *a.* plugged, clogged, obstructed

obturador *m.* plug, stopper; lens shutter ‖ ***obturador de empaque*** *m.* packer ‖ ***obturador de pared*** *m.* wall packer ‖ ***obturador del tubo revestidor*** *m.* casing packer ‖ ***obturador focal*** *m.* focal plane shutter ‖ ***obturador para pozo a dos zonas*** *m.* dual completion packer ‖ ***obturador para tubería*** *m.* tubing packer

obturar la formación *vb.* to plug a formation

obtuso *a.* obtuse, blunt

occidental *a.* western, occidental; (geol.) inferior quality gem

occidente *m.* west

oceánico *a.* marine

océano *m.* ocean ‖ ***océano Atlántico*** *m.* Atlantic Ocean ‖ ***océano Glacial Artico*** *m.* Arctic Ocean ‖ ***océano Glacial Antártico*** *m.* Antarctic Ocean ‖ ***océano Indico*** *m.* Indian Ocean ‖ ***océano Pacífico*** *m.* Pacific Ocean

oceanografía *f.* oceanography

oceanográfico *a.* oceanographic

ocelar *a.* ocellar

ocluído *a.* occluded

oclusión *f.* occlusion

ocre *m. & a.* ocher ‖ ***ocre de antimonio*** *m.* stibiconite ‖ ***ocre de molibdeno*** *m.* molibdite ‖ ***ocre rojo*** *m.* hematite

ocroíta *f.* cerite

ocrolita *f.* nadorite

octaedrita *f.* octahedrite, anatase

octaedro *m..* octahedron

octagonal *a.* octagonal

octágono *m.* octagon

octafilíta *f.* octaphyllite

octano *m.* octane

ocular *m.* eyepiece, ocular

oculto *a.* concealed, hidden

ocurrencia *f.* occurrence

Ochoense, Ochoiano *m. & a.* Ochoan

oersted, oerstedio *m.* oersted

oeste *m.* west

oferta *f.* (com.) tender, offer, sale || ***oferta y demanda*** *f.* supply and demand

oficalcita *f.* ophicalcite

oficina *f.* office

ofiolita *f.* ophiolite

ofiolítico *a.* ophiolitic

ofita *f.* ophite

ofítico *a.* ophitic

ohmio *m.* ohm

oikocrist *m.* oikocryst

ojo *m.* eye; small irregular ore pod; small lake, pond || ***ojo de agua*** *m.* spring || ***ojo de gato*** *m.* cat's eye agate || ***ojo de huracán*** *m.* eye of the storm

ojoso *a.* honeycombed structure

ola *f.* (geog.) wave, water wave || ***alcance de ola*** *m.* fetch of a wave || ***ola de bajio*** *f.* shoaling wave || ***ola de mar producido por un temblor submarino*** *m.* tsunami || ***ola de marea*** *f.* tidal wave || ***ola de superficie*** *f.* surface wave || ***ola interna*** *f.* internal wave || ***ola sísmica*** *f.* seismic wave

oleada *f.* surge, swell

oleaginoso *a.* oily

oleaje *m.* swell, surge, surf

olefina *m.* olefin

Olenkiense, Olenkiano *m. & a.* Olenkian

oleoducto *m.* pipeline || ***oleoductos mallados*** *m.* loop-gathering line || ***oleoducto tributario*** *m.* feeder line || ***oleoducto troncal*** *m.* trunk line

Olenense, Olenidense *m. & a.* Potsdam

oligisto *m.* specular hematite

Oligocénico, Oligoceno *m. & a.* Oligocene

oligoclasa *f.* oligoclase

oligoelemento *m.* trace element

oligomíctico *a.* oligomictic

oligotrofía *f.* oligotrophy

oligotrófico *a.* oligotrophic

olisoglifo *m.* olistoglyph

olistostroma *f.* olistostrome

olivina *f.* olivine

olivinita *f.* olivinite

olivino *m.* olivine

olor *m.* smell, odor

olla *f.* whirlpool, cauldron || ***olla de gigante*** *f.* pothole || ***olla de remolineo*** *f.* pothole

ombrogénuo *a.* ombrogenous

omfacita *f.* omphacite

omisión *f.* stratigraphic omission

onda *f.* (phys., física) wave, seismic or electromagnetic wave || ***longitud de onda*** *f.* wavelength || ***onda acústica*** *f.* acoustic wave || ***onda compresional*** *f.* compressional wave || ***onda de cizalla*** *f.* shear wave || ***onda de condensación*** *f.* compressional wave || ***onda de dilatación*** *f.* dilational wave || ***onda elástica*** *f.* elastic wave || ***onda primaria*** *f.* primary wave || ***onda sísmica*** *f.* seismic wave || ***onda sonora*** *f.* sound wave || ***onda superficial*** *f.* surface wave

ondeante *a.* undulating

ondícula sísmica *f.* seismic wavelet

óndula, ondulita *f.* ripple mark

ondulación *f.* undulation, crinkle || ***ondulación precontinental*** *f.* continental rise

ondulado *a.* undulate, undulated, wavy, rippled, crenulated

ondulante *a.* undulating, wavy

ondulitas *f.* ripples || ***ondulitas superpuestas*** *f.* climbing ripples

ónice, ónique, ónix *m. & a.* onyx

Onondagiense, Onandagiano *m. & a.* Onondagan

Ontariense, Ontariano *m. & a.* Ontarian

ontogénesis, ontogenía *f.* ontogenesis

onza *f.* ounce = 28.35 grams

ooide *m.* ooid

oolita *f.* oolite

oolítico *a.* oolitic

oolitización *f.* oolitization

oolito *m.* oolith

opacidad *f.* opacity

opacita *f.* opacite

opaco *a.* opaque

opalado *a.* opalized

opalescencia *f.* opalescence

opalescente *a.* opalescent, opaline

opalino *a.* opalescent

opalizado *a.* opalized

ópalo *m.* opal

opción *f.* option

operación *f.* operation || ***operación continua*** *f.* continuous operation || ***operación de ensayo*** *f.* test run || ***operación de prueba*** *f.* test run || ***operación por lotes*** *f.* batch runs, batching || ***operación por tandas*** *f.* batching

operacional *a.* operational

operaciones mineras *f.* mining operations

operado *a.* operated

operador *m.* operator

operar *vb.* to operate

operario *m.* workman, laborer, miner

Opoca *f.* Opoka

optálico *a.* caustic

óptica *f.* optics

óptico *a.* optic, optical

optimización *f.* optimization

óptimo *a.* best, optimal

opuesto *a.* opposite

oquedad *f.* pore, cavity, vug; alcove

orbículo *m.* orbicule

órbita *f.* orbit

orden *m.* order, sequence, instruction || ***orden de sucesión*** *m.* sequence || ***orden estratigráfico*** *m.* stratigraphic sequence

ordenada *f.* ordinate

ordenador *m.* computer, data processor || ***ordenador doméstico*** *m.* personal computer, PC

ordenanzas de minería *f.* mining laws

ordenar *vb.* to arrange

ordinario *a.* ordinary, common

ordinarios *m.* low-grade ores

ordinograma *m.* flowchart

ordoñezita *f.* ordonezite

Ordovicense, Ordoviciano, Ordovícico *m. & a.* Ordovician

orear *vb.* to weather

Orelliense, Orelliano *m. & a.* Orellian

orfeíta *f.* orpheite

orgánico *a.* organic

organigrama *m.* chart, organization chart, flowchart

organimorfo *a.* organoform

organismo *m.* organism

organización *f.* organization || ***organización del aprovisionamiento y del transporte*** *f.* logistics

organógeno *a.* organogenic

organosedimentario *a.* organosedimentary

orientación *f.* direction, orientation, strike, bearing || ***orientación indiferente*** *f.* random orientation || ***orientación preferida*** *f.* preferred orientation

orientado *a.* oriented

oriental *a.* eastern, oriental; genuine gem

orientar *vb.* to orient

oriente *m.* east

orificio *m.* opening, orifice || ***orificio de purga*** *m.* drain opening || ***orificio de salida*** *m.* outlet || ***orificio estrangulador*** *m.* choke

origen *m.* origin, source

orilla *f.* bank, shore, limit, edge, margin; selvage, rim

orillo *m.* selvage

orín *m.* rust

Oriskaniense, Oriskaniano *m. & a.* Oriskanian

orizita *f.* epistilbite

orla *f.* border, fringe || ***orla continental*** *f.* continental fringe || ***orla freática*** *f.* capillary fringe

oro *m.* gold || ***oro de aluvión*** *m.* placer gold || ***oro de los tontos*** *m.* fool's gold, pyrite || ***oro de veta*** *m.* vein gold || ***oro en barras*** *m.* gold bullion || ***oro en polvo*** *m.* gold dust || ***oro filoniano*** *m.* vein gold || ***oro libre*** *m.* free gold || ***oro nativo*** *m.* native gold || ***oro puro*** *m.* fine gold

oroclino *m.* orocline

orocrático *a.* orocratic

orogen *m.* orogenic belt, orogen

orogénesis *m.* orogeny, orogenesis

orogenético *a.* orogenetic, orogenic

orogénia *f.* orogenesis, orogeny

orogénico *a.* orogenic

orógeno *m.* orogen, orogenic belt

orografía *f.* orography

orográfico *a.* orographic

oroseíta *f.* iddingsite

orpimente *m.* orpiment

ortita *f.* orthite

ortoandesita *f.* orthoandesite, orthopyroxene andesite

ortobromita *f.* embolite

ortoclasa *f.* orthoclase

ortoclástico *a.* orthoclastic

ortoclasita *f.* orthoclasite, granite, seyenite

ortoclorita *f.* orthochlorite

ortoconglomerado *m.* orthoconglomerate

ortocuarcita *f.* orthoquartzite

ortocuarcítico *a.* orthoquartzitic

orto-eje *m.* orthoaxis

ortoestratigrafía *f.* orthostratigraphy

ortofelsita *f.* orthophyre

ortófiro, ortófido *m.* orthophyre

ortofírico *a.* orthophyric

ortogneis *m.* orthogneiss

ortogonal *a.* orthogonal

ortográfico *a.* orthographic

ortogranito *m.* kaligranite

ortomagmático *a.* orthomagmatic

ortométrico *a.* orthometric

ortomórfico *a.* conformal

ortomorfismo *m.* conformity

ortoneis *f.* orthogneiss

ortopiroxenita *f.* orthopyroxenite

ortopiroxeno *m.* orthopyroxene

ortorrómbico *a.* orthorhombic

ortosa *f.* orthoclase || ***ortosa sódica*** *f.* sanidine

ortosilicato *m.* orthosilicate

ortosita *f.* orthosite

ortotéctico *a.* magmatic

ortotectonismo *m.* alpinotype tectonics

oruga *f.* bulldozer

oscilación *f.* oscillation, vibration; yaw || ***oscilación libre*** *f.* free oscillation

oscilar *vb.* to oscillate

osciloscopio *m.* oscilloscope

oscurecer *vb.* to obscure

osificación *f.* ossification

osificar *vb.* to ossify

osificarse *vb.* to become ossified

osmio *m.* osmium

osmiridio *m.* osmiridium

ósmosis *f.* osmosis

osteolita *f.* osteolite

ostrácodo *m.* ostracod

otero *m.* hill, knoll, mesa

otoño *m.* autumn

otorgante *m.* grantor

ovillo *m.* skein

ovoide, ovoideo, oviforme *a.* ovoid

oxácido *m.* oxyacid

oxálico *a.* oxalic

oxamita *f.* oxammite

Oxfordiano, Oxfordiense *m. & a.* Oxfordian

oxhidrilo *a.* hydroxyl

oxiacetileno *m.* oxyacetylene

oxidable *a.* oxidizable

oxidación *f.* oxidation

oxidado *a.* oxidized

oxidante *a.* oxidizing

oxidar *vb.* to oxidize

oxidato *m.* oxidate

óxido *m.* oxide || ***óxido carbónico*** *m.* carbon monoxide || ***óxido cúprico*** *m.* tenorite || ***óxido de aluminio*** *m.* alumina || ***óxido de calcio*** *m.* quicklime, lime || ***óxido de carbono*** *m.* carbon monoxide || ***óxido de cinc*** *m.* calamine, zincite || ***óxido de cobre*** *m.* cuprite, tenorite || ***óxido de estaño*** *m.* cassiterite || ***óxido de hierro*** *m.* iron oxide || ***óxido de manganeso*** *m.* manganese oxide, pyrolusite || ***óxido de magnesio*** *m.* magnesia || ***óxido de mercurio*** *m.* hydrargyrite || ***óxido de titanio*** *m.* titanium dioxide || ***óxido de urano*** *m.* uraninite || ***óxido de zinc*** *m.* calamine, zincite || ***óxido férrico*** *m.* ferric oxide

oxigenación *f.* oxygenation

oxigenado *a.* oxygenated

oxígeno *m.* oxygen

oxihornblenda *f.* oxyhornblende

oximagnita *f.* maghemite

oximesostasis *f.* oxymesostasis

ozono *m.* ozone

ozoquerita *f.* native paraffin

P

pabellón *m.* pavilion of a gemstone

paco *m.* [Bol., Pe.] gossan; [Ch.] silver chloride ore

pack *m.* pack ice

padre del sal *m.* crystal selenite

padrón *m.* register of a filed claim

pagador *m.* paymaster

pagaré *m.* promisory note, IOU

pago *m.* payment || ***pago adelantado*** *m.* cash advance

paigeita *f.* vonsenite

paila *f.* pan

país *m.* country || ***país marginal*** *m.* borderland

paisaje *m.* countryside, landscape || ***paisaje exhumado*** *m.* resurrected landscape

paja quemada *f.* [Pe.] Jamesonite

pajuela *f.* ore that glitters

pala *f.* shovel, spade || ***pala mecanica*** *f.* backhoe

palacra, palacrana *f.* gold nugget; gold ingot

paladio *m.* palladium

palagonita *f.* palagonite

palagonización *f.* palagonization

palaita *f.* hureaulite

palasita *f.* pallasite

palasomo *m.* palasome

Palatiense, Palatiano *m. & a.* Palatinian, Pfalzian

paleador *m.* shoveller

paleaduro, paleaje *m.* shovelling

palear *vb.* to shovel

paleoacuífero *m.* paleoaquifer

paleoambiente *m.* paleoenvironment

paleoautóctono *m.* paleoautochthon

paleobatimétrico *a.* paleobathymetric

paleobiología *f.* paleobiology

paleobotánica *f.* paleobotany

paleocanal *m.* paleochannel

paleocarst *m.* paleokarst

Paleocénico, Paleoceno *m. & a.* Paleocene

paleoclima *m.* paleoclimate

paleoclimático *a.* paleoclimatic

paleoclimatología *f.* paleoclimatology

paleodrenaje *m.* paleodrainage

paleoecuador *m.* paleoequator

paleofitología *f.* paleophytology

Paleógeno *m. & a.* Paleogene

paleogeografía *f.* paleogeography

paleogeográfico *a.* paleogeographic

paleogeológico *a.* paleogeologic

paleogeomorfología *f.* paleogeomorphology

paleohidrología *f.* paleohydrology

Paleolítico *m. & a.* Paleolithic

paleomagnético *a.* paleomagnetic

paleomagnetismo *m.* paleomagnetism

paleontología *f.* paleontology

paleontológico *a.* paleontologic

paleontólogo *m.* paleontologist

paleorelieve *m.* paleorelief
paleosol *m.* paleosol, hard ground
paleotectónico *a.* paleotectonic
paleotermal *a.* paleothermal
paleotermometría *f.* paleothermometry
paleotopografía *f.* paleotopography
paleoviento *m.* paleowind
Paleozóico *m. & a.* Paleozoic
palero *m.* shoveller
paleta *f.* small shovel, trowel
paligorskita *f.* palygorskite
palimpsesto *m.* palimpsest
palingénesis *f.* palingenesis
palinología *f.* palynology
palínomorfo *m.* palynomorph, spore
palinspástico *a.* palinspastic
palma *f.* palm
palmo *m.* one quarter of a vara
palse *m.* pingo
paludal *a.* paludal
palúdico *a.* paludal, swamp
palustre *a.* swampy
pallaco *m.* [Bol., Pe.] ore found in tailings; tailings
pallador *m.* [Bol., Pe.] ore sorter
pallaquear, pallar *vb.* to sort ore from dumps
palliri *m.* ore sorter
pallón *m.* assay button
pampa *f.* grassy plain, pampas
panautomórfico *a.* panidiomorphic
pandeamiento *m.* buckling
pandeo *m.* buckle, bend, sag, bulge
pandermita *f.* priceite
pandilla *f.* gang, work crew
pando *a.* bulged, shallow
panel *m.* panel, display || ***panel explicativo*** *m.* display panel || ***panel solar*** *m.* solar panel
panetita *f.* panethite
panidiomorfo *a.* idiomorphic
panino *m.* vein material
paniso, panizo *m.* [Bol., Pe.] altered rock, gangue, gouge; [Ch., Pe.] trachytic country rock
Pannoniense, Pannoniano *m. & a.* Pannonian
panorama *f.* panorama
pantalla *f.* video screen, screen; (comp.) screen [Bol.] dike || ***pantalla de cristal líquido*** *f.* liquid crystal display || ***pantalla de tubo de rayos catódicos*** *f.* CRT, cathode ray tube screen
pantano *m.* swamp, marsh, [U.K.] fen
pantanoso *a.* swampy
papa *m.* gold nugget; nodule
papilla *f.* slurry
paquer *m.* packer
paquete *m.* complex, series; package || ***paquete de estratos*** *m.* sedimentary complex || ***paquete de mantos*** *m.* thrust complex
par *m.* pair, couple || ***par estereoscópico*** *m.* stereo pair || ***par térmico*** *m.* thermocouple
parabituminoso *a.* parabituminous
Paracelsiano, Paracelsense *m. & a.* Paracelsian
paraclasa *f.* fault
paraclinal *a.* paraclinal
paraconcordancia *f.* paraconformity
paraconcordante *a.* paraconformable
paracrisotil *f.* parachrysotile
paracronología *f.* parachronology
parada *f.* stop; [Méx.] work gang, crew || ***parada a la carga*** *f.* crew paid by the load || ***parada de hacienda*** *f.* crew paid by the day || ***parada de obra*** *f.* crew paid by the day
paradiagenético *a.* paradiagenetic
paradigma *m.* paradigm
paradiscordancia *f.* paraunconformity
paraesquisto *m.* paraschist
paraestratigrafía *f.* parastratigraphy
paraestratigráfico *a.* parastratigraphic
parafina *f.* paraffin, wax
parafínico *a.* paraffinic
paragénesis *m.* paragenesis
paragenético *a.* paragenetic
parageosinclinal *m.* parageosyncline
paragneiss *m.* paragneiss

paragolpe *m.* buffer, bumper, shock absórber
paragón *m.* paragon (variety of diamond)
paraje *m.* place, spot
paralaje *f.* parallax
paralelo *m. & a.* parallel of lattitude; gallery driven parallel to the main gallery; parallel
parálico *a.* paralic
paralóctono *m.* parallochthon
parallamas *f.* flame arrester
paramagnético *a.* paramagnetic
paramagnetismo *m.* paramagneitsm
paramar *m.* season of wind and snow
paramear *vb.* to drizzle
paramento *m.* face of hewn stone
paramera *f.* desert; altiplano
paramétrico *a.* parametric
parámetro *m.* parameter
páramo *m.* altiplano, high plain, [U.K.] moor
paramorfismo *m.* paramorphism
paramorfo *m.* paramorph
paraóndula *f.* para-ripple
parar *vb.* to stop || ***parar una pérdida*** *vb.* to stop a leak
pararizadura *f.* para-ripple
pararrayos *m.* lightning rod, lightning arrester
parasecuencia *f.* parasequence
parasítico *a.* parasitic
parásito *m.* parasite
paratectónico *m. & a.* paratectonics; paratectonic
paratelurita *f.* paratellurite
paratipo *m.* paratype
parautóctono *m. & a.* parautochthon; parautochthonous
paravan *m.* paravane
paravolastonita *f.* parawollastonite
parcela *f.* lot, plot, parcel
parcial *a.* partial
parcionero *m.* mining partner
parche *m.* patch of rock, patch
parchetita *f.* parchettite, leucite trachyandesite
pardo *m.* earth-color
pardusco *a.* earth-colored
pared *f.* wall, cliff || ***pared colgante*** *f.* hanging wall || ***pared de base*** *f.* footwall || ***pared de seguridad*** *f.* safety wall
paredón *m.* cliff, escarpment
paria *f.* cinnabar
paro *m.* stop, labor stoppage, shutdown
paroxismo *m.* paroxysm; *a.* paroxysmal
parque *m.* park
parragón *m.* standard silver for assayers
parranthina *f.* scapolite
parte *f.* part, share, portion || ***parte aguas*** *f.* drainage divide || ***parte del propietario*** *f.* landowner royalty || ***parte diario*** *f.* daily report || ***parte diario del perforador*** *f.* driller's log
parteaguas *f.* drainage divide
par térmico *m.* thermocouple
partes contratantes *f.* contracting parties
participación *f.* (com.) share, interest; working interest || ***participación del concesionario*** *f.* working interest || ***participación en el mercado*** *f.* market share
partícula *f.* particle || ***partícula alfa*** *f.* alpha particle || ***partícula de fango*** *f.* mud particles
partida *f.* lot, shipment, consignment
partido *m.* [Méx.] division of ore between owners, investors, workers
partido *a.* cracked, split, parted
partidor *m.* ore sorter
partir *vb.* to split, divide, crack
partridgeita *f.* bixbyite
parva *f.* smelter mixture or charge
Pasadeniense, Pasadeniano *m. & a.* Pasadenan
pasaje *m.* passage, straits
pasar la cuenta *vb.* to bill, invoice
pasarse *vb.* to occur
paseo de tierras *m.* transport on the backs of laborers
pasivo *m.* liability, debt || ***pasivos devengados*** *m.* accrued liabilities

paso *m.* pass; underground passage, drift, or gallery; pace, linear measure equivalent to two steps || ***paso real*** *m.* main gallery

pasta *f.* paste; (min.) matrix, groundmass; silver ore; gold or silver amalgam; gold or silver bullion || ***pasta mineral*** *f.* groundmass, matrix

pasto *m.* pasture, grass

pata de pero *f.* dogleg, well deviation

patentado *a.* proprietary, patented

patentar *vb.* to patent

patente *f.* patent; grant, permit, license; concession; [Bol., Ch.] annual mine tax

paternoita *f.* kaliborite

pátina *f.* patina || ***pátina del tiempo*** *f.* weathering || ***pátina desértica*** *f.* desert varnish

patio *m.* yard, court, floor || ***patio de tanques*** *m.* tank farm

patrón *m.* pattern, framework; template; standard

patrones estratales *m.* stratal patterns

paurocristalino *a.* paurocrystalline

pauta *f.* norm

pavimento *m.* pavement || ***pavimento del desierto*** *m.* desert pavement

pavonado *m.* [Pe.] tetrahedrite

pectolita *f.* pectolite

pechblenda *f.* pitchblende

pecho de paloma *m.* bornite, peacock ore

pedálfer *m.* pedalfer

pedazo *m.* piece, bit, part, fragment

pedemontano *m.* piedmont

pedernal *m.* flint, chert

pedestal *m.* butte

pedido *m.* order, request

pedimentación *f.* pedimentation

pedimento *m.* & *a.* pediment

pedión *f.* pedion

pedir *vb.* to request; to order

pedogénesis *f.* pedogenesis

pedología *f.* pedology

pedomórfosis *f.* pedomorphosis

pedragal *m.* stony ground; quarry

pedregoso *a.* stony, rocky

pedregullo *m.* gravel || ***pedregullo de ladera*** *m.* [Arg.] talus

pedrejal *m.* quarry

pedrera *f.* quarry

pedrería *f.* precious stones

pedrero *m.* stone cutter; scree, talus

pedriscal *m.* stony or rocky ground

pedrizo *a.* stony, rocky

pedrosita *f.* hornblendite

pedrusco *m.* rough stone, boulder

pega *f.* blast

pegado *a.* stuck

pegador *m.* blasting foreman

pegadura *f.* dirt, mud, smudge on mine walls

pegajoso *a.* gummy, tacky

pegamento *m.* glue

pegar *vb.* to fire the loaded drill holes

pegmatita *f.* pegmatite

pegmatítico *a.* pegmatitic

pegmatitización *f.* pegmatitization

pegmatoide *m.* pegmatoid

pegmatofírico *a.* granophyric

pegmatofiro *a.* granophyre

pegmatolita *f.* orthoclase

peladilla *f.* quartz pebble

peladura *f.* window

pelágico *a.* pelagic

Peleano, Pelense *m.* & *a.* Peelean

pelecípodo *m.* pelecypod

peletoidal *a.* peletoidal, peletoid

peletoide *m.* peletoid

película *f.* film || ***película aérea*** *f.* aerial film || ***película color*** *f.* color film || ***película de aceite*** *f.* oil film || ***película infrarroja*** *f.* infrared film || ***película interfacial*** *f.* interfacial film || ***película pancromática*** *f.* panchromatic film

peligro *m.* danger, hazard

pelita *f.* shale

pelítico *a.* pelitic

pelitomórfico *a.* pelitomorphic

pelo *m.* opaque streak in a gemstone

pelodita *f.* pellodite, pelodite

peloide *m.* peloid

pelolítico *a.* argillaceous

pelotilla *f.* pellet

peltre *m.* spelter

pella *f.* mass of native metal; amalgam; pellet || ***pella fecal*** *f.* fecal pellet, casting

penacho *m.* plume

pendiente *f.* slope, gradient, grade, declivity; hanging wall, roof, pendant; pitch, dip || ***pendiente continental*** *f.* continental slope, shelf || ***pendiente de escombros de talud*** *f.* talus slope || ***pendiente de un río*** *f.* fall of a river || ***pendiente estructural*** *f.* dip slope || ***pendiente homoclinal*** *f.* homoclinal dip || ***pendiente inestable*** *f.* unstable slope || ***pendiente magnética*** *f.* magnetic gradient || ***pendiente regional*** *f.* regional dip

pendulo *m.* plum bob

peneconcordante *a.* peneconcordant

penecontemporáneo *a.* penecontemporaneous

peneplanación *f.* peneplanation

peneplanicie *f.* peneplain

peneplanización *f.* peneplanation

penesalino *a.* penesaline

penetración *f.* penetration, piercement || ***penetración del agua*** *f.* water breakthrough || ***penetración pasiva*** *f.* passive piercement

penetrante *a.* pervasive

penetrar *vb.* to penetrate

penillanura *f.* peneplain

península *f.* peninsula

peniplanicie *f.* peneplain

peniplano *m.* peneplain

Penjabiano, Penyabiano, Penjabiense, Penyabiense *m. & a.* Punjabian

pennina *f.* penninite

Penokeiense, Penokeiano *m. & a.* Penokean

Pensilvaniano, Pensilvánico, Pensilvaniense, Pensylvaniano *m. & a.* Pennsylvanian

pentacíclico *a.* pentacyclic

pentagonal *a.* pentagonal

pentágono *m.* pentagon

pentano *m.* pentane

pentlandita *f.* pentlandite

Penutiense, Penutiano *m. & a.* Penutian

penumbra *f.* penumbra

peña *m.* boulder, rock, cobble; wall rock

peñascal *m.* rocky ground, gravel

peñasco *m.* large rock, boulder, crag, stack, cliff; klippe; escarpment

peñol, peñon *m.* large rock, boulder, rock mass

peón *m.* common laborer, worker, roustabout || ***peón de cuadrilla*** *m.* roustabout, roughneck || ***peón de perforación*** *m.* roughneck, roustabout || ***peón suelto*** *m.* roustabout, roughneck

peonaje *m.* work gang

pepa de oro *f.* nugget

pepena *f.* smelting ore, cobbed ore

pepenado *m.* [Sp.] dressed ore

pepenador *m.* ore sorter, cobber

pepenar *vb.* to sort ore

peperita *f.* volcanic sand

pepino *m.* karst tower

pepita *f.* nugget

peracidita *f.* silexite

peralcalino *a.* peralkaline

peraluminoso *a.* peraluminous

percolación *f.* percolation, seep, seeapage, filtration

percolar *vb.* to percolate

percristalino *a.* percrystalline

percusión *f.* percussion

perder(se) *vb.* to become lost, pinch out; to leak, to lose

pérdida *f.* loss due to treatment; leak, leakage; (legal) forfeiture || ***pérdida de circulación*** *f.* circulation loss || ***pérdida de gas*** *f.* gas leak || ***pérdida de lodo*** *f.* mud losses

pérdidas *f.* (com.) losses

perdigón *m.* iron pellet

perenne *a.* perenial

perfeccionar *vb.* to follow up, reexamine

perfil *m.* profile, cross section, section; log || *perfil columnar m.* log; columnar section || *perfil de calibración m.* caliper log || *perfil de densidad m.* density log || *perfil de equilibrio m.* graded profile || *perfil de inducción m.* induction log || *perfil de perforación m.* well log || *perfil de pozo m.* well log || *perfil de velocidad m.* velocity log || *perfil sísmico de reflexión m.* seismic reflection profile || *perfil de suelo m.* soil profile || *perfil de velocidad vertical m.* vertical velocity profile || *perfil edáfico m.* soil profile || *perfil eléctrico m.* electric log || *perfil estratigráfico m.* columnar section || *perfil geológico m.* geologic cross section || *perfil litológico m.* lithologic section || *perfil según muestras m.* sample profile || *perfil tipo, típico m.* type section || *perfil transversal m.* cross section

perfilado *a.* streamlined

perfiladora *f.* grader

perfilaje *m.* logging || *perfilaje continuo m.* continuous profiling || *perfilaje de rayos gama f.* gamma ray logging || *perfilaje neutrónico f.* neutron log

perfilar *vb.* to log

perforabilidad *f.* drillability

perforable *a.* drillable

perforación *f.* drill hole; perforation || *perforación a múltiples zonas f.* multizone completion || *perforación a núcleos f.* core drilling || *perforación a percusión f.* percussion drilling, churn drilling || *perforación bajo agua f.* submarine drilling || *perforación con cable f.* cable drilling || *perforación con diamante f.* diamond drilling || *perforación con diámetro reducido f.* slim hole drilling || *perforación dirigida f.* directional drilling || *perforación con extracción de testigos f.* core drilling || *perforación de cateo f.* boring, exploratory drilling || *perforación de disparo f.* shot hole || *perforación de ensayo f.* wildcat || *perforación de exploración f.* exploration drilling, wildcat || *perforación de prueba f.* test hole || *perforación de reconocimiento f.* exploratory drilling || *perforación desviada f.* deviated hole || *perforación direccional f.* directional drilling || *perforación guiada f.* directional drilling || *perforación inicial f.* spud || *perforación por rotación f.* rotary drilling || *perforación reducida f.* slim hole || *perforación rotativa f.* rotary drilling

perforado *a.* perforated; drilled, bored

perforador *m. & a.* driller; boring, drilling || *perforador neumático m.* air hammer, jackhammer

perforadora *f.* borer, driller, rig, drill || *perforadora a balancín f.* spudder || *perforadora a percusión f.* percussion drill, jackhammer || *perforadora de tubería de revestimiento f.* casing perforator

perforando *m.* drilling, boring

perforante *a.* drilling, boring

perforar *vb.* to drill, bore

perhialina *f.* perhyaline

periferia *f.* periphery

periclasa *f.* periclase

periclinal *m. & a.* pericline; periclinal

peridotita *f.* peridotite

peridoto *m.* peridot, olivine

peirférico *a.* peripheral

perigénico *a.* perigenic

periglaciario *a.* periglacial

perímetro *m.* perimeter

periodicidad *f.* periodicity

período *m.* period, cycle, stage || *período de amortización m.* amortization period || *período de glaciación m.* glacial period || *período de operación m.* operating cycle || *período de paro m.* shutdown time, turnaround time

peritéctico *a.* peritectic, reaction

perito *m.* expert; appraiser; skilled worker

perjuicio *m.* damage, injury

perknita *f.* ultramafic rock

perla *f.* pearl

perlado *a.* pearly

perliforme *a.* bead-like

perlita *f.* perlite

perlítico *a.* perlitic

perlofita *f.* perloffite

permanente *a.* permanent

permeabilidad *f.* permeability || ***permeabilidad absoluta*** *f.* absolute permeability

permeable *a.* permeable

permeación *f.* permeation

permeámetro *m.* permeameter

Permiano, Pérmico *m. & a.* Permian

permineralización *f.* permineralization

permisible *a.* permissible, allowable

permiso *m.* license, lease, permit || ***permiso de cateo*** *m.* exploration license

permitir *vb.* to permit, grant, allow, let

Permocarbonífero *a.* Permo-Carboniferous

permutación *f.* permutation

perno *m.* (min.) spike

perofskita, perovskita, perowskita *f.* perovskite

peróxido *m.* peroxide

perpendicular *a.* perpendicular, normal, at right angles

personal *m.* crew; personnel

perspectiva *f.* perspective, outlook; prospect

pertenencia *f.* mining claim; possession

pértiga *f.* rod (meas.)

pertita *f.* perthite

pertítitco *a.* perthitic

pertosita *f.* perthosite, syenite

perturbación *f.* perturbation, dislocation

perturbado *a.* disturbed

pesa *f.* weight

pesado *a.* heavy; dense, of high specific gravity

pesca *f.* (perf.) fishing for tools down the drill hole || ***pesca caños*** *f.* collar socket || ***pesca cuchara*** *f.* bailer grab, boot jack || ***pesca válvulas*** *f.* valve spear

pescado *m.* (perf.) fish

pescador *m.* (perf.) fishing tool

pescar *vb.* (perf.) to fish

peso *m.* weight; monetary unit || ***peso atómico*** *m.* atomic weight || ***peso avoirdupois*** *m.* avoirdupois weight (1 pound = 0.454 kg) || ***peso bruto*** *m.* gross weight, dead weight || ***peso espicífico*** *m.* specific gravity || ***peso joyería*** *m.* troy weight || ***peso molar*** *m.* molar weight || ***peso molecular*** *m.* molecular weight || ***peso muerto*** *m.* dead weight || ***peso neto*** *m.* net weight

pestaña *f.* ledge

pétanque, petlanque *f.* tetrahedrite, ruby silver

pétreo *a.* of stone, stony, rocky

petrificación *f.* petrification

petrificado *a.* petrified, lithified

petrificar *vb.* to petrify

petroclástico *a.* detrital

petrofábrico *m. & a.* fabric, petrofabric

petrofísica *f.* petrophysics

petrogénesis *f.* petrogenesis

petrografía *f.* petrography

petrográfico *a.* petrographic

petrógrafo *m.* petrographer

petrolato *m.* petrolatum

petroleno *m.* petrolene

petróleo *m.* petroleum,. hydrocarbons, oil (véase aceite) || ***petróleo con azufre*** *m.* sour crude || ***petróleo con gas*** *m.* live oil || ***petróleo crudo*** *m.* crude oil || ***petróleo de lutitas*** *m.* shale oil || ***petróleo desgasificado*** *m.* dead oil || ***petróleo en bruto*** *m.* crude oil || ***petróleo húmedo*** *m.* wet oil || ***petróleo intemperizado*** *m.* weathered oil || ***petróleo lampante*** *m.* kerosene, paraffin oil || ***petróleo ligero*** *m.* light oil || ***petróleo liviano*** *m.* light oil || ***petróleo muerto*** *m.* dead oil || ***petróleo oreado*** *m.* weathered crude oil || ***petróleo original in-situ*** *m.* original oil in place (OOIP) || ***petróleo para calderas*** *m.* bunker oil || ***petróleo para calefacción*** *m.* heating oil || ***petróleo parafínico*** *m.* waxy oil || ***petróleo pesado*** *m.* heavy crude || ***petróleo sulfuroso*** *m.* sour crude oil

petrolero *m.* oilman; tanker; *a.* oil, petroleum; oil-bearing

petrolífero *a.* petroliferous, oil-bearing

petrología *f.* petrology

petrológico *a.* petrologic

petrólogo *m.* petrologist

petromorfo *m.* petromorph

petroquímica *f.* petrochemistry

petroquímico *a.* petrochemical

petrosilex *m.* flint, amausite

pez *f.* tar, pitch

Pfalziense, Pfalziano *m. & a.* Pfalzian

Piacentino *m. & a.* Piacenzian

piamonte *m.* piedmont, foothills

piamontita *f.* piedmontite

picacho *m.* pointed peak, summit; spur

picada *f.* survey line

picador *m.* pick miner

picadura *m.* pitting

picapedrero *m.* stonecutter

picaporro *m.* pick

picar *vb.* to cut stone, to hew, work with a pick or hammer and gad

picnita *f.* pycnite (var. of topaz)

pico *m.* peak, summit; pickaxe, miner's pick

picota *f.* peak

picrita *f.* picrite

picrítico *a.* picritic

picrocromita *f.* picrochromite, magnesiochromite

picurita *f.* pitch coal

pie *m.* (meas.) foot = 0.3048 m.; base of a mountain || ***pie-acre*** *m.* acre-foot || ***pie cúbico*** *m.* cubic foot || ***pie de monte*** *m.* piedmont, foothills

piedra *f.* rock, stone || ***piedra bruta*** *f.* country rock || ***piedra córnea*** *f.* chert, flint || ***piedra de afilar*** *f.* whetstone, grinding stone || ***piedra de esmeril*** *f.* grinding stone || ***piedra de mano*** *f.* hand specimen || ***piedra de sapo*** *f.* soapstone || ***piedra imán*** *f.* lodestone || ***piedra preciosa*** *f.* precious stone || ***piedra tallada*** *f.* cut stone || ***piedra triturada*** *f.* crushed stone

piedrín *m.* crushed stone

piés cúbicos *m.* cubic feet

pieza *f.* piece, part, member

piezoclasa *f.* compression joint

piezocristalización *f.* piezocrystallization

piezoelectricidad *f.* piezoelectricity

piezoeléctrico *a.* piezoelectric

piezométrico *a.* piezometric, equipotential

pigeonita *f.* pigeonite

pila *f.* basin, trough, tank, pothole; evaporation tank; fountain; battery; pile, heap || ***pila termoeléctrica*** *f.* thermocouple

pilar *m.* horst, tower, pinnacle, spire, pillar || ***pilar de tierra*** *m.* hoodoo || ***pilar tectónico*** *m.* horst, uplifted block

pileta *f.* mine sump; crucible

pilita *f.* pilite

pilón *m.* stone trough

pilotáxico *a.* pilotaxitic

pilotaxítico *a.* pilotaxitic, felty

pinacoide *m. & a.* pinacoid; pinacoidal

pináculo *m.* pinnacle, spire, pillar, needle

pinado *a.* pinnate

pingo *m.* drill hole terminated prematurely

pinguita *f.* nontronite

pinta *f.* pint = 0.473 liters

pintar *vb.* to exhibit indications of ore

pínula *f.* sight of a surveying instrument

pinza *f.* fishing tool

piña *f.* cone-shaped ingot of silver; [And.] bullion

piocha *f.* pick

piotina *f.* saponite

pipa *f.* pipe

pipeta *f.* pipette

pique *m.* bottom, ground; [And.] shaft || ***pique comunicante*** *m.* winze || ***pique interno*** *m.* blind shaft

piquero *m.* miner

piqueta *f.* mattox, pickaxe

piquete *m.* surveyor's stake; small prospect work

piquetilla *f.* gad; small pickaxe

piralmandita *f.* pyralmandite

piralspita *f.* pyralspite

piramidal *a.* pyramidal

pirámide *m.* pyramid

pirargirita *f.* pyargyrite

pireneita *f.* melanite

pirex *m.* pyrex

piriforme *a.* pear-shaped

pirita *f.* pyrite; sulfides ‖ ***pirita arsenical*** *f.* arsenopyrite ‖ ***pirita blanca*** *f.* marcasite ‖ ***pirita capilar*** *f.* millerite ‖ ***pirita cobriza*** *f.* chalcopyrite ‖ ***pirita de azufre*** *f.* iron pyrite ‖ ***pirita de carbón*** *f.* scud ‖ ***pirita de cobalto*** *f.* linnaeite ‖ ***pirita de cobre*** *f.* chalcopyrite ‖ ***pirita de cresta de gallo*** *f.* marcasite ‖ ***pirita de estaño*** *f.* stannite ‖ ***pirita de fierro*** *f.* iron pyrite ‖ ***pirita de hierro*** *f.* iron pyrite ‖ ***pirita de manganeso*** *f.* hauerite ‖ ***pirita ferroniquelífera*** *f.* pentlandite ‖ ***pirita magnética*** *f.* pyrrhotite ‖ ***pirita marcial*** *f.* iron pyrite ‖ ***pirita prismática*** *f.* marcasite ‖ ***pirita roja*** *f.* niccolite

pirítico *a.* pyritic

piritización *f.* pyritization

piritizar *vb.* to pyritize

piritoedro *m.* pyritohedron

piritoso *a.* pyrite-bearing

piroaurita *f.* pyroaurite

pirobelonita *f.* pyrobelonite

pirobitumen *m.* pyrobitumen

piroclástico *a.* pyroclastic

piroclasto *m.* pyroclast

pirocloro *m.* pyrochlore

pirofana *f.* pyrophane, fire opal

pirofilita *f.* pyrophyllite

pirogenación *f.* carbonization

pirogénesis *f.* pyrogenesis

pirogénico, pirógeno *a.* pyrogenic

pirólisis *f.* pyrolysis

pirolita *f.* pyrolite

pirolítico *a.* pyrolytic

pirolusita *f.* pyrolusite

piromelana *f.* brookite

piromerida *f.* pyromeride, nodular rhyolite

pirometamorfismo *m.* pyrometamorphism

pirometasomatismo *m.* pyrometasomatism

pirometría *f.* pyrometry

pirómetro *m.* pyrometer

piromorfita *f.* pyromorphite

pironafta *f.* pyronaphtha

piropisita *f.* pyropissite

pirópo *m.* precious stone; pyrope

piroretinita *f.* pyroretinite

pirosfero *m.* pyrosphere

pirosmalita *f.* pyrosmalite

pirostibina *f.* kermesite

pirostilpnita *f.* pyrostilpnite

piroxena *f.* pyroxene

piroxenita *f.* pyroxinite

piroxeno *m.* pyroxene

piroxenoide *m.* pyroxenoid

pirquín *m.* [Bol., Ch.] contract mining

pirquinero *m.* [Bol., Ch.] contract miner

pirrotina *f.* pyrrhotite

pisada fósil *f.* ichnite

piso *m.* floor, level; stage; footwall; age ‖ ***piso de un galería*** *m.* sill

pisolita *f.* pisolite, pisolith

pisolítico *a.* pisolitic

pista *f.* track; magnetic tape track

pistola *f.* small drill hole ‖ ***pistola perforadora*** *f.* perforating gun

pistón de achique *m.* (perf.) swab a well

pistonear *vb.* to swab

piticita *f.* pitticite

pivote *m.* stock

pixel *m.* pixel, picture element

pizarra *f.* slate; schist

pizarral *m.* slate quarry

pizarreño *a.* slatey

pizarrería *f.* slate quarry

pizarrosidad *f.* slaty, slaty cleavage

pizarroso *a.* slatey

placa *f.* plate, slab, sheet ‖ ***placa de agua*** *f.* water leg ‖ ***placa de deslizamiento*** *f.* slide block ‖ ***placa de presión*** *f.* pressure plate ‖ ***placa litosférica*** *f.* lithospheric plate ‖ ***placa para desenrosque de trépanos*** *f.* bit breaker ‖ ***placa tectónica*** *f.* tectonic plate

placer *m.* placer ‖ ***placer eluvial*** *m.* dry digging

plagioclasa *f.* plagioclase

plagioclasita *f.* anorthosite

plagionita *f.* plagionite

Plaisanciano, Plaisanciense *m. & a.* Plaisancian

plan *m.* program, plan; layout; bottom level in a mine, [C.A., Car., And.] plain; *a.* [C.A., Car., And.] flat

plana *f.* [Arg.] plain

planación *f.* planation

planctón *m.* plankton

planctónico *a.* planktonic

plancha *f.* plate or sheet of metal; bar, ingot; pig

plancheta *f.* plane table

planchón *m.* barge

planeamiento *m.* planning

planear *vb.* to plan

planeta *f.* planet

planicie *f.* plain || ***planicie abisal*** *f.* abyssal plain || ***planicie aluvial*** *f.* alluvial plain || ***planicie costanera*** *f.* coastal plain || ***planicie costera*** *f.* coastal plain || ***planicie de abrasión*** *f.* denudation plane || ***planicie de lodo*** *f.* mudflat || ***planicie troncal*** *f.* peneplain

planificación *f.* planning

planilla *f.* payroll

planimetría *f.* planimetry, survey, mapping

planimétrico *a.* planimetric

planímetro *m.* planimeter

planización de pedimentos *f.* pediplanation

plano *m.* map, plan, drawing; *a.* planar, flat, level || ***plano aluvial*** *m.* flood plain || ***plano axial, axil*** *m.* axial plane || ***plano basal*** *m.* footwall flat || ***plano colgado*** *m.* hanging wall flat || ***plano de clivaje*** *m.* cleavage plane || ***plano de cota zero*** *m.* datum plane || ***plano de crucero*** *m.* cleavage plane || ***plano de deslizamiento*** *m.* glide plane || ***plano de diaclasa*** *m.* joint plane || ***plano de escurrimiento*** *m.* thrust plane || ***plano de estratificación*** *m.* bedding plane || ***plano de falla*** *m.* fault plane || ***plano de fractura*** *m.* fracture plane || ***planos de fricción*** *m.* slickensides || ***plano de junta*** *m.* joint plane || ***plano de localización*** *m.* index map || ***plano de nivel*** *m.* datum || ***plano de referencia*** *m.* datum || ***plano de resbalamiento*** *m.* slickenside || ***plano focal*** *m.* focal plane || ***plano óptico*** *m.* optic plane

planta *f.* floor, level; horizontal cut; map, plan; plant, factory, facility || ***planta compresora*** *f.* compressor plant || ***planta de concentración*** *f.* concentration works || ***planta de coque, de coqueo*** *f.* coking plant || ***planta de craqueo*** *f.* cracking plant || ***planta de extracción*** *f.* extraction plant || ***planta de gas*** *f.* gas plant || ***planta de recirculación*** *f.* recycling plant || ***planta de recuperación de ácidos*** *f.* acid recovery plant || ***planta de tratamiento ácido*** *f.* acid treatment plant || ***planta intermedia de refuerzo*** *f.* booster station || ***planta para tratamiento de agua*** *f.* water treatment plant || ***planta piloto*** *f.* pilot plant

plantar *vb.* to erect a mill, machinery, or plant

planteaje, planteamiento, plantel *m.* plant, mill

planteo *m.* layout

plantilla *f.* template, pattern || ***plantilla normal*** *f.* standard gauge

plasma *f.* plasma

plasticidad *f.* plasticity

plástico *a.* plastic

plastotipo *m.* plastotype

plata *f.* silver; silver ore || ***plata agria*** *f.* argentite, polybasite, stephanite || ***plata antimonial*** *f.* discrasite || ***plata azul acercado*** *f.* polybasite || ***plata azul de catorce*** *f.* selbite || ***plata blanca*** *f.* native silver || ***plata bruto*** *f.* native silver || ***plata ceniza*** *f.* silver chloride || ***plata córnea*** *f.* cerargyrite || ***plata córnea amarilla*** *f.* iodyrite || ***plata córnea blanca*** *f.* cerargyrite || ***plata córnea verde*** *f.* embolite || ***plata dúctil*** *f.* argentite || ***plata en barra*** *f.* silver bullion || ***plata gris*** *f.* argentite || ***plata labrada*** *f.* wrought silver || ***plata mixta*** *f.* electrum || ***plata nativa*** *f.* native silver || ***plata natural*** *f.* native silver || ***plata negra*** *f.* argentite || ***plata prismática*** *f.* stephanite || ***plata parda*** *f.* cerargyrite || ***plata pella*** *f.* silver amalgam || ***plata roja clara*** *f.* proustite || ***plata roja oscura*** *f.* pyrargyrite || ***plata verde*** *f.* embolite || ***plata vítrea*** *f.* argentite

plataforma *f.* platform; microscope stage; roadbed; shelf || ***plataforma aeroportada*** *f.* airborne platform, plane || ***plataforma autoelevadiza*** *f.* jackup rig || ***plataforma continental*** *f.* continental rise, shelf || ***plataforma costera*** *f.* coastal plain || ***plataforma de abrasión*** *f.* wave-cut platform || ***plataforma de cables en tensión*** *f.* tension leg platform || ***plataforma de corona*** *f.* crow's nest || ***plataforma de perforación*** *f.* drilling platform, rig || ***plataforma de invierno*** *f.* winter berm || ***plataforma del astillero*** *f.* monkey board || ***plataforma del enganchador*** *f.* derrickman's platform, monkey board || ***plataforma del torrero*** *f.* monkey board || ***plataforma de verano*** *f.* summer berm || ***plataforma epicontinental*** *f.* continental slope, shelf || ***plataforma espacial*** *f.* space-based platform, satellite || ***plataforma externa*** *f.* external platform, outer shelf || ***plataforma flotante*** *f.* floating platform || ***plataforma interna*** *f.* platform interior || ***plataforma montada en gatos mecánicos*** *f.* jack-up rig || ***plataforma para caños*** *f.* pipe rack || ***plataforma replegable*** *f.* jack-up rig || ***plataforma semisumergible, semisumersible*** *f.* semi-submersible rig || ***plataforma submarina*** *f.* continental shelf || ***plataforma sumergible*** *f.* submersible rig

platina *f.* platinum ore; native platinum; microscope stage

platino *m.* platinum

plato para lavar mineral *m.* dish, plate; gold pan || ***plato de burbujeo*** *m.* bubble plate

platón *m.* gold pan

play *m.* exploration play, trend, concept

playa *f.* beach, shore, strand; river bank || ***playa alta*** *f.* backshore || ***playa de barrera*** *f.* barrier beach || ***playa de marea*** *f.* tidal flat || ***playa levantada*** *f.* raised beach || ***retro playa*** *f.* backshore

playero *m.* [Bol., Pe.] gold washer

playo *a.* shallow

plazo *m.* term, time range || ***corto plazo*** *m.* short-term || ***largo plazo*** *m.* long-term

pleamar *f.* high tide, flood tide, highstand || ***pleamar media*** *f.* mean high water

plegable *a.* pliable, folding, collapsible

plegadizo *a.* pliable, folding, collapsible

plegado *m.* folding; *a.* folded

plegadura *f.* folding, plication

plegamiento *m.* folding, fold, plication || ***plegamiento cabrío*** *m.* chevron folding || ***plegamiento de flujo*** *m.* flow folding || ***plegamiento en echelón*** *m.* en echelon folding || ***plegamiento entrecruzado*** *m.* crossfolding || ***plegamiento escalonado*** *m.* en echelon folds || ***plegamiento flexo-deslizamiento*** *m.* flexural-slip folding || ***plegamiento imbricado*** *m.* imbricate folding || ***plegamiento minúsculo*** *m.* crenulation, wrinkle || ***plegamiento por cizalla*** *m.* shear folding || ***plegamiento por deslizamiento*** *m.* slip folding || ***plegamiento por flexión de fallas*** *m.* fault-bend folding || ***plegamiento por flexión y deslizamiento*** *m.* flexural-slip folding || ***plegamiento por flujo*** *m.* flexural flow folding || ***plegamiento por propagación de fallas*** *m.* fault-propagation folding || ***plegamiento ptigmático*** *m.* ptygmatic folding || ***plegamiento similar*** *m.* similar folding || ***plegamiento verdadero*** *m.* flexural folding

plegar *vb.* to fold

Pleistocénico, Pleistoceno *m. & a.* Pleistocene

plena carga *f.* full load

plenargirita *f.* matildite

pleno diámetro *m.* full hole

pleocristalino *a.* holocrystalline

pleocróico, pleocroisto *a.* pleochroic

pleocroísmo, pleocromatismo *m.* pleochroism

pleocroísto *a.* pleochroic

pleonasta *f.* ceylonite

plicación *f.* plication

pliegue *m.* fold || ***pliegue abanico*** *m.* fan fold || ***pliegue abierto*** *m.* open fold || ***pliegue abovedado*** *m.* dome fold || ***pliegue acostado*** *m.* recumbant fold || ***pliegue anticlinal*** *m.* anticline || ***pliegue apretado*** *m.* tight fold || ***pliegue armónico*** *m.* competent fold || ***pliegue arqueado*** *m.* arcuate fold || ***pliegue***

***asimétrico** m.* asymmetric fold || ***pliegue bucle** m.* buckle fold || ***pliegue buzante** m.* plunging fold || ***pliegue cabrío** m.* chevron fold || ***pliegue cerrado** m.* isoclinal fold, closed fold || ***pliegue concéntrico** m.* concentric fold || ***pliegue corrido** m.* thrust fold || ***pliegue de arrastre** m.* drag fold || ***pliegue de chevron** m.* chevron fold || ***pliegue de compactación** m.* supratenuous fold || ***pliegue de cubierta** m.* sheath fold || ***pliegue de despegue** m.* detachment fold || ***pliegue de fluencia** m.* flow fold || ***pliegue diapírico** m.* diapiric fold || ***pliegue disarmónico** m.* disharmonic fold || ***pliegue elongado** m.* elongate fold || ***pliegue en escalón** m.* en echelon fold || ***pliegue escalonado** m.* en echelon fold || ***pliegue flexurado por falla** m.* fault-bend fold || ***pliegue forzado** m.* forced fold || ***pliegue inclinado** m.* inclined fold || ***pliegue invertido** m.* overturned fold || ***pliegues imbricados** m.* imbricate folds || ***pliegue isoclinal** m.* isoclinal fold || ***pliegue kink** m.* kink fold || ***pliegue largo** m.* elongated fold || ***pliegue monoclinal** m.* monocline || ***pliegue parado** m.* upright fold, normal fold || ***pliegue paralelo** m.* parallel fold || ***pliegue por propagación** m.* fault-propagation fold || ***pliegue recostado** m.* recumbent fold || ***pliegue recto** m.* symmetric fold || ***pliegue recumbente** m.* recumbent fold || ***pliegue semejante** m.* similar fold || ***pliegue simétrico** m.* symmetric fold || ***pliegue similar** m.* similar fold || ***pliegue sinclinal** m.* syncline || ***pliegue suave** m.* gentle or broad fold || ***pliegue supratenue** m.* supratenuous fold || ***pliegue tumbado** m.* overturned fold || ***pliegue vertical** m.* upright fold || ***pliegue volcado** m.* overturned, inverted fold || ***pliegue yacente** m.* recumbent fold

Pliensbaquiano, Pliensbaquense *m. & a.* Pliensbachian

Pliniano, Pliniense *m. & a.* Plinian

Pliocénico, Plioceno *m. & a.* Pliocene

plisamiento *m.* microfolding

plomada *f.* plumb, plumb bob, plumb line

plombagina *f.* graphite

plomífero, plomilloso *a.* lead-bearing

plomizo *a.* leaden

plomo *m.* lead; lead ore || ***plomo abigarrado** m.* pyromorphite || ***plomo amarillo** m.* wulfenite || ***plomo blanco** m.* cerussite || ***plomo negro** m.* graphite || ***plomo pardo** m.* pyromorphite || ***plomo rojo** m.* crocoite || ***plomo verde** m.* pyromorphite

plomoso *a.* lead-bearing; plumose

pluma *f.* plume, hot spot

plúmbeo *a.* leaden

plurivalencia *f.* polyvalence

plutón *m.* pluton, intrusion, intrusive

plutónico *a.* plutonic

plutonio *m.* plutonium

plutonismo *m.* plutonism, vulcanism

pluvial *a.* pluvial

pneumatolítico *a.* pneumatolytic

población *f.* city, town; population || ***población de referencia** f.* reference population || ***población tipo** f.* type population

pobre *a.* barren, sterile, unmineralized; low grade, lean, poor

pobremente *a.* poorly

pocero *m.* well sinker; shaft digger

poceta *f.* sump, pool

pocillo *m.* small well or drill hole

poco profundo *a.* shallow

poder *m.* power; thickness; power of attorney || ***poder de resolución** m.* resolving power

podolita *f.* carbonate-apatite

podrido *a.* rotten, decomposed

podsol *m.* podsol

poiquilítico *a.* poikilitic

poiquiloblástico *a.* poikiloblastic

poise *m.* poise

polar *a.* polar

polaridad *f.* polarity

polarimetría *f.* polarimetry

polarímetro *m.* polarimeter

polarización *f.* polarization

polarizado *a.* polarized

polarizador *m. & a.* polarizer; polarizing

polea *f.* pulley, block, sheave

polen *m.* pollen

polenita *f.* pollenite, olivine phonolite

poliargirita *f.* polyargyrite
polibasita *f.* polybasite
policíclico *a.* polycyclic
policónico *a.* polyconic
policraso *m.* polycrase
policristal *m.* polycrystal
policromo *a.* polychrome, multicolored
policuarzo *m.* polyquartz
polidimita *f.* polydymite
polie *m.* polje
polietileno *m.* polyethylene
poligénico *a.* polycyclic
poligonal *a.* polygonal
polígono *m.* polygon || ***polígono de congelación*** *m.* frost polygon || ***polígono de piedras*** *m.* stone polygon
polihalita *f.* polyhalite
polikinemático *a.* polykinematic
polimerización *f.* polymerization
polimerizado *a.* polymerized
polímerizar *vb.* to polymerize
polímero *m.* polymer
polimetálico *a.* polymetallic
polimetamorfismo *m.* polymetamorphism
polimíctico *a.* polymictic
polimínerálico *a.* polymineralic
polimodal *a.* polymodal
polimorfía, polimorfismo *f.* polymorphism
polimorfo *m. & a.* polymorph; polymorphic
pólipo *m.* polyp
polisintético *a.* polysynthetic
polisulfuro *m.* polysulfide
politipismo *m.* polytypism
politipo *m.* polytype
póliza de seguro *f.* insurance policy
polo *m.* pole || ***polo geográfico*** *m.* geographic pole || ***polo magnético*** *m.* magnetic pole
polonio *m.* polonium
polucita *f.* polucite
polvillo *m.* fine dust; tailings
polvillos *m.* rich concentrates
polvo *m.* dust, powder
pólvora *f.* gunpowder, powder || ***pólvora para voladura*** *f.* blasting powder
polvorilla *f.* argentite; polybasite
polvorín *m.* powder magazine
pomáceo *a.* pumaceous
pómez *m. & a.* pumice
poner a cero *vb.* to reset
poner en bombeo *vb.* to put on pump
poner en producción *vb.* to put on production, to bring in a well
pongo *m.* gorge
poniente *m.* west
Pontiano, Pontiense *m. & a.* Pontian
Pontico *m. & a.* Pontic
pontón *m.* pontoon, floating bridge
población *f.* statistical population
poquedad *a.* paucity, scarcity
por año *a.* per annum, per year
porcelainita, porcelanita *f.* porcellanite
porcelana *f.* porelain
porcelanáceo *a.* porcellaneous
porcentaje *f.* percentage; rate || ***porcentaje de mol*** *f.* mole percent
porcentil *f.* percentile
por ciento *m.* percent
por día *m.* per diem; *a.* per day
por encima *p.* above
porfídico *a.* porphyritic
porfidita *f.* porphyry
pórfido *m.* porphyry
porfidoblástico *a.* porphyroblastic
porfidoblasto *m.* porphyroblast
porfiroclástico *a.* porphyroclastic
porfírico *a.* porphyritic
pórfirio *m.* porphyry
porfirita *f.* porphyry
porfirítico *a.* porphyritic
porfiroide *m.* porphyroid; *a.* porphyritic
poro *m.* pore || ***radio de la garganta del poro*** *m.* pore throat radius
porosidad *f.* porosity || ***de baja porosidad*** *a.* tight || ***porosidad efectiva*** *f.* effective porosity || ***porosidad primaria*** *f.* primary porosity || ***porosidad secundaria*** *f.* secondary porosity || ***porosidad total*** *f.* absolute porosity || ***sin porosidad*** *a.* tight
porosímetro *m.* porosimeter

poroso *a.* porous; vuggy

por peso *a.* by weight

porra *f.* sledge hammer

portada *f.* arc, arch, warp

portal *m.* rock arch, natural arch

portamira *m.* rodman

portaobjetos *m.* stage, slide

portátil *a.* portable

portezuelo *m.* pass, gap || ***portezuelo de acción eólica*** *m.* wind gap || ***portezuelo de agua*** *m.* water gap || ***portezuelo de viento*** *m.* wind gap

portillo *m.* gap, opening, pass

Portlandiano, Portlandiense *m. & a.* Portlandian

portón de glaciar *m.* glacial outlet

poruña *f.* [Arg., Ch.] gold pan

posfosa *f.* back-deep

posglacial *a.* Holocene; postglacial

posición *f.* position

posicionamiento *m.* position, positioning

poskinemático *a.* post-tectonic

posorogénico *a.* post-orogenic

pospaís *m.* hinterland

postarrecife *m. & a.* backreef

poste *a. & p.* post, after, behind

posterior *a. & p.* posterior, behind, rear

postglacial *a.* postglacial

post-litoral *a.* backshore

postor *m.* bidder

postpaís *m.* hinterland

potamógeno *a.* potamic

potasa *f. & a.* potash

potásico *a.* potassic

potasio *m.* potassium || ***potasio argón*** *m.* K–Ar, potassium–argon

potencia *f.* thickness; power; strength || ***potencia de régimen*** *f.* rated horsepower || ***potencia efectiva*** *f.* effective power || ***potencia nuclear*** *f.* nuclear power

potencial *m. & a.* potential || ***potencial de gravitación*** *m.* gravitational potential || ***potencial eléctrico*** *m.* electrical potential || ***potencial espontáneo*** *m.* spontaneous potential || ***potencial nominal*** *m.* rated potential, rated capacity

potenciométrico *a.* potentiometric

potenciómetro *m.* potentiometer

potente *a.* powerful, strong; thick

Potsdamiano, Potsdamiense *m. & a.* Potsdamian

powelita *f.* powellite

pozo *m.* well, bore, drill hole; shaft, pit || ***pozo abajo*** *a.* downhole || ***pozo abandonado*** *m.* abandoned well || ***pozo abierto*** *m.* open hole || ***pozo a bomba*** *m.* pumping well || ***pozo adyacente*** *m.* offset well || ***pozo agotado*** *m.* dead well, depleted well || ***pozo artesiano*** *m.* artesian well || ***pozo brotante*** *m.* flowing well, gusher || ***pozo cárstico*** *m.* yama || ***pozo casi agotado*** *m.* stripper well || ***pozo cerrado*** *m.* shut-in well || ***pozo contrarestante*** *m.* offset well || ***pozo compensador*** *m.* offset well || ***pozo de agua*** *m.* water well || ***pozo de alivio*** *m.* relief well || ***pozo de avancada*** *m.* step-out well || ***pozo de cateo*** *m.* wildcat well, prospect pit || ***pozo de delineación*** *m.* delineation well || ***pozo de desarrollo*** *m.* development well || ***pozo de descarga*** *m.* disposal well || ***pozo de diámetro reducido*** *m.* slim hole || ***pozo de ensayo*** *m.* test hole || ***pozo de exploración*** *m.* wildcat || ***pozo de explotación*** *m.* development well, infill well || ***pozo de explosión*** *m.* shot hole || ***pozo de límite*** *m.* offset well || ***pozo de mina*** *m.* mine shaft || ***pozo de prueba*** *m.* test well || ***pozo de ventilación*** *m.* air shaft || ***pozo descubridor*** *m.* discovery well || ***pozo desviado*** *m.* deviated well || ***pozo direccional*** *m.* directional well, slant hole || ***pozo dirigido*** *m.* directional well, slant hole || ***pozo exploratorio*** *m.* exploratory well, (perf.) wildcat || ***pozo eruptivo*** *m.* gusher || ***pozo franco*** *m.* open hole || ***pozo horizontal*** *m.* horizontal well || ***pozo improductivo*** *m.* dry hole, (perf.) duster || ***pozo inyector*** *m.* injection well || ***pozo mermado*** *m.* (perf.) stripper, stripper well || ***pozo muerto*** *m.* abandoned well || ***pozo obligatorio*** *m.* obligation well || ***pozo parado*** *m.* shut-in well || ***pozo pobre*** *m.* (perf.) stripper, stripper well || ***pozo principal*** *m.* main (min.) shaft || ***pozo productivo*** *m.* producing well || ***pozo propietario*** *m.* tight hole || ***pozo saltante*** *m.* flowing well || ***pozo seco*** *m.* dry hole || ***pozo sin entubar*** *m.* open hole || ***pozo surgente*** *m.* flowing well, (perf.) gusher

pozolano *a.* pozzolan
práctica *f.* practice, method, practice session
practicable *a.* workable, feasible
practicante *m.* trainee
practicar *vb.* to practice
práctico *a.* practical, expert, trained
pradera *f.* meadow
pradería *f.* meadowlands, grasslands
prado *m.* meadow
Praetigliense, Praetigliano *m. & a.* Praetiglian
praseodimio *m.* praseodymium
prasio *m.* green quartz, prase
prasma *f.* dark green agate
prealpino *a.* prealpine
pre-apilamiento *a.* (seis.) pre-stack
Precambriano, Precámbrico *m. & a.* Precambrian
precio *m.* price
precioso *a.* precious
precipicio *m.* precipico, cliff
precipitación *f.* precipitation
precipitado *m. & a.* precipitate; precipitated
precipitar *vb.* to precipitate
precisión *f.* precision, accuracy
preciso *a.* precise, exact
precursor *m.* preliminary seismic wave; precursor
predazita *f.* predazzite
predicción *f.* prediction
predecir *vb.* to predict
prediagénesis *f.* prediagenesis
predicha *vb.* predicted
predio minero *m.* mining property
prefosa *f.* foredeep
prehnita *f.* prehnite
premontaña *f.* foothills ·
prender fuego *vb.* to fire drill holes or shot holes
preorogénico *a.* pre-orogenic
preparación microscópica *f.* slide
preparar *vb.* to prepare, train; to prime || ***preparar minerales*** *vb.* to dress ores || ***preparar para la explotación*** *vb.* to block out ore
presa *f.* dam, weir; reservoir || ***presa de lodo*** *f.* sump hole || ***presa de tierra y escollera*** *f.* earthfill dam
presencia *f.* presence, occurrence
presentar *vb.* to present, show, display
presentarse *vb.* to occur, be found
preservación *f.* preservation
preservar *vb.* to preserve
presión *f.* pressure; oil, gas, or water drive || ***presión absoluta*** *f.* absolute pressure || ***presión a pozo abierto*** *f.* open hole pressure || ***presión a pozo cerrado*** *f.* shut-in pressure || ***presión atmosférica*** *f.* atmospheric pressure || ***presión capilar*** *f.* capillary pressure || ***presión cerrada*** *f.* shut-in pressure || ***presión confinante*** *f.* confining pressure || ***presión crítica*** *f.* critical pressure || ***presión de burbuja*** *f.* bubble point pressure || ***presión de confinamiento*** *f.* confining pressure || ***presión de fluidos en los poros*** *f.* pore fluid pressure || ***presión de flujo*** *f.* fluid pressure, flowing pressure || ***presión de fondo*** *f.* bottom-hole pressure || ***presión de fondo fluyendo*** *f.* flowing bottom-hole pressure || ***presión de formación*** *f.* formation pressure || ***presión de la tubería*** *f.* tubing pressure || ***presión de surgencia*** *f.* flowing pressure || ***presión de trabajo*** *f.* working pressure || ***presión de vapor*** *f.* vapor pressure || ***presión diferencial*** *f.* differential pressure || ***presión efectiva*** *f.* working pressure || ***presión encerrada*** *f.* shut-in pressure || ***presión en la cabeza de cañaría*** *f.* casinghead pressure || ***presión estática*** *f.* static pressure || ***presión fluctuante*** *f.* fluctuating pressure || ***presión hermética*** *f.* shut-in pressure || ***presión hidrostática*** *f.* hydrostatic pressure || ***presión litostática*** *f.* lithostatic pressure || ***presión momentánea*** *f.* pressure transient || ***presión osmótica*** *f.* osmotic pressure || ***presión parcial*** *f.* partial pressure
presupuesto *m.* budget; cost estimate
pretectónico *a.* pretectonic
pretratamiento *m.* pretreatment
preventor de erupciones *m.* blowout preventer

previsible *a.* predictable

Priaboniano *m. & a.* Priabonian

prima *f.* bonus

Primario *m. & a.* Paleozoic; primary

primavera *f.* spring

primer plano *m.* foreground

primera pendiente *f.* slope, first derivative

primer meridiano *m.* prime meridian

primeros auxilios *m.* first aid

primitivo *a.* primitive, primary, basic

primordial *a.* primordial, essential

prioridad *f.* priority

priorizar *vb.* to prioritize

prisma *f.* prism || ***prisma de acrecentamiento*** *f.* accretionary prism || ***prisma de cuarzo*** *f.* quartz wedge || ***prisma de Nicol*** *f.* Nicol prism

prismático *a.* prismatic

prismoide *m. & a.* prismoid; prismoidal

probabilidad *f.* probability

probable *a.* probable, likely

probado *a.* proven, proved

probador *m.* tester, sampler || ***probador de fondo*** *m.* bottom sampler

probar *vb.* to test

probeta *f.* test tube || ***probeta graduada*** *f.* graduate, graduated cylinder

problema *m.* problem || ***problema de los tres puntos*** *m.* three-point problem

procedencia *f.* occurrence, provenance

procedimiento *m.* process, method, procedure

procesar *vb.* to process

proceso *m.* process; processing || ***proceso de datos*** *m.* (comp.) data processing || ***procesos de la vida*** *m.* life processes || ***proceso en tiempo real*** *m.* real-time processing || ***proceso por lotes*** *m.* (comp., ref.) batch processing

proclorita *f.* ripidolite

producción *f.* production, yield || ***producción acumulada*** *f.* cumulative production || ***producción anual*** *f.* annual production || ***producción compartida*** *f.* shared production || ***producción en bruto*** *f.* gross production || ***producción final*** *f.* ultimate production || ***producción inicial*** *f.* initial production (IP) || ***producción neta*** *f.* effective output || ***producción potencial*** *f.* draw-down potential || ***producción secundaria*** *f.* secondary recovery || ***producción total*** *f.* ultimate recovery || ***producción última*** *f.* ultimate recovery

produciendo *a.* producing

producir *vb.* to produce

productividad *f.* productivity

productivo *a.* productive, producing

producto *m.* product, return, yield, output; material, commodity || ***producto de carga*** *m.* feedstock || ***producto de condensación*** *m.* condensate yield || ***producto sin valor*** *m.* waste product

productor *m.* producer

productos de cabeza *m.* tops

productos de erosión *m.* erosion products

productos químicos intermedios *m.* chemical intermediates

profesionista *m. & a.* professional

profundidad *f.* depth || ***profundidad de cementación*** *f.* setting point || ***profundidad de clavamiento*** *f.* setting point || ***profundidad del foco*** *f.* depth of focus || ***profundidad de entubación*** *f.* casing point || ***profundidad focal*** *f.* focal depth || ***profundidad total*** *f.* total depth, TD

profundización *f.* deepening

profundizar *vb.* to deepen

profundo *a.* deep || ***poco profundo*** *a.* shallow

progradación *f.* progradation

programa *m.* program, schedule; computer program || ***programa de ubicaciones*** *m.* rig schedule || ***programa piloto*** *m.* pilot program

programación *f.* programming

programador *m.* (comp.) programmer

programar *vb.* to plan

progresar con la perforación *vb.* to make hole

progreso *m.* progress, development

prolado *a.* prolate

prolífico *a.* prolific

prolongación *f.* prolongation

proluvio *m.* proluvium
promediar *vb.* to average
promedio *m.* average; middle
prominencia *f.* (geog.) knob, hill, rise
promontorio *m.* promontory, headland
promover *vb.* to promote
pronosticar *vb.* to predict, forecast
pronóstico *m.* prediction, forecast
propagación *f.* propagation; (seis.) diffusion
propaganda *f.* advertisement
propano *m.* propane
propiedad *f.* property; ownership || ***propiedad física*** *f.* physical property || ***propiedad minera*** *f.* mining claim || ***propiedad particular*** *f.* private property || ***propiedad privada*** *f.* private property
propiedades de rocas *f.* rock properties
propiedades ópticas *f.* optical properties
propiedades químicas *f.* chemical properties
propiedades térmicas *f.* thermal properties
propietario *m.* owner; *a.* proprietary
propileno *m.* propylene
propilita *f.* propylite
propilitización *f.* propylitization
proponente *m. & f.* bidder
proporción *f.* ratio, proportion || ***proporción del carbono*** *f.* carbon ratio || ***proporción de movilidad*** *f.* mobility ratio
proporcional *a.* proportional
propósito *m.* purpose, objective, intention
propuesta *f.* (com.) proposal, application, tender
propulsión *f.* propulsion, drive
prorata *f.* (com.) share, quota
proratear *vb.* to share, apportion, prorate
prorrateo *m.* proration
prórroga *f.* time extension, renewal
prorogar *vb.* to renew, extend
prospección *f.* prospecting, survey || ***prospección geofísica*** *f.* geophysical survey
prospectar *vb.* to prospect
prospecto *m.* prospect; (com.) prospectus
prospector *m.* prospector
protección contra la grabación *f.* (comp.) write protection
Proterozóico *m. & a.* Proterozoic
protoclástico *a.* protoclastic
protocuarzita *f.* protoquartzite
protogénico *a.* protogenic
protomena *f.* protore
protomilonita *f.* protomylonite
protomineral *m.* protore
protón *m.* proton
protuberancia *f.* boss, protuberance, bulge, swell
proustita *f.* proustite
provecho *m.* profit
provincia *f.* province || ***provincia fisiográfica*** *f.* physiographic province || ***provincia geológica*** *f.* geologic province || ***provincia metalogénica*** *f.* metallogenic province
provisión *f.* supply
provitriano *m.* provitrian
proximal *a.* proximal
próximo *a. & p.* near, close; proximal, nearby; next
proyección *f.* map projection || ***proyección buzamiento hacia abajo*** *f.* down-plunge projection || ***proyección cristalina*** *f.* crystal projection || ***proyección descendente*** *f.* down-plunge projection || ***proyección en el hemísfero inferior*** *f.* lower hemisphere projection || ***proyección estereográfica*** *f.* stereographic projection || ***proyección isométrica*** *f.* isometric projection || ***proyección Mercator*** *f.* Mercator projection || ***proyección ortográfica*** *f.* orthographic projection || ***proyección policónica*** *f.* polyconic projection
proyectar *vb.* to project
proyectil *m.* projectile || ***proyectil volcánico*** *m.* volcanic bomb, ejecta
proyecto *m.* project; plan, design
prueba *f.* test, proof; preprint || ***a prueba de fuego*** *a.* fireproof || ***a prueba de fugas*** *a.* leak-proof || ***prueba de incremento de presión*** *f.* pressure buildup test (PBUT) || ***prueba de independencia*** *f.* test of independence || ***prueba de ji cuadrado*** *f.* chi-

squared test || ***prueba de presión momentánea*** *f.* pressure transient test || ***prueba de razón versimilitud*** *f.* likelihood ratio test || ***prueba de significación*** *f.* significance test || ***prueba de student*** *f.* student t test || ***prueba no destructiva*** *f.* nondestructive test || ***prueba sin sesgo*** *f.* unbiased test

psamita *f.* psammite

psaturosa *f.* stephanite

psefita *f.* rudite

pseudo *a.* pseudo (véase seudo)

pseudobrecha *f.* pseudobreccia

pseudoclorita *f.* swelling chlorite, pseudochlorita

pseudocristal *m.* pseudocrystal

pseudocarst *m.* pseudokarst

pseudoestratificación *f.* pseudostratification

pseudofósil *m.* pseudofossil

pseudomórfico *a.* pseudomorphous, pseudomorphic

pseudomorfismo *m.* pseudomorphism

pseudomorfo *m.* pseudomorph

pseudoporfírico *a.* pseudoporphyritic

pseudoporfiroblástico *a.* pseudoporphyroblastic

pseudowavelita *f.* crandallite

psicrómetro *m.* psychrometer || ***psicrómetro giratorio*** *m.* sling psychrometer

psilomelano *m.* psilomelane

psitacinita *f.* mottramite

Pteroceriano, Pteroceriense *m. & a.* Pterocerian

pudinga *f.* conglomerate, [Sp.] pebble

Puerciense, Puerciano *m. & a.* Puercan

puente *m.* bridge; plugging a well; headframe; pillar of ore

puertas *f.* massive barren rock or horses in veins

puerto *m.* pass, gap

puesta en producción *f.* bringing in a well

puesto del torrero *m.* monkey board, derrickman post

pulaskita, pulasquita *f.* pulaskite

pulgada *f.* inch = 2.54 cm; pinch

pulido *a.* polished; slickensided

pulidor *m.* polisher, buffer, grinder

pulidora *f.* polisher, buffer, grinder

pulimentación *f.* polishing

pulimento *m.* polish, shine, polishing, grinding

pulir *vb.* to polish, grind, burnish

pulpería *f.* company store

pulsación *f.* pulse, pulsation; (elect.) spike

pulverización *f.* pulverization

pulverizador *m.* pulverizer

pulverizar *vb.* to pulverize

púmice *m.* pumice

pumicita *f.* pumice

pumpeliita *f.* pumpellyite

Punjabiano, Punjabiense, Punyabiano, Punyabiense, Penjabiano, Penyabiano *m. & a.* Punjabian

punta *f.* source; point; sharp end of tool || ***punta de tierra*** *f.* headland

puntal *f.* shore, strut, prop, spur; dredge; spud

puntero *m.* stonemason's chisel

punterola *f.* small pick

puntiagudo *a.* pointed

punto *m.* point || ***punto acotado*** *m.* bench mark || ***punto brillante*** *m.* bright spot || ***punto candente*** *m.* hot spot || ***punto cedente*** *m.* yield point || ***punto común profundo*** *m.* common depth point (CDP) || ***punto de bifurcación*** *m.* branch point || ***punto de burbuja, de burbujeo*** *m.* bubble point || ***punto de cesión*** *m.* yield stress || ***punto de comprobación*** *m.* control point || ***punto de concentración del esfuerzo*** *m.* stress concentration point || ***punto de condensación*** *m.* condensation point, dew point || ***punto de congelación*** *m.* freezing point || ***punto de control*** *m.* control point || ***punto de deformación*** *m.* yield point || ***punto de derrame*** *m.* spill point || ***punto de derretimiento*** *m.* melting point || ***punto de ebullición*** *m.* boiling point || ***punto de enturbamiento*** *m.* cloud point || ***punto de escurrimiento*** *m.* melting point, pour point || ***punto de fluidez*** *m.* pour point || ***punto de fragüe*** *m.* setting point || ***punto de fusión***

m. melting point || ***punto de inflamación*** *m.* flash point || ***punto de la explosión*** *m.* shot point || ***punto de marca*** *m.* survey station || ***punto de neblina*** *m.* cloud point || ***punto de niebla*** *m.* dew point || ***punto de observación*** *m.* viewpoint || ***punto de opacidad*** *m.* cloud point || ***punto de origen*** *m.* focus, focal point; origin || ***punto de referencia*** *m.* reference point || ***punto de rocío*** *m.* dew point || ***punto de tiro*** *m.* (seis.) shotpoint || ***punto de turbidez*** *m.* cloud point || ***punto de vaporización*** *m.* vaporizing point || ***punto de vista*** *m.* viewpoint || ***punto eutéctico*** *m.* eutectic point || ***punto fijo*** *m.* control point || ***punto mínimo de fluidez*** *m.* pour point || ***punto muerto*** *m.* breakeven point || ***punto nodal*** *m.* nodal point

puntoagudo *a.* taper-pointed

puntual *a.* pointed

punza caños *m.* casing perforator

punzado *a.* perforated

punzador *m.* perforator

punzar *vb.* to perforate, punch

punzón *m.* gad, punch

punzonador *m.* perforator, perforating gun

punzonar *vb.* to perforate pipe, to punch

puquio *m.* [Ch., Pe.] spring

Purbeckiano, Purbeckiense *m. & a.* Purbeckian

pureza *f.* purity

purga *f.* draining, venting, purging, blowing off

purgador *m.* blowoff valve, drain cock, petcock, bleeder

purgar *vb.* to bleed off

purificación *f.* purification

purificado *a.* purified

purificador *m.* purifier

purificar *vb.* to purify

puro *a.* pure, refined; native

putrefacción *f.* putrefaction

puzol *m.* pozzolana

puzolana *f.* puzzolana

PV (presión de vapor) *f.* steam pressure

Q

quark *m.* quark

quebrada *f.* ravine, gully, narrow pass, gorge, draw; fissure or break; stream, brook, [U.S.] arroyo, gulch || ***quebrada seca*** *f.* dry stream

quebradizado *a.* fractured; brittle, friable

quebradizo *a.* brittle, fragil

quebrado *a.* broken, fractured, rough, uneven; bankrupt, insolvent

quebrador *m.* crusher; ore sorter

quebradora *f.* crusher

quebradura *f.* fracture, break, crack, fissure; ravine

quebraja *f.* crack, split, flaw, break

quebrajar *vb.* to crack, split

quebrajoso *a.* brittle, fragil

quebrantable *a.* brittle

quebrantador *m. & a.* stone crusher; crushing

quebrantadura *f.* fracture, rupture

quebrantamiento *m.* fracture, rupture, shattering

quebrantar *vb.* to break, fracture, crack, crush, split, spall

quebrantarse *vb.* to break, crack, split

quebrantura *f.* ravine

quebrar *vb.* to break, fracture; to cob ore

quebrarse *vb.* to break, to be broken

quedar determinado *vb.* to be delineated in space

queenstownita *f.* Darwin glass

quelidonia *f.* chelidon

quema *f.* retorting of amalgam; burning, roasting ore

quemadero *m.* burning pit; retort furnace for amalgam

quemado *m.* porcellanite; *a.* porcellanous

quemador *m.* burner || ***quemador Bunsen*** *m.* Bunsen burner

quemahlita *f.* kemahlite

quemar *vb.* to roast or calcine ore; to retort amalgam; to burn

quemazón *m.* outcrop

queme *m.* retorting of amalgam

quenita *f.* kenyite

quentalenita *f.* kentallenite

queramiquita *f.* keramikite

queramohalita *f.* alunogen

querargira, querargirita *f.* cerargyrite

queratófido, queratófiro, queratoporfirita *m.* keratophyre

quermesita *f.* kermesite

querógeno *m.* kerogen

querol *m.* kerol

querosene, queroseno *m.* kerosene

queroteno *m.* kerotene

quersantita *f.* kersantite

quesilita *f.* azurite

quetenita *f.* botryogen

quevellita *f.* kvellite, ultrabasite

quiastolita *f.* chiastolite

quiebra *f.* fault, fracture, break, fissure, crack; breaking, crushing; bankruptcy, failure

quieselgur *m.* kieselguhr

quietud *f.* calm, quiescence, still stand

quijo *m.* gangue, matrix; gold or silver ore pocket

quilatador *m.* assayer

quilatar *vb.* to assay

quilate *m.* carat (205 mg)

quilaveita *f.* kilaueite

quilita *f.* kylite

quilo, quilogramo *m.* kilogram

quilolino *m.* kiloline

quilómetro *m.* kilometer

quimbalete *m.* crude ore

quimberlita *f.* kimberlite

química *f.* chemistry; chemist

químico *m. & a.* chemical

quine *m.* olivine diabase

quintal *m.* hundredweight, 100 libras (45.4 kg)

quirunavaarita *f.* magnetite

quiselgur *m.* kieselguhr

quitacostra *f.* scale remover

quitaherrumbre *f.* rust remover

quitar *vb.* to throw out, discard, remove; to extract, separate

quitina *f.* chitin

quitinoso *a.* chitinous

quitón *m.* chiton

R

rabdomancia *f.* dowsing

rabión *m.* rapids; riffle

rabiones *m.* cataracts

racimo de menas *m.* ore cluster

radar *m.* radar

radiación *f.* radiation

radiado *a.* radiate

radial *a.* radial

radián *m.* radian

radiancia *f.* radiance

radiante *a.* radiant

radiar *vb.* to radiate

radical *m. & a.* radical || ***radical libre*** *m.* free radical

radio *m.* radio; radius; radium || ***radio de drene*** *m.* drainage radius

radioactividad *f.* radioactivity

radioactivo *a.* radioactive

radiodetector *m.* radioactivity detector

radiogénico *a.* radiogenic

radiográfico *a.* x-ray || ***difracción radiográfica*** *f.* x-ray diffraction

radioisótopo *m.* radioisotope

radiolario *m.* radiolarian

radiolarita *f.* radiolarite

radiólisis *f.* radiolysis

radioluminiscencia *f.* radioluminescence

radiométrico *a.* radiometric

radiómetro *m.* radiometer

radioquímica *f.* radiochemistry

radón *m.* radon

raedura *f.* scraping

raer *vb.* to scrape, abrade

rafa *f.* subsidence

rafaelita *f.* grahamite, asphaltite

ráfaga *f.* gust

raiz *f.* root || ***raiz cuadrada*** *f.* square root || ***raiz cúbica*** *f.* cube root

raja *f.* cut; crack, split

rajadizo *a.* fissile

rajado *a.* cut, cracked, split

rajadura *f.* fissure, crack

rajar *vb.* to cut, split, crack

rajatubos *m.* casing splitter

rajo *m.* stope; shallow open cut || ***rajo abierto*** *m.* glory hole, open cut || ***rajo a cielo abierto*** *m.* open cut

ralo *a.* thin, sparse

rama *f.* branch; tributary; branch vein; main gallery

ramal *m.* secondary branch; secondary gallery; branch, branch vein

ramalearse *vb.* to split off, branch off

rambla *f.* dry ravine, dry stream bed; sandy place

ramificación *f.* division of a vein into branches; branching

ramificado *a.* branching, braided

ramificarse *vb.* to split up or divide into branches

ramita de filón *f.* branch or leader vein

ramito *m.* stringer

ramosita *f.* scoria

rampa *f.* ramp; working face; ascent; slope || ***rampa basal*** *f.* footwall ramp || ***rampa colgada*** *f.* hanging wall ramp || ***rampa frontal*** *f.* frontal ramp || ***rampa lateral*** *f.* lateral ramp

rampla *f.* ramp, shallow incline

ramplón *a.* upright

ramsayita *f.* lorenzenite

Rancholabreiense, Rancholabreano *m. & a.* Rancholabrean

rango *m.* range, scope

rankinita *f.* rankinite

ranquilita *f.* haiweeite

ranura *f.* groove, notch, slot || ***ranura de deslizamiento*** *f.* striae

ranurar *vb.* to groove

raña *f.* fanglomerate

rapasa *f.* easily worked stone

rapidez *f.* rate || ***rapidez de crecimiento*** *f.* growth rate

rápido *a.* quick

rápidos *m.* ravine, chute, rapids

ras *m.* level, even

rasa *f.* high mountain plain

rascar *vb.* to rework an abandoned mine

rasgado *a.* ripup || ***clastos rasgados*** *m.* ripup clasts

rasgar *vb.* to tear, rip

rasgo *m.* feature, trait, characteristic; tear || ***rasgo abierto*** *m.* open cut || ***rasgo sobresaliente*** *m.* landmark

rasgón volcánico *m.* volcanic rent, fissure

rasguño *m.* scratch

raspa *f.* rasp, scraper

raspador *m.* scraper

raspadura *f.* streak; abrasion; scrapings; rake

raspaje *m.* gall

raspar *vb.* to abrade; to scrape, scour

raspatubos *m.* casing scraper

rastra *f.* rake

rastreador *m.* tracer

rastro *m.* track, trace; rake

rastros *m.* shows, traces || ***rastros de gas*** *m.* gas shows || ***rastros de petróleo*** *m.* hydrocarbon shows

ratón *m.* (comp.) mouse

ratonera *f.* rathole

raudal *m.* cataract, torrent, flood

raudales *m.* rapids

raudas *f.* rapids

Rauraciano, Rauraciense *m. & a.* Rauracian

raya *f.* line; (petrol.) streak; scratch; [Méx.] day's pay

rayado *a.* striped, banded, grooved, scored

rayano *a.* bordering, adjacent

rayo *m.* ray, beam; lightning; radius || ***rayo catódico*** *m.* cathode ray || ***rayo de barrido*** *m.* heading || ***rayo gamma*** *m.* gamma ray || ***rayo x*** *m.* X-ray

razón *m.* ratio; rate; reason || ***razón de correlación*** *m.* correlation coefficient || ***razón social*** *m.* corporate name, company address

reabsorción, resorción *f.* resorption

reacción *f.* reaction, response || ***reacción al esfuerzo aplicado*** *f.* strain || ***reacción en cadena*** *f.* chain reaction || ***reacción química*** *f.* chemical reaction

reaccionar *vb.* to react

reacondicionamiento *m.* workover, reconditioning, reworking

reacondicionar *vb.* to workover, recondition, overhaul

reactancia *f.* reactance

reactante *m.* reactant

reactivación *f.* reactivation

reactividad *f.* reactivity, being a reagent

reactivar *vb.* to reactivate

reactivo *m.* reagent; *a.* reactive

reactor *m.* reactor, reaction chamber

reajustar *vb.* to overhaul

reajuste *m.* overhaul

real de fontanero *m.* measure of the rate of water flow

realce *m.* raised working face; luster, significance; enhancement || ***realce lateral*** *m.* overhand stoping

realzar *vb.* to enhance, highlight || ***realzar la galería*** *vb.* to back stope

rearmar *vb.* to reassemble

rebaja *f.* discount, rebate; deduction

rebajamiento *m.* planation

rebajar *vb.* to underhand stope; to reduce, lower, discount, abate, lessen

rebaje *m.* butting against; cut

rebanada *f.* slice of rock or mineral

rebanado *m.* slicing

rebasar *vb.* to exceed, surpass

reborde *m.* ledge; rim, border, edge; dike

rebosadero *m.* weir, dam; overflow; gossan

rebosamiento *m.* overflow

rebosar *vb.* to overflow

rebotear *vb.* to rebound

rebote *m.* rebound || ***rebote isostático*** *m.* isostatic rebound

recámara tubular *f.* lubricator

recarga *f.* recharge || ***recarga de agua subterránea*** *f.* groundwater recharge

recargar *vb.* to recharge

recargo de precio *m.* overcharge

recebo *m.* road gravel; screenings

receptáculo *m.* reservoir

receptor *m.* receiver, geophone

reciente *a.* recent

recial *m.* rapids

recibo *m.* receipt, bill, account

reciclado *a.* recycled

reciclaje *m.* recycling || ***reciclaje de gas*** *m.* cycling of gas

reciclar *m.* to recycle

recio *a.* coarse

recipiente *m.* recipient, vessel, container || ***recipiente a presión*** *m.* pressure vessel

recirculación *f.* recirculation, recycling, cycling

recirculado *a.* recirculated

reclamación *f.* reclamation; complaint, claim

reclamo *m.* claim

recobramiento *m.* recovery

recobrar *vb.* to recover

recobro *m.* recovery || ***recobro secundario*** *m.* secondary recovery

recocción *f.* annealing

recocer *vb.* to anneal; to cook

recocido *a.* annealed

recodo *m.* river bend

recogemuestras *f.* sample catcher

recoger *vb.* to collect; to gather

recogida de la velocidad *f.* velocity pick

recolección de muestras *f.* sampling

recolectar *vb.* to collect

recompensa *f.* reward

recompresión *f.* repressuring; recompression

reconcentrados *m.* concentrates
reconcentrar *vb.* to concentrate
reconocer *vb.* to survey, reconnoiter, scout, examine; to search, explore, prospect
reconociendo *m.* prospecting, exploration
reconocimiento *m.* reconnaissance, survey, exploration, prospecting; inspection; recognition; acknowledgment || ***reconocimiento aéreo*** *m.* aerial reconnaissance || ***reconocimiento a plancheta*** *m.* plane table survey || ***reconocimiento del subsuelo*** *m.* subsurface survey || ***reconocimiento geofísica*** *m.* geophysical survey
reconsequente *a.* resequent
reconstrucción *f.* reconstruction, recompletion
reconstructivo *a.* reconstructive
reconversión *f.* restructuring, reorganization, streamlining
recopilación *f.* compilation, summry
recopilador *m.* compiler, (seis.) recorder
recopilar *vb.* to compile
recorrer *vb.* to examine, survey, traverse
recorrido *m.* path, route
recorrimiento *m.* traversing
recortar *vb.* to trim
recortes *m.* cuttings
recostado *a.* recumbent
recostar *vb.* to dip
recristalización *f.* recrystallization
rectángulo *m.* rectangle
rectificación *f.* rectification; correction
rectificador *m.* stripper tower; reamer
rectificar *vb.* to straighten; rectify; to balance; to rebore, to mill
rectilíneo *a.* rectilinear
recto *a.* straight
recubrimiento *m.* covering
recubrir *vb.* to coat, cover
recuento *m.* count
recuesto *m.* inclination, dip; slope
recumbente *a.* recumbent
recuperable *a.* recoverable, reclaimable, retrievable
recuperación *f.* recovery || ***recuperación asistida*** *f.* enhanced recovery || ***recuperación final*** *f.* ultimate recovery || ***recuperación primaria*** *f.* primary recovery || ***recuperación secundaria*** *f.* secondary recovery || ***recuperación terciaria*** *f.* tertiary recovery || ***recuperación total*** *f.* ultimate recovery || ***recuperación última*** *f.* ultimate recovery
recuperador *m.* reclaimer
recuperar *vb.* to retrieve, recover
recurrente *a.* recurrent
recurso *m.* resource || ***recurso no renovable*** *m.* nonrenewable resource || ***recursos comprobados*** *m.* proven resources
rechazo *m.* offset, displacement, shift || ***rechazo horizontal*** *m.* heave || ***rechazo vertical*** *m.* throw
red *f.* net, grid; network, stockwork; crystal structure || ***red alveolar*** *f.* network || ***red atómica*** *f.* atomic structure || ***red cristalográfica*** *f.* crystal lattice || ***red de cañerías*** *f.* pipeline network, piping system || ***red de control*** *f.* control net, control grid || ***red de fallas*** *f.* fault system, fault network || ***red de fracturas*** *f.* fracture network || ***red de triangulación*** *f.* triangulation net || ***red de triángulos*** *f.* triangulation net || ***red estereograma*** *f.* stereographic net || ***red filoniana*** *f.* stockwork
redeposición *f.* redeposition
redepositado *a.* redeposited, reworked
redepositar *vb.* redeposit
redepósito *m.* redeposition
redistilación *f.* rerunning
rédito *m.* interest, income, revenue
redoma *f.* flask, beaker
redondeado *a.* rounded
redondear *vb.* to round, round off
redondez *f.* roundness
redondo *a.* round
Redoniense, Redoniano *m. & a.* Redonian
redruthita *f.* chalcocite
reducción *f.* reduction, decrease
reducido *a.* reduced
reducir *vb.* (chem.) to reduce

reductor *m.* reducing agent; choke; *a.* reducing
reelaborado *a.* reworked
reembolso *m.* reimbursement
reemplazable *a.* removable, replaceable
reemplazamiento *m.* replacement
reemplazar *vb.* to replace, displace
reemplazo *m.* replacement
reentrante *m. & a.* reentrant
reentubar *vb.* to rerun casing
referencia cero de sondeo *f.* (naut.) sounding datum reference
referencia *f.* reference
refinación *f.* refining || ***refinación a base de disolventes*** *f.* solvent extraction process || ***refinación ácida*** *f.* acid refining process || ***refinación por lotes*** *f.* batch distillation
refinado *m.* refined product; *a.* refined
refinador *m.* refiner
refinadura *f.* refining
refinamiento *m.* refining || ***refinamiento por fundición*** *m.* smelting
refinar *vb.* to refine
refinería *f.* refinery
refino *m.* refinery, refining
reflectancia *f.* reflectance
reflectividad *f.* reflectivity
reflector *m.* reflector || ***reflector lento*** *m.* slow reflector
reflejado *a.* reflected
reflejar *vb.* to reflect
reflejo *m.* reflectance; *a.* reflected
reflección, reflexión *f.* reflection
reflujo *m.* ebb tide; reflux
refogar *vb.* to retort amalgam
refoliación *f.* refoliation
reformación *f.* (ref.) reformation
reformar *vb.* to reform, improve, reorganize
refracción *f.* refraction
refractar *vb.* to refract
refractario *m.* refraction; *a.* refractory
refractividad *f.* refractivity
refractivo *m.* refractive
refractometría *f.* refractometry
refractómetro *m.* refractometer
refringencia *f.* refractivity, refringence
refringir *vb.* to refract
refuerzo *m.* reinforcement; abutment, backing, bracing
Refugiense *m. & a.* Refugian
regadura *f.* undermining; undercut
regalía *f.* royalty
regar *vb.* to irrigate
regata *f.* irrigation canal
regato *m.* pool; stream
regeneración *f.* regeneration
regenerado *a.* regenerated
régimen *m.* rate || ***régimen de entrada*** *m.* input rate || ***régimen estacionario*** *m.* steady state
región *f.* region, area || ***región de hundimiento*** *f.* graben || ***región frontera*** *f.* foreland; frontier || ***región interior*** *f.* hinterland
regional *a.* regional
registración *f.* registration
registrar *vb.* (seis.) to image; to file, to register, to log, record
registrador *m.* recorder
registradora *f.* recorder
registro *m.* chart, electric or well log, record; logging || ***registro acústico*** *m.* acoustic log || ***registro con despliegue en abanico*** *m.* fan shooting || ***registro de buzamientos*** *m.* dipmeter || ***registro de cabecera*** *m.* header record || ***registro de pozo*** *m.* well log || ***registro diario de perforación*** *m.* driller's log || ***registro eléctrico*** *m.* electric log || ***registro sísmico*** *m.* seismic recorder || ***registro vertical*** *m.* vertical shooting
regla *f.* ruler, rule, scale; pattern; survey rod || ***regla de cálculo*** *f.* slide rule || ***regla de las fases*** *f.* phase rule || ***regla de uves*** *f.* rule of v's || ***regla práctica*** *f.* rule of thumb
reglón *m.* vertical survey rod
regmático *a.* regmatic
regola *f.* undercut
regolita *f.* regolith
regolito *m.* regolith

regradación *f.* regradation

regresión *f.* regression || ***regresión lineal*** *f.* linear regression

regresivo *a.* regressive

reguero *m.* undercut || ***reguero de bloques*** *m.* boulder train

regulación *f.* regulation; adjustment, control || ***regulación estructural*** *f.* structural control

regulado *a.* controlled

regulador *m.* regulator, controller || ***regulador de caudal*** *m.* flow regulator || ***regulador de flujo*** *m.* flow regulator || ***regulador de presión*** *m.* pressure regulator

regular *a.* regular, methodical, orderly; ordenary; even; good, fair

regular *vb.* to regulate, adjust, govern

rehabilitación *f.* rehabilitation, reopening || ***rehabilitación de pozos*** *f.* workover job, reconditioning of wells

rehabilitar *vb.* to recondition, overhaul

reja *f.* grating; grate, grid

rejalgar *m.* realgar

rejilla *f.* lattice, grid; ripple

rejuvenecimiento *m.* rejuvenation

relación *f.* ratio, proportion, rate; relation; report || ***relación arenisca–lutita*** *f.* sand–shale ratio || ***relación gas–petróleo*** *f.* gas–oil ratio || ***relación hidráulica*** *f.* hydraulic ratio

relacionado *a.* related, allied

relajación *f.* relaxation

relativo *a.* relative

relaves *m.* tailings; residues from washing of ore

relevamiento *m.* topographic survey

relices de fallas *m.* slickensides

relicto *a.* relict || ***relicto de erosión*** *m.* butte, erosional remnant || ***relicto exterior*** *m.* outlier || ***relicto interior*** *m.* inlier

relieve *m.* topographic relief || ***relieve superficial*** *m.* surface texture

reliquia *f.* relict

reliz *m.* vein wall; [Méx.] landslide || ***reliz del alto*** *m.* hanging wall || ***reliz del bajo*** *m.* footwall

reluctancia *f.* reluctance || ***reluctancia magnética*** *f.* magnetic reluctance

relumbroso *a.* shining, glistening

rellenar *vb.* to fill, backfill

relleno *m.* filling, filler, fill; backfill; pad || ***relleno de cascajo*** *m.* gravel packing || ***relleno de cauce*** *m.* cut and fill, scour and fill || ***relleno de valle*** *m.* valley fill || ***relleno exterior*** *m.* overburden

remanente *m.* remnant, remainder; *a.* remaining, relict

remanso *m.* pool, backwater

remate *m.* pinchout, thinning out, wedge-out, top, abutment

remesa *f.* remittance

remisión *f.* shipment

remoción *f.* denudation || ***remoción de la tubería*** *f.* laying down pipe || ***remoción en masa*** *f.* mass wasting

remojar *vb.* to slake

remolenita *f.* atacamite

remolino *m.* whirlpool, eddy, swirl; pocket of ore

remoto *a.* remote

remover *vb.* to remove

removible *a.* removable

remuestrear *vb.* to resample

rendición *f.* yield

rendimiento *m.* yield, return, revenue, earnings, performance; output, production rate || ***rendimiento final*** *m.* ultimate yield || ***rendimiento por acre*** *m.* per acre yield || ***rendimiento total*** *m.* overall efficiency

rendija *f.* crevice

rendir *vb.* to yield, produce

renglón *m.* line; part of income; row

renio *m.* rhenium

renovable *a.* renewable

renovación *f.* renewal, replacement

renovar *vb.* to renew

renta *f.* (com.) rent; income; interest, return || ***rentas brutas*** *f.* (com.) gross income

rentabilidad *f.* return, yield, profitability

rentable *a.* profitable

rentar *vb.* to yield, produce; to rent

renuncia de una concesión *f.* relinquishment

reodid *m.* rheidity

reología *f.* rheology

reológico *a.* rheologic

reomórfico *a.* rheomorphic

reostato *m.* rheostat

reotropismo *m.* rheotropism

repacho *m.* slope

reparación *f.* repair, reparation, restoration, remediation || ***reparación mecánica de un pozo*** *f.* well repair

reparar *vb.* to fix, repair, restore, mend; to rework

reparo *m.* restoration, repair

repartamiento *m.* allotment

repartible *a.* distributable

reparto de la demanda *m.* prorationing

repasar *vb.* to review; to recondition, overhaul

repaso *m.* review; recycle, recycling; revision

repecho *m.* slope, declivity

reperforación *f.* workover

repetabilidad *f.* repeatability

repetición *f.* repetition, repeatability

repetir *vb.* to rerun

Repettiense, Repettiano *m. & a.* Repettian

(se) repite *a.* recurrent

replanteo *m.* layout

repleción *f.* silting

replegamiento *m.* refolding, convolute folding

réplica de un terremoto *f.* aftershock

reponer *vb.* to replace reserves

reposadores *m.* settling tanks

reposición *f.* reposition, replacement

represa *f.* dam, reservoir, weir

represar *vb.* to dam

representación *f.* representation || ***representación de un registro*** *f.* (seis.) play back || ***representación gráfica*** *f.* plotting, plot || ***representación visual*** *f.* display

representante *m. & f.* representative

representar *vb.* to represent; to express, symbolize

representativo *m.* representative

reprocesamiento *m.* recycling

reproducción *f.* reproduction

reproducibilidad *f.* reproducibility

reptación *f.* creep || ***reptación del suelo*** *f.* soil creep

reptante *a.* creeping

repulsión *f.* repulsion

requiebro *m.* crushing of ore

resaca *f.* backwash, surf; oxbow lake

resalte, resalto *m.* ledge, projection, salient; offset; rebound

resalto de pendiente *m.* nickpoint

resbaladero *m.* ore chute

resbalado *a.* slumped; overlapping

resbalamiento *m.* overlap, slippage, slip

resbalar *vb.* to slump, slip, slide

resbalarse *vb.* to run side-by-side

resbalo *m.* steep hill; abrupt slope

resbalón *m.* slip, fault; landslide; overlap || ***resbalón flexible*** *m.* flexural slip

resbaloso *a.* slippery

rescatador *m.* ore buyer

rescate *m.* ore purchase; purchased ores; ransom, recovery

resección del meandro *f.* meander cutoff

reseñar *vb.* to review

reservas *f.* reserves || ***reservas comprobadas estimadas*** *f.* estimated proved reserves || ***reservas probadas*** *f.* proven reserves || ***reservas recuperables*** *f.* recoverable reserves || ***reservas remanentes*** *f.* remaining reserves

reservar *vb.* to reserve

resecuente *a.* resequent

resedimentación *f.* resedimentation

resedimentado *a.* resedimented

resguardo *m.* collateral

residual *a.* residual, relict, remnant

resíduo *m.* residuum, residue, remainder, sludge || ***resíduo granítico*** *m.* granite wash || ***resíduo insoluble*** *m.* insoluble residue || ***resíduos*** *m.* tailings, fines || ***resíduos de meteorización*** *m.* weathering residue || ***resíduos sólidos*** *m.* solid wastes

resina *f.* resin, epoxy

resinoso *a.* resinous

resistencia *f.* resistance, strength || ***resistencia a la abrasión*** *f.* resistance to abrasion || ***resistencia a la compresión*** *f.* crushing strength || ***resistencia a la flexión*** *f.* bending strength || ***resistencia al aplastamiento*** *f.* compressive strength || ***resistencia a la rotura*** *f.* ultimate strength || ***resistencia a la ruptura*** *f.* rupture strength || ***resistencia a la tensión*** *f.* tensile strength || ***resistencia al cizallamiento*** *f.* shear strength || ***resistencia al corte*** *f.* shear strength || ***resistencia cohesiva*** *f.* cohesive strength || ***resistencia fundamental*** *f.* fundamental strength || ***resistencia tensora*** *f.* tensile strength

resistente *m.* resistant, strong

resistividad *f.* resistivity || ***resistividad de la tierra*** *f.* soil resistivity || ***resistividad eléctrica*** *f.* electric resistivity

resolución *f.* resolution, (seis.) aliasing; decree

resonancia *f.* resonance

resorber *vb.* to reabsorb

resorción, reabsorción *f.* resorption

respaldar *vb.* (comp.) to back up (a disk)

respaldo *m.* wall of a vein; country rock

respiración *f.* breathing

respiradero *m.* vent valve, breather, air inlet, vent || ***respiradero volcánico*** *m.* volcanic vent

respirador *m.* respirator

responsabilidad *f.* responsibility, liability || ***responsabilidad ilimitada*** *f.* unlimited liability

respuesta *f.* response

resquebrajar *vb.* to breach

resquebrajo *a.* cracked, desiccated

restablecer *vb.* to restore

restauración *f.* restoration

restaurado *a.* restored

restaurar, restaurarse *vb.* to restore

restinga *f.* bank, shoal, sandbar, beach ridge, barrier beach, bar

restita *f.* restite

restitución *f.* replacement; restoration, plotting on film or paper

restituído *a.* restored, plotted

restituidor *m.* film or paper plotter

restos *m.* residue, remains

restricciones *f.* constraints

restringido *a.* restricted

resultados *m.* results

resumen *m.* abstract, summary

resumidero *m.* sump

resumir *vb.* to sum up

resurgencia *f.* resurgence

resurgente *a.* resurgent

retallo *m.* ledge

retardo *m.* retardation, lag

retención *f.* retention

retenedor *m.* retainer

reticulado *a.* reticulated, gridded

reticular *a.* net-like

retículo *m.* lattice; cross-hairs

Retiano, Retiense *m.* & *a.* Rhaetian, Rhaetic

retinita *f.* retinite, pitchstone

retirar *vb.* to retire, remove

retiro *m.* withdrawal

retiros *m.* retirements

retorcer *vb.* to kink, to twist

retorcido *a.* kinked, twisted

retorcimiento *m.* kinking, twisting

retorta *f.* retort

retrabajar *vb.* to rework

retractable *a.* retractable

retraso *a.* retardation, delay

retratar *vb.* to photograph

retroceso *m.* recession, backwash || ***retroceso de las laderas*** *m.* slope retreat

retrocondensación *f.* retrograde condensate

retrodeformable *a.* retrodeformable

retrodiagénesis *f.* hypergenesis

retroexcavador *m.* backhoe

retrogradación *f.* retrogradation

retrógrado *a.* retrograde

retrogresión *f.* retrogression

retrogresivo *a.* retrogressive

retrolavado *m.* backwash

retrolectura *f.* backsight

retrometamorfismo *m.* retrograde metamorphism

retroplaya *f.* backshore

retrotaponamiento *m.* plugging back; backplug

retrotaponar *vb.* to plug back

retrovisual *f.* backsight

retuerta *f.* bend in a river

Retziano, Retziense *m. & a.* Retzian

reunir *vb.* to assemble, join, unite

revelador *m.* developer

revelar *vb.* to develop

revenidero *m.* altered rock

revenimiento *m.* subsidence, cave-in

reverberar *vb.* to roast

revendedor *m. & f.* dealer, retailer

revenido *a.* annealed

revenir *vb.* to anneal

reventadero *m.* rough, uneven ground; outcrop

reventar *vb.* to burst, to crack

reventón *m.* massive native silver; outcrop; blowout, explosion

reverberación *f.* roasting in a reverberatory furnace

reverberante *a.* reverberatory

reverberar *vb.* to roast or calcine ore

reversible *a.* reversible

reversión *f.* reversal

revestido *a.* lined

revestidor *m.* casing, liner

revestimiento *m.* facing, supportive cover, grout, coating

revestir *vb.* to coat, to line

revisar *vb.* to review

revocar con inyección *vb.* to mud off

revolución *f.* revolution; orogeny

revoque *m.* mud cake, filtrate

rezaga *f.* ore left in a mine, waste rock

rezago *m.* ore left in a mine, waste rock

rezumadero *m.* seep

rezumar(se) *vb.* to ooze, exude, seep, bleed; percolate

Rhodaniense, Rhodaniano *m. & a.* Rhodanian

ría *f.* creek; estuary

riacolita *f.* sanidine

riacho *m.* creek, stream, rivulet

riachuelo *m.* rivulet, creek, [U.S.] bayou

riada *f.* flood, overflow

Riazaniense, Riazaniano *m. & a.* Ryazanian

ribazo *m.* mound, hill, sloping bank

ribera *f.* river bank; seashore, strand || ***ribera posterior*** *f.* backshore

rico *a.* rich

riebeckita *f.* riebeckite

riego *m.* irrigation

riel *m.* small ingot, metal bar

riesgo *m.* risk

rifle *m.* riffle

rigidez *f.* rigidity

rígido *a.* rigid, brittle

rimaya *f.* bergschrund, crevasse

rimero *m.* heap, pile

rimpilita *f.* rimpylite

rincón *m.* inside corner, nook; open-ended basin on a mountain; recess or hollow in a cliff; re-entrant

rinkita, rinkolita *f.* mosandrite

riñon *m.* nodular ore

río *m.* river, stream || ***barranca de río*** *f.* river bank || ***boca de río*** *f.* river mouth || ***desembocadura de río*** *f.* river mouth || ***río anastomosado*** *m.* braided stream || ***río antecedente*** *m.* antecedent stream || ***río consecuente*** *m.* consequent stream || ***río en equilibrio*** *m.* graded stream || ***río inciso*** *m.* incised river || ***río intermitente*** *m.* intermittent stream || ***río obsecuente*** *m.* obsequent stream || ***río subsecuente*** *m.* subsequent stream || ***río superpuesto*** *m.* superposed stream || ***río temporario*** *m.* intermittent stream || ***río transitorio*** *m.* intermittent stream || ***serpenteo de río*** *m.* stream meandering

riodacita *f.* rhyodacite

riolita *f.* rhyolite || ***riolita alcalina*** *f.* alkali rhyolite

riostra *f.* vertical brace

ripio *m.* rubble; spalled rock; ore fines, drill cuttings

riqueza *f.* riches; richness; wealth

risco *m.* crag, cliff, escarpment, bluff

rítmicamente *adv.* rhythmically

rítmico *a.* rhythmic

rítmita *f.* rythmite

ritmo *m.* rate ‖ ***ritmo de crecimiento*** *m.* growth rate

rizadura *f.* ripple, ripple mark

rizo del meandro *m.* meander loop

robado *a.* robbed for pillars

robusto *a.* robust

roca *f.* rock ‖ ***roca aborregada*** *f.* roches moutonées ‖ ***roca alcalina*** *f.* alkali rock ‖ ***roca almacén*** *f.* reservoir rock ‖ ***roca almacenadora*** *f.* reservoir ‖ ***roca alterada*** *f.* altered rock ‖ ***roca alveolar*** *f.* honeycombed rock ‖ ***roca apartada*** *f.* outlier ‖ ***roca atrapada*** *f.* horse ‖ ***roca cataclástica*** *f.* cataclastic rock ‖ ***roca clástica*** *f.* clastic rock ‖ ***roca córnea*** *f.* hornfels ‖ ***roca cristalina*** *f.* crystalline rock, hard rock ‖ ***roca de basamento*** *f.* basement rock ‖ ***roca de caja*** *f.* wall rock ‖ ***roca de cubierta*** *f.* cover rock, caprock ‖ ***roca de espaldos*** *f.* country rock ‖ ***roca de flujo*** *f.* flowstone ‖ ***roca de fondo*** *f.* bedrock ‖ ***roca de lecho*** *f.* bedrock ‖ ***roca del lugar*** *f.* country rock ‖ ***roca de mineral de hierro*** *f.* ironstone ‖ ***roca descompuesta*** *f.* altered rock ‖ ***roca efusiva*** *f.* extrusive rock ‖ ***roca entera*** *f.* & *a.* whole rock ‖ ***roca extraña*** *f.* outlier ‖ ***roca filoniana*** *f.* dike rock ‖ ***roca firme*** *f.* bedrock ‖ ***roca foliada*** *f.* foliated rock ‖ ***roca generadora*** *f.* source rock ‖ ***roca huesped*** *f.* host rock ‖ ***roca ígnea*** *f.* igneous rock ‖ ***roca intrusiva*** *f.* intrusive rock ‖ ***roca jabón*** *f.* soapstone ‖ ***roca madre*** *f.* source rock; gangue, country rock ‖ ***roca metamórficas*** *f.* metamorphic rock ‖ ***roca petrolífera*** *f.* source rock ‖ ***roca picada*** *f.* crushed rock ‖ ***roca piroclástica*** *f.* pyroclastic rock ‖ ***roca plutónica*** *f.* plutonic rock ‖ ***roca productiva*** *f.* reservoir rock, pay interval ‖ ***roca recipiente*** *f.* reservoir rock ‖ ***roca que cubre*** *f.* caprock ‖ ***roca reservoro*** *f.* reservoir rock ‖ ***roca salina*** *f.* evaporite ‖ ***roca sedimentária*** *f.* sedimentary rock ‖ ***roca sello*** *f.* seal ‖ ***roca sobre carga*** *f.* overburden ‖ ***roca sólida*** *f.* bedrock ‖ ***roca vecina*** *f.* country rock, wall rock ‖ ***roca verde*** *f.* greenstone ‖ ***roca viva*** *f.* bedrock ‖ ***roca volcánica*** *f.* volcanic rock

rocalla *f.* rubble, talus; striated pebbles

rocalloso, rocoso *a.* rocky, stony

roce *m.* rubbing, friction, attrition

rocío *m.* dew

rodado *m.* gravel, boulder, float; *a.* round

rodar *vb.* to roll

rodear *vb.* to surround

rodesita *f.* rhodesite

rodillo *m.* (geol.) roller ‖ ***rodillo de arcilla*** *m.* shale roller ‖ ***rodillo de sal*** *m.* salt roller

rodio *m.* rhodium

rodita *f.* rhodite

rodocrosita *f.* rhodocrosite

rodonita *f.* rhodonite

roedura *f.* erosion; corrosion, attrition

roela *f.* gold or silver nugget

roeperita *f.* roepperite

roer *vb.* to erode

roeslerita *f.* roesslerite, rösslerite

rogianita *f.* roggianite

rombal, rómbico, rombiforme *a.* rhombic

romboédrico *a.* rhombohedral

romboedro *m.* rhombohedron

romboidal *a.* rhomboidal, rhombic

romboide *m.* rhomboid

romboideo *a.* rhomboid

romo *a.* blunt

rompehielos *m.* icebreaker

rompeolas *m.* breakwater, jetty

rompepiedras *m.* stone crusher

romper *vb.* to break, tear, fracture, shatter, crack, crush; to penetrate, pierce; to break

rompiente *m.* (naut.) breaker; reef, shoal

rompimiento *m.* break, fracture; breakthrough

roqueda *f.* rocky place

roquedal *m.* rocky place

roquedo *m.* crag, rock
roqueño *a.* rocky
rosario *m.* chain || *en rosario a.* en echelon
rosa de los vientos *f.* compass rose
rosca *f.* whorl, convolution
roscoelita *f.* roscoelite
roseta *f.* rosette
rosicler *m.* ruby silver, proustite
rosolita *f.* landerite
rotación *f.* rotation
rotacional *a.* rotational
rotativo, rotatorio *a.* rotary, rotational
roto *a.* broken
rotura *f.* failure, breakage || ***rotura de cizalla*** *f.* shear failure || ***rotura por tracción*** *f.* pull-apart
roza *f.* undercut
rozadora *f.* coal mining machine
rozadura *f.* abrasion; undercut; gall || ***rozaduras*** *f.* abrasions
rozamiento *m.* friction, rubbing
rozar *vb.* to undercut, cut away, wear, rub, scrape
rubelita *f.* rubellite
rubí *m.* ruby || ***rubí de bohemia*** *m.* rose quartz || ***rubí estrellado*** *m.* star ruby || ***rubí oriental*** *m.* corundum
rubidio *m.* rubidium
rubio *m.* limonite, brown hematite
rudístido *m. & a.* rudist; rudistid
rudita *f.* rudite
rugosidad *f.* rugosity, wrinkle
rugoso *a.* rugose, wrinkled, rough
rumbadero *m.* ore chute
rumbo *m.* strike, trend, direction, azimuth, bearing
rumbón *m.* weathering
runita *f.* graphic granite
Rupeliano, Rupeliense *m. & a.* Rupelian
ruptura *f.* rupture, break, parting || ***ruptura de pendiente*** *f.* break in slope
Rusciniense, Rusciniano *m. & a.* Ruscinian
ruta *f.* route, road, track
rutenio *m.* ruthenium
ruteo *m.* roadbuilding
rutilado *a.* rutilated
rutilo *m.* rutile

S

Saaliense, Saaliano *m. & a.* Saalian
sábado *m.* high grade ore
sábana *f.* sheet, cover, blanket; level plain, savannah, flat land; mine on a river bank
sabanera *f.* diorite
Sabiniense, Sabiniano *m. & a.* Sabinian
sablón *m.* coarse sand
sábulo *m.* fine gravel
sabuloso *a.* sandy, gritty
saca *f.* extraction, withdrawal
sacabarrena *f.* drill extractor
sacada de la tubería *f.* tripping out
sacador *m.* sampler
sacamuestras *f.* sampler; core barrel || ***sacamuestras de pared*** *f.* sidewall sampler
sacanúcleos *m.* core barrel
sacar *vb.* to extract, remove, take out
sacaróideo *a.* saccharoidal
sacatestigos *m.* core barrel
saco *m.* sack, bag
sacudida *f.* shaking, shock, jarring || ***sacudida sísmica*** *f.* earthquake
sacudidora *f.* shaker
sacudir *vb.* to shake
saetín *m.* chute, headframe, flume
sagenítico *a.* rutilated
Saheliano, Saheliense *m. & a.* Sahelian
Sajoniano, Sajónico, Sajoniense, Saxoniense *m. & a.* Saxonian
sajonita *f.* saxonite
Sakmariense, Sakmariano *m. & a.* Sakmarian
sal *f.* salt, halite || ***sal amarga*** *f.* epsomite || ***sal común*** *f.* halite, rock salt || ***sal de barco*** *f.* impure salt || ***sal de corazón*** *f.* pure salt || ***sal de roca*** *f.* rock salt || ***sal gema*** *f.* halite, rock salt || ***sal pedres, piedra*** *f.* rock salt

salacuifer *m.* saline aquifer

sala *f.* room || ***sala de bombas*** *f.* pump room || ***sala de conferencias*** *f.* conference room || ***sala de control*** *f.* control room || ***sala de gobierno*** *f.* control room || ***sala de proyectos*** *f.* drafting room || ***sala de trabajo*** *f.* work room

salada *f.* salt flats

saladar *m.* salt marsh; dry lake, salar, salt mine

salado *m.* salt flats; *a.* brackish, salty, saline

salar *m.* salt marsh; salt mine; salt flat, salt pan, [U.S.] playa

salario *m.* salary

salbanda *f.* selvage; thin clay between a vein and wallrock; vein wall; slickensides

saldar *vb.* (com.) to balance an account

saldo *m.* (com.) balance || ***saldo acreedor*** *m.* credit balance || ***saldo deudor*** *m.* debit balance || ***saldo líquido*** *m.* net balance

salegar *m.* salt lick

salero *m.* salt mine

salida *f.* outcrop; leak, seep, outlet, output || ***salida impresa*** *f.* readout, printout

saliente *m.* salient, projection; *a.* salient, outstanding

salina *f.* salt pan, salt mine; salt marsh

salinas *f.* salt works; salt marsh

salinero *a.* salty, salt

salinidad *f.* salinity

Saliniense, Saliniano *m. & a.* Salinan

salinización *f.* salinization

salino *a.* saline, salty

salir *vb.* to go out || ***salir de madre*** *vb.* overflow the banks

salitral *m. & a.* salt deposit, saltpeter deposit; saltpetrous

salitre *m.* saltpeter, niter

salitrera *f.* saltpeter deposit

salitroso *a.* saltpetrous

salma *f.* ton

salmoita *f.* tarbuttite

salmuera *f.* brine

salobral *a.* saline

salobre *a.* brackish, briny, salty

salobreño *a.* salty

salobridad *f.* saltiness, brackishness

salón *m.* cavern, ore chamber, chamber

Salopiense, Salopiano *m. & a.* Salopian

salpicadura *f.* spray

salsa *f.* mud volcano

salsedumbre *f.* salinity

saltación *f.* saltation

salteado *a.* staggered

salteria *f.* colluvium

saltierra *f.* impure salt from dry lake beds

salto *m.* waterfall; fault throw; fault || ***salto de falla*** *m.* shift, offset, displacement || ***salto horizontal*** *m.* shift, offset || ***salto vertical*** *m.* throw

salud *f.* health

salvamento *m.* salvage

samario *m.* samarium

sammita *f.* psammite

sancocho *m.* [Vz.] drilling rig

saneado *a.* drained

saneamiento *m.* drainage

sanear *vb.* to drain, dry out

Sangamoniense, Sangamoniano *m. & a.* Sangamonian

sangría *f.* tunnel that joins two shafts

sanguinaria *f.* bloodstone

sanidina *f.* sanidine

Sannoisiense, Sannoisiano *m. & a.* Sannoisian

Sannoniano, Sannoniense *m. & a.* Sannonian

sano *a.* whole, virgin, unworked ground

Santoniano, Santoniense *m. & a.* Santonian

sanukitoide *m.* orthoandesite

sapantracita *f.* sapanthracite

saperita *f.* sapperite

saponificación *f.* saponification

saponificar *vb.* to saponify

saprógeno *a.* saprogenic

saprolita *f.* saprolite

saprolítico *a.* saprolitic

saprolito *m.* mantle rock, regolith

sapropel *m.* sapropel

sapropélico *a.* sapropelic

sapropelita *f.* sapropel

saprozóico *a.* saprozoic

sarabiado *a.* gneissic

Sardiense, Sardiano *m. & a.* Sardic

sardio, sardo *m.* sard

sardónica *f.* sard

sardónice, sardonio, sardónique, sardónix *m.* sardonyx

Sarmatiano, Sarmatiense *m. & a.* Sarmatian

sarro *m.* scale, incrustation

sarroso *a.* scaly, incrusted

sarta *f.* line, series, string || ***sarta de tuberías*** *f.* casing string

sassolina *f.* sassolite

satélite *m.* satelite

satelita *f.* fibrous serpentine

satrolito *m.* mantle rock, regolith

saturación *f.* saturation || ***nivel de la saturación irreducible*** *m.* level of irreducible water saturation

saturado *a.* saturated || ***bajo saturado*** *a.* undersaturated || ***super saturado*** *a.* supersaturated

saturar *vb.* to saturate

Saucesiense, Saucesiano *m. & a.* Saucesian

saukovita *f.* metacinnabar

saussuritización *f.* saussuritization

Saxoniano, Saxoniense, Sajoniense, Sajoniano, Sajónico *m. & a.* Saxonian

scachita *f.* scacchite

scorodita *f.* scorodite

Scytiano, Scytiense *m. & a.* Scythian

schapbachita *f.* matildite

scheelita *f.* scheelite

scheibeita *f.* phoenicochroite

schlieren *m.* schlieren

schoenita *f.* picromerite

schroeckingerita *f.* schroeckingerite

secadal *m.* [U.K.] moorland

secador *m.* dryer

secano *m.* sandbar

secar *vb.* to dry

secas *f.* refined silver

sección *f.* section, cross section || ***sección aumentada*** *f.* expanded section || ***sección axial*** *f.* axial section || ***sección balanceada*** *f.* balanced section || ***sección condensada*** *f.* condensed section || ***sección delgada*** *f.* thin section || ***sección geológica*** *f.* geologic column || ***sección perdida*** *f.* missing section || ***sección polida*** *f.* polished section || ***sección pulimentada*** *f.* polished section || ***sección restaurada*** *f.* restored section || ***sección típica*** *f.* type section || ***sección transversal*** *f.* cross section

seco *a.* dry, barren; dry well

secreción *f.* secretion

sector *m.* section, part, sector; area, zone

Secuaniano, Secuaniense *m. & a.* Sequanian

secuencia *f.* sequence || ***secuencia de deformación*** *f.* deformation sequence || ***secuencia estratigráfica*** *f.* stratigraphic sequence || ***fuera de secuencia*** *f.* out of sequence

secundario *a.* secondary; supergene

sedentario *a.* sedentary

sedimentación *f.* sedimentation

sedimentar(se) *vb.* to settle, to deposit

sedimentario *a.* sedimentary

sedimento *m.* sediment || ***sedimentos clásticos*** *m.* clastic sediments || ***sedimentos consolidados*** *m.* consolidated sediments || ***sedimentos eólicos*** *m.* aeolian sediments || ***sedimentos marinos*** *m.* marine sediments || ***sedimentos subyacentes*** *m.* underlying sediments || ***sedimentos suprayacentes*** *m.* overlying sediments

sedimentología *f.* sedimentology

segmento *m.* segment

segmentación *f.* segmentation

segmentado *a.* segmented

segregación *f.* segregation

segregar *vb.* to segregate

seguir *vb.* to follow

(dar) seguimiento *vb.* to follow up

seguridad *f.* security; safety || ***seguridad de funcionamiento*** *f.* reliability

seguro *m.* (com.) insurance, security, retainer

selección *f.* (seis.) pick

seleccionado *m.* sorting; *a.* sorted, selected || ***bién seleccionado*** *a.* well-sorted || ***pobremente seleccionado*** *adv.* poorly sorted

seleccionar *vb.* to select, sort

selector de campo *m.* range selector || ***selector de campo de modo de funcionamiento*** *m.* mode switch

selenio *m.* selenium

selenita *f.* selenite

selva *f.* jungle; rainforest; woods, forest

sellar *vb.* to seal

sello *m.* seal

semiantracita *f.* semianthracite

semiárido *a.* semiarid

semicontemporáneo *a.* penecontemporaneous

semicristalino *a.* semicrystalline

semisumergible *a.* semisubmersible drill rig

semivariograma *m.* semivariogram

senda *f.* trail, track

sendero *m.* path, trail

senectud *f.* old age

senil *a.* old

sensible *a.* sensitive

seno *m.* trough

Senoniano, Senoniense *m. & a.* Senonian

sensibilidad *f.* sensitivity; sensibility

sensor *m.* sensor || ***sensor de presión*** *m.* pressure sensor || ***sensor electromagnético*** *m.* electromagnetic sensor || ***sensor solar*** *m.* sun sensor

sensores remotos *m.* remote sensing

sentado *a.* sessile

sentido *m.* sense, meaning

señal *f.* sign, signal || ***señal de artifacto*** *f.* aliasing || ***señal de olas*** *f.* ripple marks || ***señal electromagnética*** *f.* electromagnetic signal

señalado *a.* marked

señalar *vb.* to mark, to denote

señalizador *m.* (survey) flag; signpost

separación *f.* separation, spacing; segregation || ***separación del asfalto*** *f.* deasphalting || ***separación estratigráfica*** *f.* stratigraphic separation, stratigraphic throw

separado *a.* separate; reprint

separador *m.* separator, spacer, spreader, trap || ***separador ciclonal*** *m.* cyclone separator || ***separador de agua*** *m.* steam scrubber, steam trap

separar *vb.* to separate

sepelio *m.* burial

sepia *f.* sepia

sepiolita *f.* sepiolite

septa *f.* septum

septentrional *a.* northern

septo *m.* septum

sepultado *a.* buried

sepultar *vb.* to bury

sepultura *f.* burial

sequía *f.* drought

sericita *f.* sericite

serie *f.* series || ***serie alcalina-cálcica*** *f.* calc-alkali series || ***serie homóloga*** *f.* homologous series

serpentear *vb.* to meander

serpenteo *m.* meandering

serpentina *f.* serpentine

serpentinita *f.* serpentinite

serración *f.* serration

serrado *a.* serrate

serranía *f.* mountainous country, mountain range, ridge

serrano *a.* hilly

servicio *m.* service, duty

Servicio Geológico *m.* Geological Survey

servidor *m.* server

servidumbre de paso *f.* right-of-way

servir de anfitrión *vb.* (min.) to host

sesgado *m.* skew; *a.* skewed, askew, biased

sesgo *m.* obliqueness, slope, skew; *a.* biased

sésil *a.* sessile

seudobrecha *f.* pseudobreccia

seudocarst *m.* pseudokarst

seudoclorita *f.* pseudochlorite

seudoestratificación *f.* pseudostratification

seudofósil *m.* pseudofossil

seudomórfico *a.* pseudomorphic, pseudomorphous

seudomorfismo *m.* pseudomorphism
seudomorfo *m.* pseudomorph
seudoporfirítico *a.* pseudoporphyritic
seudoporfiroblástico *a.* pseudoporphyroblastic
seudorégimen estacionario *m.* pseudo-steady state
seudotaquilita *f.* pseudotachylite
seudovitrinita *f.* pseudovitrinite
seudowavelita *f.* crandallite
shonkinita *f.* shonkinite
sial *m.* sial, upper crust
siálico *a.* sialic
sialita *f.* clay mineral sialite
Siciliano, Siciliense *m. & a.* Sicilian
siderita, siderosa *f.* siderite
siderugia *f.* iron and steel industry
siderúrgico *a.* pertaining to iron and steel
Siegeniano, Siegeniense *m. & a.* Siegenian
sienita *f.* syenite || ***sienita alcalina*** *f.* alkali syenite
sienodiorita *f.* syenodiorite
sierra *f.* sierra, mountains, mountain range
sierranita *f.* sierranite
siferna *f.* siferna, sima
sifón *m.* siphon
sigmoideo *a.* sigmoid
signatorio *m.* signatory
significante *a.* significant
signo óptico *m.* optic sign
sil *m.* yellow ochre
silcreta *f.* silcrete
Silesiense, Silesiano *m. & a.* Silesian
sílex *m.* flint, silex, chert; silica
silicado *a.* silicate
silicato *m.* silicate || ***silicato aislado*** *m.* nesosilicate || ***silicato de aluminio*** *m.* aluminosilicate || ***silicato en andamiaje*** *m.* tectosilicate || ***silicato en anillo*** *m.* cyclosilicate || ***silicato en cadena*** *m.* inosilicate || ***silicato en hoja*** *m.* phyllosilicate
sílice *f.* silica
silíceo *a.* siliceous
siliciclástico, silicoclastico *m.* siliciclastic
silícico *a.* silicic
silicificación *f.* silicification
silicificado *a.* silicified
silicificar *vb.* to silicify
silicio *m.* silicon
silicoso *a.* siliceous
silicotélico *a.* telehemic
silificación *f.* silicification
silimanita *f.* sillimanite
silo *m.* sill
silomelana *f.* psilomelane
Siluriano, Silúrico *m. & a.* Silurian
silvita, silvina *f.* sylvite
silvanita *f.* sylvanite
silla *f.* (geog., geol.) saddle
sillar *m.* ash-flow tuff
sillería *f.* ashlar stone
sima *f.* sima, basaltic layer; cave, cavern, shaft, chimney, grotto, chasm, abyss, pit, fissure
simático *a.* simatic
simbiosis *f.* symbiosis
símbolo *m.* symbol
simetría *f.* symmetry
simétrico *a.* symmetric
simictita *f.* diamictite
simplectita *f.* symplectite
simulación *f.* simulation
simular *vb.* to simulate
simulado *a.* simulated
simulador *m.* simulator
simultaneidad *f.* synchronism
sinantéctico *a.* synantectic
sincinemático *a.* synkinematic
sinclinal *m. & a.* syncline; synclinal || ***sinclinal compuesto*** *m.* synclinorium || ***sinclinal marginal*** *m.* rim syncline
sinclinorio *m.* synclinorium
sin compactar *a.* uncompacted
sincroneidad *f.* synchroneity
sincronía *f.* simultaneity, synchroneity
sincrónico, síncrono *a.* synchronous
sincronismo *m.* synchronism
sincronización *f.* synchronization

sincronizadamente *adv.* simultaneously
sincronizador *m.* timer
sincronizar *vb.* to synchronize
sinchisita *f.* synchysite
sindeposición *f.* syndeposition
sindeposicional *a.* syndepositional
sindicación *f.* unionization
sindical *a.* union, trade union
sindicalismo *m.* trade unionism
sindicalizar, sindicar *vb.* to unionize
sindicato *m.* trade union
sindicatura *f.* trade union
sineclisa *f.* synclise
Sinemuriano, Sinemuriense *m. & a.* Sinemurian
sinéresis *f.* syneresis
sinestral *a.* right-lateral, sinistral
sinforma *f.* synform
sinformal *a.* synformal
singénesis *f.* syngenesis
singenético, singénico *a.* syngenetic
singonia cristalográfica *f.* crystal system
Siniense, Siniano *m. & a.* Sinian, Siphean
siniestro *a.* sinistral, right-lateral
sinkinemático *a.* synkinematic
sinóptico *a.* synoptic
sinorogénico *a.* synorogenic
sin porosidad *a.* tight, without porosity
sinsedimentario *a.* synsedimentary
sintectónico *a.* syntectonic
sinter *f.* sinter
síntesis *m.* synthesis
sintético *a.* synthetic
sintetizar *vb.* to synthesize
sinuoso *a.* sinuous, winding, wavy
sinuosidad *a.* sinuosity, wavyness
sirca *f.* [Ch.] vein
sismicidad *f.* seismicity
sísmica *f.* seismic || ***sísmica de dos dimensiones*** *f.* 2D seismic || ***sísmica de tres dimensiones*** *f.* 3D seismic
sísmico *a.* seismic
sismita *f.* seismite
sismo *m.* seism; earthquake || ***sismo final*** *m.* aftershock || ***sismo preliminar*** *m.* foreshock || ***sismo precursor*** *m.* foreshock
sismoescopo *m.* seismoscope
sismógrafo *m.* seismograph || ***sismógrafo de reflexión*** *m.* reflection seismograph || ***sismógrafo de refracción*** *m.* refraction seismograph || ***sismógrafo sintético*** *m.* synthetic seismograph
sismograma *m.* seismogram
sismología *f.* seismology
sismológico *a.* seismological, seismic
sismólogo *m.* seismologist
sismometría *f.* seismometry
sismómetro *m.* seismometer
sismondina *f.* sismondine
sistema *m.* system, process || ***sistema cristalino*** *m.* crystal system || ***sistema de circulación*** *m.* circulation system || ***sistema de coordenadas*** *m.* coordinate system || ***sistema de diaclasas*** *m.* joint system || ***sistema de fallas*** *m.* fault system || ***sistema de información geográfica (SIG)*** *m.* geographic information system (GIS) || ***sistema de oleoductos recolectores*** *m.* loop-gathering system || ***sistema de petroleo*** *m.* petroleum system || ***sistema de posicionamiento global*** *m.* global positioning system (GPS) || ***sistema de pozo inyector y cuatro productores*** *m.* five spot flooding system || ***sistema fluvial*** *m.* river system || ***sistema hidrográfico*** *m.* river system || ***sistema MTU*** *m.* UTM coordinate system || ***sistema petrolero*** *m.* petroleum system
sistemático *a.* systematic
sitaparita *f.* bixbyite
sitio *m.* site, location, place || ***sitio de ganado mayor*** *m.* 1 square league, or 4338.11 acres || ***sitio de ganado minor*** *m.* 3333.33 acres
situación *f.* location, site
situados *m.* silver shipments
situar *vb.* to locate
Skiddaviano, Skiddaviense *m. & a.* Skiddavian
Skytiense, Skytiano *m. & a.* Skythian
smirnovita *f.* thorutite
Smithiense, Smithiano *m. & a.* Smithian

smithsonita *f.* smithsonite

sobrante *a.* surplus, extra

sobre *p.* over || ***sobrecalentado*** *a.* superheated || ***sobrecapa*** *f.* overburden || ***sobrecarga*** *f.* overcharge, overload; surcharge || ***sobrecolgante*** *m.* overhang || ***sobrecorrido*** *m.* overthrust || ***sobrecorrimiento*** *m.* overthrust || ***sobrecubrimiento*** *m.* overlap || ***sobreempinado*** *a.* oversteepened || ***sobreempuje*** *m.* overthrust || ***sobreescurrimiento*** *m.* overthrust || ***sobreguía*** *f.* intermediate level of a mine || ***sobreimponer(se)*** *vb.* to superimpose || ***sobreimposición*** *f.* superimposition, superposing || ***sobreimpresión*** *f.* overprint || ***sobreimprimir*** *vb.* to overprint || ***sobremedido*** *a.* oversize || ***sobrepeso*** *a.* overweight || ***sobreponer*** *vb.* to overlap, to overlay || ***sobreposición*** *f.* overlap || ***sobreproducción*** *f.* overproduction || ***sobrepuesta en película*** *f.* film overlay || ***sobrepuesto*** *a.* superimposed, overlapping || ***sobresabana*** *f.* alluvium || ***sobresaliente*** *a.* overhanging, projecting; outstanding || ***sobresalir*** *vb.* to overhang, project, jut out || ***sobresaturación*** *f.* oversaturation || ***sobresaturado*** *a.* supersaturated || ***sobresaturar*** *vb.* to oversaturate || ***sobrestante*** *m.* foreman, site manager, boss, supervisor || ***sobreyacente*** *a.* overlying, draping || ***sobreyacer*** *vb.* to drape, overlie

sobreprecio *m.* premium

sobrestante *m.* foreman, boss, supervisor

sócalo *m.* basement

socava, socavación *f.* undermining

socavamiento *m.* undermining, washout

socavar *vb.* to undermine, undercut, dig under

socavón *m.* excavation; gallery, tunnel, adit; subsidence; cave || ***socavón á hilo de veta*** *m.* drift || ***socavón crucero*** *m.* crosscut

socavonero *m.* tunnel driver

socaz *m.* tailrace

sociedad anónima (S.A.) *f.* corporation (Inc.)

socio *m. & f.* partner; companion, member, associate

soda *f.* soda || ***soda cáustica*** *f.* caustic soda

sodaclaso *m.* albite

sodalita *f.* sodalite

sódico *a.* sodic

sodio *m.* sodium

sofocado *a.* (min.) abandoned and/or flooded

sofocar *vb.* to extinguish, asphyxiate

sofocarse *vb.* to flood

software *m.* software

sol *m.* sun

solapado *a.* overlapped, overlapping

solapante *a.* overlapping

solapar *vb.* to overlap

solapo *m.* overlap

solar *m.* lot, site, plot; *a.* solar, sun

soldado *a.* welded

soldadura *f.* coalescence, fusion, weld || ***soldadura después de sal*** *f.* salt weld

soldar *vb.* to weld

soldarse *vb.* to coalesce

solera *f.* sill; [Méx.] flagstone

soleta *f.* clay or shale parting in coal

solevamiento, solevantamiento *m.* uplift

solevantado *a.* upthrown

solfatara *f.* solfatara

solfatárico *a.* solfataric

solidificación *f.* solidification

solicitación *f.* request; (mech.) stress

solicitar *vb.* to apply; to request, seek, attract, pursue

solicitud *f.* diligence, care; (com.) application, request

solidez *a.* solidity

solidificación *f.* solidification

solidificar *vb.* to solidify

sólido *a.* solid

solidus *m.* solidus

soliflucción, solifluxión *f.* solifluction

solsticio *m.* solstice || ***solsticio de invierno*** *m.* winter solstice || ***solsticio de verano*** *m.* summer solstice

soltar *vb.* to release, loosen, let go

solubilidad *f.* solubility

soluble *a.* soluble

solución *f.* solution || ***solución coloidal*** *f.* colloidal solution || ***solución de sólidos*** *f.* solid solution || ***solución diluída*** *f.* dilute solution || ***solución hidrotermal*** *f.* hydrothermal solution || ***solución molar*** *f.* molar solution || ***solución no única*** *f.* non-unique solution || ***solución saturada*** *f.* saturated solution || ***solución sobresaturada*** *f.* supersaturated solution || ***solución sólida*** *f.* solid solution

soluto *m.* solute

solvación *f.* solvation

solvente *m.* solvent

sombra *f.* shadow, shade || ***sombra de deformación*** *f.* strain shadow

sombreado *m.* shading, hachure; *a.* hachured, shaded

sombrear *vb.* to shade

sombrero *m.* hat || ***sombrero de alteración*** *m.* gossan || ***sombrero de hierro*** *m.* gossan || ***sombrero protector*** *m.* safety hat, hard hat

somerizando *m.* shoaling

somerización *f.* shoaling

somero *m.* (naut.) shoal; *a.* shallow

sonar *m.* sonar

sonda *f.* (elect.) sonde; (perf.) drill, probe; (geog.) sound, strait

sondador *m.* echo sounder

sondaje *m.* boring, drilling; sounding

sondar, sondear *vb.* to drill, bore; to take soundings

sondeo *m.* drilling, boring, borehole; sounding || ***sondeo acústico*** *m.* echo-sounding, acoustic sounding

sónico *a.* sonic

sonógrafo *m.* sonograph

soplado *m.* deep fissure; open vein, pocket || ***soplado por el viento*** *a.* windblown

soplador *m.* blower

sopladura *f.* flaw

soplar *vb.* to blow, blow away, blow off

soplete *m.* blowpipe, blowtorch

sopletear con arena *vb.* to sandblast

soplo *m.* blast || ***soplo inverso*** *m.* blow-back

soportar *vb.* to endure

soporte *m.* support, stand || ***soporte lógico*** *m.* software

sorbencia *f.* seepage

sordawalita *f.* tachylyte

sordina *f.* sound damper

soroche *m.* argentiferous galena

soroque *m.* gangue, matrix

sorosilicato *m.* sorosilicate

sosa *f.* soda ash, soda; natrona

sostén *m.* support, stand

sótano *m.* [Mex.] vertical cave, karst chimney; basement

sotavento *a.* leeward, lee

soterrado *a.* buried, underground, blind

soterramiento *m.* burial

soterraño *a.* subterranean, underground

soterrar *vb.* to bury

sotlanque *m.* chalcopyrite

sotominero *m.* mine subforeman

soyota, soyote *f.* mineralized vug or cavity

Sparnaciense, Sparnaciano *m. & a.* Sparnacian

Spatiense, Spatiano *m. & a.* Spathian

spencita *f.* tridymite Y

spessartina *f.* spessartite

spilita *f.* spilite

Staffordiano, Staffordiense *m. & a.* Staffordian

Stampiense, Stampiano *m. & a.* Stampian

standar(d) *a.* standard

status *m.* status

Stefaniano, Stefaniense *m. & a.* Stephanian

stephanita *f.* stephanite

sterrettita *f.* kolbeckite

stipoverita *f.* stishovite

stratificacion *f.* stratification

stress *m.* stress

stromeyerita *f.* stromeyerite

Struniano, Struniense *m. & a.* Strunian

Styriano, Styriense *m. & a.* Styrian

suampo *m.* swamp

suave *a.* gentle, smooth

suaveando *m.* swabbing

suavear *vb.* to swab

suavedad *f.* softness, smoothness
suavizado *a.* smoothed
suavizador de agua *m.* water softener
subacuático, subácueo *a.* underwater, subaqueous
subaéreo *a.* subaerial
subafloramiento *m.* subcrop
subágueo *a.* subaqueous
subalcálico *a.* subalkalic
subalcalino *a.* subalkaline
subaluvial *a.* suballuvial
subalpino *a.* subalpine
subaluminoso *a.* subaluminous
subangular *a.* subangular
subarriendo *m.* sublease
sub-bentonita *f.* metabentonite
sub-bituminoso *a.* sub-bituminous
subclase *m.* subclass
subcontinente *m.* subcontinent
subcortical *a.* subcrustal
subcuenca *f.* sub-basin
subducción *f.* subduction
subestrato *m.* substrate, substratum; sole
subestructura *f.* substructure
subetapa *f.* substage
subgrafíta *f.* metagraphite
subgrauwacke *m.* subgreywacke
subgrupo *m.* subgroup
subhedrón *m.* subhedron
subhedral *a.* subhedral
Subherciniano, Subherciniense *m. & a.* Subhercinian
subhorizonte *m.* substratum
subida *f.* ascent; upgrade
subidiomórfico *a.* hypidiomorphic, subautomorphic
subir *vb.* to arise, to rise, to go up
subyacente *a.* subjacent, beneath
sublevación *f.* uplift
sublimación *f.* sublimation
sublimado *m.* sublimate || ***sublimado amarillo*** *m.* orpiment || ***sublimado rojo*** *m.* realgar
sublimar *vb.* to sublimate
sublitarenita *f.* sublitharenite
sublitoral *a.* neritic, sublittoral, foreshore
submaduro *a.* submature
submarea *a.* subtidal
submarino *a.* submarine, subaqueous
submaturado *a.* submature
submetálico *a.* submetallic
submuestreo *m.* subsampling
subofítico *a.* subophitic
subperíodo *m.* subperiod
subpiso *m.* substage
subpresión *f.* underpressure
subproducto *m.* byproduct
subrampa *f.* subincline
subsaturación *f.* undersaturation
subsaturado *a.* undersaturated
subsaturar *vb.* to undersaturate
subsecuente *a.* subsequent, next
subsidencia *f.* subsidence
subsidiar *vb.* to subsidize
subsidiario *a.* subsidiary
subsidio *m.* subsidy
substancia *f.* substance (véase sustancia)
substitución *f.* substitution || ***substitución metasomática*** *f.* metasomatism
substituto *m.* substitute
substrato *m.* substrate
subsuelo *m.* subsoil; subsurface
subterráneo *a.* underground, subterranean
subterrano *m.* underground
subvolcán *m.* subvolcano
subvolcánico *a.* subvolcanic
subyacente *a.* underlying
subyacer *vb.* to underlie
succinita *f.* succinite
succino *m.* amber
succión *f.* suction
sucesión *f.* succession, series
suciedad *f.* dirt
sucio *a.* dirty, sloppy
sucursal *f.* branch office
sud *m.* south
Sudamérica *f.* South America

Sudético *m. & a.* Sudetic

suela *f.* sole

sueldo *m.* salary, wages

suelo *m.* soil, land; ground; surface || ***suelo alcalino*** *m.* alkaline soil || ***suelo autóctono*** *m.* residual soil || ***suelo franco*** *m.* loamy soil || ***suelo impermeable*** *m.* impervious soil, impermeable soil || ***suelo laterítico*** *m.* laterite || ***suelo podsol*** *m.* podsol || ***suelo residual*** *m.* residual soil || ***suelo turboso*** *m.* bog soil

suelto *a.* loose, unconsolidated

sujeta a *vb.* to be subjected to, to undergo

sujetar *vb.* to subject to

sulfatera *f.* sulphate deposit

sulfatita *f.* sulphatite

sulfato *m.* sulfate

sulfidización *f.* sulfurization

sulfito *m.* sulfite

sulfonato *m.* sulfonate

sulfosale *m.* sulfosalt

sulfuración *f.* sulfuration

sulfurado *a.* sulfuretted

sulfúrico *a.* sulfuric

sulfurización *f.* sulfurization

sulfuro *m.* sulfur; sulfide, sulfide ore || ***sulfuro de cadmio*** *m.* greenockite || ***sulfuro de cobre*** *m.* chalcocite || ***sulfuro de hidrógeno*** *m.* hydrogen sulfide || ***sulfuro de plata*** *m.* argentite || ***sulfuro de plomo*** *m.* galena || ***sulfuro rojo*** *m.* realgar

sulfuros en masa *m.* massive sulfides

sulfuroso *a.* sulfurous

suma *f.* sum, total

sumario *m.* abstract

sumergencia *f.* submergence

sumergible *m.* submersible rig; *a.* submersible

sumergido *a.* submerged, underwater, drowned

sumergimiento *m.* submergence

sumergir *vb.* to submerge, drown

sumergirse *vb.* to plunge

sumersión *f.* submersion, immersion

sumidero *m.* sinkhole, sump, catch basin

suministrar *vb.* to supply

suministro *m.* supply

sumir *vb.* to submerge

sundtita *f.* andorite

superantracita *f.* superanthracite

superávit *m.* surplus

superetapa *f.* superstage

superficial *a.* superficial, surficial

superficie *f.* surface; area, extent; acreage || ***de superficie*** *f.* surface || ***en superficie*** *a.* at the surface || ***superficie axial*** *f.* axial surface, axial plane || ***superficie de deslizamiento*** *f.* translation plane, slickenside || ***superficie de diaclasa*** *f.* joint surface || ***superficie de equilibrio*** *f.* isostatic equilibrium surface || ***superficie de erosión*** *f.* erosion surface || ***superficie de estratificación*** *f.* bedding plane || ***superficie de fracturación*** *f.* fault plane, fracture || ***superficie de inundación*** *f.* flooding surface || ***superficie de inundacion máxima*** *f.* maximum flooding surface || ***superficie equipotencial*** *f.* equipotential surface || ***superficie freática*** *f.* phreatic surface, water table || ***superficie lístrica*** *f.* listric surface || ***superficie piezométrica*** *f.* piezometric surface || ***superficies S*** *f.* S planes, S surfaces || ***superficie terrestre*** *f.* earth's surface || ***superficie topográfica*** *f.* topographic surface

supergás *m.* liquified petroleum gas (LPG)

supergénico, supergeno *a.* supergene

supergrupo *m.* supergroup

superintendente *m.* superintendent, supervisor

superior *a.* upper; better

superperiodicidad *f.* superperiodicity

superplasticidad *f.* superplasticity

superponer *vb.* to superimpose, superpose

superposición *f.* superposition; overlap, onlap || ***superposición transgresiva*** *f.* onlap

superpresión *f.* overpressure

superpuesto *m.* overlap; *a.* superimposed, supplementary

supersaturación *f.* supersaturation

supersaturado *a.* supersaturated

supersaturar *vb.* to supersaturate

superserie *f.* superseries
supersistema *m.* supersystem
supervisar *vb.* to monitor, supervise
suplementario *a.* subsidiary
supracortical *a.* supracrustal
supralitoral *a.* supratidal
suprayacente *a.* overlying
suprayacer *vb.* to overlie
sur *m. & a.* south
surcado *a.* grooved, furrowed
surcamiento *m.* furrowing
surco *m.* groove, furrow; sill || ***surco de alud*** *m.* avalanche scar || ***surco de erosión*** *m.* erosion groove
surgencia *f.* upwelling; resurgence, flow || ***surgencia imprevista de presión*** *f.* pressure kick || ***surgencia libre*** *f.* open flow || ***surgencia por inyección de gas*** *f.* gas lift || ***surgencia regulada*** *f.* controlled flow, restricted flow
surgente *a.* flowing, upwelling
surgiendo *a.* flowing
surgir *vb.* to flow || ***surgir sin control*** *vb.* to flow wild
surtido *m.* sorting; *a.* sorted
surtidor *m.* filling station || ***surtidor de gasolina*** *m.* gasoline pump
susceptibilidad *f.* susceptibility
susexita *f.* sussexite
suspender *vb.* to suspend, cut off, stop
suspensión *f.* suspension
suspendido *a.* suspended, abandoned
sustancia *f.* substance (véase substancia) || ***sustancia radioactiva*** *f.* radioactive material
sustitución *f.* substitution
sutura *f.* suture, suture zone
suturado *a.* sutured

T

tabaire *m.* soft limestone used for bricks
Tabianiano, Tabianiense *m. & a.* Tabianian
tabilla *f.* cleat, slab
tabique *m.* partition, baffle, brick
tabla *f.* rock slab; table, list; side of a vein
tablazo *m.* mesa; flat stretch of a river
tablero *m.* board, panel || ***tablero de comprobación*** *m.* exploration permit map || ***tablero de instrumentos*** *m.* instrument panel
tabulación *f.* list, tabulation
tabular *a.* tabular; *vb.* to tabulate
tacana *f.* silver ore
tacano *m.* silver ore
taco *m.* wad, plug, chock; pipe cleaning pig
Tacónico *a.* Taconic
Taconiano, Taconiense *m. & a.* Taconian
tactita *f.* tactite, skarn
tafrogénesis *f.* taphrogenesis
tafrogenía *f.* taphrogeny
tafrogénico *a.* taphrogenic, rift-like
tafrogeosinclinal *m.* taphrogeosyncline
Taghanicano, Taghanicense *m. & a.* Taghanican
tagilita *f.* pseudomalachite
taja *f.* slice
tajada *f.* slice, slab
tajadera *f.* cold chisel
tajado *m.* slicing
tajamar *m.* breakwater, jetty; [Arg.] basin
tajear *vb.* [Pe.] to stope
tajo *m.* cut, opening, trench, slit, excavation; working face || ***tajo abierto*** *m.* open cut, open pit || ***tajo largo*** *m. & a.* longwall
talacha *f.* [Méx.] pickaxe, pick, mattock; basic work
taladrador *m.* driller; *a.* drilling, piercing, (perf.) boring
taladradora *f.* drill
taladrar *vb.* to drill, (perf.) to bore
taladro *m.* drill, auger; rig; borehole; drill bit (véase barrena, trepano)
talasocracia *f.* highstand period
talasocrático *a.* characterized by high stands
talasogénesis *f.* epirogenic processes that form ocean basins
talcita *f.* talcite; damourite

talco *m.* talc

talcoso *a.* talcose, talcous

talio *m.* thallium

talpetate *m.* [U.S.] caliche, hardpan

talquita *f.* talc schist

talud *m.* talus, slope || ***talud continental*** *m.* continental slope || ***talud de corte*** *m.* backslope || ***talud de reposo*** *m.* natural slope, angle of repose || ***talud de ribera*** *m.* floodplain || ***talud interior*** *m.* beach foreslope || ***talud litoral*** *m.* litoral plain || ***talud arrecifal*** *m.* reef talus

taludar *vb.* to slope

talweg *m.* thalweg

talla *f.* cutting or dressing of stone or gems

tallado *m.* cutting or dressing stone or gems; *a.* cut or dressed (stone)

tallar *vb.* to work; mill, carve, dress stone

taller *m.* shop, workshop, mill, factory

tamaño *m.* size; dimension; volume; capacity || ***tamaño de grano*** *m.* grain size

tambor *m.* barrel, drum, reel

tamiz *m.* sieve, sifter || ***tamiz vibratorio para lodos*** *m.* vibrating mudscreen, shale shaker

tamizador *m.* strainer

tamizando *m.* sieving

tamizar *vb.* to sift, sieve, screen

tampón *m.* (chem.) buffer

tanatocenosis *m. & f.* thanatocoenosis

tanda *f.* layer; series; work-shift; amount of work done, batch, turn, task, rotation, gang of men

Taneciano, Taneciense *m. & a.* Thanetian

tangencial *a.* tangential

tangente *f. & a.* tangent

tanque *m.* tank, reservoir, pit, pool || ***tanque a presión*** *m.* pressure tank || ***tanque asentador*** *m.* settling tank || ***tanque de almacenamiento*** *m.* storage tank, stock tank || ***tanque de combustible*** *m.* fuel tank || ***tanque de decantación*** *m.* settling tank || ***tanque de inyección*** *m.* mud pit, slush pit || ***tanque de lodo devuelto*** *m.* mud return tank || ***tanque de techo pontón*** *m.* floating-roof tank || ***tanque depurador de aire*** *m.* scrubber || ***tanque igualador*** *m.* surge tank, balance tank || ***tanque remolque*** *m.* tank trailer || ***tanque separador*** *m.* separating tank || ***tanque séptico*** *m.* septic tank

tantalio, tántalo *m.* tantalum

tantalita *f.* tantalite

tantear *vb.* to estimate, to work out, size up, measure; to test, examine, study

tanteo *m.* rough estimate; test; rough calculation

tapa *f.* hanging wall, roof; lid, cover

tapado *m.* buried treasure; *a.* plugged, capped; concealed, covered, blind || ***tapado por arena*** *a.* sanded up

tapar *vb.* to cover, plug

tapar(se) con arena *vb.* to sand up

tapón *m.* plug || ***tapón arcilloso*** *m.* clay plug || ***tapón de cementación*** *m.* cementing plug || ***tapón de tubería de revestimiento*** *m.* casing plug || ***tapón volcánico*** *m.* volcanic plug

taponado *a.* plugged, clogged

taponamiento *m.* closure, stoppage, plug

taponar *vb.* to plug, clog

taquihidrita *f.* tachyhydrite

taquilita *f.* tachylite

taquimetría *f.* tachymetry, stadia survey

taquímetro *m.* stadia

taquisismo *m.* rapid differential movement of lithosphere

tara *f.* tare weight

tardiglacial *m.* glacial stage

tardio *a.* late

tarea *f.* task, job

tarjeta *f.* card

tarifa *f.* tariff || ***tarifa de fletes*** *f.* freight rates || ***tarifa de oleoductos*** *f.* pipeline rates

tarima para tuberías *f.* pipe rack

Tartariense, Tartariano *m. & a.* Tartarian

tasa *f.* appraisal, valuation; tax; rate, rating || ***tasa de flujo*** *f.* flow rate || ***tasa de reposición*** *f.* replacement rate || ***tasa de sondeo*** *f.* sampling rate || ***tasa de utilización*** *f.* utilization rate

tasación *f.* appraisal, valuation; calculation

tasar *vb.* to appraise, tax, value; to regulate, fix

tascina *f.* naumannite

tasco *m.* cupel, crucible; fireclay

tasconio *m.* argillaceous earth used for crucibles

tasquil *m.* chip, shard

tavistockita *f.* carbonate-apatite

taxitas *f.* taxites

taxonomía *f.* taxonomy

Tayloriano, Tayloriense, Taylorense *m. & a.* Tayloran

tea *f.* flare, torch

tecali *m.* alabaster, "Mexican onyx"

tecla *f.* (comp.) key

teclado *m.* (comp.) keyboard

teclear *vb.* (comp.) to key in, to keypunch

tecnecio *m.* technetium

técnica *f.* technique

técnico *m.* technician

tecnología *f.* technology || ***tecnología informática*** *f.* information technology || ***tecnología más avanzada*** *f.* state-of-the-art, leading edge, high tech

tecpatl *m.* [Méx.] flint; green jade

tectita *f.* tektite

tectofacies *f.* tectofacies

tectogénesis *f.* tectonics

tectógeno *m.* tectogene

tectónica *f.* tectonic, tectonics || ***tectónica de gravedad*** *f.* gravity tectonics || ***tectónica de placas*** *f.* plate tectonics || ***tectónica salina*** *f.* salt tectonics

tectónico *a.* tectonic

tectonismo *m.* diastrophism

tectonita *f.* tectonite

tectonización *f.* tectonization

tectonofísica *f.* tectonophysics

tectonósfera *f.* tectonosphere

tectósfera *f.* tectosphere

tectosilicato *m.* tectosilicate

techo *m.* caprock, roof; headwall || ***techo de galería*** *m.* roof, hanging wall

tefra *f.* tephra

tefrita *f.* tephrite

tefrocronología *f.* tephrochronology

tefroíta *f.* tephroite

teja *f.* shingle

tejido *m.* web || ***tejido para filtro*** *m.* filter screen

tejo *m.* ingot

tejón *m.* wedge, plate, or cake of gold

tela metálica de colador sacudidor *f.* shaker screen

teledetección *f.* remote sensing

telemedir *vb.* to telemeter

telémetro *m.* telemeter, rangefinder

teleobjetivo *m.* telephoto lens

teleregistrador *m.* remote recorder

telescópio *m.* telescope || ***telescópio refractor*** *m.* refracting telescope || ***telescópio reflector*** *m.* reflecting telescope

telesismo *m.* distant earthquake

teletermal *a.* telethermal

telinita *f.* telinite

telurato *m.* tellurate

telúrico *a.* telluric

telúrido *f.* telluride

telurio, teluro *m.* tellurium

telurita *f.* tellurite

telogenético *a.* telogenetic

temático *a.* thematic

temblor *m.* earthquake, tremor || ***temblor final*** *m.* aftershock || ***temblor previo*** *m.* foreshock

temiskamita *f.* maucherite

témpano *m.* ice floe || ***témpano de hielo*** *m.* iceberg

temperatura *f.* temperature || ***temperatura absoluta*** *f.* absolute temperature || ***temperatura crítica*** *f.* critical temperature || ***temperatura de congelación*** *f.* freezing temperature || ***temperatura de ebullición*** *f.* boiling temperature .|| ***temperatura de encendio*** *f.* ignition temperature || ***temperatura de fondo de pozo*** *f.* bottom hole temperature || ***temperatura de fusión*** *f.* melting temperature || ***temperatura de precipitación de sólidos en petróleo*** *f.* cloud point || ***temperatura de superficie*** *f.* skin temperature, surface temperature || ***temperatura de vaporización*** *f.* vaporization temperature

tempestad *f.* storm, tempest

templa *f.* [Pe.] batch

templado *a.* temperate

templar *vb.* to temper, quench

temple *m.* temper, tempering

temporal *m.* gale, storm; *a.* temporary

temprano *a.* early

tendencia *f.* trend; tendency; drift

tender *vb.* to lay down, stretch, extend, lay out || ***tender cañería*** *vb.* to lay pipe

tendido *m.* dip; *a.* laid, lying down, flat || ***tendidos*** *m.* riffles

tenedor de libros *m.* bookkeeper

teneduría de libros *f.* bookkeeping

tener en reserva *vb.* to stand by, to hold in reserve

tennantita *f.* tennantite

tenor *m.* fineness

tenorita *f.* tenorite

tensión *f.* tension, stress || ***tensión de rotura*** *f.* breaking stress || ***tensión de vapor*** *f.* vapor pressure || ***tensión interfacial*** *f.* interfacial tension || ***tensión superficial*** *f.* surface tension

teñido *a.* stained

teodolito *m.* theodolite, transit

teorema *m.* theorem

teoría *f.* theory

tepetate *m.* barren rock, [Méx.] residual volcanic soil; [U.S.] caliche, hardpan

teposteles *m.* polybasite

tequio *m.* taxes, duties, or royalty on ore

teralita *f.* theralite

terapiés cubicos *m.* TCF (trillion cubic feet)

terbio *m.* terbium

Terciario *m.* & *a.* Tertiary

tercio *m.* half a mule load

terma *f.* therm

termal, térmico *a.* thermal

termas *f.* hot springs

térmico *a.* thermal

termío *m.* therm

terminación *f.* completion || ***terminación a dos zonas*** *f.* dual completion || ***terminación de pozo*** *f.* well completion || ***terminación multiple*** *f.* multiple completion

terminado *a.* finished, completed

terminal *m.* (elect., ref.) terminal; *a.* terminal, final || ***terminal de computadora*** (comp.) *m.* computer terminal || ***terminal informático*** (comp.) *m.* computer terminal

terminar *vb.* finish || ***terminar un pozo*** *vb.* to complete a well

términos *m.* terms

termistor *m.* thermistor

termocarst *m.* thermokarst

termoclasa *f.* crack due to change in temperature

termoclina *f.* thermocline

termoclinal, termoclino *m.* thermocline

termocopla *f.* thermocouple

termodinámico *a.* thermodynamic

termoelástico *a.* thermoelastic

termogenético, termogénico *a.* thermogenic, thermogenetic

termógeno *m.* thermogene

termogénesis *f.* thermogenesis

termografía *f.* thermography

termoluminiscencia *f.* thermoluminescence

termometría *f.* thermometry || ***termometría de inclusión de fluidos*** *f.* fluid inclusion thermometry

termonatrita *f.* thermonatrite

termoósmosis *f.* thermo-osmosis

termopar *m.* thermocouple

termopermutador *m.* heat exchanger

termoplástico *a.* thermoplastic

termósfera *f.* thermosphere

termostato *m.* thermostat

ternario *a.* ternary

terra rosa *f.* terra rossa

terracota *f.* terra cotta

Terraciano, Terraciense *m.* & *a.* Terracian

terraplén *m.* backfill, graded roadbed, rampart, terrace, slope, mound, embankment, earth fill, earthwork

terraplenar *vb.* to earthfill, to fill and level, to terrace

terraza *f.* terrace, embankment, bench || ***terraza aluvial*** *f.* alluvial terrace || ***terraza cortada por las aguas*** *f.* wave cut terrace || ***terraza costera*** *f.* raised beach || ***terraza de acumulación*** *f.* constructional terrace || ***terraza de deposición*** *f.* aggradational terrace || ***terraza de erosión*** *f.* erosional terrace || ***terraza de marea baja*** *f.* low-tide terrace || ***terraza estructural*** *f.* structural terrace || ***terraza fluvial*** *f.* stream terrace || ***terraza marina*** *f.* marine terrace

terremoto *m.* earthquake

terreno *m.* land, ground, terrane, terrain; plot, site || ***terreno bajo*** *m.* lowland, [U.K.] swale || ***terreno baldío*** *m.* waste land || ***terreno costroso*** *m.* scabland || ***terreno de acarreo*** *m.* transported soil || ***terreno elevado*** *m.* upland, highland || ***terreno permanentemente congelado*** *m.* permafrost terrain || ***terreno poligonal*** *m.* polygonal ground

térreo *a.* earthy

terrero *m.* heap or pile of dirt, dump; *a.* earth, earthy

terrestre *a.* terrestrial, earthly

terrígeno *a.* terrigenous

territorio *m.* territory

terrojo *m.* red earth

terrón *m.* clod

terroso *a.* earthy, brown

terruño *m.* piece of land, plot, site

teschenita *f.* teschenite

tesis *f.* thesis; theory, idea || ***tesis doctoral*** *f.* dissertation

tesorería *f.* treasury

tesorero *m.* treasurer

tesoro *m.* treasure

testero *m.* overhand stope

testificación *f.* logging

testigo *m.* core; sample, specimen; (survey, levan.) reference point || ***testigo cajón*** *m.* box core || ***testigo de erosión*** *m.* outlier || ***testigo orientado*** *m.* oriented core

Tetis, Tethys *m.* Tethys

tetracíclico *a.* tetracyclic

tetracloruro de carbono *m.* carbon tetrachloride

tetradimita *f.* tetradymite

tetraedrita, tetrahedrita *f.* tetrahedrite

tetraedro *m.* tetrahedron

tetraetilo *a.* tetraethyl || ***tetraetilo de plomo*** *m.* tetraethyl lead

tetragonal *a.* tetragonal

tetravalente *a.* tetravalent, quadrivalent

textura *f.* texture, fabric || ***textura amorfa*** *f.* amorphous texture || ***textura clástica*** *f.* clastic texture || ***textura cribosa*** *f.* sieve texture || ***textura cristalina*** *f.* crystalline texture || ***textura de avenamiento*** *f.* drainage texture || ***textura de drenaje*** *f.* drainage texture || ***textura gráfica*** *f.* graphic texture || ***textura pegmatítica*** *f.* pegmatitic texture || ***textura porfirítica*** *f.* porphyritic texture

Thanetiano, Thanetiense *m. & a.* Thanetian

thenardita *f.* thenardite

theralita *f.* theralite

tholeiitico *a.* tholeiitic

thomsonita *f.* thomsonite

thorio *m.* thorium

Thuringiano, Thuringiense *m. & a.* Thuringian

thuringita *f.* thuringite

tiempo *m.* time; weather || ***tiempo activo*** *m.* uptime || ***tiempo de acceso*** *m.* access time || ***tiempo de intercepción*** *m.* intercept time || ***tiempo de llegada*** *m.* arrival time || ***tiempo de paro*** *m.* shutdown time || ***tiempo de perforación*** *m.* drilling time || ***tiempo de propagación*** *m.* travel time || ***tiempo de reacción*** *m.* reaction time || ***tiempo de retardo*** *m.* delay time || ***tiempo doble*** *m.* (seis.) two way travel time || ***tiempo extra*** *m.* overtime || ***tiempo pasivo*** *m.* idle time, down time || ***tiempo vertical en el punto de disparo*** *m.* (seis.) uphole time

tienda *f.* tent; shop

tienta *f.* earth auger, probe

tierra *f.* earth, land, soil, dirt, ground; region || ***tierra adámica*** *f.* adamite earth || ***tierra adentro*** *f. & a.* inland; foreland || ***tierra alcalina*** *f.* alkaline earth || ***tierra alta*** *f.* highland, upland || ***tierra arriba*** *f.* upland || ***tierra de alumbre*** *f.* alum earth

|| *tierra de aluvión f.* bottom land || *tierra de batán f.* Fuller's earth || *tierra de Fuller f.* Fuller's earth || *tierra de diatomeas f.* diatomaceous earth || *tierra de porcelana f.* kaolin || *tierra diatomácea, diatomeas f.* diatomaceous earth || *tierra endurecida f.* hardpan || *tierra firme f.* solid land || *tierra floja f.* loose earth || *tierra franca f.* topsoil || *tierra rara f.* rare earth || *tierra verde f.* greenstone || *tierras bajas f.* lowlands || *tierras inundables f.* tidelands || *tierras malas f.* badlands

Tiffaniano, Tiffaniense *m. & a.* Tiffanian

Tigliano, Tigliense *m. & a.* Tiglian

tijera *f.* small ditch

til *m.* glacial till

tilita *f.* tillite, boulder clay

tinaja *f.* water pocket, pool, spring

tinajita *f.* solution pan

tinolita *f.* thinolite

tintura *f.* dye

Tioughniogense, Tioughniogiano *m. & a.* Tioughniogan

tipificación *f.* classification

tipo *m.* type, class, kind; rate, percentage || *tipo de descuento m.* discount rate

tipomórfico *a.* typomorphic

tirada aparte *f.* reprint

tiraje *m.* print run, printing, draft

tirar *vb.* to pull, to throw, to haul || *tirar un labor vb.* to prospect; to work a deposit

tiro *m.* text draft; (min.) mine shaft; (seis.) shot; (survey, levan.) course || *tiro ciego m.* winze || *tiro de alcancía m.* raise || *tiro de mina m.* shaft || *tiro de recueste m.* adit, inclined shaft || *tiro de velocidad m.* checkshot survey || *tiro en pauta m.* patterned shot || *tiro hacia abajo m.* downdraft || *tiro hacia arriba m.* updraft

tirolita *f.* tyrolite

titánico *a.* titanic

titanífero *a.* titaniferous

titanio *m.* titanium

titanita *f.* sphene

Titoniano, Tithoniano, Titoniense *m. & a.* Tithonian

titración *f.* titration

titulación *f.* assay; titration

título *m.* property title; degree; diploma, certificate || *título de propiedad m.* claim, deed || *títulos cotizables m.* marketable securities

tixotropía *f.* thixotropy

tiza *f.* chalk; gypsum

Toarciano, Toarciense *m. & a.* Toarcian

toba *f.* tuff, sinter; tufa, chalk || *toba calcárea f.* tufa, travertine || *toba de cenizas f.* volcanic ash, tuff || *toba silícea f.* siliceous sinter || *toba soldada f.* welded tuff || *toba vítrea f.* vitric tuff

tobáceo *a.* tuffaceous

tobagán *m.* chute

tobar *m.* tuff quarry

toboso *a.* tuffaceous

todouno *m.* unprocessed coal

tolerable *a.* tolerable, allowable, admissible

tolerancia *f.* allowance, clearance, tolerance

tolueno *m.* toluene.

tolva *f.* hopper, bin, ore chute

tollón *m.* glacial col

toma *f.* intake, outlet || *toma de primer plano a.* close-up

tómbolo *m.* tombolo, sandbar

tonalidad *f.* tone

tonalita *f.* tonalite

Tonawadiano, Tonawadense *m. & a.* Tonawadan

tonel *m.* barrel

tonelada *f.* ton || *tonelada bruta f.* gross ton || *tonelada corta f.* short ton (2000 lbs = 907.18 kg) || *tonelada larga f.* long ton (2240 lbs = 1016 kg) || *tonelada métrica f.* metric ton (2204.6 lbs = 1000 kg) || *tonelada neta f.* short ton, net ton

tonelaje *m.* tonnage || *tonelaje bruto m.* gross tonnage

tonelámetro *m.* metric ton

tonga *f.* layer; pile, heap; difference in level or head

tongada *f.* layer

Tongriano, Tongriense *m. & a.* Tongrian

tono *m.* tone

Tonolowayense, Tonolowayano *m. & a.* Tonolowayan

tonstein *m.* claystone

topacio *m.* topaz

topadora *f.* bulldozer

tope *m.* rich pocket of ore; top; buffer; obstacle

topo *m.* league and a half

topografía *f.* topography || ***topografía accidentada*** *f.* rough terrain || ***topografía de cuencas y cordones*** *f.* basin and range topography || ***topografía juvenil*** *f.* youthful topography || ***topografía quebrada*** *f.* broken terrain

topográfico *a.* topographic

topógrafo *m.* topographer, surveyor

topología *f.* topology

topometría *f.* topometry

toque *m.* touchstone

torbellino *m.* whirlwind

torbernita *f.* torbernite

torca *f.* sinkhole; basin, depression; sand gall

torcer *vb.* to change direction || ***torcer el rumbo*** *vb.* to change strike

torcido *a.* twisted, warped

torcimiento *m.* warp

torio *m.* thorium

torita *f.* thorite

tormenta *f.* storm || ***tormenta de arena*** *f.* sandstorm || ***tormenta de cerebros*** *f.* brainstorming || ***tormenta de nieve*** *f.* snowstorm, blizzard || ***tormenta de polvo*** *f.* dust storm || ***tormenta magnética*** *f.* magnetic storm

tormento *m.* [Méx.] light table

tornado *m.* tornado

tornasol *m.* sheen, iridescence; litmus

tornasolado *a.* iridescent

tornillo *m.* screw, adjustment knob || ***tornillo de nivelación*** *m.* levelling screw || ***tornillo macrométrico*** *m.* coarse focus || ***tornillo micrométrico*** *m.* fine focus || ***tornillo nivelador*** *m.* levelling screw

torpedeamiento *m.* artificial fracturing, blasting, shooting || ***torpedeamiento por vibración*** *m.* vibrofrac

torpedear *vb.* to blast, to shoot

torpedo *m.* shot, charge

torre *f.* tower, derrick (geog.); pinnacle, spire, pillar, needle || ***torre de desprendimiento*** *f.* stripping tower || ***torre de destilación*** *f.* distillation tower || ***torre de enfriamiento*** *f.* cooling tower || ***torre de evaporación*** *f.* evaporation tower || ***torre de extracción*** *f.* derrick, headframe; stripping tower || ***torre de fraccionamiento*** *f.* fractionating tower || ***torre de perforación*** *f.* derrick, headframe || ***torre depuradora*** *f.* scrubbing tower || ***torre de sondeo*** *f.* derrick, headframe || ***torre fraccionadora*** *f.* fractionating tower

Torrejoniano, Torrejoniense *m. & a.* Torrejonian

torrencial *a.* torrential

torrente *m.* ravine, stream

torrentera *f.* gully, ravine, chute

torrero *m.* derrickman

Torridoniano, Torridoniense *m. & a.* Torridonian

torsión *f.* torsion, twist, torque

torta *f.* cake || ***torta de filtro*** *f.* filter cake

tortilla *f.* hardpan

Tortoniano, Tortoniense *m. & a.* Tortonian

tortuosidad *f.* tortuosity

tortuoso *a.* sinuous, tortuous

tosca *f.* hardpan; clayey vein filling; talc seam; decomposed porphyry

toscamente *adv.* coarsely

tosco *a.* rough, coarse

tostación, tostadura *f.* roasting, calcination

tostador *m.* ore roasting furnace

tostar *vb.* to roast ore

total *m.* total, sum; *a.* total, whole, complete || ***total de sólidos disueltos*** *m.* total dissolved solids

Tournaisiano, Tournaisiense *m. & a.* Tournaisian

tóxico *a.* toxic

toxina *f.* toxin

traba de vapor *f.* vapor lock

trabajador *m.* worker

trabajar *vb.* to work a mine, placer, or well

trabajo *m.* work || ***trabajo bruto*** *m.* dirty work, grunt work || ***trabajo de campaña*** *m.* field work

trabajos *m.* workings

trabar *vb.* to interlock

tracción *f.* traction

traccionita *f.* tractionite

tráfico *m.* traffic

tragante *m.* sluice; flue

trama *f.* structure, framework, lattice

tramo *m.* length, range, lot, tract, stretch, span, bench

trampa *f.* (geol.) trap; (min.) sluice gate || ***trampa al vacío*** *f.* vacuum trap || ***trampa combinada*** *f.* combination trap || ***trampa de arena*** *f.* sand trap || ***trampa de gas*** *f.* gas trap || ***trampa de vapor*** *f.* steam trap || ***trampa estratigráfica*** *f.* stratigraphic trap || ***trampa estructural*** *f.* structural trap

trancada *f.* trench

tranquilidad *f.* standstill, quiescence

tranquilidadita *f.* tranquillityite

transacción *f.* transaction

transcurrente *a.* transcurrent

transductor *m.* transducer

transecto *m.* cross section, transect || ***transecto balanceado*** *m.* balanced section

transferencia *f.* transfer || ***transferencia de desplazamiento*** *f.* displacement transfer

transferidor *m.* protracter

transferir *vb.* to transfer

transfluencia *f.* transfluence

transformación *f.* transformation || ***transformación de la intensidad, el tono, y la saturación*** *f.* intensity-hue-saturation (IHS) transformation || ***transformación de Laplace*** *f.* Laplace transform

transformar *vb.* to transform, change

transgresión *f.* transgression

transgresivo *a.* transgressive, overlapping

tránsito *a.* transit

translúcido, transluciente *a.* translucent

transmisión *f.* transmission

transmisividad *f.* transmissivity

transmisor *m.* transmitter

transmitir *vb.* to transmit

transpaís *m.* hinterland

transparencia *f.* transparency

transparente *m. & a.* transparency; transparent

transpiración *f.* transpiration

transportador *m.* protractor

transportar *vb.* to move, transport, carry

transporte *m.* transport, transportation, rafting

transpresión *f.* transpression

transpresivo *a.* transpressive

transtensión *f.* transtension

transtensivo *a.* transtensional

transversal, transverso *m.* cross cut; *a.* transverse

tranvía *f.* tramway

trapezoedro *m.* trapezohedron

trapp *m.* trap, plateau basalt

traquiandesita *f.* trachyandesite

traquibasalto *m.* trachybasalt

traquilita *f.* trachylyte

traquiofítico *a.* trachyophitic

traquita *f.* trachyte || ***traquita alcalina*** *f.* alkali trachyte

traquítico *a.* trachytic

traquito *m.* trachyte

traquitoide *m.* trachytoid

tras *m.* tuff; tufa; *p.* behind, after

tras-arrecife *m.* back reef

traslación *f.* translation, movement across || ***traslación continental*** *f.* continental drift

traslacional *a.* translational

trasladar *vb.* to move

traslapamiento *m.* overlap

traslapar *vb.* to overlap

traslape, traslapo *m.* onlap, overlap || ***traslape hacia abajo*** *m.* downlap || ***traslape hacia abajo bidireccional*** *m.* bidirectional downlap || ***traslape regresivo*** *m.* offlap || ***traslape transgresivo*** *m.* onlap

traslúcido, trasluciente *a.* translucent

trasluz *m.* diffused light; reflected light
traspaís *m.* hinterland
trasversal *a.* transverse
tratado *a.* treated || ***tratado con ácido*** *a.* acid treated
tratado *m.* treaty || ***tratado de libre comercio (TLC)*** *m.* free trade agreement (NAFTA)
tratamiento *m.* treatment, processing || ***tratamiento ácido*** *m.* acid treatment, acidizing || ***tratamiento al calor*** *m.* heat treatment || ***tratamiento con solvente*** *m.* solvent treatment || ***tratamiento digital*** *m.* digital processing || ***tratamiento químico*** *m.* chemical treatment || ***tratamiento térmico*** *m.* heat treating
tratar *vb.* to treat
travecaja *f.* cross cut
travertino *m.* travertine
travesía *f.* traverse
traviesa *f.* cross cut
trayecto *m.* trajectory, route, way
trayectoria *f.* trajectory, path || ***trayectoria de la onda sísmica*** *f.* raypath
traza *f.* trace; streak
trazado *m.* tracing, line, traverse, path, location; *a.* plotted, traced || ***trazado de curvas de nivel*** *m.* contouring || ***trazado de reflecciones*** *m.* reflection picking
trazadora *f.* tracer
trazar *vb.* to trace, plan, locate, design, draw
trazas de fisión *f.* fission tracks
trazo *m.* sketch, plan, trace
Tremadociano, Tremadociense *m. & a.* Tremadocian
tremedal *m.* quagmire
tremolita *f.* tremolite
tremor *m.* tremor
Trempealeauiano, Trempealeauense *m. & a.* Trempealeauan
tren de olas *m.* wave train
Trentoniano, Trentoniense *m. & a.* Trentonian
trenzado *a.* braided
trepa *f.* drilling, boring
trepada *f.* climb, rise, ascent
trepado *m.* drilling, boring
trépano *m.* drill bit (véase barrena, taladro)
trepar *vb.* to climb
triangulación *f.* triangulation
triangular *a.* triangular
triángulo *m.* triangle
Trias, Triásico *m. & a.* Trias, Triassic
triaxial *a.* triaxial
triaxón *m.* triaxon
triboluminiscencia *f.* triboluminescence
tributario *m. & a.* tributary
tricíclico *a.* tricyclic
triclínico *a.* triclinic
tricroismo *m.* trichroism
tridimita *f.* trydimite
trifásico *a.* three-phase
trifilita *f.* triphylite
triedro, trihedro *a.* trihedral
trigiro *m.* ternary axis of rotation
trilobita *f.* trilobite
trillón de pies cúbicos *m.* TCF
trimestral *a.* quarterly
trimorfísmo *m.* trimorphism
trincha *f.* waste pile
trinchar *vb.* to dam
trinchera *f.* trench, ditch; rift; rock pile, ore pile; costean, costeen
trineo *m.* skid, sled
Trinitiano, Trinitiense *m. & a.* Trinitian
triple unión *f.* triple junction
tripolita *f.* diatomite
tripulación *f.* crew
triquita *f.* microlite
tritio *m.* tritium
trituración *f.* grinding, crushing
triturado *a.* ground, crushed
triturador *m.* crusher; *a.* crushing, grinding
trituradora *f.* crusher, shredder
triturar *vb.* to crush or grind
trivalente *a.* trivalent
troctolita *f.* troctolite
trona *f.* trona
troncar(se) *vb.* to truncate
tronera *f.* chimney, channel, flue

tronco *m.* stock, trunk || ***tronco alimentador** m.* feeder stock || ***tronco de montaña** m.* peneplain

tropical *a.* tropical

trópico *m.* tropic; tropics || ***Trópico de Cáncer** m.* Tropic of Cancer || ***Trópico de Capricornio** m.* Tropic of Capricorn

tropiezo *m.* obstacle, impediment

tropósfera *f.* troposphere

troquel *m.* pile driver, drop hammer

trozo *m.* fragment, bit, chunk

trueque *m.* barter, exchange

truncación *truncation*

truncado *a.* truncated, beveled

truncamiento *m.* truncation, bevelling

truncar *vb.* to truncate

trunco *m.* stock; *a.* truncated

tsunami *m.* tsunami, tidal wave

tubería *f.* pipe, tubing; drill string (véase cañería) || ***tubería colectora** f.* gathering line || ***tubería de ademe** f.* casing || ***tubería de carga** f.* penstock || ***tubería de disparo** f.* tubing, flow pipe || ***tubería de expansión** f.* expansion line || ***tubería de flujo** f.* flow line || ***tubería de producción** f.* producing drill string || ***tubería de revestimiento** f.* casing || ***tubería de transporte** f.* pipeline || ***tubería flexible** f.* flexible tubing || ***tubería plástica** f.* plastic tubing || ***tubería revestida** f.* coated pipe || ***tubería sin costura** f.* seamless tubing || ***tubería telescópica** f.* tapered tubing || ***tubería vástago** f.* drillpipe

tubo *m.* tube, pipe, conduit (véase tubería, cañería) || ***tubo ascendente** m.* riser || ***tubo calado** m.* perforated liner || ***tubo capilar** m.* capillary tube || ***tubo colador** m.* perforated pipe, liner, sand screen || ***tubo condensador** m.* condenser tube || ***tubo conducto** m.* conduit || ***tubo de admisión** m.* intake pipe || ***tubo de ensayo** m.* test tube || ***tubo de entrada** m.* penstock || ***tubo de explosión** m.* diatreme || ***tubo de humo** m.* flue || ***tubo de lava** m.* lava tube || ***tubo de perforación** m.* drill pipe || ***tubo de rayos catódicos** m.* cathode ray tube (CRT) || ***tubo de sondeo** m.* drill pipe || ***tubo de subida** m.* riser || ***tubo de ventilación** m.* ventilation pipe, vent || ***tubo distribuidor** m.* distribution line || ***tubo en U** m.* U tube || ***tubo portatestigos** m.* core barrel || ***tubo productor** m.* tubing || ***tubo revestidor** m.* liner || ***tubo sin costura** m.* seamless pipe || ***tubo sin soldadura** m.* seamless pipe || ***tubo subiente** m.* riser

tuboscópio *m.* tuboscope

tubular *a.* tubular

tucanita *f.* scarbroite

tucurubay *m.* gravel

tueste *m.* roasting of ore

tufa *f.* tuff

tufaceo *a.* tufaceous

tufita *f.* tuffite

tufítico *a.* tuffaceous

tulio *m.* thulium

tumbado *a.* recumbent

tumbar *vb.* [Méx.] to break ground, breakdown ore; to get ore

tumbe *m.* the act of breaking and removing ore

tumor *m.* dome

tundra *f.* tundra

túnel *m.* tunnel, gallery

tungstato *m.* tungstate

tungsteno *m.* tungsten, wolframite

tungstita *f.* tungstite

Tuoriano, Tuoriense *m. & a.* Tuorian

turba *f.* peat, turf || ***pantano de turba** m.* peat bog

turbación *f.* turbation

turbal *m.* peat bog

turbera *f.* peat bog || ***turbera baja** f.* [U.K.] fen soil, lowmoor

turbiedad *f.* cloudiness, turbulence

turbidez, turbiedad, turbieza *f.* turbidity

turbidita *f.* turbidite

turbina *f.* turbine

turbio *a.* turbid

turboglifo *m.* scour cast || ***turboglifo intersepto** m.* flute cast

turbonita *f.* oil shale

turbosina *f.* jet fuel

turbulencia *f.* turbulence

turbulento *a.* turbulent

turgita *f.* turgite
Turingiano, Turingiense *m. & a.* Thuringian
turingita *f.* thuringite
turmalina *f.* tourmaline
turmalinización *f.* tourmalinization
turmalita *f.* tourmalite
turno *m.* shift, crew; tour of duty
Turoliano, Turoliense *m. & a.* Turolian
Turoniano, Turoniense *m. & a.* Turonian
turquesa *f.* turquoise
Tyrrheniano, Tyrrheniense *m. & a.* Tyrrhenian
tysonita *f.* fluocerite
tyuyamunita *f.* tyuyamunite
tzinapu *m.* [Méx.] obsidian

U

uadi *m.* wadi, wash, [U.S.] arroyo, gulch
ubicación *f.* location, position, site
ubicado *a.* located
ubicar *vb.* to locate, spot a well, plot on a map
ubicarse *vb.* to be located, situated, or placed
ufertita *f.* davidite
Uintense *m. & a.* Uintan
Ulatsiense, Ulatsiano *m. & a.* Ulatsian
ulexita *f.* ulexite
Ulsteriense, Ulsteriano *m. & a.* Ulsterian
último *a.* last, final, ultimate
ultrabásico *a.* ultrabasic
ultramáfico *a.* ultramafic
ultramar *m.* lapis lazuli, lazurite
ultramilonita *f.* ultramylonite
ultrarrojo *m. & a.* infrared
ultraviolado *a.* ultraviolet
Ultravulcaniense, Ultravulcaniano *m. & a.* Ultravulcanian
ulvita *f.* ulvöspinel
umbral *m.* threshold, sill; swell
unaka *f.* monadnock
uncompahgrita *f.* uncompahgrite
unda *f.* environment
undación *f.* megafold
undulación *f.* undulation, wave
uniaxial, uniáxico *a.* uniaxial
uniclinal *a.* monoclinal
unidad *f.* unit || ***unidad bioestratigráfica*** *f.* biostratigraphic unit || ***unidad central de procesamiento*** *f.* (comp.) central processing unit (CPU) || ***unidad cronoestratigráfica*** *f.* chronostratigraphic unit || ***unidad de destilación al vacío*** *f.* vacuum distillation unit || ***unidad de disco*** *f.* (comp.) disk drive ***unidad de extracción de solventes*** *f.* solvent extraction unit || ***unidad de proceso central*** *f.* (comp.) central processing unit (CPU) || ***unidad de memoria*** *f.* (comp.) memory || ***unidad estratigráfica*** *f.* stratigraphic unit || ***unidad para acondicionamiento de agua*** *f.* water-treating unit || ***unidad para elaboración de petróleo crudo*** *f.* crude processing unit || ***unidad térmica británica*** *f.* BTU (252 g. cal.) || ***unidad térmica inglesa*** *f.* BTU
unificación *f.* unitization of an oil field
unificar *vb.* to unitize, unify
uniformar *vb.* to standardize
uniforme *a.* uniform, standard
uniformidad *f.* uniformity
uniformitarismo *m.* uniformitarianism
unión *f.* joint, coupling, connection, splice
unir(se) *vb.* to unite, coalesce, splice
uñeta *f.* stone cutter's chisel
Uraliano, Uraliense *m. & a.* Uralian
uralita *f.* uralite
uralitización *f.* uralitization
uránico *a.* uranic
uranífero *a.* uraniferous
uraninita *f.* uraninite, pitchblende
uranio *m.* uranium
urano *m.* uranium oxide
Urano *m.* Uranus
uranofano *m.* uranophane
uranolito *m.* tectite, meteorite

uranopicco *m.* uraninite, pitchblende
uranosfaerita *f.* uranosphaerite
uranotantalita *f.* samarskite
uranothallita *f.* liebigite
uranotilo *m.* uranotile
urao *m.* gaylussite
Uriconiense, Uriconiano *m. & a.* Uriconian
usado *a.* used, worn out
usamerita *f.* greywacke
usar *vb.* to use, to wear
usina *f.* factory, mill, plant, works
utahita *f.* jarosite
utahlita *f.* variscite
útiles *m.* tools, equipment || ***útiles de extracción*** *m.* fishing tools
utilidad *f.* profit, benefit; utility || ***utilidad del activo de los accionistas*** *f.* return on shareholder equity || ***utilidad del capital empleado*** *f.* return on capital employed || ***utilidad del capital propio de los accionistas*** *f.* return on shareholder equity || ***utilidad probable*** *f.* estimated profit
uvala *f.* karst valley
uvarovita, uwarowita *f.* uvarovite
uzbekita *f.* volborthite

V

V, en *f. & a.* chevron
V, en forma de *a.* V-shaped
vacia *f.* wacke
vacío *a.* empty
vaciada *f.* melt; dumping
vaciadero *m.* dump, tailings pile, spoil bank; sluiceway
vaciagris *m.* greywacke
vaciamar *f.* ebb tide
vaciante *f.* ebb tide
vaciar *vb.* to emply out, drain, hollow out, dump
vacío *m.* vacuum, void; *a.* empty
vacuo *m.* vacuum
vacuola *f.* vug, cavity
vacuómetro *m.* vacuum gage
vadear *vb.* to ford
vado *m.* ford
vagón *m.* ore car
vagonada *f.* carload
vaguada *f.* watershed; lowest part of a valley, stream channel, thalweg, waterway
vaho *m.* steam, vapor, fumes, mist
vaivén *m.* swash || ***vaivén de oleaje*** *m.* swash
val *m.* valley, [U.K.] vale
Valangiano, Valangiense *m. & a.* Valanginian
válencia *f.* valence
Valendisiano, Valendisiense *m. & a.* Valendisian
Valentiano, Valentiense *m. & a.* Llandoverian
valentinita *f.* valentinite
valía *f.* valuation, value, worth
validez *f.* validity || ***dar validez*** a *vb.* to validate
válido *a.* valid
valioso *a.* precious, valuable
Valmeyerense, Valmeyerano *m. & a.* Valmeyeran
valor *m.* value, worth || ***valor base*** *m.* base value || ***valor BTU*** *m.* BTU calorific value || ***valor calórico*** *m.* caloric value || ***valor contable*** *m.* book value
valoración *f.* valuation
valorar *vb.* to evaluate, to appraise
valores *m.* securities || ***valores vendibles*** *m.* marketable securities
valorización *f.* appreciation
valuado *a.* rated, appraised
valuar *vb.* to value, assess, appraise, rate
valva *f.* valve, shell
válvula *f.* valve || ***válvula de cierre automático*** *f.* cutoff valve || ***válvula de compuerta*** *f.* gate valve || ***válvula de descarga*** *f.* discharge valve, exhaust valve, blowoff valve || ***válvula de descarga de presión*** *f.* pressure relief valve || ***válvula de escape*** *f.* exhaust relief valve, relief valve, blowoff valve || ***válvula de seguridad*** *f.* safety valve, relief valve || ***válvula hidráulica*** *f.* hydraulic valve || ***válvula maestra*** *f.* master gate valve

valla *f.* obstacle, embankment, barrier

valle *m.* valley, [U.K.] vale || ***valle ahogado*** *m.* drowned valley, [U.K.] firth || ***valle aluvial*** *m.* alluvial valley, outwash plain || ***valle antecedente*** *m.* antecedent valley || ***valle colgante*** *m.* hanging valley || ***valle de drenaje*** *m.* water gap || ***valle de hundimiento*** *m.* rift valley || ***valle de rumbo*** *m.* strike valley || ***valle epigenético*** *m.* superimposed, or antecedent valley || ***valle estrecho entre cerros*** *m.* [U.K.] coombe, coomb || ***valle inciso*** *m.* incised valley || ***valle inundado*** *m.* drowned valley, [U.K.] firth || ***valle obsecuente*** *m.* obsequent valley || ***valle resecuente*** *m.* resequent valley || ***valle sepultado*** *m.* buried valley || ***valle subsecuente*** *m.* subsequent valley || ***valle sumergido*** *m.* submerged valley, [U.K.] firth

vallecito *m.* valley, [U.K.] dale

vanadato *m.* vanadinite; vanadate

vanadífero *a.* vanadiferous

vanadinita *f.* vanadinite

vanadio *m.* vanadium

vanado-magnetita *f.* coulsonite

vano *m.* opening

vapor *m.* vapor, steam || ***vapor vivo*** *m.* steam, live steam

vaporar *vb.* to evaporate

vaporización *f.* evaporation, vaporization

vara *f.* stick, rod, pole (véase varilla); (meas.) old unit varying between 0.8–1.1 meter or about 33 inches || ***vara de agrimensor*** *f.* measuring rod || ***vara de medir*** *f.* measuring rod || ***vara de topografía*** *f.* surveyor's rod || ***vara mágica*** *f.* divining rod || ***vara portabrújula*** *f.* Jacob's staff

varada *f.* three months work

vareo *m.* measuring in varas

vareta *f.* rod || ***vareta divinatoria*** *f.* divining rod || ***vareta mágica*** *f.* divining rod

variable *f.* variable

variación *f.* variation

variancia *f.* variance

variado *a.* varied

variante *f. & a.* variant

varianza *f.* variance

variar *vb.* to vary

variedad *a.* variety

varilla *f.* rod (véase vara) || ***varilla de bombeo*** *f.* sucker rod || ***varilla de medición*** *f.* measuring rod || ***varilla de refuerzo*** *f.* reinforcing rod || ***varilla de succión*** *f.* sucker rod || ***varilla divinatoria*** *f.* divining rod || ***varilla mágica*** *f.* divining rod || ***varilla estadimétrica*** *f.* stadia rod

variolas *f.* varioles

variolita *f.* variolite

variolítico *a.* variolitic

variómetro *m.* variometer

Varíscico, Varisco *m. & a.* Variscan

variscita *f.* variscite

varita de rabdomante *f.* divining rod

varva *f.* varve

vasija *f.* receptacle, vessel

vaso *m.* basin, reservoir; vessel

vástago *m.* rod || ***vástago de arrastre*** *m.* kelly || ***vástago de perforación*** *m.* kelly joint, kelly, drill stem

vatímetro *m.* watt meter

vatío *m.* watt

Vauclusiano, Vauclusiense *m. & a.* Vauclusian

vecindad *f.* neighborhood, surroundings, vecinity

vecino *m.* neighbor; *a.* contiguous

vector *m.* vector

vega *f.* valley, open plain, floodplain; oxbow lake

vegetación *f.* vegetation || ***vegetación mesofítica*** *f.* mesophytic vegetation || ***vegetación xerófila, xerofítica*** *f.* xerophytic vegetation

vegetal *a.* vegetable, plant

vejez *f.* old age stage

velador *m.* watchman

velaje glaciar *m.* calving of iceburgs from glaciers

velardenita *f.* gehlenite

veleta *f.* weather vane, wind vane

velo *m.* fog

velocidad *f.* velocity || ***velocidad de apilamiento*** *f.* stacking velocity || ***velocidad de avance de perforación*** *f.* drilling rate || ***velocidad de circulación*** *f.* circulation rate || ***velocidad de deformación*** *f.* strain rate || ***velocidad del intervalo*** *f.* interval velocity || ***velocidad de la lente*** *f.* lens speed

vena *f.* vein, veinlet, bed, seam (véase veta) || ***venas fibrosas antitaxiales*** *f.* antitaxial fibrous veins || ***venas fibrosas sintaxiales*** *f.* syntaxial fibrous veins

venaje *f.* quantity of water in a river

vencerse *vb.* to become due

vencimiento *m.* expiration

vendedor *m.* salesman, seller

veneno *m.* poison

venenoso *a.* poisonous

venera *f.* spring

venero *m.* sprin·· source; bed or seam; mineral deposit

venida *f.* flood, overflow of a river

venilla *f.* veinlet or stringer

venita *f.* veinlet or stringer

venoso *a.* veined

venta *f.* sale, divestment; selling; marketing || ***venta a granel*** *f.* bulk sale || ***venta a contado*** *f.* cash sale || ***venta al por mayor*** *f.* bulk sale, wholesale || ***venta al por menor*** *f.* retail sale

ventaja competitiva *f.* competitive advantage

ventana *f.* (geol.) window, fenster; inlier || ***ventana de generación de petróleo*** *f.* oil generation window || ***ventana de lectura y grabación*** *f.* read-write window || ***ventana tectónica*** *f.* fenster

ventanilla *f.* station

ventarrón *m.* gale; wide vein

ventifacto *m.* ventifact

ventilación *f.* ventilation

ventilador *m.* fan, blower, ventilator

ventisca *f.* blizzard

ventisquero *m.* glacier, snow-capped mountain, wind gap

ventosa al vacío *f.* vacuum relief valve

Venturiano, Venturiense *m. & a.* Venturian

venturímetro *m.* Venturi meter

venturina *f.* aventurine quartz

verano *m.* summer

vereda *f.* footpath, path, trail

verdadero *a.* true, actual, real

verdiones *m.* green copper stain

vergencia *f.* vergence

verificación *f.* verification

verificador *m.* tester

verificar *vb.* to test, verify

vermiculita *f.* vermiculite

vernadskita *f.* antlerite

Versiliano, Versiliense *m. & a.* Versilian

vertebrado *m.* vertebrate

vertedero *m.* spillway, drain; apron; dump

vertedor *m.* bucket; drain, spillway, outlet

verter *vb.* to pour

vertical *a.* vertical

vértice *f.* apex, vertex, crest, peak; survey corner

vertiente *m.* slope; aspect; drainage area, watershed, divide

vesícula *f.* vesicle || ***vesícula de arcilla*** *f.* clay gall

vesicular *a.* vesicular

vesículo *m.* vesicle

vestigio *m.* trace, vestige, show

vesubiana vesubianita, vesuvianita *f.* vesuvianite, leucitite

Vesuliano, Vesuliense *m. & a.* Vesulian

vesuvianita *f.* vesuvianite

veta *f.* vein, lode, mineralized bed or seam, streak (véase vena) || ***veta cuarzífera*** *f.* quartz vein || ***veta de color*** *f.* streak || ***veta de fisura*** *f.* fissure vein || ***veta de relleno de fisuras*** *f.* fissure vein || ***veta en borra*** *f.* barren vein || ***veta en frutos*** *f.* ore-carrying vein || ***veta madre*** *f.* mother lode || ***veta mineralizada*** *f.* lode || ***veta pegmatítica*** *f.* pegmatite vein || ***veta ramal*** *f.* branching vein || ***veta serpenteada*** *f.* variable strike vein || ***veta socia*** *f.* companion vein

vetarrón *m.* large vein

veteado *a.* veined

vetica, veticilla, vetilla *f.* veinlet

vetita *f.* fiber, stringer

vía *f.* gallery || ***vía de agua*** *f.* watercourse, waterway

viable *a.* viable, feasible

viaducto de caballetes *m.* trestle

viaje *m.* (perf.) tripping out || ***viaje redondo*** *m.* round trip

vibertita *f.* bassanite

vibración *f.* vibration || ***vibración de suelo*** *f.* ground vibration

vibrador *m.* vibrator, agitator

vibradoras *f.* (seis.) vibroseis

vibrar *vb.* to vibrate, oscillate

vibratorio *a.* vibrating

vibrosísmica *f.* vibroseis

vicio *m.* defect, flaw

Vicksburgiano, Vicksburgiense *m. & a.* Vicksburgian

vida *f.* life || ***vida media*** *f.* half-life

vidrio *m.* glass || ***aparatos de vidrio*** *m.* laboratory glassware || ***vidrio de aumento*** *m.* magnifying glass || ***vidrio de seguridad*** *m.* safety glass || ***vidrio para laboratorio*** *m.* laboratory glassware || ***vidrio pyrex*** *m.* Pyrex glass || ***vidrio volcánico*** *m.* obsidian, volcanic glass

vidrioso *a.* glassy, vitreous

viejo *a.* old

viento *m.* wind || ***viento del oeste*** *m.* westerlies

viga *f.* girder, beam, sill

vigía *m.* lookout, watchman

vigilancia *f.* watchfulness, vigilence

vigilar *vb.* to watch

vilemita *f.* willemite

Villafranquiano, Villafranquense *m. & a.* Villafranchian

vincularse *vb.* to be related to, bound to, tied to

vínculo *m.* tie, bond

Vindoboniano, Vindoboniense *m. & a.* Vindobonian

virazón *m.* sea breeze

virgación *f.* virgation

virgen *a.* native, virgin, pure

Virgiliano, Virgiliense *m. & a.* Virgilian

Virgloriano, Virgloriense *m. & a.* Anisian, Virglorian

Virguliano, Virguliense *m. & a.* Virgulian

viridina *f.* viridine, manganandalusite

virtual *a.* virtual

virutas *f.* cuttings

viscómetro *m.* viscometer

viscosidad *f.* viscosity

viscosímetro *m.* viscosimeter

viscoso *a.* viscous

Viseano, Visiense *m. & a.* Visean

visera *f.* visor

visibilidad *f.* visibility

visión *f.* view || ***visión estereoscópica*** *f.* stereoscopic vision || ***visión sinóptica*** *f.* synoptic view

visor *m.* optical sight

vista *f.* sight, vision, appearance, view || ***vista adelante*** *f.* foresight || ***vista aérea*** *f.* aerial view || ***vista atrás*** *f.* backsight || ***vista de frente*** *f.* front view || ***vista en perspectiva*** *f.* perspective view || ***vista en plan, en planta*** *f.* plan view || ***vista lateral*** *f.* side view

visual *f.* sight, line-of-sight; *a.* visual || ***visual adelante*** *f.* foresight || ***visual atrás*** *f.* backsight || ***visual de control*** *f.* check sight || ***visual inversa*** *f.* backsight

vítreo *a.* glassy, vitreous

vítrico *a.* vitric

vitrificación *f.* vitrification

vitrificado *a.* vitrified

vitrificar *vb.* to vitrify

vitrinita *f.* vitrinite

vitriófido *m.* vitrophyre

vitriolo *m.* vitriol || ***vitriolo azul*** *m.* copper sulfate, chalcanthite || ***vitriolo blanco*** *m.* zinc sulfate || ***vitriolo de cobre*** *m.* chalcanthite || ***vitriolo de marto*** *m.* copperas || ***vitriolo de plomo*** *m.* anglesite || ***vitriolo rojo*** *m.* botryogen

vitrita *f.* vitrian, vitrite

vitroclarita *f.* vitroclarian

vitroclástico *a.* vitroclastic

vitroclasto *m.* vitroclast
vitrofídico, vitrófido *a.* vitrophyric
vitrófiro *m.* vitrophyre
vivianita *f.* vivianite
viviente del fondo *a.* bottom-dwelling
vivo *a.* alive, live, active
volado *a.* blown, blasted
voladura *f.* blast, blasting; shot
volar *vb.* to blast; project; fly
volastonita *f.* wollastonite
volátil *a.* volatile
volatilidad *f.* volatility
volatilización *f.* volatilization
volcado *a.* overturned, inverted
volcadura *f.* tilting
volcán *m.* volcano; [And.] landslide, torrent, avalanche, fall; [Car.] pile, heap || ***volcán apagado*** *m.* extinct volcano || ***volcán compuesto*** *m.* strato-volcano || ***volcán de barro*** *m.* mud volcano || ***volcán de lodo*** *m.* mud volcano || ***volcán dormido*** *m.* dormant volcano || ***volcán en escudo*** *m.* shield volcano || ***volcán extinguido*** *m.* extinct volcano || ***volcán extinto*** *m.* extinct volcano
volcanicidad *f.* volcanism
volcánico *a.* volcanic
volcanismo *m.* volcanism
volcanita *f.* volcanite
volcanogénico *a.* volcanogenic
volcar *vb.* to invert, overturn; to dump
volframio *m.* wolframite
volframita *f.* wolframite
voltaje *m.* voltage
voltear *vb.* to dump, overturn
voltímetro *m.* voltmeter
voltio *m.* volt
volumen *m.* volume || ***volumen intersticial*** *m.* interstitial volume || ***volumen poroso*** *m.* pore volume
volumetría *f.* volumetrics
volumétrico *a.* volumetric
voluminoso *a.* large
volver a cubrir *vb.* to recover
volverse en lutita *vb.* to shale out
vorágine *f.* whirlpool
vórtice *m.* whirlpool, vortex
Vraconiano, Vraconiense *m. & a.* Vraconian
vuelco *m.* tilting, overturning, dumping || ***vuelco de la memoria*** *m.* memory dump
vuelta *f.* whorl; oxbow, meander; revolution, turn
vulcanismo *m.* volcanism
vulcanología *f.* volcanology
vulcanologista, vulcanólogo *m.* volcanologist
vulfenita *f.* wulfenite

W

Waaliano, Waaliense *m. & a.* Waalian
waca, wacka *f.* wacke
wackenrodita *f.* wachenrodite
wadi *m.* wadi
waldheimita *f.* actinolite
walkerita *f.* pectolite
walmstedtita *f.* breunerite
walpurgina *f.* walpurgite, waltherite
Waltoniano, Waltonense, Waltoniense *m. & a.* Waltonian
waluewita *f.* xanthophyllite
wallerian *m.* hornblende
wallerita *f.* greenstone
wallongita *f.* wallongongite
warrenita *f.* owyheeite, jamesonite
warringtonita *f.* brochantite
warthita *f.* blödite
warwicita *f.* warwickite
Wasatchiano, Wasatchiense *m. & a.* Wasatchian
washingtonita *f.* ilmenite
Washitano, Washitense *m. & a.* Washitan
wasita *f.* impure allanite
Waucobiano, Waucobiense, Waucobense *m. & a.* Waucoban
waudi *m.* wadi

wavelita *f.* wavellite
way *m.* oreway, chimney
Wealdiano, Wealdiense, Wealdense *m. & a.* Wealden
webnerita *f.* andorite
webskyita *f.* fibrous serpentine
wehrlita *f.* wehrlite
weibyeíta *f.* bastnesite
weinschenkita *f.* churchite
weissian *m.* scolecite
weissigita *f.* lithium allanite
Wemmeliano, Wemmeliense, Wemmelense *m. & a.* Wemmelian
Wenlockiano, Wenlockiense *m. & a.* Wenlockian
wentzelita *f.* hureaulite
Werfeniano, Werfeniense *m. & a.* Werfenian
werhemannita *f.* aluminite
wernerismo *m.* Neptunism
wernerita *f.* wernerite
werneritización *f.* dipyritization
weslienita *f.* römerite
westanita *f.* altered andalusite
Westfaliano, Westfaliense *m. & a.* Westphalian
Weybourniano, Weybourniense *m. & a.* Weybournian
whartonita *f.* nickel pyrite
Wheeleriano, Wheeleriense *m. & a.* Wheelerian
Whitbiano, Whitbense *m. & a.* Whitbian
Whitciffiano, Whitcliffiense *m. & a.* Whitcliffian
whitherita, witherita *f.* witherite
Whitneyense, Whitneyano *m. & a.* Whitneyan
wiborgita *f.* viborgite, rapakivi granite
wichtina *f.* wichtisite
Wilcoxiano, Wilcoxiense *m. & a.* Sabinian
wilhemita *f.* willemite
wiliamsita *f.* antigorite
wilinita *f.* idocrase
wiltshireíta *f.* rathite
wiluita *f.* vesuvianite
winche *m.* winch
winchellita *f.* lintonite
Wisconsiano, Wisconsinense *m. & a.* Wisconsian
wiserina *f.* anatase, xenotime
wismutaurita *f.* native gold with bismuth
wittichita *f.* wittichenite, klaprothite
wittingita *f.* altered rhodonite
wocheinita *f.* bauxite
wolastonita *f.* wollastonite
wolchita *f.* bournonite
Wolfcampiano, Wolfcampiense *m. & a.* Wolfcampian
wolfram *m.* tungsten, wolframite
wolframio *m.* wolfram
wolframífero *a.* tungsten-bearing
wolframita *f.* wolframite
wolfsbergita *f.* chalcostibite
wolftonita *f.* hetairite
wolnina *f.* barite
wolnym *m.* barite
wollongita *f.* wollongongite
Woodbiniano, Woodbiniense *m. & a.* Woodbinian
woodfordita *f.* ettringite
worobieffita *f.* cesium and lithium beryl
wulfenita *f.* wulfenite
wurtzilita *f.* asphaltic pyrobitumen, wurzilite
wurtzita *f.* wurtzite
wyomingita *f.* wyomingite

X

xalostocita *f.* landerite
xanotoxeno *m.* xanthoxenite
xanthokón *m.* variety of diamond
xantiosita *f.* xanthiosite
xantita *f.* xanthite
xantoarsenita *f.* sarkinite
xantoconita *f.* xanthoconite
xantocroíta *f.* greenockite
xantofilita *f.* clintonite
xantopirita *f.* pyrite
xantosiderita *f.* goethite
xantoxenita *f.* xanthoxenite

xenoblástico *a.* xenoblastic
xenoblasto *m.* xenoblast
xenobloque *m.* xenolith
xenocristal, xenocristo *m.* xenocryst
xenofiro *m.* xenophyre, dike
xenogénito, xenogenuoso *a.* epigenetic
xenoíquico *a.* xenoikic, poikilitic
xenolita *f.* xenolith
xenolítico *a.* xenolithic
xenolito *m.* xenolith
xenomórfico *a.* xenomorphic
xenomorfismo *m.* xenomorphism
xenomorfo *m.* xenomorph
xenón *m.* xenon
xenoporfirítico *a.* xenoporphyritic
xenotermal *a.* xenothermal
xenótima *f.* xenotime
xenotópico *m.* xenotopic
xérico *a.* xeric
xerofítico *a.* xerophytic
xerotérmico *a.* xerothermic
xihuitl *m.* [Méx.] turquoise
xilantita *f.* fossil wood, xylanthite
xileno *m.* xylene
xilito *a.* xyloid
xiloide *m.* xyloid
xilolita *f.* xilolite, xilolith
xilópalo *m.* opalized wood
xilotilo *m.* xylotile
xylenita, xylinita *f.* xylenite, xylinite
xylinoide *m.* xylinoid
xylita *f.* xylite
xylochloro *m.* green apophyllite
xylocriptita *f.* xylokryptite
xyloide *a.* xyloid
xylolitha *f.* petrified wood
xylopal *m.* opalized wood
xylotilo *m.* xylotile, fibrous serpentine
xylovitrian, xylovitriano *m.* euvitrian

Y

yacente, yaciente *adv.* lying, deposited
yacer *vb.* to be lying down, deposited || ***yacer sobre*** *vb.* to rest on
yacimiento *m.* reservoir, deposit, bed (véase almacenadora, depósito) || ***yacimiento de segregación*** *m.* segregated orebody || ***yacimiento metalífero*** *m.* orebody, mineral deposit || ***yacimiento mineral*** *m.* orebody || ***yacimiento mineral productivo*** *m.* pay rock, pay ore || ***yacimiento petrolífero*** *m.* oil pool, reservoir, oil field
Yakutense, Yakutiano *m. & a.* Yakutsk
yamagutilita *f.* oyamalite
yamasquita *f.* yamaskite
yanacona *m.* leased-mine worker
yanolita *f.* axinite
yapa *f.* mercury used to aid metal extraction
yarda *f.* yard (three feet, or 0.9144 m) || ***yarda cúbica*** *f.* cubic yard (0.7646 m^3)
yardaje *m.* yardage
Yarmouth *m. & a.* Yarmouthian
yasa *f.* dry stream
yeatmanita *f.* yeatmanite
yébel *m.* mountain, jebel
yenita *f.* ilvaite
Yeoviliano, Yeovilense *m. & a.* Yeovilian
yerba *f.* flaw in an emerald; grass, herb
yermo *m.* desert; *a.* uninhabited
yesal, yesar *m.* gypsum mine, gypsum deposit
yesera *f.* gypsum deposit
yesería *f.* gypsum kiln
yesero *m.* plasterer
yesífero *a.* gypsiferous
yesita *f.* gibbsite
yeso *m.* gypsum; plaster, chalk || ***yeso de París*** *m.* plaster of Paris || ***yeso espático*** *m.* selenite || ***yeso especular*** *m.* selenite || ***yeso mate*** *m.* plaster of Paris
yesoso *a.* gypsiferous
Yneziano, Yneziense *m. & a.* Ynezian
yodado *a.* iodized
yodar *vb.* to iodize

yodárgira, yodargirita *f.* iodargyrite
yodato *m.* iodate
yoderita *f.* yoderite
yódico *a.* iodic
yodirita *f.* iodargyrite
yodo *m.* iodine
yodobromita *f.* iodian bromargyrite, iodobromite
yodoembolita *f.* iodargyrite
yodostonolita *f.* manganese pyrochlore
yoduración *f.* iodization
yodurado *a.* iodized
yodurar *vb.* to iodize
yoduro *m.* iodide
yohnstonolita *f.* manganese pyrope
yohnstrupita *f.* mosandrite
yolita *f.* iolite
Yorkiano, Yorkiense *m. & a.* Yorkian
Ypresiano, Ypresiense *m. & a.* Ypresian
yterbio *m.* ytterbium
ytrio *m.* yttrium
ytriogranate *m.* yttrian garnet
ytroilmenita *f.* samarskite
ytterbita *f.* gadolinite
ytterita *f.* tengerite
yttrialita *f.* thorium thalenite
yttrita *f.* gadolinite
yttriumapatito *m.* yttrium apatite
yttrobetafita *f.* yttrobetafite
yttrocalcita *f.* yttrian fluorite
yttrocerita *f.* cerian fluorite
yttrocolombita *f.* samarskite
yttrofluorita *f.* yttrian fluorite
yttrogummita *f.* altered uraninite
yttrohatchettolita *f.* uranopyrochlore
yttroniobita *f.* samarskite
yttro-ortita *f.* allanite
yttropirocloro *m.* yttropyrochlore
yttrotitanita *f.* yttrium titanite
yu *m.* variety of jade
yuconita *f.* yuconite
yuksporita *f.* pectolite
yungas *f.* [And.] humid valleys east of the Andes
yusano *a.* lower, footwall
yusente *f. & a.* ebb; rising or falling with ebb tide
yusero *a.* lower, footwall
yuso *p.* below
yuxtaponer *vb.* to juxtapose
yuxtaposición *f.* juxtaposition
yuxtapuesto *a.* juxtaposed

Z

zaborra *f.* paving stone; gravel, grit; screenings
zafareche *m.* lake; reservoir
zafirita *f.* blue chalcedony
záfiro *m.* sapphire; ultramarine, lapis lazuli
zafra *f.* gangue, waste rock
zafre *m.* zaffer, cobalt oxide
zafrero *m.* common laborer, mucker
zahén *a.* of fine gold (coinage)
zahondar *vb.* to sink, to deepen
zahorra *f.* ballast stone
zala *f.* borax
zamarra *f.* slab
zamboninita *f.* nontronite
Zancleano, Zanclenense *m. & a.* Zanclean
zanja *f.* ditch, channel, trench, prospect pit || ***zanja de desagüe*** *f.* coulée
zanjadora *f.* ditch-digger, backhoe
zanjar *vb.* to dig a ditch or trench
zanjeador *m.* ditch digger
zanjear *vb.* to dig a ditch or trench
zanjeo *m.* ditch digging; underhand stoping
zanjón *m.* large or deep ditch or trench
zanoga *f.* glacial cirque
zapa *f.* sapping, undermining
zapadora *f.* excavator
zapapico *m.* pick-axe, mattock
zapar *vb.* to undermine, excavate
zaranda *f.* sieve, ore screen, riddle, mud screen || ***zaranda vibratoria*** *f.* shale shaker, vibrating screen

zarandado *a.* sieved, screened

zarandar, zarandear *vb.* to sieve, screen, to shake

zarandeo *m.* sifting, sieving, or screening

zarzo *m.* roof timber in a mine

Zemorriano, Zemorriense *m. & a.* Zemorrian

zenit *m.* zenith

zenital *a.* azimuthal

zeofilita *f.* zeophyllite

zeolita *f.* zeolite || ***zeolita azul*** *f.* lazulite || ***zeolita cúbica*** *f.* chabazite || ***zeolita de Bretaña*** *f.* laumontite || ***zeolita de Sudermania*** *f.* spodumene || ***zeolita de Suecia*** *f.* spodumene || ***zeolita del Cabo*** *f.* prehnite || ***zeolita dura*** *f.* analcime || ***zeolita harinosa*** *f.* laumontite || ***zeolita nacarada*** *f.* stilbite || ***zeolita radiada*** *f.* mesolite || ***zeolita roja*** *f.* edelforsite

zeolítico *a.* zeolitic

zeolitización *f.* zeolitization

zermattita *f.* chrysotile, antigorite

zeugira *f.* metabrushite

zeylanita *f.* ceylonite

zeyringita *f.* calcite, aragonite

zeyssatita *f.* randanite

zianita *f.* kyanite

ziguelina *f.* red cuprite; dolomite with disseminated cinnabar

zillerita *f.* tremolite-actinolite asbestos

zinalsita *f.* zinalsite

zinc *m.* zinc || ***zinc carbonatado*** *m.* smithsonite || ***zinc hidrocarbonatado*** *m.* hydrozincite, calamine || ***zinc oxidado silicífero*** *m.* calamine, hydrozincite

zincoferrita *f.* franklinite

zinconisa *f.* hydrozincite, calamine

zincosita, zinkosita *f.* zinc sulfate

zincoso *a.* zinc-bearing

zinkaragonito *m.* nicholsonite

zinkhausmannita *f.* hetairite

zinkisa *f.* tin sulfide

zinkita *f.* zincite

zinkfilita *f.* hopeite

zinopel *m.* red jasper with gold-bearing pyrite

zinque *m.* zinc

ziquitumba *f.* dewatering a mine

zircarbita *f.* zirconium carbonate

zircón *m.* zircon

zirconio *m.* zirconium

zirconita *f.* zircon

zircón-jacinto, zircón-pectolita *m.* rosembuschite

zirconolita *f.* zirkelite

zircosulfato *m.* zircosulfate

zirkofilita *f.* zircophyllite

zirlita *f.* allophane

Zoantariano, Zoantariense *m. & a.* Zoantarian

zöblitzita *f.* serpentine

zócalo *m.* basement, continental platform, base || ***zócalo continental*** *m.* continental shelf

zocollar *vb.* [Ec.] to clear land before mining

zoicita *f.* zoisite

zona *f.* zone, area, region, belt || ***zona aceite–agua*** *f.* oil–water contact || ***zona de acomodación*** *f.* accommodation zone || ***zona de agua artesiana*** *f.* zone of artesian flow || ***zona de agua estática*** *f.* static water zone || ***zona de aereación*** *f.* zone of aeriation || ***zona de agotamiento*** *f.* depletion zone || ***zona de alteración*** *f.* alteration zone || ***zona de arranque*** *f.* root zone || ***zona de asociación*** *f.* assemblage zone || ***zona de balanceo de mareas*** *f.* intertidal zone || ***zona de Benioff*** *f.* Benioff zone || ***zona de capilaridad*** *f.* capillary fringe || ***zona de captación*** *f.* catchment area || ***zona de cizallamiento*** *f.* shear zone || ***zona de contacto*** *f.* contact zone || ***zona de debilidad*** *f.* zone of weakness || ***zona de desarollo*** *f.* pay zone, development zone || ***zona de descargo*** *f.* discharge zone || ***zona de despegue*** *f.* detachment || ***zona de encarrujado*** *f.* kink zone || ***zona de evapotranspiración*** *f.* belt of soil water || ***zona de explotación*** *f.* producing zone || ***zona de fallas*** *f.* fault zone || ***zona de fracturación*** *f.* fracture zone || ***zona de fracturamiento*** *f.* fracture zone || ***zona de Fresnel*** *f.* Fresnel zone || ***zona de hundamiento*** *f.* zone of subsidence

|| ***zona de inframarea*** *f.* shoreface || ***zona de invasión*** *f.* invasion zone || ***zona de inyección*** *f.* injection zone || ***zona de las caraínas*** *f.* 30–90 m water depth || ***zona de las laminarias*** *f.* 0–30 m water depth || ***zona de las corales*** *f.* 100–500 m water depth || ***zona de meteorización*** *f.* weathering zone || ***zona de saturación*** *f.* saturation zone || ***zona de subducción*** *f.* subduction zone || ***zona de supramarea*** *f.* backshore || ***zona desviada*** *f.* bypass zone || ***zona de transferencia*** *f.* transfer zone || ***zona de transición*** *f.* transition zone || ***zona fallada*** *f.* fault zone || ***zona faunística*** *f.* faunal zone || ***zona fótica*** *f.* photic zone || ***zona fracturada*** *f.* fracture zone || ***zona interior*** *f.* hinterland || ***zona limítrofe*** *f.* frontier || ***zona mas externa*** *f.* foreland || ***zona mas interior*** *f.* hinterland || ***zona metalífera*** *f.* mineral belt || ***zona no comprobada*** *f.* untested zone || ***zona no probada*** *f.* untested zone || ***zona petrolífera*** *f.* oil zone, pay zone || ***zona porosa*** *f.* porous zone || ***zona productiva, productora*** *f.* productive interval, pay zone || ***zona sombreada*** *f.* shadow zone || ***zona templada*** *f.* temperate zone || ***zona transcurrente*** *f.* shear zone, wrench zone || ***zona triangular*** *f.* triangle zone || ***zona triturada*** *f.* shattered belt, shattered zone

zonación *f.* zonation

zonado *a.* zoned

zonal *a.* zonal

zonalidad *f.* zoning

zonamiento *m.* zoning

zonificación *f.* zoning

zonoclorita *f.* zonochlorite

zonula *f.* zonule

zonulo *m.* zonule

zoofásmico *a.* zoophasmic

zoógeno *a.* of organic origin

zooide *m.* zooid

zoolita *f.* biolith, zoolith

zopetero *m.* embankment, slope

zubía *f.* channel, pool, swamp

Zuloagense, Zuloagiano *m. & a.* Zuloagan

zumbido *m.* noise, ringing, hum, buzz || ***zumbido terrestre*** *m.* earth ringing

zunyita *f.* zunyite

zurdo *m.* shaft used to stope from below; *a.* left-handed

zurlia *f.* mellilite

zurrapa *f.* dregs, lees, sediment

zurrapiento, zurraposo *a.* muddy, turbid

REFERENCES - COMPENDIO

Allmendinger, R.W. (1986). *Técnicas Modernas de Análisis Estructural: Curso de Actualización Asociación Geológica Argentina*, Department of Geological Sciences, Cornell University, 90 pp.

Corbeil, J.C., and A. Archambault (1992). *Visual Dictionary: Facts on File*, Éditions Québec/Amérique Inc., New York, 924 pp.

Dwight, A.S. (1903). *A Glossary of Spanish–American Mining and Metallurgical Terms: Trans. Am. Inst. of Mining Engineers*, v. 32, pp. 573–603.

Galiana-Mingo, T. (1979). *La Tierra y el Espacio, Pequeño Larousse Científico*, Larousse, México, D.F.

García-Pelayo R. (1989). *Pequeño Larousse Illustrado*, Larousse, México, D.F.

Hammond Inc. (1988). *Qué es Qué: Enciclopedia visual bilingüe*, Hammond Inc., Maplewood, NJ, 435 pp.

Instituto Argentino del Petróleo (1981). *Glosario Técnico de la Industria del Petróleo Español-Inglés Inglés-Español*, Segunda Edición, Buenos Aires, 513 pp.

Long, K.R. (1991). *A Partial Glossary of Spanish Geological Terms Exclusive of Most Cognates*, U.S. Geological Survey, Open File Report 91-0579.

Orellana-Silva, E. (1967). *Diccionario Inglés-Español de Ciencias de la Tierra*, Blume, Madrid.

Petróleo Internacional (1982). *Glossary of the Petroleum Industry*, 2nd Edition, PennWell Books, Tulsa, 378 pp.

Robb, L.A. (1956) *Diccionario para Ingenieros*, Co. Editorial Continental, México, D.F., 664 pp.

Seiglie, G.A., and J.D. Weaver (1968). *Glosario de Términos Geológicos Inglés–Español*, Edición Preliminar: Universidad de Puerto Rico Facultad de Artes y Ciencias, Mayaguez, 84 pp.

Sheriff, R.E. (1973). *Encyclopedic Dictionary of Exploration Geophysics*, Society of Exploration Geophysicists, Tulsa, OK.

Smith, C. (1993). *Collins Spanish–English English–Spanish Dictionary*, 3rd Edition, Harper Collins Publishers, Glasgow, 908 pp.

Turner, J.C. (1972). *Diccionario Geológico Inglés–Español Español–Inglés: Asociación Geológica Argentina*, Buenos Aires, 288 pp.

Williams, E.B. (1978). *Diccionario Inglés–Español, Español–Inglés*, McGraw-Hill Inc., México, D.F.